TRANSPORT PHENOMENA AND UNIT OPERATIONS

TRANSPORT PHENOMENA AND UNIT OPERATIONS

A Combined Approach

Richard G. Griskey

A JOHN WILEY & SONS, INC., PUBLICATION

Solutions Manual Available for
Transport Phenomena and Unit Operations:
A Combined Approach
by Richard G. Griskey
ISBN 978-0-471-99814-3 © 2002 by John Wiley & Sons, Inc.

The Solutions Manual to the 560 problems in *Transport Phenomena and Unit Operations: A Combined Approach* is available for academic adopters of the text. To request a copy of the Solutions Manual, please contact Jonathan Rose at jrose@wiley.com.

Published by John Wiley & Sons, Inc., Hoboken, New Jersey.
Published simultaneously in Canada.

Library of Congress Cataloging-in-Publication Data is available.

ISBN-13 978-0-471-99814-3

10 9 8 7 6 5 4 3 2 1

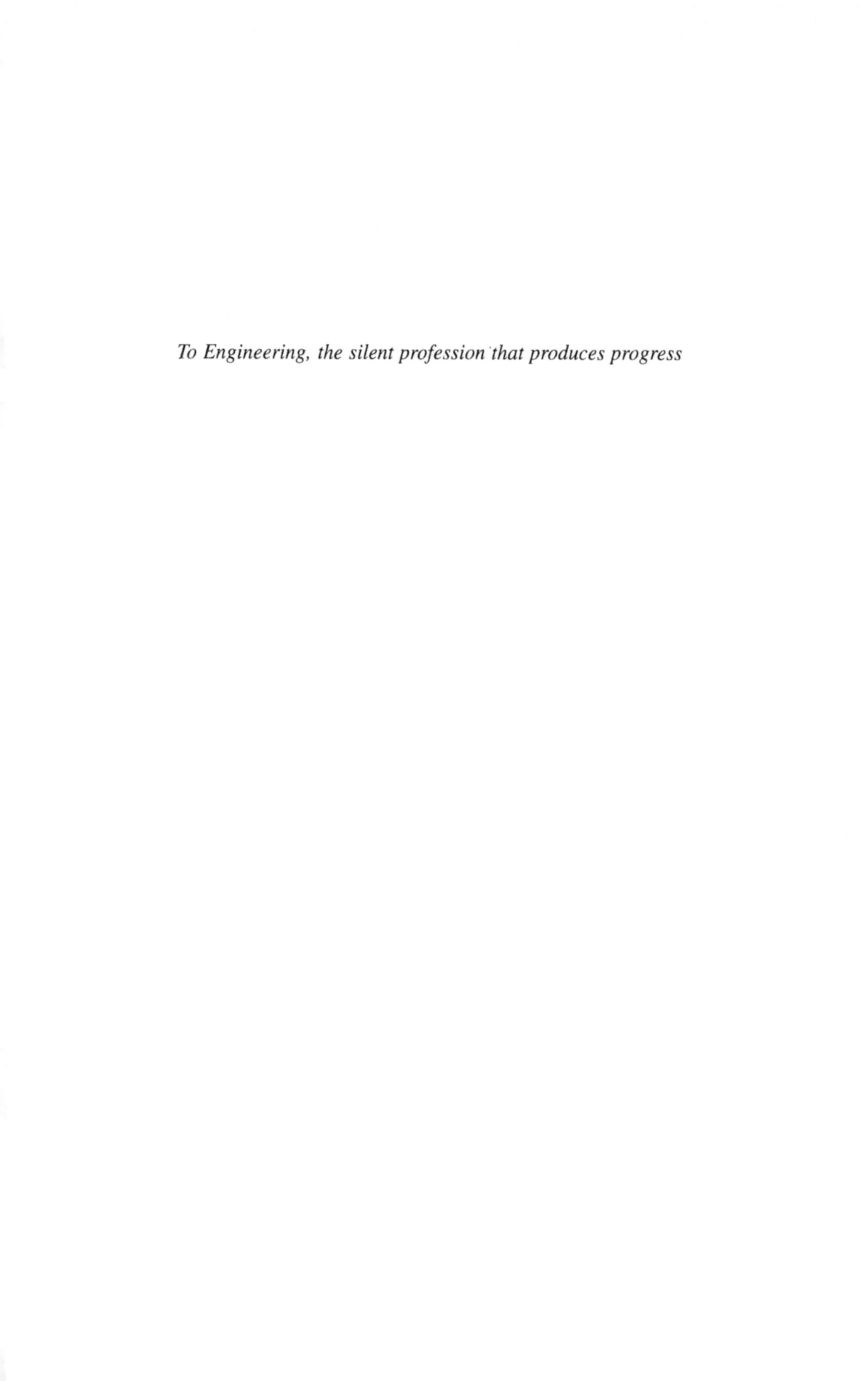

To Engineering, the silent profession that produces progress

CONTENTS

PREFACE

The question of "why another textbook," especially in the areas of transport processes and unit operations, is a reasonable one.

To develop an answer, let us digress for a moment to consider Chemical Engineering from a historical perspective. In its earliest days, Chemical Engineering was really an applied or industrial chemistry. As such, it was based on the study of definitive processes (the Unit Process approach).

Later it became apparent to the profession's pioneers that regardless of process, certain aspects such as fluid flow, heat transfer, mixing, and separation technology were common to many, if not virtually all, processes. This perception led to the development of the Unit Operations approach, which essentially replaced the Unit Processes-based curriculum.

While the Unit Operations were based on first principles, they represented nonetheless a semiempirical approach to the subject areas covered. A series of events then resulted in another evolutionary response, namely, the concept of the Transport Phenomena that truly represented Engineering Sciences.

No one or nothing lives in isolation. Probably nowhere is this as true as in all forms of education. Massive changes in the preparation and sophistication of students — as, for example in mathematics — provided an enthusiastic and skilled audience. Another sometimes neglected aspect was the movement of chemistry into new areas and approaches. As a particular example, consider Physical Chemistry, which not only moved from a macroscopic to a microscopic approach but also effectively abandoned many areas in the process.

Furthermore, other disciplines of engineering were moving as well in the direction of Engineering Science and toward a more fundamental approach.

These and other factors combined to make the next movement a reality. The trigger was the classic text, *Transport Phenomena*, authored by Bird, Stewart, and Lightfoot. The book changed forever the landscape of Chemical Engineering.

At this point it might seem that the issue was settled and that Transport Phenomena would predominate.

Alas, we find that Machiavelli's observation that "Things are not what they seem" is operable even in terms of Chemical Engineering curricula.

The Transport Phenomena approach is clearly an essential course for graduate students. However, in the undergraduate curriculum there was a definite division with many departments keeping the Unit Operations approach. Even where the Transport Phenomena was used at the undergraduate level there were segments of the Unit Operations (particularly stagewise operations) that were still used.

Experience with Transport Phenomena at the undergraduate level also seemed to produce a wide variety of responses from enthusiasm to lethargy on the part of faculty. Some institutions even taught both Transport Phenomena and much of the Unit Operations (often in courses not bearing that name).

Hence, there is a definite dichotomy in the teaching of these subjects to undergraduates. The purpose of this text is hopefully to resolve this dilemma by the mechanism of a seamless and smooth combination of Transport Phenomena and Unit Operations.

The simplest statement of purpose is to move from the fundamental approach through the semiempirical and empirical approaches that are frequently needed by a practicing professional Chemical Engineer. This is done with a minimum of derivation but nonetheless no lack of vigor. Numerous worked examples are presented throughout the text.

A particularly important feature of this book is the inclusion of comprehensive problem sets at the end of each chapter. In all, over 570 such problems are presented that hopefully afford the student the opportunity to put theory into practice.

A course using this text can take two basically different approaches. Both start with Chapter 1, which covers the transport processes and coefficients. Next, the areas of fluid flow, heat transfer, and mass transfer can be each considered in turn (i.e., Chapter 1, 2, 3, . . ., 13, 14).

The other approach would be to follow as a possible sequence 1, 2, 5, 10, 3, 6, 11, 4, 7, 8, 9, 12, 13, 14. This would combine groupings of similar material in the three major areas (fluid flow, heat transfer, mass transfer) finishing with Chapters 12, 13, and 14 in the area of separations.

The foregoing is in the nature of a suggestion. There obviously can be many varied approaches. In fact, the text's combination of rigor and flexibility would give a faculty member the ability to develop a different and challenging course.

It is also hoped that the text will appeal to practicing professionals of many disciplines as a useful reference text. In this instance the many worked examples, along with the comprehensive compilation of data in the Appendixes, should prove helpful.

Richard G. Griskey
Summit, NJ

1

TRANSPORT PROCESSES AND TRANSPORT COEFFICIENTS

INTRODUCTION

The profession of chemical engineering was created to fill a pressing need. In the latter part of the nineteenth century the rapidly increasing growth complexity and size of the world's chemical industries outstripped the abilities of chemists alone to meet their ever-increasing demands. It became apparent that an engineer working closely in concert with the chemist could be the key to the problem. This engineer was destined to be a chemical engineer.

From the earliest days of the profession, chemical engineering education has been characterized by an exceptionally strong grounding in both chemistry and chemical engineering. Over the years the approach to the latter has gradually evolved; at first, the chemical engineering program was built around the concept of studying individual processes (i.e., manufacture of sulfuric acid, soap, caustic, etc.). This approach, *unit processes*, was a good starting point and helped to get chemical engineering off to a running start.

After some time it became apparent to chemical engineering educators that the unit processes had many operations in common (heat transfer, distillation, filtration, etc). This led to the concept of thoroughly grounding the chemical engineer in these specific operations and the introduction of the *unit operations* approach. Once again, this innovation served the profession well, giving its practitioners the understanding to cope with the ever-increasing complexities of the chemical and petroleum process industries.

As the educational process matured, gaining sophistication and insight, it became evident that the unit operations in themselves were mainly composed of a smaller subset of transport processes (momentum, energy, and mass transfer). This realization generated the transport phenomena approach—an approach that owes much to the classic chemical engineering text of Bird, Stewart, and Lightfoot (1).

There is no doubt that modern chemical engineering in indebted to the transport phenomena approach. However, at the same time there is still much that is important and useful in the unit operations approach. Finally, there is another totally different need that confronts chemical engineering education—namely, the need for today's undergraduates to have the ability to translate their formal education to engineering practice.

This text is designed to build on all of the foregoing. Its purpose is to thoroughly ground the student in basic principles (the transport processes); then to move from theoretical to semiempirical and empirical approaches (carefully and clearly indicating the rationale for these approaches); next, to synthesize an orderly approach to certain of the unit operations; and, finally, to move in the important direction of engineering practice by dealing with the analysis and design of equipment and processes.

THE PHENOMENOLOGICAL APPROACH; FLUXES, DRIVING FORCES, SYSTEMS COEFFICIENTS

In nature, the trained observer perceives that changes occur in response to imbalances or driving forces. For example, heat (energy in motion) flows from one point to another under the influence of a temperature difference. This, of course, is one of the basics of the engineering science of thermodynamics.

Likewise, we see other examples in such diverse cases as the flow of (respectively) mass, momentum, electrons, and neutrons.

Hence, simplistically we can say that a *flux* (see Figure 1-1) occurs when there is a *driving force*. Furthermore, the flux is related to a gradient by some characteristic of the system itself—the *system* or *transport coefficient.*

$$\text{Flux} = \frac{\text{Flow quantity}}{(\text{Time})(\text{Area})} = (\text{Transport coefficient})(\text{Gradient}) \tag{1-1}$$

The gradient for the case of temperature for one-dimensional (or directional) flow of heat is expressed as

$$\text{Temperature gradient} = \frac{dT}{dY} \tag{1-2}$$

The flux equations can be derived by considering simple one-dimensional models. Consider, for example, the case of energy or heat transfer in a slab (originally at a constant temperature, T_1) shown in Figure 1-2. Here, one of the opposite faces of the slab suddenly has its temperature increased to T_2. The result is that heat flows from the higher to the lower temperature region. Over a period of time the temperature profile in the solid slab will change until the linear (steady-state) profile is reached (see Figure 1-2). At this point the rate of heat

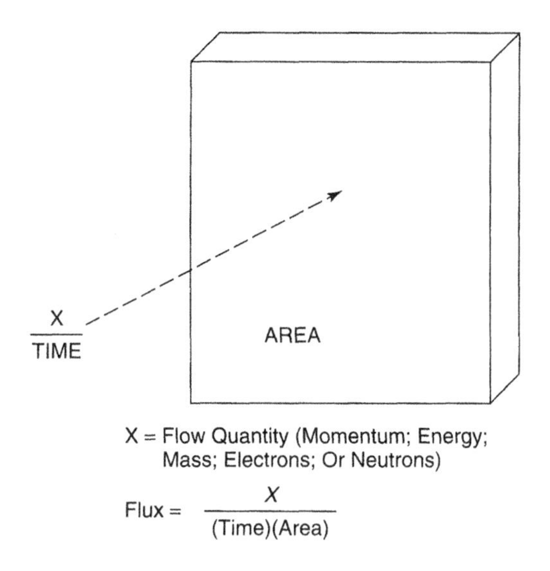

Figure 1-1. Schematic of a flux.

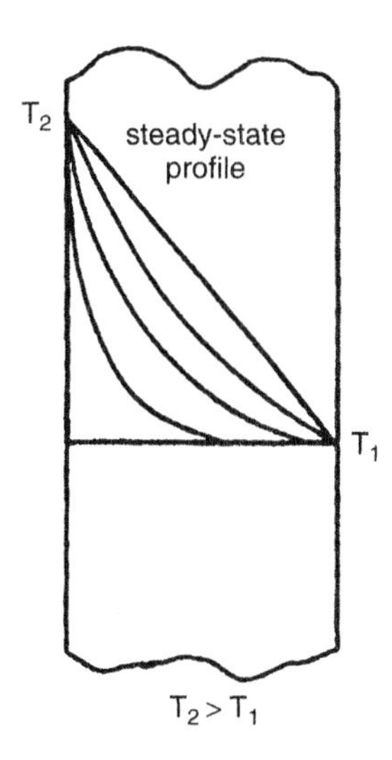

Figure 1-2. Temperature profile development (unsteady to steady state). (Reproduced with permission from reference 18. Copyright 1997, American Chemical Society.)

flow Q per unit area A will be a function of the system's transport coefficient (k, thermal conductivity) and the driving force (temperature difference) divided by distance. Hence

$$\frac{Q}{A} = k\frac{(T_1 - T_2)}{(X - O)} \tag{1-3}$$

If the above equation is put into differential form, the result is

$$q_x = -k\frac{dT}{dx} \tag{1-4}$$

This result applies to gases and liquids as well as solids. It is the one-dimensional form of Fourier's Law which also has y and z components

$$q_Y = -k\frac{dT}{dy} \tag{1-5}$$

$$q_z = -k\frac{dT}{dz} \tag{1-6}$$

Thus heat flux is a vector. Units of the heat flux (depending on the system chosen) are BTU/hr ft^2, calories/sec cm^2, and W/m^2.

Let us consider another situation: a liquid at rest between two plates (Figure 1-3). At a given time the bottom plate moves with a velocity V. This causes the fluid in its vicinity to also move. After a period of time with unsteady flow we attain a linear velocity profile that is associated with steady-state flow. At steady state a constant force F is needed. In this situation

$$\frac{F}{A} = -\mu\frac{O - V}{Y - O} \tag{1-7}$$

where μ is the fluid's viscosity (i.e., transport coefficient).

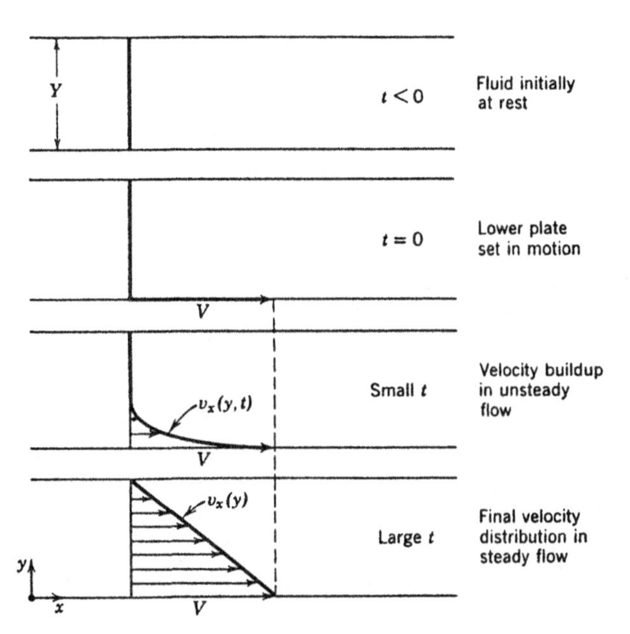

Figure 1-3. Velocity profile development for steady laminar flow. (Adapted with permission from reference 1. Copyright 1960, John Wiley and Sons.)

Hence the F/A term is the flux of momentum (because force = d(momentum)$/dt$. If we use the differential form (converting F/A to a shear stress τ), then we obtain

$$\tau_{yx} = -\mu \frac{dV_X}{dy} \tag{1-8}$$

Units of τ_{yx} are poundals/ft^2, dynes/cm^2, and Newtons/m^2.

This expression is known as Newton's Law of Viscosity. Note that the shear stress is subscripted with two letters. The reason for this is that momentum transfer is not a vector (three components) but rather a tensor (nine components).

As such, momentum transport, except for special cases, differs considerably from heat transfer.

Finally, for the case of mass transfer because of concentration differences we cite Fick's First Law for a binary system:

$$J_{A_y} = -D_{AB} \frac{dC_A}{dy} \tag{1-9}$$

where J_{A_y} is the molar flux of component A in the y direction. D_{AB}, the diffusivity of A in B (the other component), is the applicable transport coefficient.

As with Fourier's Law, Fick's First Law has three components and is a vector. Because of this there are many analogies between heat and mass transfer as we will see later in the text. Units of the molar flux are lb moles/hr ft^2, g mole/sec cm^2, and kg mole/sec m^2.

THE TRANSPORT COEFFICIENTS

We have seen that the transport processes (momentum, heat, and mass) each involve a property of the system (viscosity, thermal conductivity, diffusivity). These properties are called the transport coefficients. As system properties they are functions of temperature and pressure.

Expressions for the behavior of these properties in low-density gases can be derived by using two approaches:

1. The kinetic theory of gases
2. Use of molecular interactions (Chapman–Enskog theory).

In the first case the molecules are rigid, nonattracting, and spherical. They have

1. A mass m and a diameter d
2. A concentration n (molecules/unit volume)
3. A distance of separation that is many times d.

This approach gives the following expression for viscosity, thermal conductivity, and diffusivity:

$$\mu = \frac{2}{3\pi^{3/2}} \frac{\sqrt{mKT}}{d^2} \tag{1-10}$$

where K is the Boltzmann constant.

$$k = \frac{1}{d^2}\sqrt{\frac{K^3 T}{\pi^3 m}} \tag{1-11}$$

where the gas is monatomic.

$$D_{AB} = \frac{2}{3}\left(\frac{K^3}{\pi^3}\right)^{1/2}\left(\frac{1}{2m_A} + \frac{1}{2m_B}\right)^{1/2} \frac{T^{3/2}}{P\left(\frac{d_A + d_B}{2}\right)^2} \tag{1-12}$$

The subscripts A and B refer to gas A and gas B.

If molecular interactions are considered (i.e., the molecules can both attract and repel) a different set of relations are derived. This approach involves relating the force of interaction, F, to the potential energy ϕ. The latter quantity is represented by the Lennard-Jones (6–12) potential (see Figure 1-4)

$$\phi(r) = 4\epsilon\left[\left(\frac{\sigma}{r}\right)^{12} - \left(\frac{\sigma}{r}\right)^{6}\right] \tag{1-13}$$

where σ is the collision diameter (a characteristic diameter) and ϵ a characteristic energy of interaction (see Table A-3-3 in Appendix for values of σ and e).

The Lennard-Jones potential predicts weak molecular attraction at great distances and ultimately strong repulsion as the molecules draw closer.

Resulting equations for viscosity, thermal conductivity, and diffusivity using the Lennard-Jones potential are

$$\mu = 2.6693 \times 10^{-6}\frac{\sqrt{MT}}{\sigma^2 \Omega\mu} \tag{1-14}$$

where μ is in units of kg/m sec or pascal-seconds, T is in °K, σ is in Å, the $\Omega\mu$ is a function of KT/e (see Appendix), and M is molecular weight.

$$k = 0.8322\frac{\sqrt{T/M}}{\sigma^2 \Omega_k} \tag{1-15}$$

where k is in W/m °K, σ is in Å, and $\Omega_k = \Omega\mu$. The expression is for a monatomic gas.

$$D_{AB} = 1.8583 \times 10^{-7}\frac{\sqrt{T^3(1/M_A + 1/M_B)}}{P\sigma_{AB}^2 \Omega_{DAB}} \tag{1-16}$$

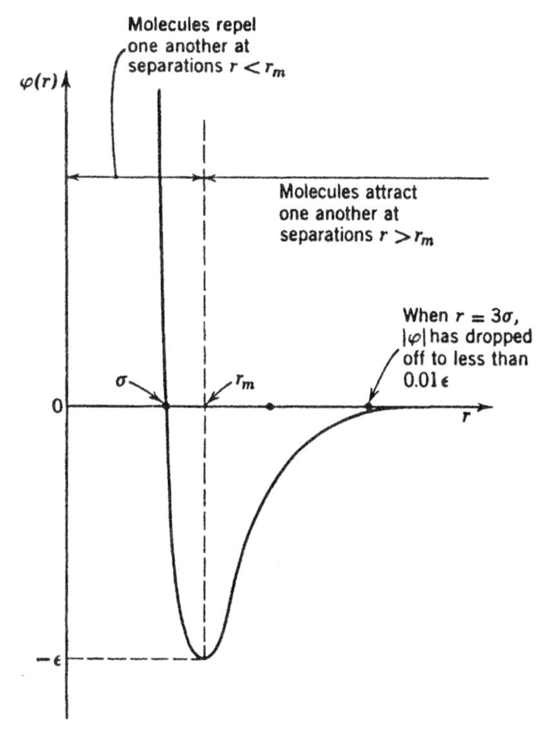

Figure 1-4. Lennard-Jones model potential energy function. (Adapted with permission from reference 1. Copyright 1960, John Wiley and Sons.)

where D_{AB} is units of m^2/sec P is in atmospheres, $\sigma_{AB} = \frac{1}{2}(\sigma_A + \sigma_B)$, $\varepsilon_{AB} = \sqrt{\varepsilon_A \varepsilon_B}$, and Ω_{DAB} is a function of KT/ε_{AB} (see Appendix B, Table A-3-4).

Example 1-1 The viscosity of isobutane at 23°C and atmospheric pressure is 7.6×10^{-6} pascal-sec. Compare this value to that calculated by the Chapman–Enskog approach.

From Table A-3-3 of Appendix A we have

$$\sigma = 5.341 \text{ Å}, \qquad \varepsilon/K = 313 \text{ K}$$

Then, $KT/\varepsilon = 296.16/313 = 0.946$ and from Table A-3-4 of Appendix B we obtain

$$\Omega_\mu = 1.634$$

$$\mu = 2.6693 \times 10^{-6} \frac{\sqrt{MT}}{\sigma^2 \Omega_\mu}$$

$$\mu = 2.6693 \times 10^{-6} \frac{\sqrt{(58.12)(296.16)}}{(5.341)^2(1.634)}$$

$$\mu = 7.51 \times 10^{-6} \text{ pascal-sec}$$

$$\% \text{ error} = [(7.6 - 7.51)/7.6] \times 100 = 1.18\%$$

Example 1-2 Calculate the diffusivity for the methane–ethane system at 104°F and 14.7 psia.

$$T = \frac{104 + 460}{1.8} \,{}^\circ\text{K} = 313^\circ\text{K}$$

Let methane be A and let ethane be B. Then,

$$\left(\frac{1}{M_A} + \frac{1}{M_B}\right) = \left(\frac{1}{16.04} + \frac{1}{30.07}\right) = 0.0956$$

From Table A in the Appendix we have

$$\sigma_A = 3.822 \text{ Å}, \qquad \sigma_B = 4.418 \text{ Å}$$

$$\frac{\varepsilon_A}{K} = 137^\circ\text{K}, \qquad \frac{\varepsilon_B}{K} = 230^\circ\text{K}$$

$$\sigma_{AB} = \tfrac{1}{2}(\sigma_A + \sigma_B) = \tfrac{1}{2}(3.822 + 4.418) \text{ Å} = 4.120 \text{ Å}$$

$$\frac{\varepsilon_{AB}}{K} = \sqrt{\left(\frac{\varepsilon_A}{K}\right)\left(\frac{\varepsilon_B}{K}\right)} = \sqrt{(137^\circ\text{K})(230^\circ\text{K})} = 177.5^\circ\text{K}$$

$$\frac{KT}{\varepsilon_{AB}} = \frac{313}{177.5} = 1.763$$

From Table A-3-4 in Appendix we have $\Omega_{DAB} = 1.125$

$$D_{AB} = \frac{1.8583 \times 10^{-7}\sqrt{T^3(1/M_A + 1/M_B)}}{P\sigma_{AB}^2\Omega_{DAB}}$$

$$D_{AB} = \frac{1.8583 \times 10^{-7}\sqrt{(313^\circ\text{K})^3(0.0956)}}{(1)\ (4.120)^2(1.125)}$$

$$D_{AB} = 1.66 \times 10^{-5} \text{ m}^2/\text{sec}$$

The actual value is 1.84×10^{-5} m^2/sec. Percent error is 9.7 percent.

TRANSPORT COEFFICIENT BEHAVIOR FOR HIGH DENSITY GASES AND MIXTURES OF GASES

If gaseous systems have high densities, both the kinetic theory of gases and the Chapman–Enskog theory fail to properly describe the transport coefficients' behavior. Furthermore, the previously derived expression for viscosity and

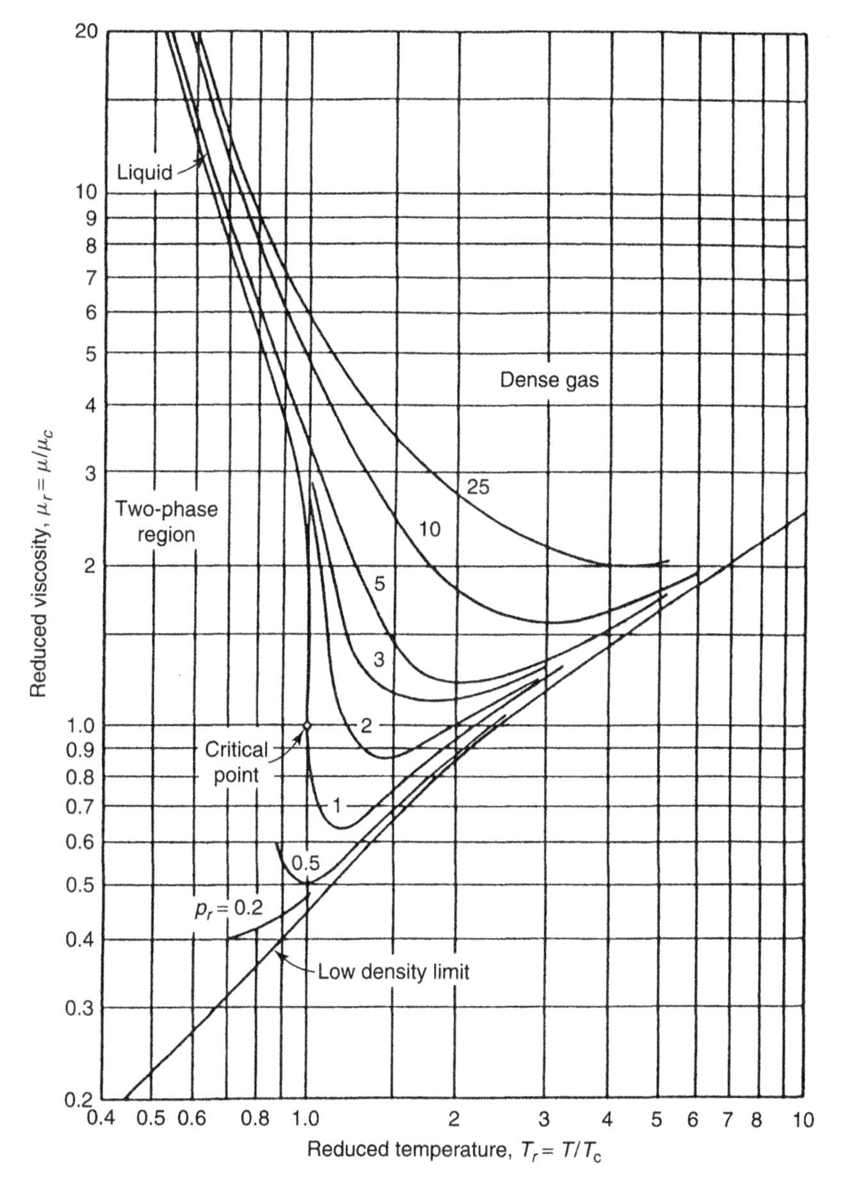

Figure 1-5. Reduced viscosity as a function of reduced pressure and temperature (2). (Courtesy of National Petroleum News.)

thermal conductivity apply only to pure gases and not to gas mixtures. The typical approach for such situations is to use the theory of corresponding states as a method of dealing with the problem.

Figures 1-5, 1-6, 1-7, and 1-8 give correlation for viscosity and the thermal conductivity of monatomic gases. One set (Figures 1-5 and 1-7) are plots of the

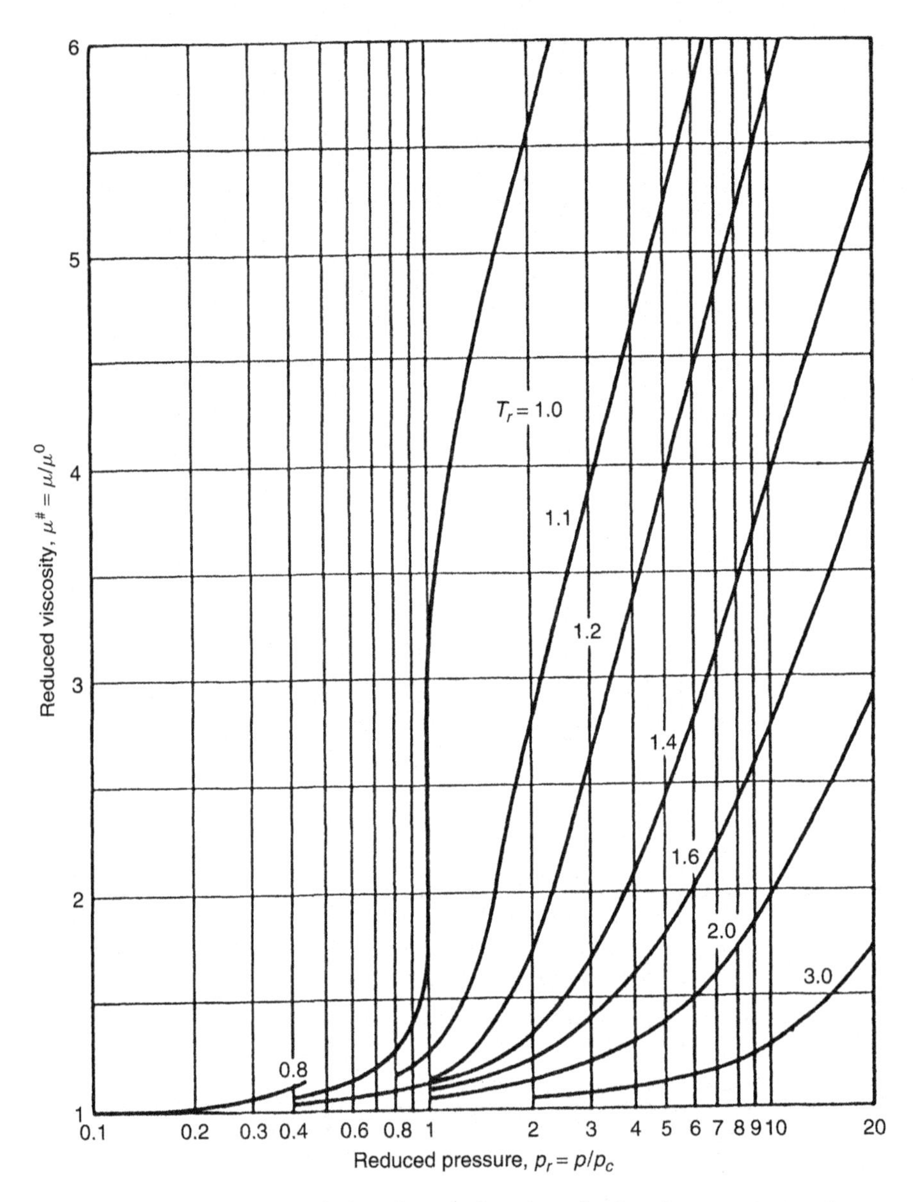

Figure 1-6. Modified reduced viscosity as a function of reduced temperature and pressure (3). (Trans. Am. Inst. Mining, Metallurgical and Petroleum Engrs *201* 1954 pp 264 ff; N. L. Carr, R. Kobayashi, D.B. Burrows.)

reduced viscosity (μ/μ_c, where μ_c is the viscosity at the critical point) or reduced thermal conductivity (k/k_c) versus (T/T_c), reduced temperature, and (p/p_c) reduced pressure. The other set are plots of viscosity and thermal conductivity divided by the values (μ_0, k_0) at atmospheric pressure and the same temperature.

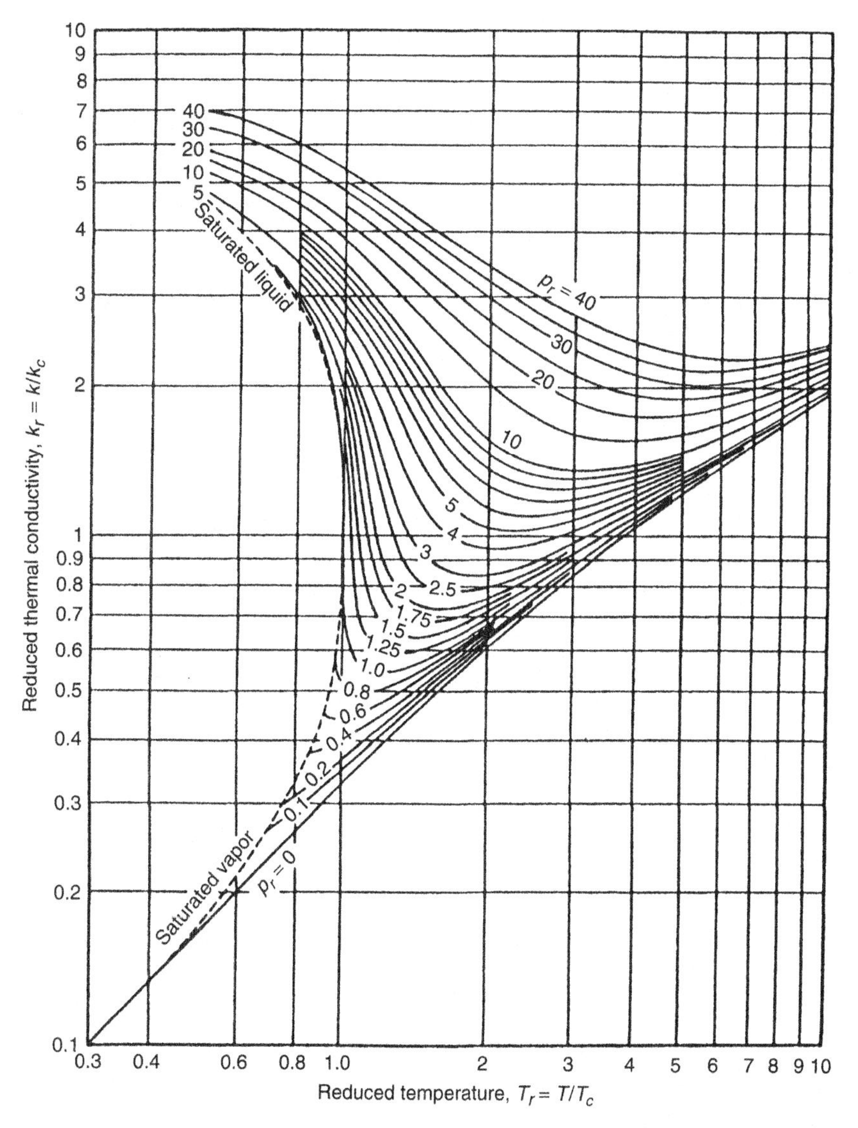

Figure 1-7. Reduced thermal conductivity (monatomic gases) as a function of reduced temperature and pressure. (Reproduced with permission from reference 4. Copyright 1957, American Institute of Chemical Engineers.)

Values of μ_c can be estimated from the empirical relations

$$\mu_c = 61.6 \frac{(MT_c)^{1/2}}{(\widehat{V}_c)^{2/3}} \tag{1-17}$$

$$\mu_c = \frac{7.70\ M^{1/2} P_c^{2/3}}{(T_c)^{1/6}} \tag{1-18}$$

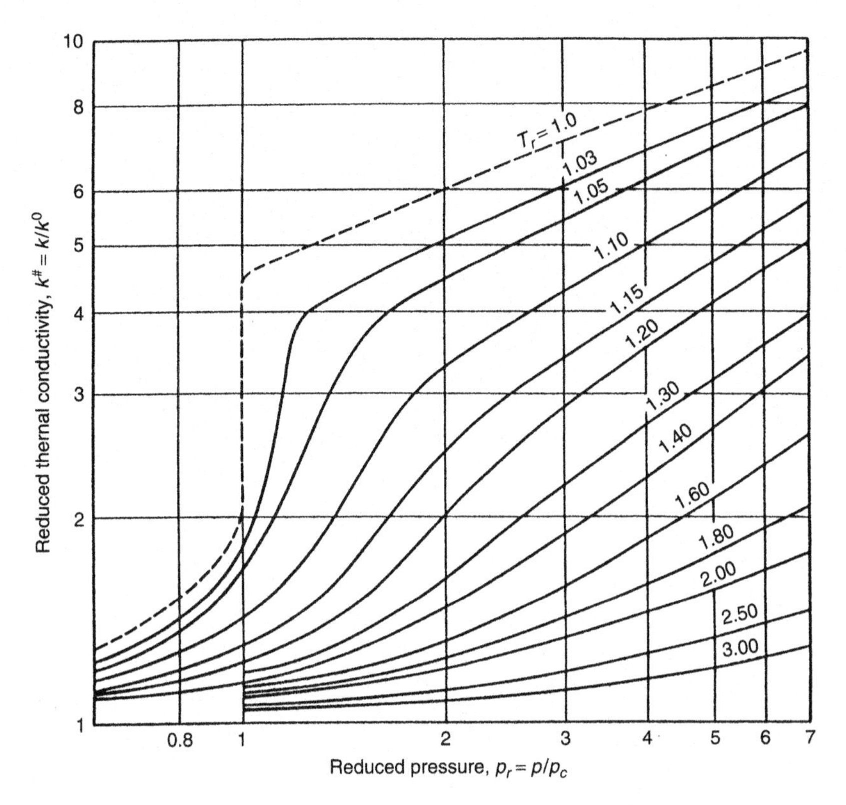

Figure 1-8. Modified reduced thermal conductivity as a function of reduced temperature and pressure. (Reproduced with permission from reference 5. Copyright 1953, American Institute of Chemical Engineers.)

where μ_c is in micropoises, T_c is in °K, P_c in atmospheres, and $\widehat{V_c}$ is in cm^3/g mole.

The viscosity and thermal conductivity behavior of mixtures of gases at low densities is described semiempirically by the relations derived by Wilke (6) for viscosity and by Mason and Saxena (7) for thermal conductivity:

$$\mu_{\text{mix}} = \sum_{i=1}^{n} \frac{y_i \mu_i}{\sum_{j=1}^{n} y_j \Phi_{ij}} \tag{1-19}$$

$$\Phi_{ij} = \frac{1}{8}\left(1 + \frac{M_i}{M_j}\right)^{-1/2} \left[1 + \left(\frac{\mu_i}{\mu_j}\right)^{1/2} \left(\frac{M_j}{M_i}\right)^{1/4}\right]^2 \tag{1-20}$$

$$k_{\text{mix}} = \sum_{i=1}^{n} \frac{y_i k_i}{\sum_{j=1}^{n} y_j \Phi_{ij}} \tag{1-21}$$

The Φ_{ij}'s in equation (1-21) are given by equation (1-20). The y's refer to the mole fractions of the components.

For mixtures of dense gases the pseudocritical method is recommended. Here the critical properties for the mixture are given by

$$P'_c = \sum_{i=1}^{n} y_i P_{ci} \tag{1-22}$$

$$T'_c = \sum_{i=1}^{n} y_i T_{ci} \tag{1-23}$$

$$\mu'_c = \sum_{i=1}^{n} y_i \mu_{ci} \tag{1-24}$$

where y_i is a mole fraction; P_{Ci}, T_{Ci}, and μ_{Ci} are pure component values. These values are then used to determine the P'_R and T'_R values needed to obtain (μ/μ_i) from Figure 1-5.

The same approach can be used for the thermal conductivity with Figure 1-7 if k_c data are available or by using a k^0 value determined from equation (1-15).

Behavior of diffusivities is not as easily handled as the other transport coefficients. The combination $(D_{AB}P)$ is essentially a constant up to about 150 atm pressure. Beyond that, the only available correlation is the one developed by Slattery and Bird (8). This, however, should be used only with great caution because it is based on very limited data (8).

Example 1-3 Compare estimates of the viscosity of CO_2 at 114.6 atm and 40.3°C using

1. Figure 1-6 and an experimental viscosity value of 1800×10^{-8} pascal-sec for CO_2 at 45.3 atm and 40.3°C
2. The Chapman–Enskog relation and Figure 1-6.

From Table A-3-3 of Appendix A, $T_c = 304.2$°K and $P_c = 72.9$ atmospheres. For the first case we have

$$T_R = \frac{313.46}{304.2} = 1.03, \qquad P_R = \frac{45.3}{72.9} = 0.622$$

so that $\mu^{\#} = 1.12$ and

$$\mu_0 = \frac{\mu}{\mu^{\#}} = \frac{1800 \times 10^{-8} \text{ pascal-sec}}{1.12} = 1610 \times 10^{-8} \text{ pascal-sec}$$

For $P = 114.6$ atm we have

$$P_R = \frac{114.6}{72.9} = 1.57, \qquad T_R = 1.03$$

so that $\mu^{\#} = 3.7$ and

$$\mu = \mu^{\#}\mu_0 = (3.7)(1610 \times 10^{-8} \text{ pascals-sec}) = 6000 \times 10^{-8} \text{ pascals-sec}$$

For the second case, from Table A-3-3 of Appendix we have

$$M = 44.01, \qquad \sigma = 3.996 \text{ Å}, \qquad \varepsilon/k = 190^\circ\text{K}$$

and

$$\frac{KT}{\varepsilon} = \frac{313.46}{190} = 1.165$$

so that from Table A-3-4 of Appendix and $\Omega_\mu = 1.264$ we obtain

$$\mu = 2.6693 \times 10^{-6} \frac{\sqrt{MT}}{\sigma^2 \Omega \mu}$$

$$\mu = 2.6693 \times 10^{-6} \frac{\sqrt{(44.01)(313.46)}}{(3.996)^2(1.264)} = 1553 \times 10^{-8} \text{ pascal-sec}$$

From Figure 1-6, $\mu^{\#}$ is still 3.7 so that

$$\mu = (3.7)(1553 \times 10 \text{ pascal-sec}) = 5746 \times 10^{-8} \text{ pascal-sec}$$

The actual experimental value is 5800×10^{-8} pascal-sec. Percent errors for case 1 and case 2 are 3.44% and 0.93%, respectively.

Example 1-4 Estimate the viscosity of a gas mixture of $CO_2(y = 0.133)$; $O_2(y = 0.039)$; $N_2(y = 0.828)$ at 1 atm and 293°K by using

1. Figure 1-5 and the pseudocritical concept
2. Equations (1-19) and (1-20) with pure component viscosities of 1462, 2031, and 1754×10^{-8} pascal-sec, respectively, for CO_2, O_2, and N_2.

In the first case the values of T_c, P_c, and μ_c (from Table A-3-3 of Appendix) are as follows:

	T_c(°K)	P_c (atmospheres)	μ_c (pascal-seconds)
CO_2	304.2	72.9	3430×10^{-8}
O_2	154.4	49.7	2500×10^{-8}
N_2	126.2	33.5	1800×10^{-8}

$$T_c' = (0.133)(304.2°\text{K}) + (0.039)(154.4°\text{K}) + (0.828)(126.2°\text{K})$$

$$T_c' = 150.97°\text{K}$$

$$P_c' = [(0.133)(72.9) + (0.039)(49.7) + (0.828)(33.5)] \text{ atm}$$

$$P_c' = 39.37 \text{ atmospheres}$$

$$\mu_c' = [(0.133)(3430) + (0.039)(2500) + (0.828)(1800)] \times 10^{-8} \text{ pascal-sec}$$

$$\mu_c' = 2044.1 \times 10^{-8} \text{ pascal-sec}$$

Then

$$T_R' = \frac{293}{150.97} = 1.94, \qquad P_R' = \frac{1}{39.37} = 0.025$$

From Figure 1-5 we have

$$\mu_R' = \frac{\mu}{\mu_c'} = 0.855$$

$$\mu = (0.855)(2044.1 \times 10^{-8} \text{ pascal-sec}) = 1747.7 \times 10^{-8} \text{ pascal-sec}$$

For case 2, let $CO_2 = 1$, $O_2 = 2$, and $N_2 = 3$. Then:

i	j	M_i/M_j	M_i/M_j	Φ_{ij}
1	1	1.00	1.00	1.00
	2	1.38	0.72	0.73
	3	1.57	0.83	0.73
2	1	0.73	1.39	1.39
	2	1.00	1.00	1.00
	3	1.14	1.16	1.04
3	1	0.64	1.20	1.37
	2	0.88	0.86	0.94
	3	1.00	1.00	1.00

$$\mu_{\text{mix}} = \sum_{i=1}^{3} \frac{Y_i \mu_i}{\sum_{j=1}^{3} y_j \Phi_{ij}}$$

$$\mu_{\text{mix}} = \left[\frac{(0.133)(1462)}{(0.76)} + \frac{(0.039)(2031)}{(1.06)} + \frac{(0.828)(1754)}{(1.05)}\right] \times 10^{-8} \text{ pascal-sec}$$

$$\mu_{\text{mix}} = 1714 \times 10^{-8} \text{ pascal-sec}$$

Actual experimental value of the mixture viscosity is 1793×10^{-8} pascal-sec. The percent errors are 2.51 and 4.41%, respectively, for cases 1 and 2.

TRANSPORT COEFFICIENTS IN LIQUID AND SOLID SYSTEMS

In general, the understanding of the behavior of transport coefficients in gases is far greater than that for liquid systems. This can be partially explained by seeing that liquids are much more dense than gases. Additionally, theoretical and experimental work for gases is far more voluminous than for liquids. In any case the net result is that approaches to transport coefficient behavior in liquid systems are mainly empirical in nature.

An approach used for liquid viscosities is based on an application of the Eyring (9,10) activated rate theory. This yields an expression of the form

$$\mu = \frac{Nh}{V} \exp\left(\frac{0.408\Delta U_{\text{vap}}}{RT}\right) \tag{1-25}$$

where N is Avogadro's number, h is Plancks constant, V is the molar volume, and ΔU_{vap} is the molar internal energy change at the liquid's normal boiling point.

The Eyring equation is at best an approximation; thus it is recommended that liquid viscosities be estimated using the nomograph given in Figure B-1 of the appendix.

For thermal conductivity the theory of Bridgman (11) yielded

$$k = 2.80 \left(\frac{\widehat{N}}{\widehat{V}}\right)^{2/3} K V_S \tag{1-26}$$

where V_S the sonic velocity, is

$$V_S = \sqrt{\frac{C_P}{C_V}\left(\frac{\partial P}{\partial \rho}\right)_T} \tag{1-27}$$

The foregoing expressions for both viscosity and thermal conductivity are for pures. For mixtures it is recommended that the pseudocritical method be used where possible with liquid regions of Figures 1-5 through 1-8.

Diffusivities in liquids can be treated by the Stokes–Einstein equation

$$\frac{D_{AB}\mu_B}{KT} = \frac{1}{4\pi R_A} \tag{1-28}$$

where R_A is the diffusing species radius and μ_B is the solvent viscosity.

Table 1-1 Association Parameters

Solvent System	ψ_B
Water	2.6
Methanol	1.9
Ethanol	1.5
Unassociated Solvents (ether, benzene, etc.)	1.0

Another approach (for dilute solutions) is to use the correlation of Wilke and Chang (12) :

$$D_{AB} = 7.4 \times 10^{-8} \frac{(\psi_B M_B)^{1/2} T}{\mu(\widehat{V}_A)^{0.6}} \tag{1-29}$$

where $\widehat{V}_A$ is the diffusing species molar volume at its normal boiling point, μ is the solution viscosity, and ψ_B is an empirical parameter ("association parameter") as shown in Table 1-1.

Liquid diffusivities are highly concentration-dependent. The foregoing expressions should therefore be used only for dilute cases and of course for binaries.

For solid systems the reader should consult the text by Jakob (13) for thermal conductivities. Diffusivities in solid systems are given by references 14–17.

Example 1-5 Compare viscosity values for liquid water at 60°C as determined by

1. The Eyring equation (the ΔU_{vap} is 3.759×10^7 J/kg mole)
2. Using the nomograph in Figure B-1 and Table B-1 of Appendix

For case 1,

$$\widehat{V} = \frac{M}{\rho} = \frac{18.02}{1000} \frac{m^3}{\text{kg mole}} = 0.01802 \frac{m^3}{\text{kg mole}}$$

$$\mu = \frac{\widehat{N} h}{\widehat{V}} \exp\left(\frac{0.408 \Delta U_{vap}}{RT}\right)$$

$$\mu = \frac{6.023 \times 10^{26}}{\text{kg mole}} \frac{6.62 \times 10^{34} \text{ J sec}}{0.01802 \text{ m}^3/\text{kg mole}} \exp\left[\frac{(0.408)(3.759 \times 10^7 \text{ m}^3/\text{kg mole}}{(333.10°\text{K})(8314.4 \text{ j/kg mole})}\right]$$

$$\mu = 0.00562 \text{ pascal-sec}$$

For case 2, the coordinates for water are (10.2, 13). Connecting 60°C with this point gives a viscosity of 4.70×10^{-4} pascal-sec.

Actual experimental viscosity value is 4.67×10^{-4} pascal-sec, which clearly indicates that the Eyring method is only an approximation.

Example 1-6 Estimate the thermal conductivity of liquid carbon tetrachloride (CCl_4) at 20°C and atmospheric pressure.

$$\rho = 1595 \text{ kg/m}^3, \qquad \frac{1}{\rho}\left(\frac{\partial \rho}{\partial P}\right)_T = 89.5 \times 10^{-11}\frac{\text{m}^2}{\text{N}}$$

$$\left(\frac{\partial P}{\partial \rho}\right)_T = \frac{1}{\rho[1/\rho(\partial\rho/\partial P)_T]} = \frac{1}{(1595 \text{ kg/m}^3)(89.5 \times 10^{-11} \text{ m}^2/\text{J})} = 7 \times 10^5 \frac{\text{m}^2}{\text{sec}^2}$$

If it is assumed that $C_P = C_V$ (a good assumption for the conditions) then

$$V_s = \sqrt{\frac{C_P}{C_V}\left(\frac{\partial P}{\partial \rho}\right)_T} = \sqrt{(1)\ (7 \times 10^5 \text{ m}^2/\text{sec}^2)} = 837 \text{ m/sec}$$

The value of V is given by

$$\widehat{V} = \frac{M}{\rho} = \frac{153.84}{1595}\frac{\text{m}^3}{\text{kg mole}} = 0.0965\frac{\text{m}^3}{\text{kg mole}}$$

$$kV = 2.80\left(\frac{\widehat{N}}{\widehat{V}}\right)^{2/3}$$

$$KV_s = 2.80\left(\frac{6.023 \times 10^{26} \text{ kg mole}}{\text{kg mole } 0.0965 \text{ m}^3}\right)^{2/3}\left(1.3805 \times 10^{-23}\frac{\text{J}}{°\text{K}}\right)(837 \text{ m/sec})$$

$$k = 0.112\frac{\text{J}}{\text{m sec °K}}$$

The experimental value is 0.103 J/m sec °K. This gives a percent error of 8.74 percent.

Example 1-7 What is the diffusivity for a dilute solution of acetic acid in water at 12.5°C? The density of acetic acid at its normal boiling point is 0.937 g/cm^3. The viscosity of water at 12.5°C is 1.22 cP.

Using the Wilke–Chang equation, we obtain

$$D_{AB} = 7.4 \times 10^{-8}(\psi_B M_B)^{1/2}\frac{T}{\mu_B \widehat{V}_A^{0.6}}$$

$$\widehat{V}_A = \frac{M}{\rho} = \frac{60}{0.937} = \frac{\text{cm}^3}{\text{g mole}} = 64.1\frac{\text{cm}^3}{\text{g mole}}$$

$$\psi_B = 2.6 \text{ (from Table 1-1)}$$

$$D_{AB} = 7.4 \times 10^{-8}(2.6 \times 18)^{1/2}\frac{287.5°\text{K}}{(1.22)(64.1 \text{ m}^3/\text{g mole})^{0.6}}$$

$$D_{AB} = 9.8 \times 10^{-6} \text{ cm}^2/\text{sec} = 9.8 \times 10^{-10} \text{ m}^2/\text{sec}$$

ONE-DIMENSIONAL EQUATION OF CHANGE; ANALOGIES

As was shown earlier, each of the three transport processes is a function of a driving force and a transport coefficient. It is also possible to make the equations even more similar by converting the transport coefficients to the forms of diffusivities. Fick's First Law [equation (1-9)] already has its transport coefficient (D_{AB}) in this form. The forms for Fourier's Law [equation (1-7)] and Newton's Law of Viscosity [equation (1-8)] are

$$\text{Thermal diffusivity} = \frac{k}{\rho C_P} = \alpha \tag{1-30}$$

$$\text{Momentum diffusivity} = \frac{\mu}{\rho} = \nu \tag{1-31}$$

Units of all the diffusivities are (length)^2/unit time (i.e., cm^2/sec, m^2/sec, and ft^2/hr). Momentum diffusivity is also known as kinematic viscosity.

If the transport coefficients are thus converted, the equations of change in one dimension then become

$$q_y = -\alpha \frac{d(\rho C_P T)}{dy} \tag{1-32}$$

$$\tau_{yx} = -\nu \frac{d(\rho V_x)}{dy} \tag{1-33}$$

$$J_{Ay} = -D_{AB} \frac{d(C_A)}{dy} \tag{1-34}$$

This result indicates that these three processes are analoguous in one-dimensional cases. Actually, heat and mass transfer will be analogous in even more complicated cases as we will demonstrate later. This is not true for momentum transfer whose analogous behavior only applies to one-dimensional cases.

SCALE-UP; DIMENSIONLESS GROUPS OR SCALING FACTORS

One of the most important characteristics of the chemical and process industries is the concept of scale-up. The use of this approach has enabled large-scale operations to be logically and effectively generated from laboratory-scale experiments. The philosophy of scale-up was probably best expressed by the highly productive chemist Leo Baekeland (the inventor of Bakelite and many other industrial products). Baekeland stated succinctly, "make your mistakes on a small scale and your profits on a large scale."

Application of scale-up requires the use of the following:

1. Geometric similarity
2. Dynamic similarity
3. Boundary conditions
4. Dimensionless groups or scaling factors.

The first of these, geometric similarity, means that geometries on all scales must be of the same type. For example, if a spherically shaped process unit is used on a small scale, a similarly shaped unit must be used on a larger scale. Dynamic similarity implies that the *relative* values of temperature, pressure, velocity, and so on, in a system be the same on both scales. The boundary condition requirement fixes the condition(s) at the system's boundaries. As an example, consider a small unit heated with electrical tape (i.e., constant heat flux). On a larger scale the use of a constant wall temperature (which is not constant heat flux) would be inappropriate.

Dimensionless groups or scaling factors are the means of sizing the units involved in scaling up (or down). They, in essence, represent ratios of forces, energy changes, or mass changes. Without them the scale-up process would be almost impossible.

Additionally, these groups are the way that we make use of semiempirical or empirical approaches to the transport processes. As we will see later, the theoretical/analytical approach cannot always be used, especially in complex situations. For such cases, dimensionless groups enable us to gain insights and to analyze and design systems and processes.

PROBLEMS

1-1. Estimate the viscosities of n-hexane at 200°C and toluene at 270°C. The gases are at low pressure.

1-2. What are the viscosities of methane, carbon dioxide, and nitrogen at 20°C and atmospheric pressure?

1-3. Estimate the viscosity of liquid benzene at 20°C.

1-4. Determine a value for the viscosity of ammonia at 150°C.

1-5. A young engineer finds a notation that the viscosity of nitrogen at 50°C and a "high pressure" is 1.89×10^{-9} pascal-seconds. What is the pressure?

1-6. Available data for mixtures of hydrogen and dichlorofluoromethane at 25°C and atmospheric pressure are as follows:

Mole Fraction Hydrogen	Viscosity of Mixture ($\times 10^5$ pascal-sec)
0.00	1.24
0.25	1.281
0.50	1.319
0.75	1.351
1.00	0.884

Compare calculated values to the experimental data at 0.25 and 0.75 mole fraction of hydrogen.

1-7. Estimate the viscosity of a 25-75 percent mole fraction mixture of ethane and ethylene at 300°C and a pressure of 2.026×10^7 pascals.

1-8. Values of viscosity and specific heats, respectively, for nitric oxide and methane are (a) 1.929×10^{-5} pascal-sec and 29.92 kilojoules/(kg mole °K) and (b) 1.116×10^{-5} pascal seconds and 35.77 kilojoules/(kg mole °K). What are the thermal conductivities of the pure gases at 27°C?

1-9. Compare values of thermal conductivity for argon at atmospheric pressure and 100°C using equations (1-11) and (1-15), respectively.

1-10. A value of thermal conductivity for methane at 1.118×10^7 pascals is 0.0509 joules/(sec m K). What is the temperature for this value?

1-11. Water at 40°C and a pressure of 4×10^{12} pascals has a density of 993.8 kg/m^3 and an isothermal compressibility $(\rho^{-1}(\partial\rho/\partial P)_T)$ of 3.8×10^{-18} pascals^{-1}. What is its thermal conductivity?

1-12. What is the thermal conductivity of a mixture of methane (mole fraction of 0.486) and propane at atmospheric pressure and 100°C?

1-13. Argon at 27°C and atmospheric pressure has values of viscosity and thermal conductivity of 2.27×10^{-5} pascal-sec and 1.761×10^{-4} Joules/(sec m °K) from each property respectively. Calculate molecular diameters and collision diameters, compare them, and evaluate.

1-14. Compute a value for D_{AB} for a system of argon (A) and oxygen (B) at 294°K and atmospheric pressure.

1-15. The diffusivity for carbon dioxide and air at 293°K and atmospheric pressure is 1.51×10^{-5} m^2/sec. Estimate the value at 1500°K using equations (1-12) and (1-16).

1-16. A dilute solution of methanol in water has a diffusivity of 1.28×10^{-9} m^2/sec at 15°C. Estimate a value at 125°C.

1-17. Estimate a value of diffusivity for a mixture of 80 mole percent methane and 20 mole percent of ethane at 40°C and 1.379×10^7 pascals.

1-18. Determine a value of D_{AB} for a dilute solution of 2,4,6-trinitrotoluene in benzene at 15°C.

1-19. Find values of σ_{AB} and ϵ_{AB} from the following data:

D_{AB} (m/sec)	T (°K)
1.51×10^{-5}	293
2.73×10^{-5}	400
5.55×10^{-5}	600
9.15×10^{-5}	800

1-20. At 25°C, estimate the diffusivity of argon (mole fraction of 0.01) in a mixture of nitrogen, oxygen, and carbon dioxide (mole fractions of 0.78, 0.205, and 0.005, respectively).

REFERENCES

1. R. B. Bird, W. E. Stewart, and E. N. Lightfoot, *Transport Phenomena*, John Wiley and Sons, New York (1960).
2. O. A. Uyehara and K. M. Watson, *Nat. Petrol. News Tech. Section* **36**, 764 (1944).
3. N. L. Carr, R. Kobayashi, and D. B. Burroughs, *Am. Inst. Min. Met. Eng.* **6**, 47 (1954).
4. E. J. Owens and G. Thodos, *AIChE J.* **3**, 454 (1957).
5. J. M. Lenoir, W. A. Junk, and E. W. Comings, *Chem. Eng. Prog.* **49**, 539 (1953).
6. C. R. Wilke, *J. Chem. Phys.* **17**, 550 (1949).
7. E. A. Mason and S. C. Saxena, *Physics of Fluids* **1**, 361 (1958).
8. J. C. Slattery and R. B. Bird, *AIChE J.* **4**, 137 (1958).
9. Glasstone, K. J. Laidler, and H. Eyring, *Theory of Rate Processes*, McGraw-Hill, New York, (1941).
10. J. F. Kincaid, H. Eyring, and A. W. Stearn, *Chem. Rev.* **28**, 301 (1941).
11. P. W. Bridgman, *Proc. Am. Acad. Arts Sci.* **59**, 141 (1923).
12. C. R. Wilke and P. Chang, *AIChE J.* **1**, 264 (1955).
13. M. Jakob, *Heat Transfer*, Vol. I, John Wiley and Sons, New York (1949), Chapter 6.
14. R. M. Barrer, *Diffusion in and Through Solids*, Macmillan, New York (1941).
15. W. Jost, *Diffusion in Solids, Liquids and Gases*, Academic Press, New York (1960).
16. P. G. Shewman *Diffusion in Solids*, McGraw-Hill, New York (1963).
17. J. P. Stark, *Solid State Diffusion*, John Wiley and Sons, New York, (1976).
18. R. G. Griskey, *Chemical Engineering for Chemists*, American Chemical Society, Washington, D.C. (1997).

2

FLUID FLOW BASIC EQUATIONS

INTRODUCTION

In the beginning, the chemical industry was essentially a small-scale batch-type operation. The fluids (mainly liquids) were easily moved from one vessel to another literally by a "bucket brigade" approach. However, after a time both the increasing complexity of the processes and the desire for higher production levels made it necessary for industry to find ways to rapidly and efficiently transport large quantities of fluids.

This need led to the sophisticated and complicated fluid transportation systems in place in today's chemical and petroleum process industries. These system are characterized by miles of piping (Figure 2-1), complicated fittings, pumps (Figure 2-2), compressors (Figure 2-3), turbines, and other fluid machinery devices. As such, today's engineers must be highly skilled in many aspects of the flow of fluids if they are going to be competent in the analysis design and operation of modern chemical and petroleum processes.

This chapter will introduce the student to some of the important aspects of fluid flow (momentum transfer). Later, other subjects will be introduced to give the fledgling engineer the competence required to meaningfully deal with this overall area of momentum transport and fluid flow.

The subjects in this chapter will include fluid statics, fluid flow phenomena, categories of fluid flow behavior, the equations of change relating the momentum transport, and the macroscopic approach to fluid flow.

Figure 2-1. Petrochemical plant complex. (Shell Chemical.)

FLUID STATICS

Before considering the concept of fluids in motion, it is worthwhile to examine the behavior of fluids at rest or the subject of fluid statics.

An important equation relating to fluid statics is the barometric equation

$$\frac{dP}{dZ} = \rho g \tag{2-1}$$

where P is the pressure, Z is the vertical distance, ρ is the fluid density, and g is the acceleration of gravity. Consider the barometric equation with respect to the world itself. If, for example, we find ourselves in the surface of the North Atlantic Ocean, then we know that the pressure is atmospheric. On the other hand, if we visit the final resting place of the Titanic on the ocean floor, we would find that the pressure is many times atmospheric with the difference due to the effect predicted by the barometric equation for ocean water. Likewise, if we would go to California, we would find that the pressure in Death Valley (282 feet below sea level) is higher than that at the top of Mount Whitney (elevation 14,494 feet). Once again the difference in pressure would be governed by the barometric equation.

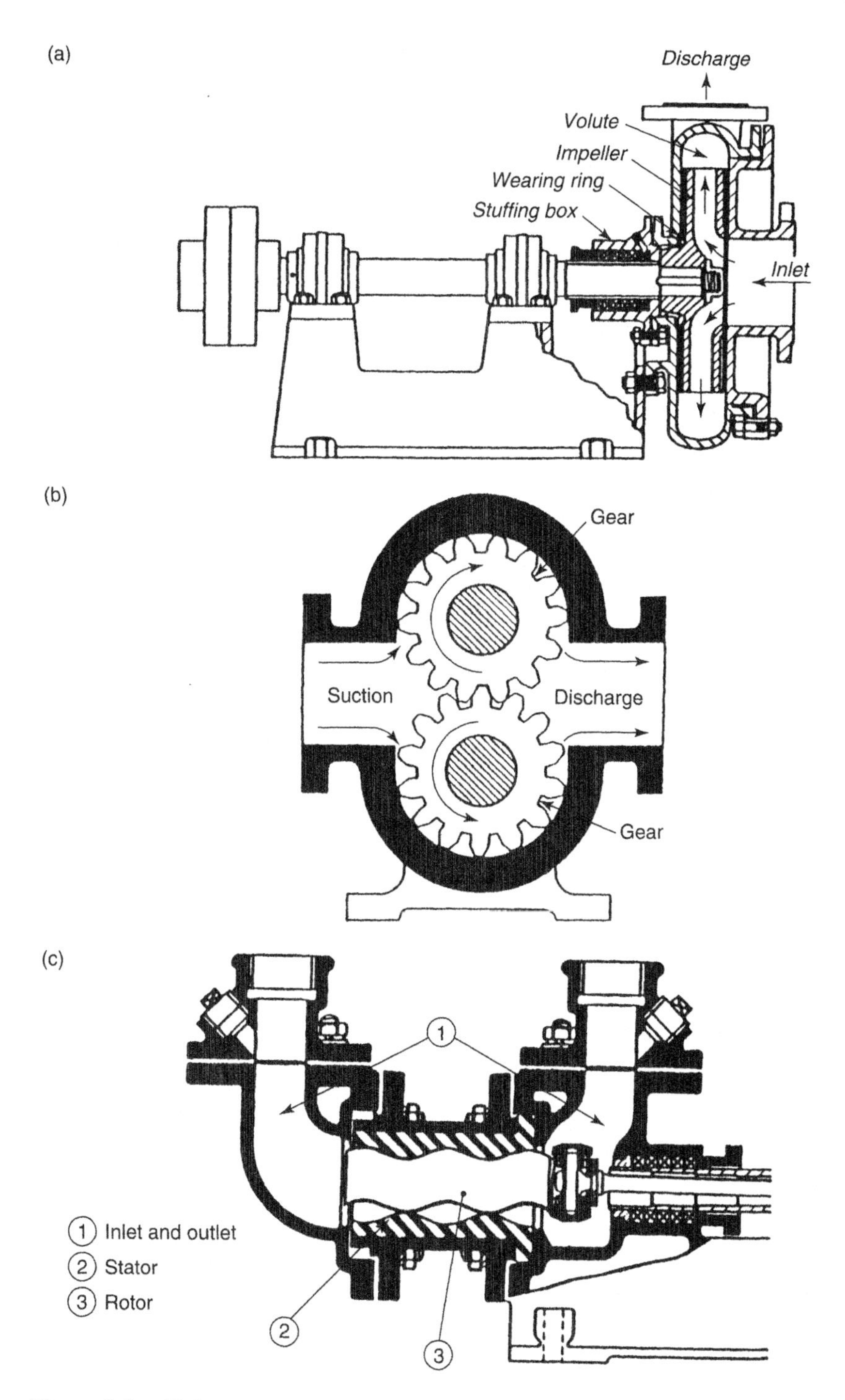

Figure 2-2. Various process pumps. (a) Centrifugal; (b) gear; (c) Moyno. (1, 2).

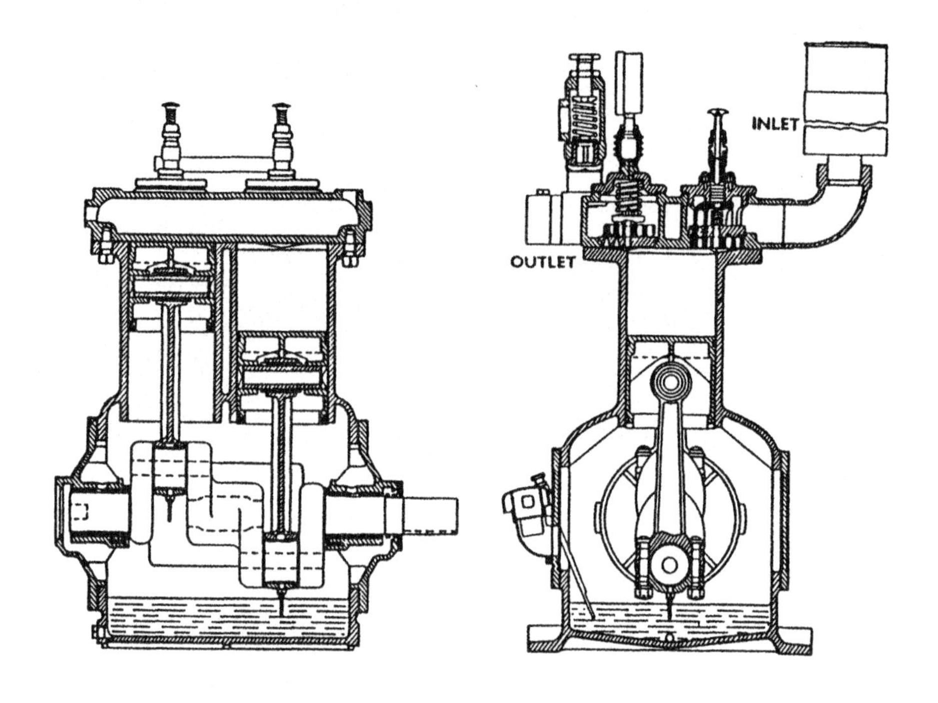

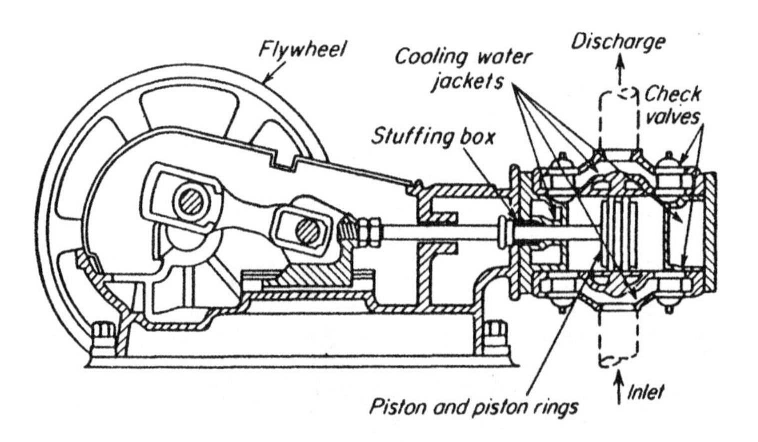

Figure 2-3. Reciprocating compressors (1, 2)

We can also apply the barometric equation to directions other than those that are directly vertical. In such cases we use trigonometry to correct the equation

$$\frac{dP}{dl} = \rho g \cos \Theta \tag{2-2}$$

where Θ is the angle between the vertical Z and the dimension l.

In using the barometric equation, we also must consider the usage of the terms absolute, gauge, and atmospheric pressure. The interrelation between these is given in equation (2-3):

$$P_{\text{absolute}} = P_{\text{gauge}} + P_{\text{atmospheric}} \tag{2-3}$$

$P_{\text{atmospheric}}$ is the ambient pressure at that point where we make the reading. The P_{gauge} is the pressure read by some measuring device for a vessel or a container. See that such readings must have the atmospheric pressure added to them to get the absolute pressure value. For example, if the reading gave a gauge pressure of 1.26×10^5 N/m^2, the atmospheric pressure of 1.01×10^5 N/m^2 must be added to it to give the absolute pressure of 2.27×10^5 N/m^2.

Fluid statics can be involved in process operations in a number of ways. One such case is in the measurement of pressure differentials in a system. This is illustrated in Example 2-1.

Example 2-1 Suppose a manometer is used to measure a pressure differential in a pipe with a flowing fluid "x" at room temperature as shown in Figure 2-4. The manometer reads a differential height of 1.09 feet. The liquid in the pipe has a density of 78.62 ft^3/lb mass. Mercury (density of 848.64 ft^3/lb mass) is the manometer fluid. What is the pressure measured?

In this example we will use a detailed approach in order to demonstrate how the barometric equation works. Further more, the units used will be English in order to help us illustrate some important facts relating to units.

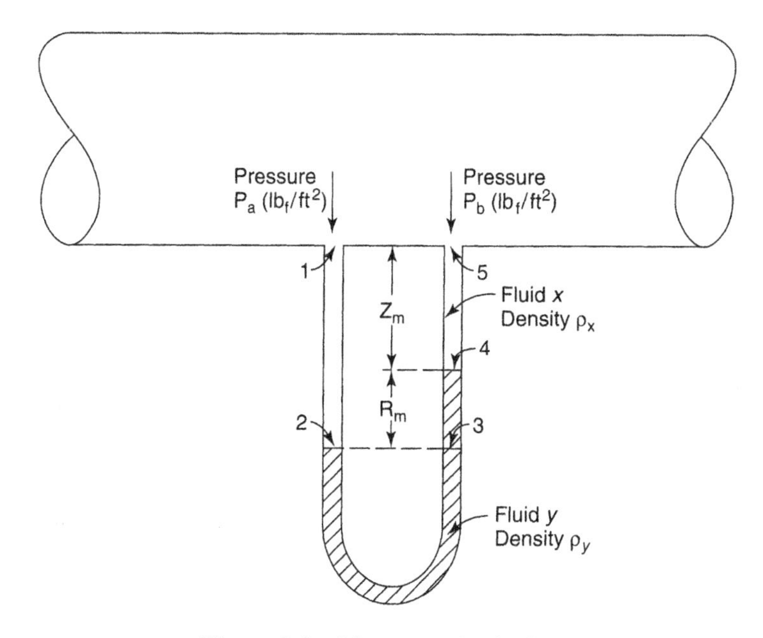

Figure 2-4. Manometer in pipeline.

The overall pressure differential in the system shown in Figure 2-4 is $(P_1 - P_5)$. In order to derive the expression for it, let us use the barometric equation step by step.

If we first move from point 1 to point 2 [the interface between the flowing (x) and manometer (y) fluids], we see that

$$P_2 = P_1 + g(Z_m + R_m)\rho_x$$

Then, at point 4 we see the other interface between the two fluids:

$$P_4 = P_2 - gR_m\rho_Y$$

$$P_4 = P_1 + g(Z_m + R_m)\rho_x - gR_m\rho_Y$$

and

$$P_4 = P_1 + gR_m(\rho_x - \rho_Y) + gZ_m\rho_x$$

Finally at point 5 we see that

$$P_5 = P_4 - gZ_m\rho_x$$

$$P_5 = P_1 + gR_m(\rho_x - \rho_y) + gZ_m\rho_x - gZ_m\rho_x$$

$$(P_1 - P_5) = gR_m(\rho_Y - \rho_x)$$

Or if R_m is taken as the height h of the manometer fluid, we obtain

$$(P_1 - P_5) = gh(\rho_y - \rho_x)$$

Substituting the given values yields

$$(P_1 - P_5) = (32.2 \text{ ft/sec}^2)(1.09 \text{ ft})[848.64 - 78.62] \text{ lb mass/ft}^3 1/g_c$$

See that the factor g_c is needed to derive the pressure (lbf/ft^2) in this case. In the metric system we do not consciously employ g_c because it is unity. In the English system, however, g_c is not unity.

To see this, consider Newton's Law that relates force, mass, and acceleration:

$$\text{Force} = \text{Mass} \times \text{Acceleration}$$

This in symbolic form is actually

$$F = \frac{1}{g_c} ma$$

Table 2-1 g_c Values for Various Unit Systems

Mass	Length	Time	g_c	Force
gram	cm	sec	$\frac{1 \text{ g cm}}{\text{sec}^2 \text{ dyne}}$	dyne
kg	meter	sec	$\frac{1 \text{ kg m}}{\text{sec}^2 \text{ newton}}$	newton
lb mass	feet	sec	$32.2\frac{\text{lb mass ft}}{\text{lb forcesec}^2}$	lb force
slug	feet	sec	$1\frac{\text{slug ft}}{\text{lb forcesec}^2}$	lb force

The g_c is a universal constant that depends on the specified unit system. The difference can be seen in Table 2-1.

As was noted the value of g_c is unity for the metric system (centimeter–gram–second, c.g.s.; and meter–kilogram–second, m.k.s.). However, in the English system if lb mass is employed, then a value of 32.2 lb mass ft/lb force sec^2 must be used. See that the use of the slug as a unit of mass also gives a g_c of unity. However, this is a unit that is rarely employed at any time in engineering (or even scientific) usage.

Returning to the result of Example 2-1, we have $(P_1 - P_5) = (32.2 \text{ ft/sec}^2)$ (1.09 ft) [848.64 – 78.62] lb mass/ft^3 1/32.2 lb mass ft/lb force sec^2 $(P_1 - P_5)$ = 839.3 lb force/ft^2 or 5.82 lb force/in^2.

Another aspect of fluid statics that will be utilized later in this text is the Principle of Archimedes relating to buoyant force.

Buoyant force for an object floating in a liquid is

$$F_{\text{B}} = [\rho_{\text{liquid}}\, g\, V_{\text{liquid}} + \rho_{\text{air}}\, g\, V_{\text{air}}] \tag{2-4}$$

where

V_{liquid} = the volume of liquid displaced by the object

V_{air} = the volume of air displaced

FLUID DYNAMICS — A PHENOMONOLOGICAL APPROACH

If a fluid is put into motion, then its behavior is determined by its physical nature, the flow geometry, and its velocity. In order to obtain a phenomenological view of the process of fluid flow, consider the pioneering experiments done by Osborne Reynolds in the nineteenth century on Newtonian fluids (those that obey Newton's Law of Viscosity).

Reynolds, considered by many to be the founder of modern-day fluid mechanics, injected a dye stream into water flowing in a glass tube. At certain fluid

Figure 2-5. Reynolds experiment—streamline flow.

velocities he found that the dye stream moved in a straight line (see Figure 2-5). This behavior was found to occur over the entire cross section of the tube for a given overall flow rate.

Reynolds also found that the dye streams' velocity was the same for a given radial distance from the tube center (or wall). Hence, at a particular circumference the velocity had the same value. Furthermore, he found that the maximum fluid velocity occurred at the tube's center line and then decreased as the radius approached the radius of the tube wall.

These behavior patterns led Reynolds to conclude that such flows were streamline (i.e., the dye stream showed a straight line behavior with a given velocity at a circumference). Furthermore, since the velocity moved from a maximum at the tube center to a minimum at the wall, the fluid itself moved in shells or lamina (see Figure 2-6). Because of these patterns of behavior he termed such flows as *streamline* and *laminar.*

Reynolds, being an experimentalist, went further by increasing the flow rate. In so doing, he first observed that the stream line began to move in a sinuous or oscillating pattern (Figure 2-7) and ultimately developed into a chaotic pattern of eddies and vortices. The chaotic flow was termed *turbulent* flow, and the intermediate range was called *transition* flow.

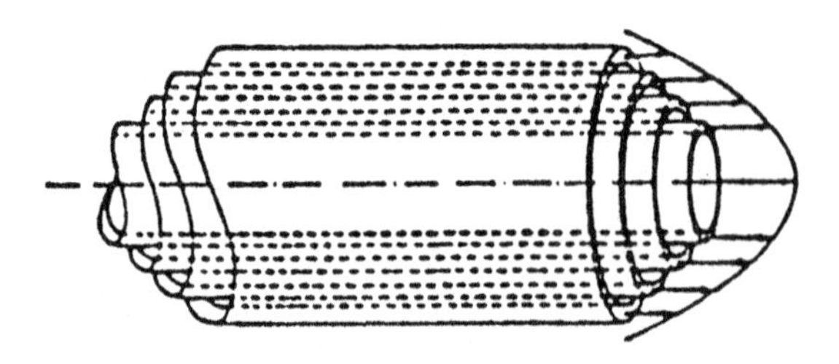

Figure 2-6. Laminar flow schematic.

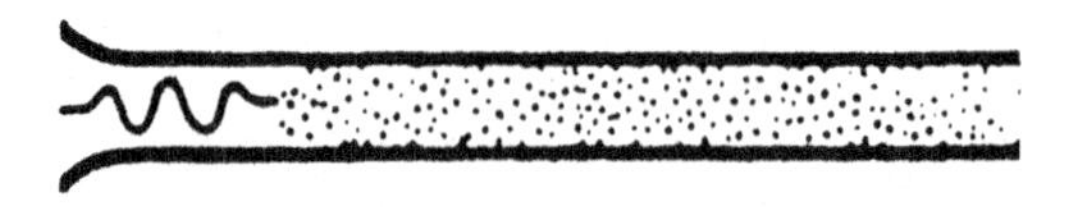

Figure 2-7. Reynolds experiment—transition and turbulent flow.

In considering these flows, Reynolds noted that two principal forces occurred in such cases. One was inertia forces,

$$\text{Inertia forces} = \rho \overline{V}^2 \tag{2-5}$$

where $\overline{V}$ is the fluid's *average velocity.*

The other was viscous forces,

$$\text{Viscous forces} = \frac{\mu \overline{V}}{D} \tag{2-6}$$

where μ is the fluid's viscosity and D is the tube diameter.

It was further postulated that the ratio of these two forces could give an important phenomenological insight into fluid flow. The ratio of the inertial forces to the viscous forces gives a dimensionless grouping

$$\frac{\text{Inertia forces}}{\text{Viscous forces}} = \frac{\rho \overline{V}^2}{\dfrac{\mu \overline{V}}{D}} = \frac{D \overline{V} \rho}{\mu} \tag{2-7}$$

This ratio is called the Reynolds number (Re).

The significance of the Reynolds number can best be realized by considering the behavior shown in Figures 2-5 and 2-7. For the laminar region (lower flow rates) the viscous forces predominate, giving low Reynolds number values. As flow rates increase, inertia forces become important until in turbulent flow these forces predominate.

Interestingly, in tube flow the Reynolds number clearly indicates the range of a given type of flow. For example, for Reynolds numbers up to 2100 the flow is laminar; from 2100 to about 4000 we have transition flow; and from 4000 on up, the flow is turbulent. Actually, it is possible to extend laminar flow beyond 2100 if done in carefully controlled experiments. This, however, is not the usual situation found in nature. Furthermore, it should be mentioned that the boundary between transition and turbulent flow is not always clearly defined. The ranges given above are considered to hold for most situations that would be encountered.

We can further delineate the differences between laminar and turbulent flow by considering the shape of the relative velocity profiles of each for tube flow (see Figure 2-8). In considering this figure, remember that the values of velocity are relative ones (hence the center line velocities for laminar and turbulent are not the same). Furthermore, see that the fluid velocity is taken to be zero at the tube wall. This convention means that a molecular layer of the fluid has a zero velocity. Using this convention enables the engineer to more easily deal with fluid mechanics in a reasonable and effective manner.

As can be seen from the shapes of the curves in Figure 2-8, there is a considerable difference between laminar and turbulent flow. The shape of the former is a true parabola. This parabolic shape is a characteristic of laminar flow. Turbulent

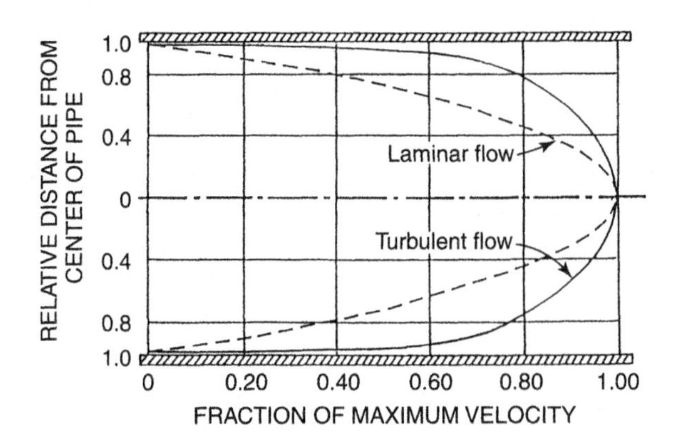

Figure 2-8. Velocity distributions in a pipe. (Adapted from references 1 and 9.)

flow, on the other hand, is blunter in shape. The ultimate beyond both laminar and turbulent flow is ideal or plug flow where all the velocities across the tube cross section are the same.

The plug or ideal flow concept is used in a number of applications. Particular examples are such systems as flow reactors or flow-through packed beds. Plug flow cases will be considered later in the text.

CLASSIFICATION OF FLUID BEHAVIOR

We have seen that many fluids obey Newton's Law of Viscosity. In these fluids, which are called Newtonian, the viscosity is a property of the system. As such, it depends on the substance or substances in the system, temperature, and pressure but not on the velocity gradient, which is the rate of shear or on the time parameter.

In nature there are many other fluid systems in which this is not the case. Further more, there are other cases where other factors influence the flow behavior. This means that there are a number of categories of fluid behavior. These are

1. Newtonian
2. "Simple" non-Newtonian (the viscosity is a function of shear rate)
3. "Complex" non-Newtonian (the behavior is a function of both rate of shear and the time parameter)
4. Fluids influenced by external force fields
5. Fluids that are noncontinuous
6. Relativistic fluids

The *"simple" non-Newtonian* are those fluids in which the rate of shear influences the flow behavior (see Figure 2-9). As can be seen, the straight line (i.e.,

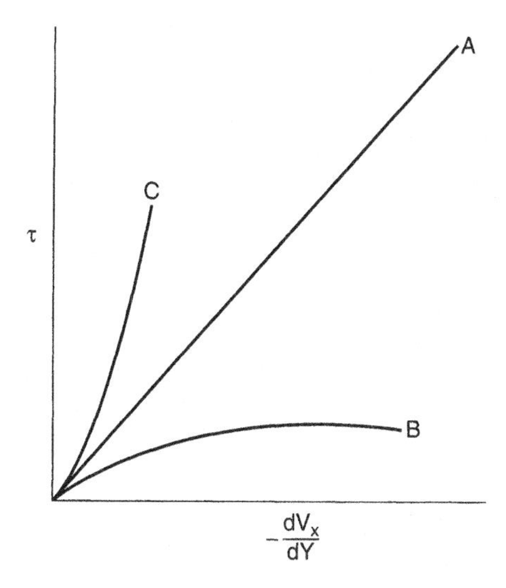

Figure 2-9. Comparison shear stress versus shear rate for Newtonian (A), shear-thinning (B), and shear-thickening (C) fluids.

constant viscosity) behavior of the Newtonian is not matched by two categories of fluids (shear-thinning and shear-thickening).

In order to get a better feel of these fluids, let us rewrite Newton's Law of Viscosity so that we have an apparent viscosity μ_{app}:

$$\tau_{yx} = -\mu_{\text{app}}\left(\frac{dV_x}{dy}\right) \tag{2-8}$$

and

$$\mu_{\text{app}} = \frac{\tau_{yx}}{\dfrac{-dV_x}{dy}} \tag{2-9}$$

Hence, if we plot the behavior of μ_{app} on a log–log plot, we obtain Figure 2-10.

As can be seen, the μ_{app} for a Newtonian is a constant. However, the other fluids show a decreasing apparent viscosity with increasing shear rate (shear-thinning fluid) or an increasing apparent viscosity with an increasing shear rate (shear-thickening fluid). Both of these fluid types have other popular names. Shear-thinning fluids are also called *pseudoplastic*, whereas shear-thickening fluids are termed dilatant. The latter name is somewhat unfortunate because it confuses the flow behavior of shear-thickening fluids with the concept of volumetric dilation (a different phenomenon which can or cannot occur in shear-thickening fluids).

Incidentally the overall science that considers flow and deformation of fluids (as well as solids) is termed *rheology*. It emanates from the Greek philosopher Heraclitus, who wrote "panta rhei," translated as "everything flows."

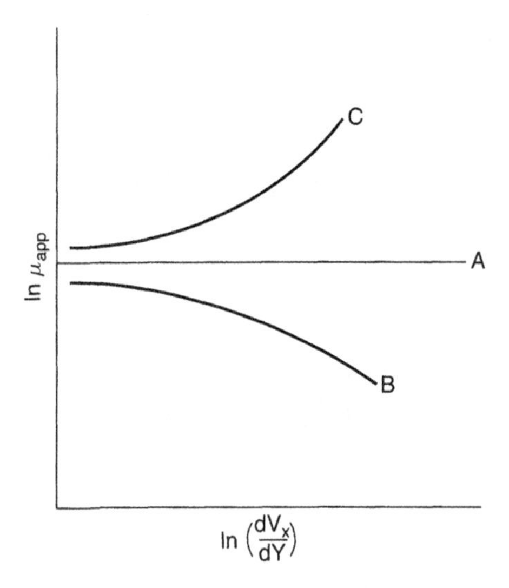

Figure 2-10. Logarithm apparent viscosity versus logarithm shear rate for Newtonian (A), shear-thinning (B), and shear-thickening (C) fluids.

It is obvious that the "simple" non-Newtonians cannot be treated by Newton's Law of Viscosity. As such, other approaches must be taken which lead to rheological constitutive equations that can be quite complicated. The simplest ones are equation (2-8) (the apparent viscosity expression) and the Ostwald–De Waele Power Law

$$\tau_{yx} = -k\left|\frac{dV_x}{dy}\right|^{n-1}\frac{dV_x}{dy} \tag{2-10}$$

where $|dV_x/dy|$ is the absolute value of the rate of shear. The K term is the consistency index, while n is the flow behavior index.

It is also possible to derive the expression

$$\tau_{yx} = K\left(\frac{d\gamma}{dt}\right)^n \tag{2-11}$$

where $d\gamma/dt$ is the rate of strain and $(d\gamma/dt = -dV_x/dy)$. The power law is not applicable over the entire range of shear rate behavior but rather only where $\log(\tau_{yx})$ versus $\log(d\gamma/dt)$ is a straight line. Also, note that the flow behavior index, n, is the slope of such a plot. The various systems are described in Table 2-2.

More on the behavior of "simple" non-Newtonian will be covered later in the text. However, one additional point of interest is the velocity profile (i.e., Figure 2-8) if a pseudoplastic fluid flowing in a circular tube is not a parabola (not even in laminar flow) but rather a blunted profile (as with Newtonian turbulent flow).

Table 2-2 System Values for *n*

Fluid Type	n	Examples
Shear-thinning pseudoplastic	<1.0	Polymer solutions; polymer melts; foods
Newtonian	1.0	Water; organic fluids
Shear-thickening; dilatant	>1.0	Continuous and dispersed phases (water–sand; cement, water–corn starch, etc.)

Complex non-Newtonians are those in which the time parameter becomes a factor. In essence,

$$\tau = \phi\left(\frac{d\gamma}{dt}, t\right) \tag{2-12}$$

A way of considering the behavior of these fluids is to first reflect that in a Newtonian fluid

$$\tau = \mu \frac{d\gamma}{dt} \tag{2-13}$$

with μ being independent of rate of shear. Next, if the behavior of an ideal elastic solid is considered (see Figure 2-11), we see that Hooke's Law applies:

$$\tau = G\gamma \tag{2-14}$$

Furthermore, it is possible to use mechanical analogs to represent the behavior of both of these systems. In the case of Hooke's Law we can use a spring (Figure 2-12) which, when distended by a stress, returns to its original shape and thereby releases the work previously done on it.

The Newtonian fluid can be represented by a dashpot (Figure 2-12) in which a piston placed in a liquid is attached to move over a pulley. When a stress is

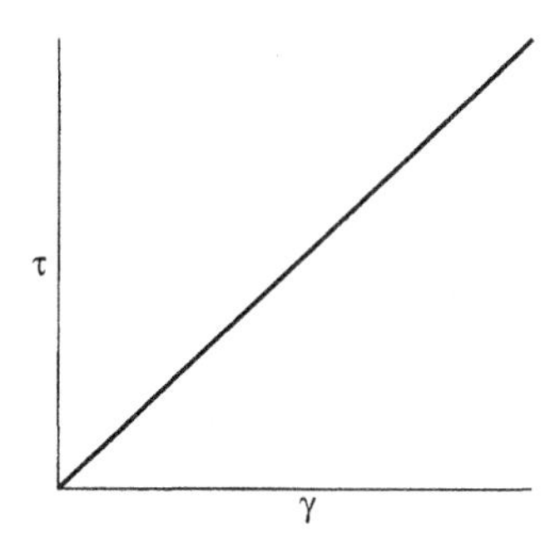

Figure 2-11. Hooke's Law behavior.

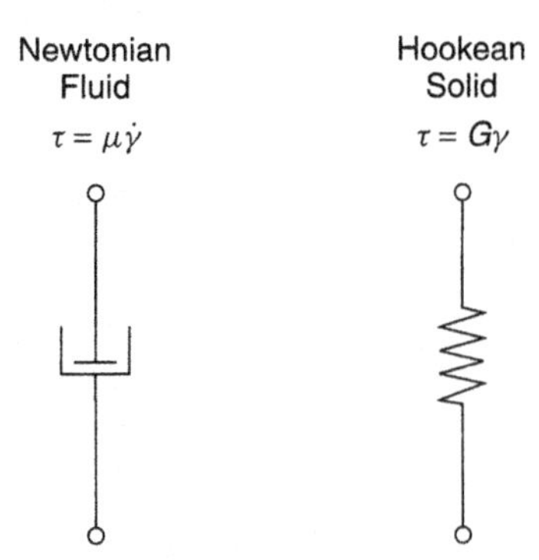

Figure 2-12. Ideal elastic solid spring model. Newtonian liquid dashpot model. (Adapted from reference 4 with permission of J. M. McKelvey.)

exerted, the piston moves to a new location. If the stress is removed, the piston stays in place and does not return to its original position. Further more, the work used to move the piston is not regained.

In nature there are fluids that have the characteristics of both the Newtonian fluid and the elastic solid. A simple representation of such a *viscoelastic* fluid is shown in Figure 2-13. Here we see that the application and release of stress does not allow the piston to stay in place but rather to recoil.

A more intensive and detailed treatment of viscoelasticity gives us the concept of relaxation time, which means, of course, that time is a parameter. Note that the viscoelastic fluid will behave as a simple non-Newtonian after some time (after the elastic effects have taken place). Typical viscoelastic fluids are certain polymer melts or polymer solutions.

Two other fluid types in which time is a parameter are those that have been categorized as time-dependent or more specifically as thixotropic or rheopectic.

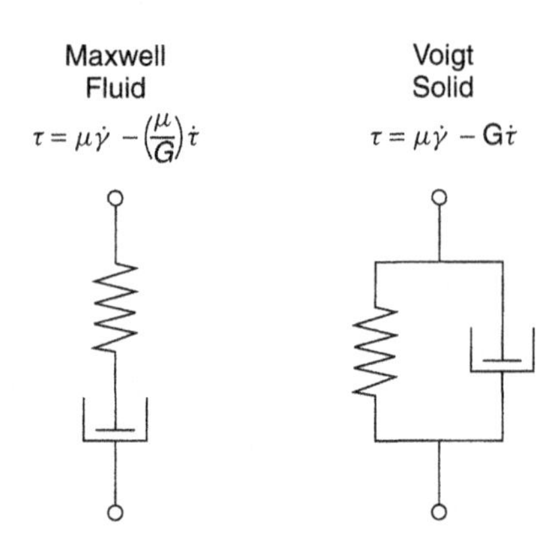

Figure 2-13. Viscoelastic models. (Adapted from reference 4 with permission of J. M. McKelvey.)

These fluids can be thought of as fluids in which

$$\tau = \phi\left(\frac{d\gamma}{dt}, t\right) \tag{2-15}$$

as shown in Figures 2-14 and 2-15.

Here, it can be seen that the apparent viscosity increases with time for the rheopectic fluid and decreases with time for the thixotropic. Examples of the latter are paints and inks, while rheopectic fluids include various lubricants.

Fluids influenced by external force fields are best exemplified by systems in which electrically conducting fields play a role (for example, magnetohydrodynamics). In such cases

$$\tau = \phi\left(\frac{d\gamma}{dt}, \text{ field}\right) \tag{2-16}$$

In essence, in order to handle such systems the fluid mechanical behavior must be spliced with Maxwell's equations for electromagnetic fields.

Originally used for such cases as plasmas and ionized gases, it now has possibilities elsewhere (i.e. electro-viscous fluids). An interesting aspect of such fluids is that their velocity profile behavior is similar to that of "simple" non-Newtonian in that they are blunted.

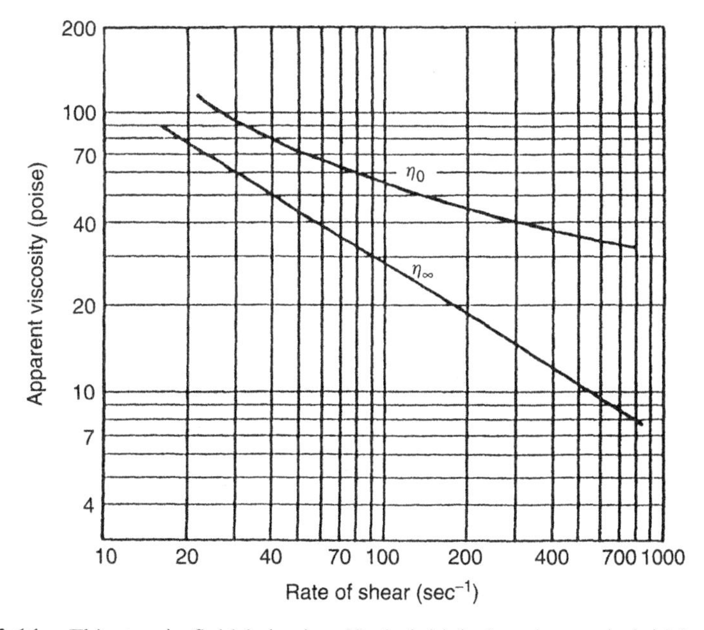

Figure 2-14. Thixotropic fluid behavior. No is initial viscosity: η_0 is initial viscosity; η_∞ is viscosity after infinite time. (Adapted with permission from reference 5. Copyright 1963, John Wiley and Sons.)

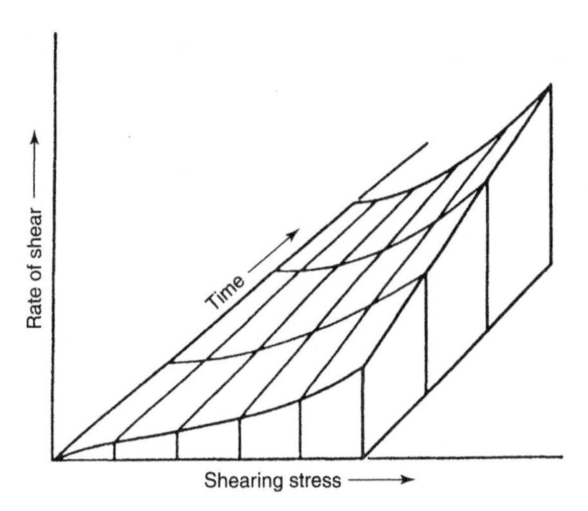

Figure 2-15. Rheopectic fluid behavior. (Adapted with permission from reference 5. Copyright 1963, John Wiley and Sons.)

In treating the fluid systems described previously we have assumed that the fluid is a continuum. There are cases where this is no longer correct. Specifically consider such cases as fogs, aerosols, sprays, smoke and rarefied gases (i.e., such as might be encountered extraterrestrially). All of these are *noncontinuum fluids* and as such mandate that special techniques be used.

Finally, we have the case where a fluid could approach the speed of light. This would invalidate Newtonian mechanics and as such mean that a form of *relativistic fluid mechanics* be used.

MULTIDIMENSIONAL FLUID FLOW—THE EQUATIONS OF CHANGE

We have considered a one-dimensional flow case for a Newtonian fluid (Newton's Law of Viscosity) as well as a phenomenological consideration of fluid dynamics (the Reynolds experiment, the Reynolds number, velocity profiles). Now, let us direct our attention to the concepts of the multidimensional cases.

First, consider the conservation of mass. We can write in word form that for a stationary element of volume we have

$$\text{Mass rate in} - \text{Mass rate out} = \text{Mass accumulation rate} \tag{2-17}$$

If this balance is carried out on the element shown in Figure 2-16, we obtain, by letting $\Delta x \Delta y \Delta z$ approach zero, the differential equation

$$\frac{\partial \rho}{\partial t} = -\left(\frac{\partial}{\partial x}\rho v_x + \frac{\partial}{\partial y}\rho v_y + \frac{\partial}{\partial z}\rho v_z\right) \tag{2-18}$$

which is the *equation of continuity.*

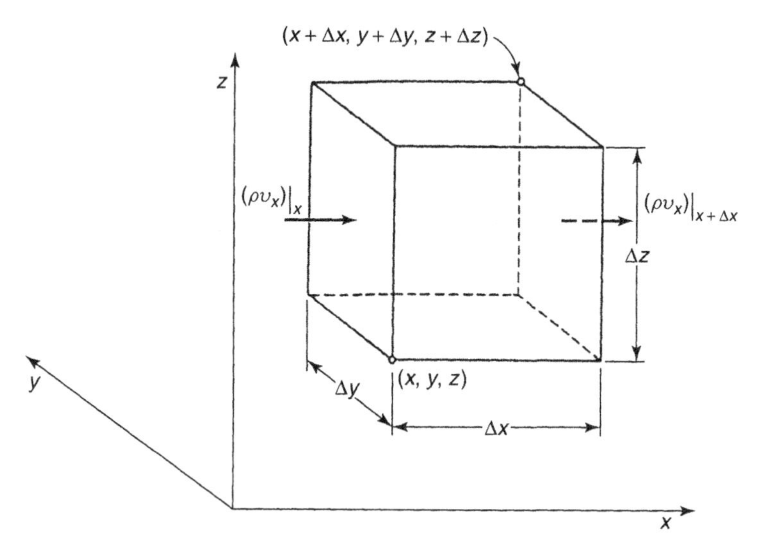

Figure 2-16. Volume element for mass balance. (Adapted with permission from reference 6. Copyright 1960, John Wiley and Sons.)

Furthermore, if the above is differential and separated into density and velocity derivatives, we obtain

$$\frac{\partial \rho}{\partial t} + v_x \frac{\partial \rho}{\partial x} + v_y \frac{\partial \rho}{\partial y} + v_z \frac{\partial \rho}{\partial z} = -\rho \left(\frac{\partial v_x}{\partial x} + \frac{\partial v_y}{\partial y} + \frac{\partial vz}{\partial z} \right) \tag{2-19}$$

in which

$$-\rho \left(\frac{\partial v_x}{\partial x} + \frac{\partial v_y}{\partial y} + \frac{\partial v_z}{\partial z} \right) = -\rho(\nabla \cdot v) \tag{2-20}$$

where ∇ is a vector operator and the term in the parentheses on the right-hand side of equation (2-20) is the dot product or divergence of velocity.

Then, for a fluid of constant density (i.e., incompressible) we have

$$(\nabla \cdot v) = 0 \tag{2-21}$$

Note that while true incompressibility cannot be attained, there are many instances where constant density is very closely approximated. Such an assumption is valid for many engineering situations and is therefore a useful concept.

A momentum balance for a volume element (similar to the one shown in Figure 2-17) would yield

(Momentum rate in) − (Momentum rate out)
+ (Sum of all forces acting on the system)
= (Momentum accumulation rate)

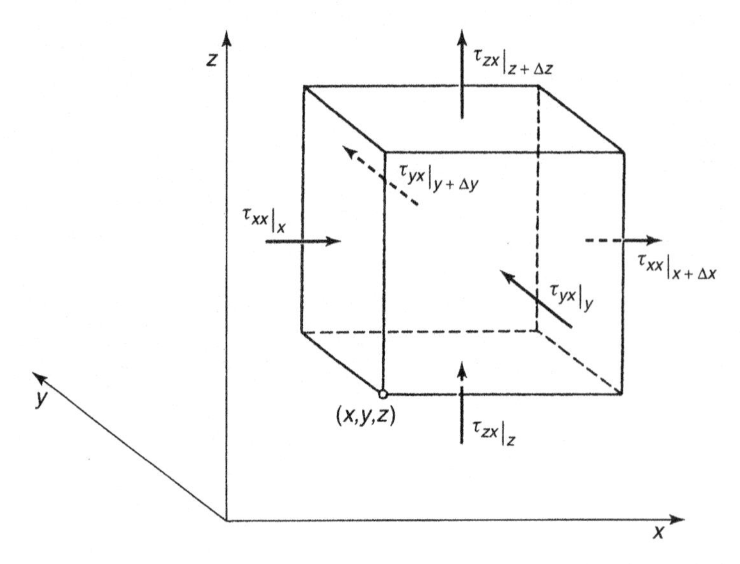

Figure 2-17. Volume element for momentum balance. (Adapted with permission from reference 6. Copyright 1960, John Wiley and Sons.)

The momentum rate expressions would include both a convective (for example, $\rho v_x \dfrac{\partial v_x}{\partial x}$) and a molecular transport term (involving shear stress).

The force term would include both pressure and gravity forces. Once again using the approach (see Figure 2-17) for the equation of continuity, let $\Delta X \Delta Y \Delta Z$ zero. This gives the following for the x component of the motion equation:

$$\rho\left(\frac{\partial v_z}{\partial t} + v_x\frac{\partial v_x}{\partial x} + v_y\frac{\partial v_x}{\partial y} + v_z\frac{\partial v_x}{\partial z}\right) = -\frac{\partial p}{\partial x} - \left(\frac{\partial \tau_{xx}}{\partial x} + \frac{\partial \tau_{yx}}{\partial y} + \frac{\partial \tau_{zx}}{\partial z}\right) + \rho g_x \tag{2-22}$$

In a similar way we can find the y and z components.

$$\rho\left(\frac{\partial v_y}{\partial t} + \mathrm{v_X}\frac{\partial v_y}{\partial x} + v_Y\frac{\partial v_y}{\partial y} + v_z\frac{\partial v_y}{\partial z}\right) = -\frac{\partial p}{\partial y} - \left(\frac{\partial \tau_{xy}}{\partial x} + \frac{\partial \tau_{yy}}{\partial y} + \frac{\partial \tau_{zy}}{\partial z}\right) + \rho g_Y \tag{2-23}$$

$$\rho\left(\frac{\partial v_z}{\partial t} + v_x\frac{\partial v_z}{\partial x} + v_Y\frac{\partial v_z}{\partial y} + v_z\frac{\partial v_z}{\partial z}\right) = -\frac{\partial p}{\partial z} - \left(\frac{\partial \tau_{xz}}{\partial x} + \frac{\partial \tau_{yz}}{\partial y} + \frac{\partial \tau_{zz}}{\partial z}\right) + \rho g_z \tag{2-24}$$

For the special case of constant density and viscosity, equations (2-22), (2-23), and (2-24) become, respectively,

$$x \text{ component}: \quad \rho\left(\frac{\partial v_x}{\partial t} + v_x\frac{\partial v_x}{\partial x} + v_y\frac{\partial v_x}{\partial y} + v_z\frac{\partial v_x}{\partial z}\right) = -\frac{\partial p}{\partial x} + \mu\left(\frac{\partial^2 v_x}{\partial x^2} + \frac{\partial^2 v_x}{\partial y^2} + \frac{\partial^2 v_x}{\partial z^2}\right) + \rho g_x \quad (2\text{-}25)$$

$$y \text{ component}: \quad \rho\left(\frac{\partial v_y}{\partial t} + v_x\frac{\partial v_y}{\partial x} + v_y\frac{\partial v_y}{\partial y} + v_z\frac{\partial v_y}{\partial z}\right) = -\frac{\partial p}{\partial y} + \mu\left(\frac{\partial^2 v_y}{\partial x^2} + \frac{\partial^2 v_y}{\partial y^2} + \frac{\partial^2 v_y}{\partial z^2}\right) + \rho g_y \quad (2\text{-}26)$$

$$z \text{ component}: \quad \rho\left(\frac{\partial v_z}{\partial t} + v_x\frac{\partial v_z}{\partial x} + v_y\frac{\partial v_z}{\partial y} + v_z\frac{\partial v_z}{\partial z}\right) = -\frac{\partial p}{\partial z} + \mu\left(\frac{\partial^2 v_z}{\partial x^2} + \frac{\partial^2 v_z}{\partial y^2} + \frac{\partial^2 v_z}{\partial z^2}\right) + \rho g_z \quad (2\text{-}27)$$

Note that equations (2-22)–(2-24) represent a general case (i.e., can be used, for example, for simple non-Newtonians), while equations (2-25)–(2-27) are applicable only to Newtonian fluids with constant densities.

It is also possible to derive both the equation of continuity and the equation of motion in cylindrical (r, θ, z) and spherical (r, θ, ϕ) coordinates. There are summarizeu in Tables C-1, C-2, C-3 of the Appendix.

Example 2-2 Consider a Newtonian fluid flowing in a circular tube at constant temperature (Figure 2-18). The fluid (in laminar flow) is in steady-state flow and has a fully developed velocity profile. What is the velocity profile across the tube?

For this case use Figure 2-18 with z being the axial dimension and r the radial dimension (i.e., cylindrical coordinates). If constant density and viscosity are assumed, we use the z-component portion of the equation of motion (Table C-2 of Appendix). The r and θ component are both zero because there is only a V_Z (velocity in the axial direction) and there are no pressure gradients in either the r or θ coordinates.

Now by considering the Equation of Continuity and the Z component of the Motion Equation, we will obtain the equation applicable to this situation. In carrying out this analysis, note that it is important to consider each term and to justify its nonapplicability.

First, for the Equation of Continuity we have

$$\frac{\partial \rho}{\partial t} + \frac{1}{r}\frac{\partial}{\partial r}(\rho r v_r) + \frac{1}{r}\frac{\partial}{\partial \theta}(\rho v_\theta) + \frac{\partial}{\partial z}(\rho v_z) = 0$$

Steady-state density constant — No V_r — No V_θ

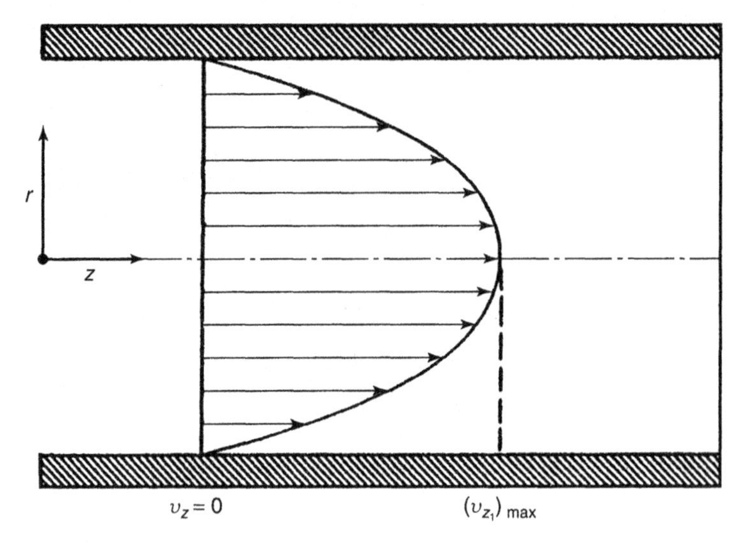

Figure 2-18. Laminar flow in a tube. (Adapted with permission from reference 6. Copyright 1960, John Wiley and Sons.)

Because density is constant, we have

$$\frac{\partial v_z}{\partial z} = 0$$

Also note that

$$\frac{\partial^2 v_z}{\partial z^2} = 0$$

Next, consider the z component of the Motion Equation:

$$\rho\left(\cancel{\frac{\partial v_z}{\partial t}} + v_r \cancel{\frac{\partial v_z}{\partial r}} + \frac{v_\theta}{r}\cancel{\frac{\partial v_z}{\partial \theta}} + v_z \cancel{\frac{\partial v_z}{\partial z}}\right) = -\frac{\partial p}{\partial z}$$

Steady state — No v_r — No v_θ — Equation of Continuity

$$+\mu\left[\frac{1}{r}\frac{\partial}{\partial r}\left(r\frac{\partial v_z}{\partial r}\right) + \frac{1}{r^2}\cancel{\frac{\partial^2 v_z}{\partial \theta^2}} + \cancel{\frac{\partial^2 v_z}{\partial z^2}}\right] + \rho g_z$$

No θ dependence — Equation of Continuity

so that

$$0 = \frac{-\partial P}{\partial z} + \frac{\mu}{r}\frac{\partial}{\partial r}\left(r\frac{\partial v_z}{\partial r}\right) + \rho g_z$$

Next, define a term $\overline{P}$ such that

$$\overline{P} = P - \rho g_z$$

For a z distance, $Z = L$, $(\partial p/\partial z - \rho Z g_z)$ can be rewritten $(\Delta P/L - \rho g_z)$, but

$$\frac{\Delta P/L}{L} - \rho g_z \frac{P_L - P_0 - \rho g_z L}{L} = \frac{\overline{P}_L - \overline{P}_0}{L}$$

Hence,

$$-\frac{\overline{P}_0 - \overline{P}_L}{L} = \frac{\mu}{r}\frac{\partial}{\partial r}\frac{r\partial v_z}{dr}$$

Then,

$$\frac{(\overline{P}_0 - P_L)r}{\mu} dr = d\frac{r dv_z}{dr}$$

Note that we have written the above as ordinary differentials because only r is involved.

Integrating and using the boundary conditions

$$r = 0, \frac{r dv_z}{dr} = 0$$

$$r = 0, \frac{r dv_z}{dr} = \frac{r\partial v_z}{\partial r}$$

gives

$$\frac{-(\overline{P}_0 - \overline{P}_L)r}{2\mu L} = \frac{dv_z}{dr}$$

Again integrating and using the boundary conditions

$$r = R, \qquad V_z = 0$$

$$r = r, \qquad V_z = V_z$$

gives

$$\frac{\overline{P} - \overline{P}_L}{4\mu L}[-r^2 + R^2] = V_z$$

or

$$V_z = \frac{\overline{P}_0 - \overline{P}_L}{4\mu L}\left[1 - \left(\frac{r}{R}\right)^2\right]$$

Remembering that phenomenonologically at $r = 0$ the velocity is a maximum, we have

$$(V_z)_{\max} = \frac{\overline{P}_0 - \overline{P}_L}{4\mu L}$$

Hence,

$$V_z = (V_z)_{\max}\left[1 - \left(\frac{r}{R}\right)^2\right]$$

Note that this is the equation of a parabola the relation predicted in the phenomenological studies covered earlier.

Example 2-3 Find the shear stress profile for the case of Example 2-2.

As in the previous example, only the Equation of Continuity and the z component of the Equation of Motion apply. Hence, from the Equation of Continuity we obtain

$$\frac{\partial V_z}{\partial Z} = 0$$

Using the z component (Table C-2 of Appendix) for cylindrical coordinates and analyzing the terms gives

$$\rho\left(\frac{\partial v_z}{\partial t} + v_r\frac{\partial v_z}{\partial r} + \frac{v_\theta}{r}\frac{\partial v_z}{\partial \theta} + v_z\frac{\partial v_z}{\partial z}\right) = -\frac{\partial p}{\partial z}$$

Steady state — No v_r — No v_θ — Equation of Continuity

$$-\left(\frac{1}{r}\frac{\partial}{\partial r}(r\tau_{rz}) + \frac{1}{r}\frac{\partial \tau_{\theta z}}{\partial \theta} + \frac{\partial \tau_{zz}}{\partial z}\right) + \rho g_z$$

No θ dependence — No z dependence

$$0 = \frac{-\partial P}{\partial z} - \frac{1}{r}\frac{\partial}{\partial r}(r\tau_{rz}) + \rho g_z$$

$$0 = \frac{\partial P}{\partial Z} - \rho g_z \frac{-1}{r}\frac{\partial}{\partial r}(r\tau_{rz})$$

Making use of $\overline{P} = P - \rho g_z$ and shifting to ordinary differentials, we obtain

$$-\frac{\overline{P}_0 - \overline{P}_L}{L} = \frac{-1}{r}\frac{d}{dr}(r\tau_{r_z})$$

Then by integrating we obtain

$$\tau_{rz} = \frac{\overline{P}_0 - \overline{P}_L}{2L}r + \frac{C_1}{r}$$

But $C_1 = 0$ because τ_{rz} would be infinite at $r = 0$.

$$\tau_{rz} = \frac{(\overline{P}_0 - \overline{P}_L)r}{2L}$$

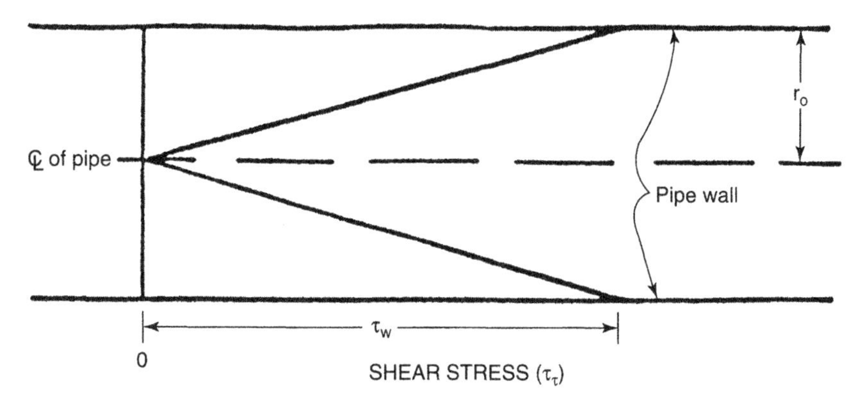

Figure 2-19. Shear stress profile in a tube. (Reproduced with permission from reference 9. Copyright 1997, American Chemical Society.)

This gives a linear relation between shear stress and pressure change as shown in Figure 2-19.

FLUID FLOW—THE MACROSCOPIC APPROACH

One of the interesting aspects of science in general and the engineering sciences in particular are the interrelationship and analogies that exist between them. We have seen examples of the latter in the case of the one-dimensional equations of change and the use of mechanical analogs in describing the Newtonian fluid and the elastic solid. The former is vividly shown in the interaction of thermodynamics with fluid flow.

The First Law of Thermodynamics for a closed system (constant mass) is

$$\Delta U = Q - W \tag{2-28}$$

where U is the internal energy, W is the work, and Q is the heat. A more general version of this equation is

$$\Delta E = Q - W \tag{2-29}$$

where E is all the energy types involved (i.e., internal, kinetic, potential, electrical, etc.).

If the First Law is derived for an open system (constant volume; see Figure 2-20) with a steady-state situation (mass flow rates in and out are equal), then the result is the First Law of Thermodynamics for an open system of a macroscopic balance:

$$\Delta\left(\frac{P}{\rho} + gZ + \frac{\overline{V}^2}{2}\right) = \frac{-dW_s}{dm} - \left(\Delta u - \frac{dQ}{dm}\right) \tag{2-30}$$

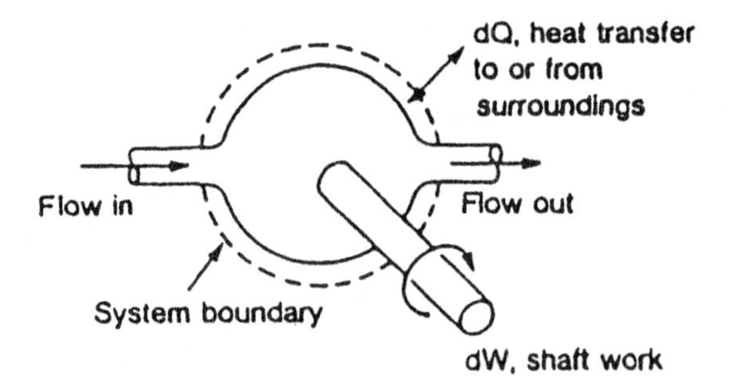

Figure 2-20. First law flow system. (Adapted from reference 3.)

In equation (2-30), P is pressure, ρ is density, Z is height above a datum plane, V is the average velocity of flow, W_S is shaft work [all work other than injection or $(\Delta P/\rho)$ work], u is the internal energy charge, Q is heat, and m is mass. The equation basically balances injection work $(\Delta P/\rho)$, potential energy $(g\Delta Z)$, and kinetic energy $\Delta V^2/2$ against shaft work, internal energy, and heat.

All of the quantities in equation (2-30) (pressure, density, etc.) *except* internal energy and heat can be directly measured. If the latter items could be handled, equation (2-30) would then be extremely useful for many engineering applications.

The method that is used is to consider an incompressible fluid (good approximation for most liquids and also for gases under certain conditions), and we can equate the internal energy and heat combination to a friction heating term:

$$\left(\Delta u - \frac{dQ}{dm}\right) = F_h \tag{2-31}$$

The concept of the friction heating can best be comprehended by realizing that the energy put into the fluid to get it to flow is converted to a nonuseful form.

This gives us the form shown below:

$$\Delta\left(\frac{P}{\rho} + gz + \frac{V^2}{2}\right) = \frac{-dW_s}{dm} - F_h \tag{2-32}$$

the Bernoulli Equation with frictional heating. It represents the starting point for the consideration of many problems in the area of fluid mechanics.

It is apparent that the frictional heating term, F_h, will have to be dealt if the Bernoulli Equation is to be applied. This subject will be deferred until the next chapter. However, it should be recognized that there are instances where fluid mechanics situations can be treated by the Bernoulli Equation if F_h is taken to be zero. Some of these will be discussed in the following sections.

Example 2-4 What is the volumetric flow rate for the tank (h = 3.048 m; exit cross-sectional area of 0.279 m^2) shown in Figure 2-21 if F_h is taken to be zero.

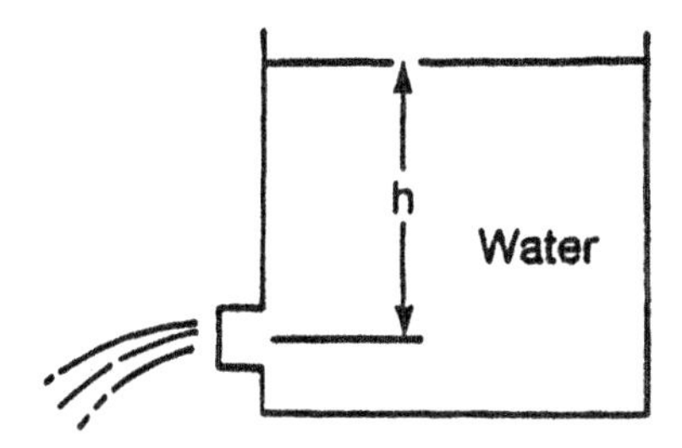

Figure 2-21. Tank flow exit. (Reproduced with permission from reference 9. Copyright 1997, American Chemical Society.)

Our starting point is the Bernoulli Equation. As pointed out in Example 2-2, a systematic and orderly approach should always be taken. An analysis of the Bernoulli Equation (inlet, 1, is liquid surface in the tank; outlet, 2, is the exit)

$$\Delta\left[\cancel{\frac{P}{\rho}} + gz + \frac{V^2}{2}\right] = \cancel{\frac{-dW_s}{dm}} - \cancel{F_h}$$

Inlet and outlet at atmospheric pressure; No shaft work performed; Frictionless

yields

$$g(Z_2 - Z_1) + \frac{V_2^2}{2} = 0$$

Also note that the average velocity from the exit (V_2) will be much greater than the velocity V_1. For this reason, V_1 can be neglected.

$$g(Z_2 - Z_1) + \frac{\overline{V}_2^2}{2} = 0$$

$$-g(h) + \frac{\overline{V}_2^2}{2} = 0$$

$$\overline{V}_2 = \sqrt{2gh}$$

$$\overline{V}_2 = \sqrt{2 \times 9.8 \text{ m/sec}^2 \times 3.048 \text{ m}}$$

$$\overline{V}_2 = 7.77 \text{ m/sec}$$

Now, the volumetric flow rate, Q, is defined as $Q = V$ (conduit cross-sectional area)

$$Q = (7.77 \text{ m/sec} \times 0.279 \text{ m}^2) = 2.17 \text{ m}^3/\text{sec}$$

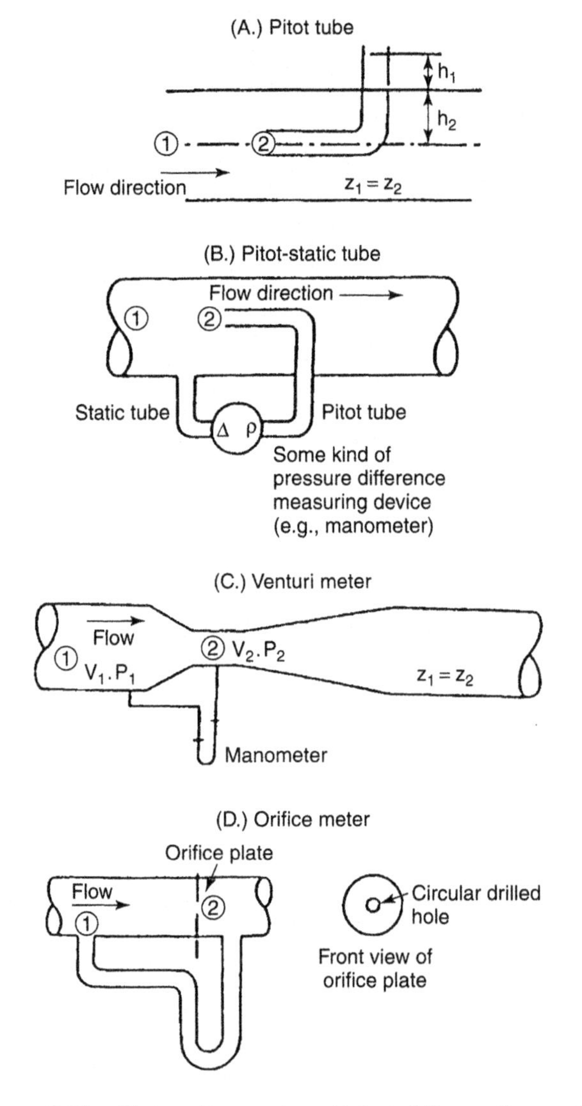

Figure 2-22. Various flow meters. (Adapted from reference 3.)

The frictionless form of the Bernoulli equation is the basis for many fluid-flow measuring devices. These include the pitot tube, the pitot-static tube, the Venturi meter, and the orifice meter (see Figure 2-22).

In general the equation form is

$$\frac{P_2 - P_1}{\rho} + \frac{V_2^2 - V_1^1}{2} = 0 \tag{2-33}$$

this neglects friction, height effects, and shaft work.

For of Figure 2-22A (the pitot tube), the equation is

$$V_1 = \sqrt{2gh_1} \tag{2-34}$$

Likewise, for Figure 2-22B (the pitot-static tube), the equation is

$$V_1 = \frac{\sqrt{2\Delta P}}{\rho} \tag{2-35}$$

The Venturi meter Figure 2-22C is handled by

$$V_2 = \left[\frac{2(P_1 - P_2)/\rho}{1 - \dfrac{A_2^2}{A_1^2}}\right]^{1/2} \tag{2-36}$$

The ratio of the areas occurs because volumetric flow, Q, is constant but the V and A values at points 1 and 2 are different.

The flow rate calculated from equation (2-36) is usually higher than that observed. This is due to two factors: friction heating and nonuniform flow. In order to compensate for this, an empirical coefficient C_v (the coefficient of discharge) is introduced into equation (2-36):

$$V_2 = C_v\left[\frac{2(P_1 - P_2)}{\rho\left(1 - \dfrac{A_2^2}{A_1^2}\right)}\right]^{1/2} \tag{2-37}$$

Equation (2-37) also applies to the fourth (Figure 2-22D) fluid-flow measuring device, the orifice plate. C_v behavior is, however, considerably different for Venturi and orifice meters (see Figures 2-23 and 2-24).

Example 2-5 What is the velocity as measured by an orifice plate (0.06-m diameter) in a 0.305-m-diameter pipe with the measured pressure drop being 75,150 n/m^2?

Equation (2-37) is applicable:

$$V_2 = C_v\left[\frac{2(P_1 - P_2)}{\rho\left(1 - \dfrac{A_2^2}{A_1^2}\right)}\right]^{1/2}$$

Note that all of terms except C_v and V_2 are known. The C_v presents a problem because equation (2-37) requires that a value for V_1 be known in order to find C_v.

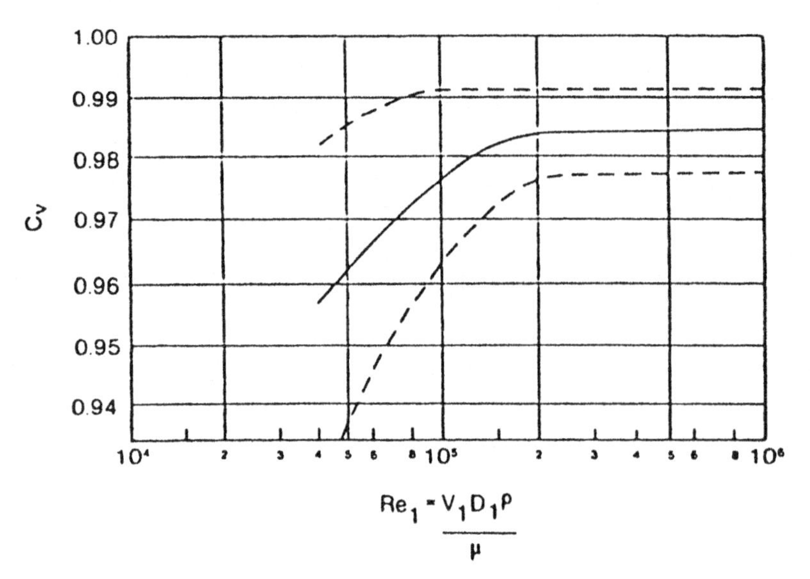

Figure 2-23. C_v data for Venturi meters. (Adapted with permission from reference 7. Copyright 1959, American Society of Mechanical Engineers.)

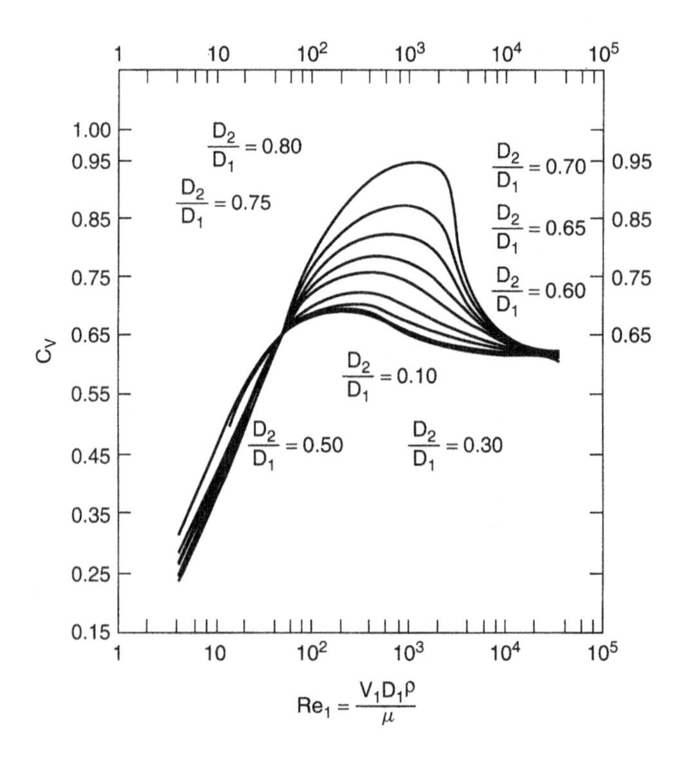

Figure 2-24. C_v Data for orifice meters. (Adapted from reference 8.)

V_1 cannot be determined unless V_2 is known (by using the volumetric flow rate $V_1 A_1 = V_2 A_2$). However, neither V_1 nor V_2 is known, which gives the initial aspect of an unsolvable problem.

However, if we consider the behavior of the curve of $D_2/D_1 = 0.20$ in Figure 2-24, we see from a Reynolds number of about 40 to 40,000 that

$$0.60 \leq C_v \leq 0.70$$

Therefore, as a starting point, we assume that $C_v = 0.65$. Using this value, we calculate V_2 and in turn V_1:

$$\overline{V}_2 = 0.65 \left[\frac{2(75{,}150 \text{ n/m}^2)}{999.6 \text{ kg/m}^3 \; 1 - \dfrac{\pi^2 D_2^4}{\pi^2 D_1^4}} \right]^{1/2}$$

$$\overline{V}_2 = 8 \text{ m/sec}$$

$$\overline{V}_1 = \frac{\overline{V}_2}{A_2} A_1 = 8 \text{ m/sec} \frac{(0.06/\text{m})^2}{(0.305 \text{ m})^2} = 0.321 \text{ m/sec}$$

If this V_1 value is used for a Reynolds number, we find that $C_v \neq 0.65$. Taking the C_v for the V_1, we repeat (i.e., carry out a trial and error solution or an iteration).

By so doing until the C_v value checks with V_1, we find $V_1 = 0.305$ m/sec.

PROBLEMS

2-1. An incompressible fluid is flowing at steady state in the annular region (i.e., torus or ring between two concentric cylinders). The coaxial cylinders have an outside radius of R and inner radius of λR. Find:

(a) Shear stress profile

(b) Velocity profile

(c) Maximum and average velocities

2-2. Repeat problem 2-1 for flow between very wide or broad parallel plates separated by a distance 2h.

2-3. Suppose an incompressible fluid flows in the form of a film down an inclined plane that has an angle of Θ with the vertical. Find the following items:

(a) Shear stress profile

(b) Velocity profile

(c) Film thickness

(d) Volumetric flow rate

2-4. An incompressible fluid flows upward through a small circular tube and then downward on the outside of the tube in a falling film. The tube radius is R. Determine:

(a) The falling film's velocity distribution

(b) Volumetric flow rate for the falling film

2-5. What are the shear stress and velocity distributions for an incompressible fluid contained between two coaxial cylinders (inner cylinder of radius λR stationary; outer cylinder of R moving with an angular velocity of ω. Also find the torque needed to turn the outer cylinder.

2-6. A cylindrical container of radius R with a vertical axis contains an incompressible liquid of constant density and viscosity. If the cylinder rotates about its axis with an angular velocity of ω, what is free liquid surface's shape? (*Hint*: Take r and Θ components of gravity to be zero.)

2-7. In wire coating operations a solid cylinder is pulled through a larger coaxial cylinder in which a fluid is contained. If the larger cylinder's inside radius is R and the solid cylinder's outside radius is λR, what is the fluid's velocity profile and the net volumetric fluid flow?

2-8. Obtain the velocity profile and volumetric flow rate for a non-Newtonian fluid obeying the Ostwald–De Waele Power Law in a circular tube.

2-9. A lubricant flows radially between two parallel circular disks from a radius r_1 to another r_2 because of ΔP.

(a) What is the velocity profile?

(b) Find the volumetric flow rate.

2-10. Assume that two fluids are flowing simultaneously between two broad parallel flat plates. The system is adjusted so that each fluid fills half of the space between the plates. Fluid A (more dense) has a viscosity of μ_A, and fluid B's viscosity is μ_B. Find the velocity distribution under a pressure drop of $P_L - P_O$. Also determine the shear stress profile.

2-11. Compare a thin annular ring flow situation to that for a very thin slit. What do you conclude?

2-12. A Stormer viscometer has two concentric cylinders (outer stationary, inner moving). Measurement of viscosity is done by measuring the rate of rotation of the inner cylinder under a given torque. Obtain the velocity distribution for v_θ in terms of the applied torque.

2-13. A tank of water is completely immersed in a tank of gasoline. Water flows out of an opening in the tank bottom. The height of the water above the opening is 10 meters. Specific gravity of gasoline is 0.7. What is the flow velocity of the water.

2-14. Water is flowing through a diffuser in which cross-sectional area 2 (exit) is 3 times the entrance cross section. What is the pressure change if entrance average velocity is 3 m/sec? Indicate all assumptions.

2-15. A venturi meter is used to measure the flow rate of water at 21°C. The pipe and nozzle diameters are 0.3048 m and 0.1524 m, respectively. Pressure drop is measured at 6895 Pa. What is the volumetric flow rate in the system?

2-16. Two limiting incompressible flow cases for the annulus are represented by flow through a slit and flow through a circular tube. The former is for the case of a large core (i.e., inner tube), whereas the latter represents a small core. For a 0.0254 m tube find the range of cores for each case that will be within 15 percent of the pressure drop values for the annulus.

2-17. Find the volumetric flow of an Ostwald–De Waele power fluid through a narrow slit.

2-18. Helium in a large tank escapes vertically through a hole in the tank's top at a velocity of 38.7 m/sec. The helium in the tank is above air in the same tank. What is the height of the helium in the tank?

2-19. What is the velocity distribution of an incompressible fluid in laminar flow between two cylinders (inner R, outer λR) rotating at ω_1 and ω_2, respectively.

2-20. Find the pressure distribution for an incompressible fluid at constant temperature flowing in the r direction between two spherical porous concentric spheres (inner radius R, outside radius KR).

2-21. A wetted wall tower is a piece of process equipment that uses a liquid film of δ meters thickness in laminar flow in the axial (Z direction). Find the velocity profile in the falling film.

2-22. The velocity of a motorboat is to be measured using a pitot tube device. If the maximum velocity of the boat is 7.7 m/sec, how high must the pitot tube be?

2-23. In open tanks with an exit nozzle a distance Z below the liquid surface, it is possible to obtain a relation between V, the exit velocity, and Z the height of liquid. Obtain data (in the form of a plot) for Z values up to 300 meters.

2-24. A paper cup designed in the shape of a truncated cone (top diameter of 0.06 m, bottom diameter of 0.0254 m and height of 0.08 m) filled with water suddenly develops a leak in the bottom. How long will it take for the cup to empty? (Assume that the entire bottom gives way.)

2-25. A vertical process unit (cylindrical, 2-m diameter, 4-m height) is cooled by water sprayed on top and allowed to flow down the outside wall. What is the thickness of the water layer for a flow rate of 0.002 m^3/sec at 40°C.

2-26. A transfer line used to move various grades of a fluid is 0.05 m in diameter and 10 m long. The line is first used for the commercial grade and then for ultrapure material. How much of the latter must be pumped before the exiting fluid contains not more than 1% of the commercial material? Assume that all properties are the same and that the flow is laminar.

2-27. A packed column is filled with solid objects (surface area of packing per unit volume of packing is 190 m^2/m^3). The water rate through the column (mass velocity) is 0.041 kg/m sec. Using the concept of flow over inclined surfaces, estimate the holdup in the column (i.e., m^3 of liquid per m^3 of packing). Assume half of the packing is wetted.

REFERENCES

1. W. L. McCabe and J. C. Smith, *Unit Operations of Chemical Engineering*, McGraw-Hill, New York (1967).
2. J. M. Coulson and J. F. Richardson, *Chemical Engineering*, Vols. 1 and 2, Pergamon, London (1978).
3. N. DeNevers, *Fluid Mechanics*, Addison-Wesley, Reading, MA (1970).
4. J. M. McKelvey, *Polymer Processing*, John Wiley and Sons, New York (1962).
5. J. R. Van Wazer, J. W. Lyons, K. Y. Kim and R. E. Colwell, *Viscosity and Flow Measurement*, Interscience, New York (1963).
6. R. B. Bird, W. E. Stewart and E. N. Lightfoot, *Transport Phenomena*, John Wiley and Sons, New York (1960).
7. *Fluid Meters: Their Theory and Applications*, ASME International, New York (1959).
8. G. L. Tuve and R. E. Sprenkle, *Instruments* **6**, 201 (1933).
9. R. G. Griskey, *Chemical Engineering for Chemists*, American Chemical Society, Washington, D.C. (1997).

3

FRICTIONAL FLOW IN CONDUITS

INTRODUCTION

Previously it was noted that the Bernoulli Equation with frictional heating was the starting point for many engineering calculations relating to analysis and design. However, in order to use this equation we must be able to satisfactorily handle F_h, frictional heating.

It will be our purpose in this chapter to develop those techniques that are needed to determine F_h. We will consider both laminar and turbulent flows, conduits with circular and noncircular cross sections, complex piping situations, expansions, and contractions. The overall result will be the ability to handle effectively many of the situations that can and do confront the engineering practitioner.

FRICTIONAL HEATING IN LAMINAR FLOW

For a fluid in laminar flow in a horizontal circular conduit we have

$$V_Z = \frac{(P_0 - P_L)}{4\mu L}\left(1 - \left(\frac{r}{R}\right)^2\right) \tag{3-1}$$

as per Example 2-2 (note that because the tube is horizontal, $\overline{P} = P$).

The volumetric flow for such a case is

$$Q = \int V_Z dA \tag{3-2}$$

or

$$Q = \frac{(P_0 - P_L)}{L}\frac{\pi}{\mu}\frac{D^4}{128} \tag{3-3}$$

Now if the Bernoulli Equation is applied to a length of the tube, L, then we obtain

$$\Delta\left(\frac{P}{\rho} + gZ + \frac{V^2}{2}\right) = \frac{-dW_s}{m} - F_h \tag{3-4}$$

No z change — No kinetic energy change — No shaft work

so that

$$F_h = \frac{P_0 - P_L}{\rho} \tag{3-5}$$

From equation (3-3), after multiplying both sides by ρ, we obtain

$$\frac{P_0 - P_L}{\rho} = QL\frac{\mu}{\rho}\frac{128}{\pi D^4} \tag{3-6}$$

so that

$$F_h = QL\frac{\mu}{\rho}\frac{128}{\pi D^4} \tag{3-7}$$

We can also show for a vertical tube — indeed, for a tube at any slope — that

$$F_h = QL\frac{\mu}{\rho}\frac{128}{\pi D^4} \tag{3-8}$$

FRICTIONAL HEATING IN TURBULENT FLOW — THE FRICTION FACTOR

In turbulent flow the fluid has a chaotic pattern as shown by the Reynolds experiment. Because of the nature of turbulent flow, the velocity will actually fluctuate (see Figure 3-1). It is possible to time-smooth these fluctuations such that

$$V_Z = \text{time-smoothed velocity} = \frac{1}{\Delta t}\int_t^{t+\Delta t} V_Z dt \tag{3-9}$$

Furthermore, the velocity is therefore

$$V_Z = \overline{V}_Z + V_z^1 \tag{3-10}$$

where V_Z' is the fluctuating velocity as shown in Figure 3-1.

The fluctuating velocity is a useful parameter in characterizing two important aspects of turbulent flow behavior, namely the intensity and scale of turbulence.

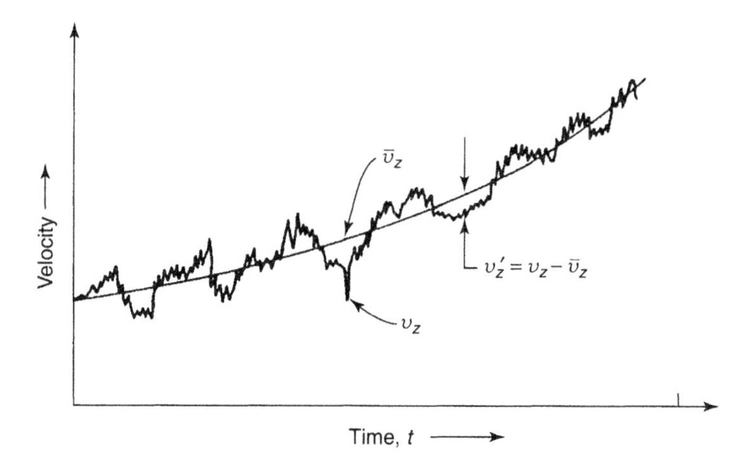

Figure 3-1. Instantaneous, time-smoothed, and fluctuating velocities. (Adapted with permission from reference 1. Copyright 1960, John Wiley and Sons.)

Intensity deals with the speed of rotation of the eddies and the energy contained in an eddy of a given size. Scale of turbulence measures the size of the eddies. Equations for each are given below:

$$\text{Intensity of turbulence (in percent)} = \frac{\sqrt{\overline{V_i'^2}}}{\langle \overline{V}_i \rangle}(100) \tag{3-11}$$

where $\langle \overline{V}_i \rangle$ is the average velocity (of the time-smoothed velocity $\overline{V}_i$) in the i direction and V_i' the fluctuating velocity in the i direction.

$$\text{Scale turbulence of} = \int_0^\infty R_V \, dj \tag{3-12}$$

where j is a coordinate not in the direction of flow and

$$R_V = \frac{\overline{(V_i')_1 (V_i')_3}}{\overline{(V_i')_1^2}\ \overline{(V_i')_3^2}} \tag{3-13}$$

The $(V_i')_1$ and $(V_i')_3$ represent fluctuating velocity values at j coordinates of 1 and 3.

Typical values of the intensity of turbulence range from 1 to 10 percent in flow in circular conduits. A scale of turbulence value for air flowing in a tube at 12 m/sec is 0.01 meter (average measure of eddy diameter). Now in order to carry out the *frictional flow* analysis as with laminar flow, we must first substitute $\overline{V}_Z + V_z'$ everywhere in the motion equation that V_Z appears. If the overall resultant equation is time-smoothed, there is no effect on $\overline{V}_Z$. Furthermore, by the nature of the fluctuating velocity we have

$$\overline{V}'_Z = 0 \tag{3-14}$$

However, there, are additional terms of the form $\rho\overline{V'_Z V'_Z}$ which are not zero. These terms, components of the turbulent momentum flux, are called Reynolds stresses. Their nature is such that they must be handled semiempirically. As such, therefore, the approach used for frictional heating in laminar flow cannot be used.

What is done for frictional heating in turbulent flow is to use an approach that relies on the scaling factor or dimensionless group approach. This can be handled in a number of ways.

One approach is to visualize frictional heating in a long, smooth tube as being proportional to the stress in the fluid (a function of velocity gradient) and the velocity itself. If the ratio of V/D is used for the gradient, then

$$F_h \sim \frac{V^2}{D} \tag{3-15}$$

Next, F_h should also be proportional to the tube length so that

$$F_h \sim \frac{LV^2}{D} \tag{3-16}$$

The proportionality is changed to an equality by using an empirical function, f (the friction factor):

$$F_h = \frac{2fLV^2}{D} \tag{3-17}$$

Experimentation shows that for long, smooth tubes we have

$$f = \Theta\left(\frac{DV\rho}{\mu}\right) = \Theta(\text{Re}) \tag{3-18}$$

where Re is the Reynolds number.

The concept of the friction factor can be also developed by alternative techniques. One approach is to define F_{kinetic} as

$$F_{\text{kinetic}} = AKf \tag{3-19}$$

where F_{kinetic} is the force associated with the fluid's kinetic behavior, A is a characteristic area, K is a characteristic kinetic energy per unit volume, and f is the friction factor.

Then,

$$F_{\text{kinetic}} = (\pi DL)\left(\frac{1}{2}\rho V^2\right)f \tag{3-20}$$

Also,

$$F_{\text{kinetic}} = \int_0^L \int_0^{2\pi} \left(-\mu \frac{\partial V_Z}{\partial r}\right)\bigg|_{r=R} R d\theta\, dZ \tag{3-21}$$

Ultimately for a long, smooth tube it can be shown that

$$f = \Theta(\text{Re}) \tag{3-22}$$

Still another technique is to use the Buckingham Pi Theorem, which involves taking the quantities involved and writing then in terms of the appropriate dimensions (i.e., mass, length, time) and then solving a set of relations to obtain the pertinent dimensionless group (7).

For the case of laminar flow, note that if frictional heating [equation (3-7)] is equated to equation (3-17), we obtain

$$QL\frac{\mu}{\rho}\frac{128}{D^4} = F_h = \frac{2fV^2}{D} \tag{3-23}$$

and

$$Q = V\frac{\pi D^2}{4} \tag{3-24}$$

so that

$$\frac{V\mu 16}{\rho D^2} = \frac{2fLV^2}{D} \tag{3-25}$$

and

$$f = \frac{16\mu}{DV\rho} = \frac{16}{\text{Re}} \tag{3-26}$$

Hence, the friction factor concept can be used for both laminar and turbulent flow.

EFFECT OF TUBE ROUGHNESS—THE OVERALL FRICTION FACTOR CORRELATION

The friction factor derivation culminating in the relationship of equations (3-18) and/or (3-22) specified that the tube be smooth. This raises a natural question as to what the effect of tube roughness would be on frictional heating.

Consider tubes with various roughness (see Figure 3-2). If a fluid were to flow in such tubes in laminar flow, the streamlines would conform in such a way as to minimize the effect of tube roughness on frictional heating. On the other hand,

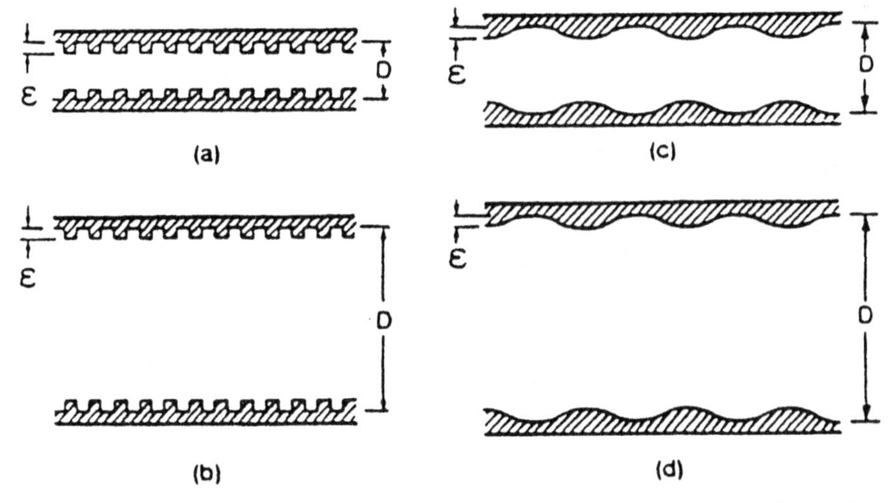

Figure 3-2. Various types of tube roughness. (Reproduced with permission from reference 8. Copyright 1997, American Chemical Society.)

in turbulent flow with its vortices and eddies there would be a perceptible effect of the roughness on the friction factor.

Hence, for cases of rough tubes with turbulent flow, an additional dimensionless group ε/D becomes important. This group represents the ratio of the average tube roughness protuberance to tube diameter.

Essentially then in such cases

$$f = \phi\left(\text{Re}, \frac{\varepsilon}{D}\right) \tag{3-27}$$

Some typical values of tube roughness are given in Table 3-1.

The overall correlation of f with Reynolds number and ε/D is shown in Figure 3-3. A few comments are in order. First of all see that, as mentioned,

Table 3-1 Values of Surface Roughness for Various Materials

Material	Surface Roughness, ε (m)
Drawn tubing (brass, lead, glass, etc.)	0.0000015
Commercial steel or wrought iron	0.0000457
Asphalted cast iron	0.0001219
Galvanized iron	0.0001524
Cast iron	0.0002591
Wood stave	0.0001829–0.0009144
Concrete	0.0003048–0.0030480
Riveted steel	0.0009144–0.0091440

Source: Reference 2.

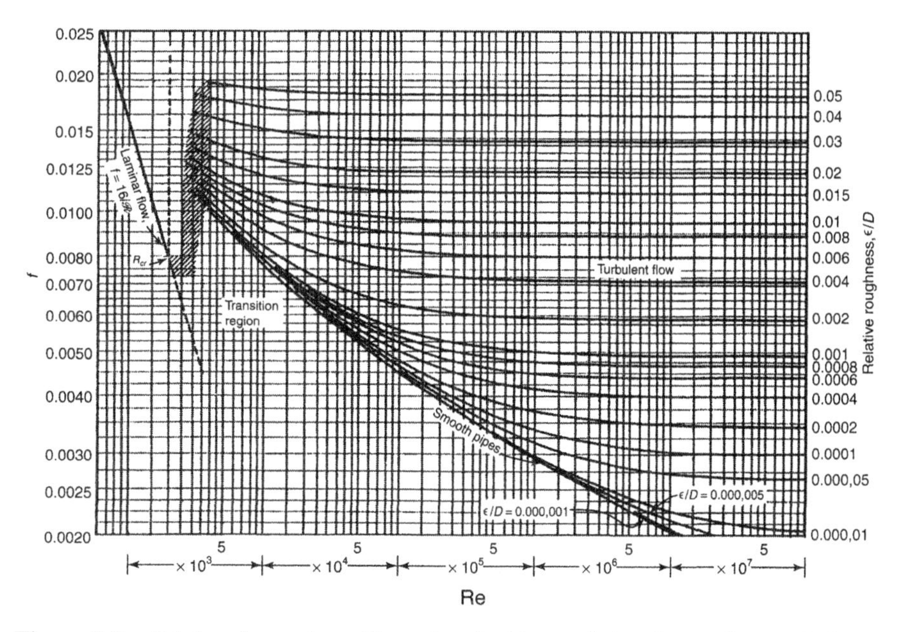

Figure 3-3. Friction factor chart. (Reproduced with permission from reference 9. Copyright 1944, American Society of Mechanical Engineers.)

there is no roughness effect in laminar flow. Furthermore, the equation of the laminar line is

$$f = \frac{16}{\text{Re}} \tag{3-28}$$

It should be understood that Figure 3-3 is only for those fluids that obey Newton's Law of Viscosity (i.e., Newtonian fluids).

It should also be pointed out that another overall f correlation exists that differs from Figure 3-3. In this other case, f is so defined that in laminar flow

$$f^1 = \frac{64}{\text{Re}} \tag{3-29}$$

The f^1 is used to distinguish the value of equation (3-26) from that of Figure 3-3. In essence, then

$$f^1 = 4f \tag{3-30}$$

Care should be taken as to which of the friction factors are being used.

The usefulness and comprehensiveness of Figure 3-3 can't be understated. These data cover (a) conduit diameters from fractional centimeter ranges up to multiples of meters, (b) viscosities from gases to highly viscous liquids, and (c) a wide range of densities. As such, it is a very useful engineering tool.

FLOW-THROUGH FITTINGS

The modern chemical or petroleum processing facility is characterized by its complex piping layout. A portion of the complexity is reflected in the presence of various types of fittings (some examples are shown in Figure 3-4).

The flow through these fittings will give rise to frictional heating that cannot directly be handled using Figure 3-3 and equation (3-17). Instead, we make use of another concept.

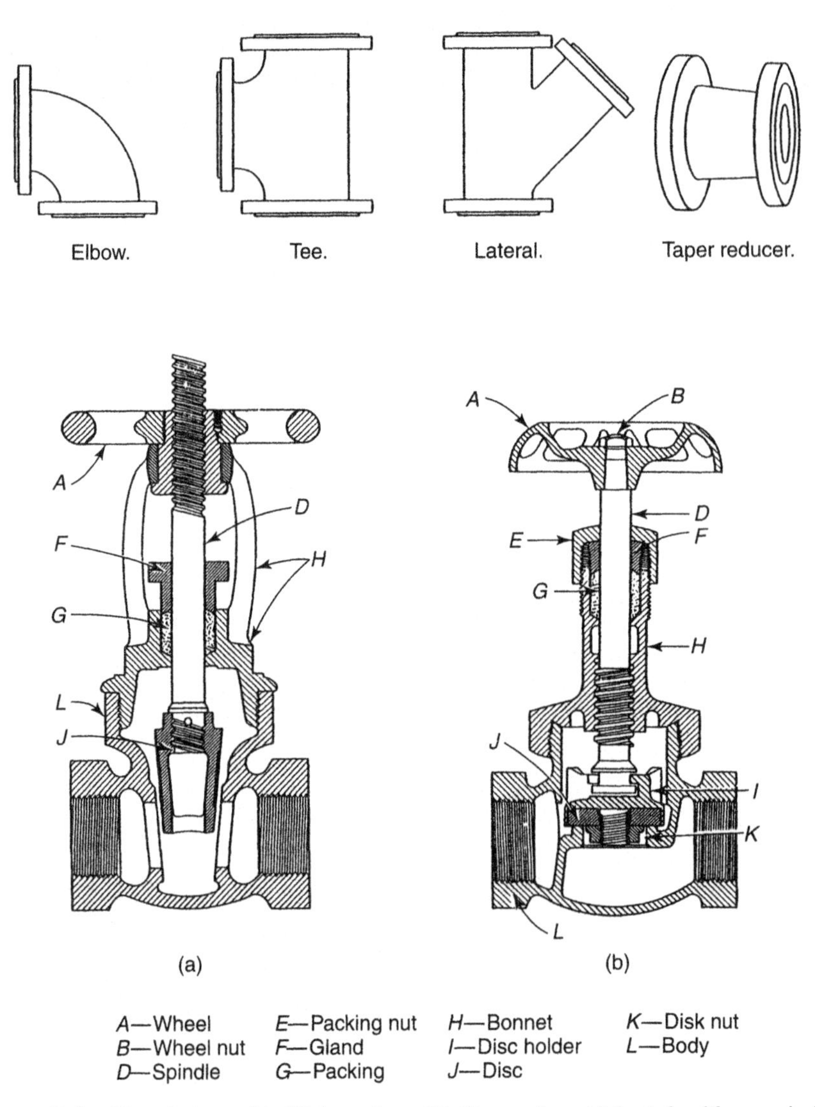

Figure 3-4. Pipe fittings. (a) Globe valve. (b) Gate valve. (Adapted with permission from reference 2. Copyright 1960, John Wiley and Sons.)

Table 3-2 Equivalent Lengths for Various Kinds of Fittings

Type of Fitting	Equivalent Length L/D (Dimensionless)
Globe valve, wide open	340
Angle valve, wide open	145
Gate valve, wide open	13
Check valve (swing type)	135
90° standard elbow	30
45° standard elbow	16
90° long-radius elbow	20

Source: Reference 5.

This leads to the idea of equivalent length for a given fitting. That is, in turbulent flow a length of a pipe of a certain, diameter will give the same frictional loss as a particular fitting. In other words, there is a length-to-diameter ratio that is characteristic of the fitting.

A listing of such values is given in Table 3-2.

Hence, if a pipeline has a tube diameter of 0.025 meters, a globe valve (wide open) will have the same effect as (340) (0.025) or 8.5 m of straight pipe. Two such valves would be the equivalent of 17.0 m.

Therefore, in order, to compute frictional heating in a pipeline, it is necessary to sum all of the (L/D) ratios for all fittings present in the system. This sum is then multiplied by the pipe diameter to yield an equivalent length of pipe that must be added to the actual pipe length. The "new" length can then be used to calculate frictional heating for the system.

EXPANSIONS AND CONTRACTIONS

Sometimes there are situations where the pipe diameters change for a particular system. Such situations result in contractions or expansions (see Figure 3-5) that cause frictional heating. In these instances, the frictional heating is related to an average velocity by a resistance coefficient K.

$$F_h = K\frac{V_1^2}{2} \tag{3-31}$$

The average velocity V_1 is that in the small-diameter tube (see Figure 3-6).

The data of Figure 3-6 show that the frictional heating is generally higher for enlargements. However, also note that the *change* of the resistance coefficient for sudden contractions is not as great as that for sudden enlargements. In other words, the K values for sudden contractions decrease only about 20 percent from a diameter ratio of 0 to one of 0.4, while those for enlargement change about 33 percent for the same diameter ratio range.

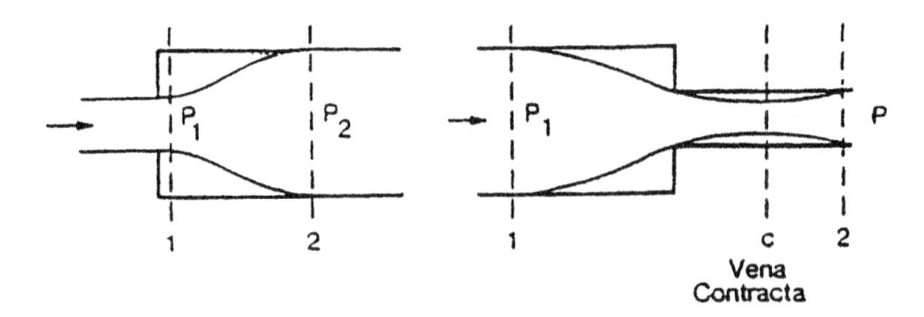

Figure 3-5. Expansion and contraction in pipe flow. (Adapted from reference 4.)

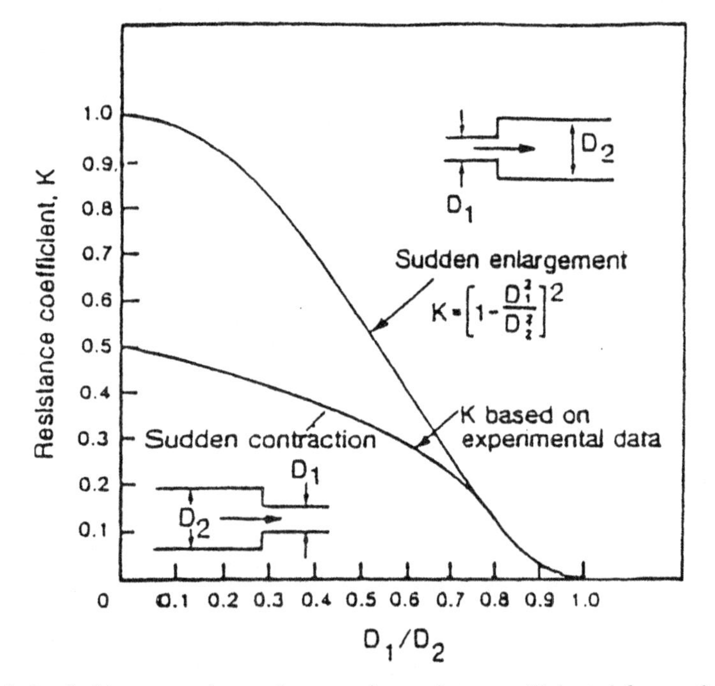

Figure 3-6. Sudden expansion and contraction resistances. (Adapted from reference 5. Permission of Crane Co. Copyright, all rights resolved.)

Generally, expansion and contraction losses are important mainly for short pipe lengths.

One other point relevant to contraction is illustrated in Figure 3-7, which shows the development of a boundary layer of fluid on the walls of the tube. Note that the fluids velocity profile forms within this boundary layer in contrast to the remainder of the fluid which remains in plug or ideal flow. Ultimately, the growing boundary layers come together in the tube center. This point is the length at which the fluid's velocity profile is completely developed.

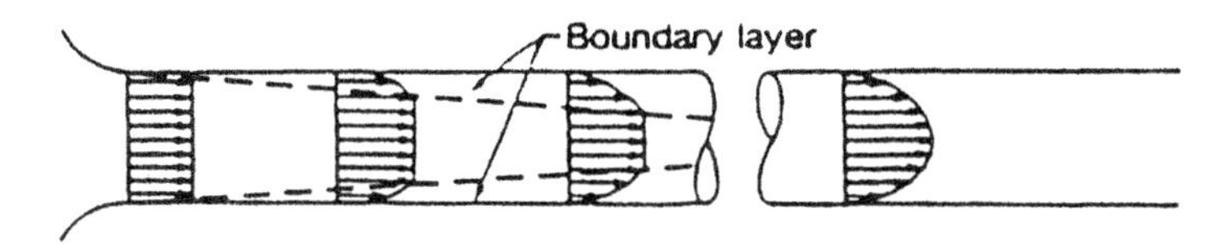

Figure 3-7. Boundary layer and velocity profile development. (Reproduced with permission from reference 8. Copyright 1997, American Chemical Society.)

This leads to the concept of an entrance length L_e needed for the full development of the velocity profile. If the fluid entering the contraction is in turbulent flow, L_e is about $50D$. For laminar flow we have

$$\frac{L_e}{D} = 0.05\text{Re} \tag{3-32}$$

If, for example, the smaller pipe were 0.05 m in diameter, the entrance length for turbulent flow would be 2.5 m; while for laminar flow with an Re of 1200, L_e would be 3 m.

FRICTIONAL HEATING IN NONCIRCULAR CONDUITS

Conduits do not always have circular cross sections. For these cases, we must have a different approach. Actually, there are two such approaches: one for laminar flow and the other for turbulent flow.

In the former case, we proceed just as before with the circular tube recognizing that we now have a different geometry. As an example, consider one-dimensional flow through a wide narrow slit (see Figure 3-8).

An analysis similar to that done for the circular tube will yield for the volumetric flow rate

$$Q = \frac{P_0 - P_L}{L}\frac{1}{\mu}\frac{1}{12}wh \tag{3-33}$$

where w and h are, respectively, the slit width and height.

For a horizontal system, we can show that equation (3-5) applies and hence

$$F_h = \frac{P_0 - P_L}{\rho} \tag{3-34}$$

Dividing both sides of equation (3-33) by density rearranging gives

$$\frac{P_0 - P_L}{\rho} = 12\frac{QL_\mu}{wh_\rho} \tag{3-35}$$

By identity

$$F_h = 12\frac{QL_\mu}{wh_\rho} \tag{3-36}$$

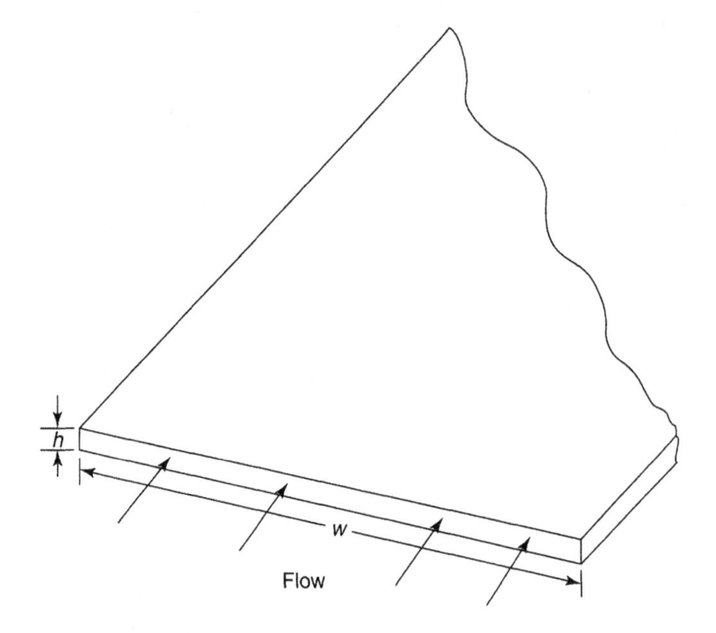

Figure 3-8. Flow between parallel plates.

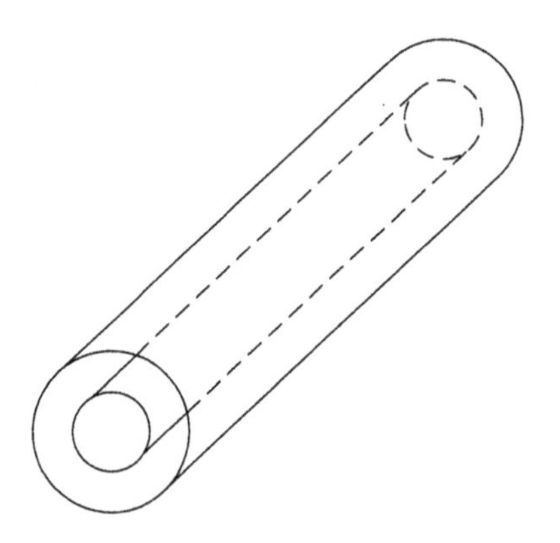

Figure 3-9. Annulus.

The same approach can be used for other cross sections. There is a difficulty with this method which is that we soon encounter mathematical complexity in obtaining the Q equations. As an example, consider the Q expression for the ring or torus of an annulus (Figure 3-9). This is given by

$$Q = \frac{P_1 - P_2}{L}\frac{1}{\mu}\frac{\pi}{128}(D_0^2 - D_1^2)\left[(D_0^2 - D_1^2) - \left[\frac{D_0^2 - D_1^2}{\ln\dfrac{D_0}{D_1}}\right]\right] \qquad (3\text{-}37)$$

Turbulent flow in noncircular cross sections cannot be handled by the laminar approach. This is, of course, because of the problems encountered with the Reynolds stresses.

Once again as with the circular tube case, we rely on an empirical approach. This involves making use of the notion that the shear stress occurring at the wall will be the same for a given fluid average velocity regardless of conduit shape. The result is the definition of a new term: the hydraulic radius, R_h, where

$$R_h = \frac{\text{Cross-sectional area perpendicular to flow}}{\text{Wetted perimeter}} \tag{3-38}$$

The usefulness of R_h becomes evident if we evaluate it for a circular tube

$$R_h = \frac{\dfrac{(\pi D^2)}{4}}{\pi D} = \frac{D}{4} \tag{3-39}$$

and

$$D = 4R_h \tag{3-40}$$

This leads to the concept that there is an equivalent or hydraulic diameter for *any cross section* such that

$$D_{\text{eq}} = D_h = 4R_h \tag{3-41}$$

It is then possible to use this D_{eq} or D hydraulic in the circular pipe expression for pressure drop:

$$\frac{\Delta P}{L} = \frac{2f_\rho V^2}{D_{\text{eq}}} \tag{3-42}$$

The question naturally arises as to whether the hydraulic radius approach could be used in laminar flow. The answer is a highly qualified yes—qualified in the sense that first of all the hydraulic radius approach gives a large deviation from the analytical approach. For example, in the case of the annulus, an error of over 40% would occur for an annulus in which $D_1 = 0.5D_2$. However, if we were, for example, to encounter a conduit of the shape shown in Figure 3-10, we would be hard pressed to find an analytical solution of the type given in equations (3-33) and (3-35).

Thus, it is recommended to use the analytical approach where possible, but to use hydraulic radius for complex cross sections while recognizing that the pressure drop could differ by an order of magnitude from the actual case.

SOLVED EXAMPLES

Example 3-1 Air is flowing through a horizontal tube of 2.54-cm diameter. What is the maximum average velocity at which laminar flow will be stable? What is the pressure drop at this velocity?

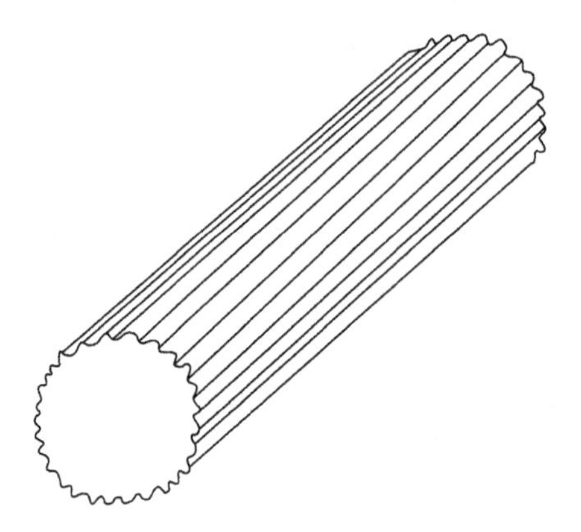

Figure 3-10. Corrugated pipe.

The transition point to unstable flow occurs at Re = 2100. Hence

$$\frac{DV_{\rho}}{\mu} = 2100$$

and

$$V = \frac{2100_{\mu}}{D_{\rho}} = \frac{(2100)\ (1.8 \times 10^{-5}) \left(\dfrac{\text{kg}}{\text{m sec}}\right)}{(0.025\ \text{m})(1.2\ \text{kg/m}^3)}$$

The properties (viscosity, density) can be obtained from the data in the appendices as illustrated there. Solving for V gives

$$V = 1.22\ \text{m/sec}$$

Using equation (3-7) together with the definition of Q[i.e., $V(\pi D^2/4)$], we obtain

$$\frac{\Delta P}{L} = \frac{128 Q\mu}{\pi D^4} = \frac{V\mu 32}{D^2}$$

and

$$\frac{\Delta P}{L} = \frac{(1.22\ \text{m/sec})(1.8 \times 10^{-5}\ \text{kg/m sec})(32)}{(0.0254\ \text{m})^2}$$

$$\frac{\Delta P}{L} = 1.041 \frac{\text{N/m}^2}{\text{m}}$$

Note that this is a very low pressure drop, which is to be expected for air and gaseous substance at a moderate Reynolds number.

Example 3-2 A pump takes water at 10°C from a large open reservoir and delivers it to the bottom of an open elevated tank (see Figure 3-11). The level of the tank averages 48.77 m above the surface of the reservoir. The pipe is 0.076 m in diameter and consists of 152.4 m (160 feet) of straight pipe, six elbows, two gate valves, and 2 tees ($L/D = 60$). The pump delivers 0.00898 m^3/sec. What is the horsepower consumed if the pump has a mechanical efficiency of 55%?

We use the Bernoulli balance developed in Chapter 2. Here we take points 1 and 2 to be the liquid surfaces in the reservoir and tank. Thus

$$\Delta\left(\frac{P}{\rho} + gZ + \frac{V^2}{2}\right) = \frac{-dW_s}{m} - F_h$$

Open to atmosphere (the P/ρ term is struck out) — No kinetic energy change (the $V^2/2$ term is struck out)

The elimination of the $\Delta P/\rho$ term results because the 152.4 m of height does not appreciably change the air's atmospheric pressure. Kinetic energy changes are also negligible because the liquid surface velocities are very low.

$$g\Delta Z + F_h = \frac{-dW_s}{dm}$$

$$F_h = \frac{4fLV^2}{2\,g_c D}$$

Next, we must determine an equivalent length that considers the fittings

$$L = (\text{height}) + (D)\left(\frac{L}{D}\right) \text{ fittings}$$

$$L = 152.4\ \text{m} + (0.076\ \text{m})(326)$$

$$L = 177.2\ \text{m}$$

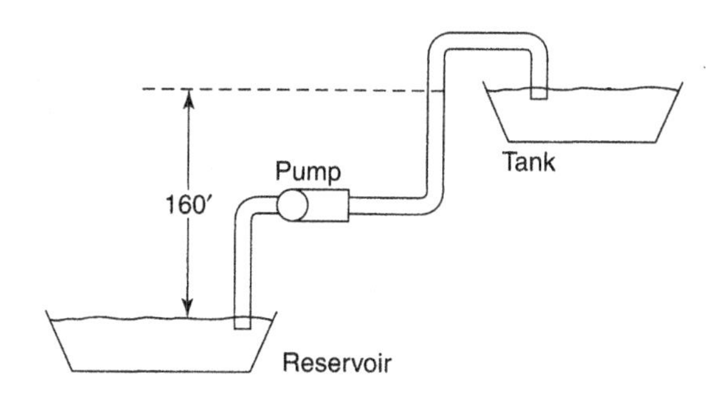

Figure 3-11. Schematic of flow in Example 3-2.

Also, the average velocity V can be determined from the volumetric flow rate

$$V = 0.00898\ \text{m}^3/\text{sec} \times \frac{(4)}{(\pi)(0.076\ \text{m})^2} = 1.98\ \text{m/sec}$$

In turn, we can find the Reynolds number:

$$\text{Re} = \frac{DV_\rho}{\mu} = \frac{(0.076\ \text{m})(1.98\ \text{m/sec})(999.6\ \text{kg/m}^3)}{(1.3 \times 10^{-3}\ \text{kg/m sec})}$$

$$\text{Re} = 115{,}700$$

Also, from Figure 3-3 we can find f (assuming a smooth tube):

$$f = 0.0046$$

Substituting into the modified Bernoulli balance and solving for shaft work, we obtain

$$\frac{-dw_s}{dm} = \frac{(4)\ (0.0046)(177.2\ \text{m})(1.98\ \text{m/sec})^2}{(2)\ (0.076\ \text{m})} + (48.77\ \text{m})(9.8\ \text{m/sec}^2)$$

Note that the units in the above are m^2/sec^2, which don't seem to make sense. Actually, we need to multiply by $1/g_c$ for the metric system (where $g_c =$ kg m/Newtons sec^2). This gives

$$\frac{-dW_s}{-dm} = 0.562\ \text{kJ/kg}$$

The work per unit mass is negative because we are putting work *into* the system. Next, converting to power (by multiplying by mass flow rate).

$$\frac{-dW_s}{dm}\frac{dm}{dt} = 0.562\ \text{kJ/kg} \times 0.00898\ \text{m}^3/\text{sec} \times 999.6\ \text{kg/m}^3$$

$$P = 5.05\ \text{kW}$$

Finally, we have to take into account the efficiency of the pump. This efficiency actually represents the irreversibility of the work performed (i.e., work was calculated for a reversible process):

$$\text{Actual power consumed} = \frac{5.05\ \text{kW}}{0.55} = 9.18\ \text{kW}$$

Example 3-3 How many gallons of water at 20°C can be delivered through a 400-m length of smooth pipe (0.15-m diameter) with a pressure difference of 1720 N/m^2?

In order to work this problem, we can use a trial-and-error method. That is, assume a flow rate, get f, and compute pressure drop. When the calculated pressure drop matches the actual value, then flow rate is correct.

We have an alternate method available that utilizes Figure 3-12. This is a plot of f versus Re f cross-plotted from the data of Figure 3-3. The function of Re f eliminates V:

$$\mathrm{Re}\sqrt{f} = \frac{dP}{\mu}\left(\frac{\Delta P D}{2L\rho}\right)^{1/2}$$

Hence, we can compute $\mathrm{Re}\sqrt{f}$ and then read off the value of f:

$$\mathrm{Re}\sqrt{f} = \frac{(0.15\ \mathrm{m})(998\ \mathrm{kg/m^3})}{(0.001\ \mathrm{kg/m\ sec})}\left(\frac{(1720\ \mathrm{N/m^2})(0.15\ \mathrm{m})}{(400\ \mathrm{m})(998\ \mathrm{kg/m^3})(2.0)}\right)^{1/2}$$

$$\mathrm{Re}\sqrt{f} = 2695$$

Then, from Figure 3-12 we have

$$f = 0.0057$$

$$\mathrm{Re} = \frac{\mathrm{Re}\sqrt{f}}{\sqrt{f}} = \frac{2695}{\sqrt{0.0057}} = 35{,}700$$

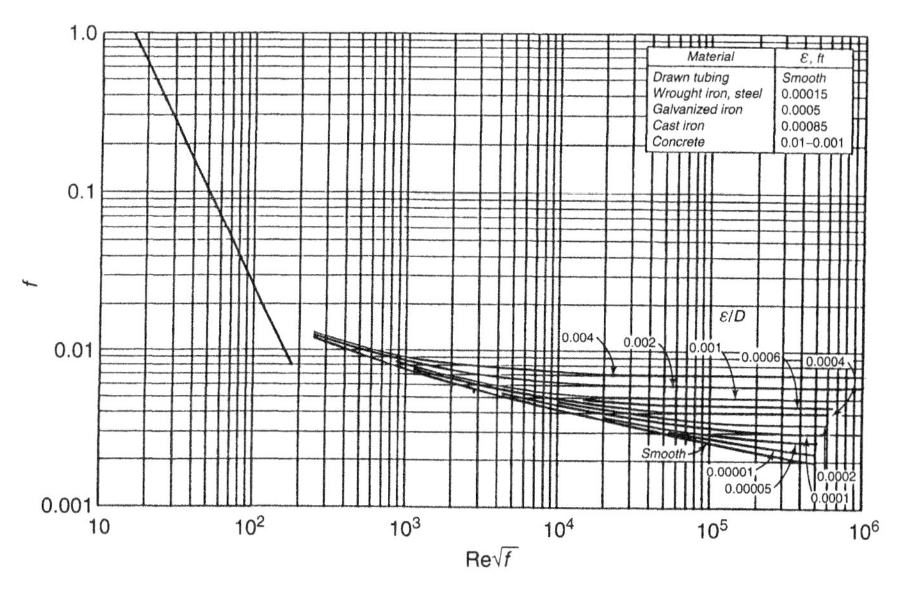

Figure 3-12. Modified friction factor plot. (Reproduced with permission from reference 8. Copyright 1997, American Chemical Society.)

and

$$\frac{DV\rho}{\mu} = 35{,}700$$

$$V = \frac{(35{,}700)(0.001\ \text{kg/m sec})}{(0.15\ \text{m})(998\ \text{kg/m}^3)}$$

$$V = 0.238\ \text{m/sec}$$

$$Q = \frac{\pi D^2 V}{4} = \frac{0.15\ \text{m}}{4} 0.238\ \text{m/sec} = 0.0042\ \text{m}^3/\text{sec}$$

$$Q = 67\ \text{gallons/minute}$$

Example 3-4 A large, high-pressure chemical reactor contains water at an absolute pressure of 1.38×10^7 N/m^2 and a temperature of 20°C. A 0.07-m inside diameter line connected to it ruptures at a point 3 m from the reactor (see Figure 3-13). What is the flow rate from the break?

Note that this represents an unsteady-state flow case. However, we can treat it as a pseudo-steady-state flow by realizing that the *initial* flow will be the maximum outflow.

Using the Bernoulli balance and taking the liquid surface and the pipe break as the relevant limits, we obtain

$$\Delta\left(\frac{P}{\rho} + gZ + \frac{V^2}{2}\right) = \frac{-dW_s}{m} - F_h$$

No Z effect   No shaft work

The potential energy effect is neglected because of the magnitude of the other effects:

$$\Delta\left(\frac{P}{\rho}\right) + \frac{V_2^2}{2} = -F_h$$

Neglect V_1

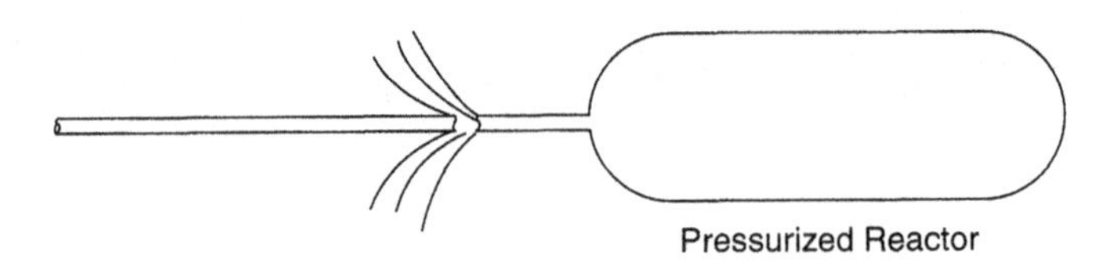

Figure 3-13. Schematic of Example 3-4.

This is so because the liquid surface velocity V_1 will be much lower than V_2 the velocity of the fluid out at the pipe break, but

$$F_h = \underset{\substack{\text{Entrance}\\ \text{effect}}}{\frac{KV_2^2}{2}} + \underset{\substack{\text{Pipe}\\ \text{effect}}}{4f\frac{L}{D}\frac{V_2^2}{2}}$$

We take the entrance effort to include only the contraction (from tank to pipe) because the expansion (from the broken pipe) takes place *past* boundary 2.

Substituting and solving for V_2, we obtain

$$V_2 = \left[\frac{2(\Delta P)/\rho}{1 + K + 4f(L/D)}\right]^{1/2}$$

We take the value of K from Figure 3-6 to be 0.5. The justification for this is that pipe diameter is much smaller than the tank and, further more, that K changes slowly for contractions.

For the pipe we have $\varepsilon/D = 0.000043/0.07 = 0.00063$, using an ε value for commercial pipe.

Now assuming Re to be large for such an ε/D, and noticing that for such cases the f versus Re relation is nearly flat, we estimate f to be about 0.0043 from Figure 3-3. Then

$$V_2 = \left[\frac{2(1.38 - 0.01) \times 10^7 \text{ N/m}^2}{998\dfrac{\text{kg}}{\text{m}^3}\left(1 + 0.5 + \dfrac{4 \times 0.0043 \times 3 \text{ m}}{0.07 \text{ m}}\right)}\right]^{1/2}$$

$$V_2 = 110 \text{ m/sec}$$

Next, we check our assumption for the f versus Re behavior.

The Reynolds number would then be

$$\text{Re} = \frac{(0.07 \text{ m})(110 \text{ m/sec})(998 \text{ kg/m}^3)}{(0.001 \text{ kg/m sec})} = 7.7 \times 10^6$$

Hence, at this Reynolds number we would be on the flat portion of the line, and the assumed f value is correct.

Also,

$$Q = (110 \text{ m/sec})\frac{\pi}{4}(0.07 \text{ m})^2 = 0.423 \text{ m}^3/\text{sec}$$

or

$$Q = 6800 \text{ gallons/minute}$$

Example 3-5 Find the hydraulic radius for each of the following: (1) a filled equilateral triangle (side $= a$; altitude $= h$); (2) an open semicircle (diameter D); (3) a torus of an annulus (outside diameter D_0; inside diameter D_1); (4) a square (side $= S$); (5) a semicircle with the top closed.

For the equilateral triangle we have

$$R_h = \frac{\frac{1}{2}ah}{3a} = \tfrac{1}{6}\ h$$

For the open semicircle we have

$$R_h = \frac{\frac{1}{2}(\pi D^2/4)}{\frac{1}{2}\pi D} = \frac{D}{4}$$

For annulus we have

$$R_h = \frac{\dfrac{\pi D_0^2}{4} - \dfrac{\pi D_1^2}{4}}{\pi D_0 + \pi D_1}$$

$$R_h = \frac{1}{4}\frac{D_0^2 - D_1^2}{(D_0 + D_1)} = \frac{1}{4}(D_0 - D_1)$$

Next, for the square we have

$$R_h = \frac{S^2}{4S} = \frac{S}{4}$$

Finally, for the closed semicircle we have

$$R_h = \frac{\frac{1}{2}(\pi D^2/4)}{\frac{1}{2}(\pi D + D)}$$

$$R_h = \frac{\frac{1}{2}(\pi D^2/4)}{\frac{1}{2}(\pi D + 2D)}$$

$$R_h = \frac{\pi D^2}{4(\pi D + 2D)} = \frac{\pi D}{4(\pi + 2)}$$

Example 3-6 A volumetric flow of air (14.2 m^3/ min) is to be moved from an air conditioner to a building 243.8 m away. The air is at a temperature of 5°C and a pressure of 689.5 N/m^2. Building pressure is 0 N/m^2. What would be the dimension of a smooth square duct used to transport the air?

This is a trial-and-error solution. For a first stab use the square dimension to be 0.3 m.

The average velocity in the duct is then

$$V = \frac{14.2\ \text{m}^3}{\text{min}}\frac{\text{min}}{60\ \text{sec}}\frac{1^2}{0.3}$$

$$V = 2.63\ \text{m/sec}$$

Next, the hydraulic radius for a square is $S/4$ (see Example 3-5). Hence

$$\text{Dequivalent} = 4\frac{S}{4} = S$$

The Reynolds number is

$$\text{Re} = \frac{SV\rho}{\mu}$$

$$\text{Re} = \frac{(0.3\ \text{m})(2.63\ \text{m/sec})(1.282\ \text{kg/m}^3)}{1.7 \times 10^{-5}\ \text{kg/m sec}}$$

$$\text{Re} = 59{,}400$$

For this Re the f value (Figure 3-3) is 0.005:

$$\Delta P = \frac{2fLV^2\rho}{gcD}$$

$$\Delta P = \frac{2(0.005)(243.8\ \text{m})(2.63\ \text{m/sec})^2(1.28\ \text{kg/m}^3)}{\dfrac{\text{kg/m}}{\text{sec}^2\ \text{N}}0.3\ \text{m}}$$

$$\Delta P = 71.95\ \text{N/m}^2$$

In order to accelerate the trial-and-error process, we can make use of an approximation. This involves holding density friction factor, length, and volumetric flow rate constant so that

$$\Delta P = \frac{C_1 V^2}{D}$$

where $C_1 = 2fL\rho/gc$.

Furthermore, since Q volumetric flow rate is

$$Q = \frac{\pi D^2}{4}\ V$$

we obtain

$$V = \frac{4Q}{\pi D^2} = \frac{C_2}{D^2}$$

because 4, π, and Q are constants.

This gives

$$\Delta P = \frac{C_3}{D^5}$$

Then, substituting for this case we obtain

$$S = (0.3 \text{ m}) \left(\frac{71.95 \text{ N/m}^2}{689.5 \text{ N/m}^2} \right)^{1/5}$$

$$S = 0.191 \text{ m}$$

Then for this S we have

$$V = \frac{14.2 \text{ m}^3}{\text{min}} \frac{\text{min}}{60 \text{ sec}} \frac{1}{0.191 \text{ m}}$$

$$V = 6.49 \text{ m/sec}$$

Also,

$$\text{Re} = (59{,}400) \frac{0.191 \text{ m}}{0.3 \text{ m}} \frac{6.49 \text{ m/sec}}{2.63 \text{ m/sec}}$$

$$\text{Re} = 93{,}300$$

This gives an f value of 0.0044:

$$\Delta P = 71.95 \text{ N/m}^2 \frac{0.0044}{0.005} \frac{6.49 \text{ m/sec}^2}{2.63 \text{ m/sec}} \frac{0.3 \text{ m}}{0.191 \text{ m}}$$

$$\Delta P = 605.5 \text{ N/m}^2$$

One additional permutation is necessary which gives an S value of 0.188 m. See that the approximation works reasonably well even though f actually changes.

PROBLEMS

3-1. The vapor at the top of a vacuum distillation column is at 0.0227 bar and 68°C. This vapor flows at a rate of 6.785×10^{-3} kg/sec to a condenser at 0.0200 bar. The pipe connecting the top of the column to the condenser has a roughness factor of 0.00457 cm (i.e., $\epsilon = 0.0045$ cm) and a diameter of 4.445 cm. What length of pipe is needed?

$$\text{Viscosity vapor} = 1.4 \times 10^{-5} \frac{\text{kg}}{\text{m sec}}$$

$$\text{Vapor molecular wt} = 67.7$$

3-2. A high-viscosity liquid ($\mu = 0.25$ kg/m sec) is to be pumped at 0.1878 m^3/min through a 5.08-cm-diameter line for a distance of 487.68 m. Can this be done with a gear pump that is capable of a maximum ΔP of 8.274 bar?

$$\text{Fluid density} = 0.8814 \frac{\text{g}}{\text{cm}^3}$$

3-3. What power per unit width is needed to pump a viscous fluid ($\mu = 0.025$ pascal-sec; $\rho = 880$ kg/m^3) through a slit (opening of 0.008 m; length of 4 m) at a flow rate (per unit width) of 0.03 m^3/sec.

3-4. Water at 5°C is to flow through a commercial steel pipe ($L = 300$ m) at a rate of 9.46×10^{-3} m^3/sec. Available pressure drop is 0.59 atmosphere. What pipe diameter should be used?

3-5. Water (5°C) is supplied to a factory (by gravity) from a large reservoir 100 m above it. The supply line (0.15-m diameter) is 300 m long with five gate valves, one globe valve, three standard 90° elbows, and two 45° elbows. Although the water supply is adequate, future needs will make it necessary to double the amount needed. How can this be done if

(a) A pump is installed

(b) A larger-diameter pipe is used

3-6. Water at 15°C is pumped through an annular section (i.e., torus) at a rate of 0.015 m^3/sec. The inner and outer diameters of the annulus are 0.08 m and 0.19 m. The inlet is 1.52 m lower than the outlet. Length of the conduit is 6.2 m. If the pressures at the pump inlet and annulus outlet are the same, what is the power required for the pump?

3-7. Two large water tanks are connected by a 610-m-long pipe. Water levels in both tanks are the same and the tanks are open to the atmosphere. If 0.0126 m^3/sec at 16°C are pumped from one tank to the other, what power is required? What is the pressure drop across the pump?

3-8. A pipeline 32 km long delivers 0.09 m^3/sec with a pressure drop of 3.45 megapascals. What would be the capacity of a new system be if a parallel identical line were laid along the last 19.2 km of the original system (neglect the effect of the connection to the 19.2-km line). Pressure drop would be unchanged. Also assume laminar flow.

3-9. Water ($\mu = 0.0013$ pascal seconds; $\rho = 1000$ kg/m^3) is to be pumped at a rate of 0.00125 m^3/sec through a steel pipe 0.025 m in diameter and 30 m long to a tank 12 m above the starting point. Pump efficiency is 60%. What power is needed? What type of pump should be used?

3-10. A liquid ($\mu = 0.001$ pascal seconds, $\rho = 4700$ kg/m^3) flows through a smooth pipe of unknown diameter at a mass flow rate of 0.82 kg/sec. The pressure drop for the 2.78-km-long pipe is 1262 pascals. What is the pipe's diameter?

3-11. A system consists of water flowing through a 0.10-m-diameter pipe at a rate of 2 m/sec. At a point downstream the pipe divides into a 0.10-m diameter main and a 0.025-m-diameter bypass. The equivalent length of the bypass is 10 m, and the length of the main is 8 m. What fraction of the water flows through the bypass? Neglect entrance and exit losses.

3-12. A petrochemical fluid ($\mu = 0.0005$ pascal-sec; $\rho = 700$ kg/m^3) is pumped through a 100-m pipeline (diameter of 0.15 m). The pressure drop is 7×10^4 pascals. A break in the pipeline causes the fluid to be redirected through a line of 70 m of 0.2-m diameter followed by 50 m of a 0.1-m-diameter pipe. The pump used with the original line can develop 300 pascals pressure. Will it be sufficient for the makeshift system? Take ϵ to be 5×10^{-5} m.

3-13. A system moves 6×10^{-4} m/sec of water at 47°C through a 0.04-m-diameter pipe ($\epsilon = 0.0002$ m) in a horizontal distance of 150 m and vertically up 10 m. The pipeline has valves and fittings that account for 260 pipe diameters and a heat exchanger. The pump uses 128 W to move the fluid. What is the equivalence of the heat exchanger in pipe diameters for its equivalent pressure drop?

3-14. A new material is being used for a specialty pipe. When water is pumped through the pipeline made of this material (diameter = 0.08 m), a friction factor of 0.0070 is found for a $\overline{V}$ of 12 m/sec. What is the ϵ (roughness of the material)?

3-15. A 0.08-m-diameter pipe 150 m long connects two tanks. The level in one tank is 6 m above the level in the other tank. However, the pressure in the second tank is 0.68 atmosphere greater than the pressure in the first tank (i.e., one with higher level). How much fluid ($\mu = 0.1$ pascal-sec; $\rho = 850$ kg/m^3) will flow in the pipeline? In which direction does the fluid flow?

3-16. Water at 20°C flows through 30 m of horizontal 0.2-m pipe (4.6×10^{-5} m roughness) with a pressure change of 2100 pascals.
What is the water's mass flow rate?

3-17. Air at 60°C flows through a horizontal duct system (rectangular-shaped). At first the flow goes through a duct 0.35 m high and 0.08 m wide. Then it branches into two lines. The first is a continuation of the original duct (i.e., in line) 6 m in length with a height half of the original duct but with the same width. The second branch with the same dimensions as the first branch is connected by a 90° elbow, rises vertically 1.5 m and then with another 90° elbow runs horizontally for 1.5 m. The width of the second branch is 0.08 m, but its final depth is unknown. What is the value of this depth for a flow of 0.17 m/sec if all pressure drops are the same? For purposes of this problem, assume that a K value of 0.2 can be used for the elbows. Also take f to be 0.005.

3-18. Mercury at 20°C flows in a horizontal 0.02-m-diameter pipe ($\epsilon = 1 \times 10^{-4}$ m) that is 15 m long. If the pressure drop is 100 kPa, what is the mercury's average velocity, volumetric flow rate, and mass flow rate? Mercury properties are viscosity of 1.54×10^{-3} pascal-sec and density 13,500 kg/m^3.

3-19. Water at 15°C is pumped from a reservoir to the top of a mountain (1220 m high) through a 0.15-m-diameter pipe that is 1500 m long at a velocity of 3 m/sec. Assume that the pump used has an efficiency of 60 percent and that electric power cost is 4 cents per kilowatt hour. What is the hourly cost of pumping the water?

3-20. A process unit for purifying salt water is based on the reverse osmosis principle. In such a device, hollow fibers are used that retain the salt but permit water to diffuse out. For a unit using 900,000 hollow fibers (diameters of 85 and 42×10^{-6} m, respectively) that are 0.9 m long, the volumetric output (feed pressure of 2.86×10^5 pascals) is 8.76×10^{-5} m^3/sec. What is the pressure drop in an individual fiber from inlet to outlet?

3-21. A condenser consisting of 400 tubes, 4.5 m long and a diameter of 0.01 m, uses a water flow rate of 0.04 m^3/sec. What is the power of pumping required and the pressure drop in the unit? Assume that a K of 0.4 is the value for contraction at the entrance of the condenser tubes.

3-22. An acid solution ($\mu = 0.065$ pascal-sec; $\rho = 1530$ kg/m^3) is to be pumped through lead pipe (diameter of 0.025 m) and raised to a height of 25 m. The pipe is 30 m long and includes two 90° bends (i.e., elbows). For a pump efficiency of 50 percent, what is the power requirement?

3-23. Two tanks are connected with a 0.08-m-diameter pipe 190 m long. The pipeline contains six elbows, four gate valves, and one globe valve. The pump in the system is equipped with a bypass line that is usually closed. The system fluid can range from 800 to 850 kg/m^3 in density and from 1.6×10^{-3} to 4.25×10^{-3} pascal-sec. Tank A's pressure has minimum and maximum values of 1.59×10^5 and 2.41×10^5 pascal. The corresponding values for tank B are 4.28×10^5 and 6.6×10^5 pascals. The highest and lowest liquid levels in tank A (measured from a datum plane) are 13 and 2.4 m. Values for tank B are 39 and 30 m. If the flow rate range is 0.01 m^3/sec to 0.0064 m^3/sec, what is the size of pump required?

3-24. If the pump in problem 3-23 is shut and the bypass opened, what will be the direction of flow? What will the minimum and maximum flow rate values be?

3-25. A blower is used to move air through an air-conditioning duct (0.10 m high, 0.2 m wide, 10 m long). If the fan uses 30 k/W of power, find the volume rate of flow and the pressure immediately downstream from the blower.

3-26. Pipe roughness is found to increase according to the formula

$$\epsilon = \epsilon_0 + C_1\, t$$

where t is in years, ϵ_0 is 2.6×10^{-4} m, and C_1 is 0.00001 m per year. If the pressure drop is always kept at 15 kPa, find the volumetric flow rates for water (20°C) for 2, 10, and 20 years.

3-27. A fan moves air (15°C, (atmospheric pressure) through a rectangular duct (0.2 by 0.3 m) that is 50 m long at a flow rate of 0.6 m^3/sec. What power is required?

3-28. A pipe (0.6 m in diameter, ϵ of 9×10^{-5} m) is used to move water at a velocity of 4.5 m/sec. Can the capacity be increased by inserting a smooth liner that reduces the diameter to 0.58 m? What would the change in pressure drop be?

3-29. A 5-m-diameter cylindrical tank has water at room temperature flowing out through a steel pipe 90 m long whose diameter is 0.2 m. The pipe is fixed to the base of the tank. How long would it take for the level to drop from 3 m to 1 m above the exit?

3-30. Two tanks containing water are connected with a horizontal pipe 0.075 m in diameter and 300 m long. The bottoms of both tanks are at the same level. Tank A is 7 m in diameter and has a depth of 7 m. Tank B (5 m in diameter) has a depth of 3 m. How long will it take for the liquid level in Tank A to fall to 6 m?

3-31. Air at room temperature and atmospheric pressure flows through a long rectangular duct (cross section of 0.3 by 0.45 m) at a velocity of 15 m/sec. If the pressure drop per unit length is 1.289 Pa/m, what is the duct's surface roughness?

3-32. Water at 15°C is to be moved at a rate of 28 m/sec in an opening (at the top) square duct (0.3 by 0.3 m). What slope should be used to cause this flow?

3-33. What differences in the result of problem 3-7 would occur if

(a) The pipe line also contained two globe valves, and nine 90° elbows

(b) Expansion and contraction losses are considered

3-34. Water (20°C) flowing at a rate of 10 m^3/sec has its flow split into two horizontal lines. One branch is 100 m long with a 0.2-m diameter. The second branch, which contains a half-opened gate valve, is 200 m long (diameter of 0.25 m). After passing through the branches, flow resumes as before in a single pipe. What are the flows and pressure drops through each branch?

3-35. Water (20°C) flowing at a rate of 0.25 m^3/sec is split equally into two branches before resuming its flow in a single pipe. Branch A is 30 m long and has a diameter of 0.15 m. The second branch, B, has a diameter of 0.075 m and a length of 30 m. It is also equipped with a pump. Size the pump for branch B.

3-36. Water at 30°C flows at a rate of 0.01 m^3/sec through a 0.075-m-diameter pipe to a 5-m length of 0.04-m-diameter pipe and thence back to 0.075-m-diameter pipe. Find the pressure drop between two points in each section of the 0.075-m-diameter pipe as well as the entire length of the 0.04-m pipe. The total length of 0.075-m pipe is 0.5 m.

3-37. Water is pumped at a rate of 0.05 m^3/sec through a 0.015-m-diameter pipe 300 m long to a reservoir 60 m higher. The pump requires 800×10^3 N/m^2. If the pipe roughness increases by a factor of ten, what is the reduction of flow rate?

3-38. A petroleum product ($\rho = 705$ kg/m^3; $\mu = 0.0005$ pa-sec) is pumped 2 km to storage tanks through a 0.15-m-diameter pipeline (roughness of 4×10^{-6} m) at a rate of 0.04 m/sec. The pump for the system has a mechanical efficiency of 50 percent. If the pump impeller is damaged to the extent that its delivery pressure is halved, what will be the amount of flow rate reduction?

3-39. Two storage tanks X and Y contain an organic fluid ($\rho = 870$ kg/m^3; $\mu = 0.0007$ pascal-sec.) discharge through 0.3-m-diameter pipes 1.5 km long to a junction Z. The fluid then moves through 0.8 km of a 0.5-m-diameter pipe to another storage tank K. Levels in X and Y are initially 10 and 7 m above that of K. What is the initial rate of flow?

3-40. Water at 20°C is pumped at a rate of 2.36×10^{-3} m^3/sec to the top of an experimental unit (4.5 m high). Frictional losses in the 0.05-m-diameter pipe are 2.39 J/kg. If the pump used can develop net power of 93 watts, what height is needed in the water supply tank?

3-41. A pump moves a fluid ($\rho = 1180$ kg/m^3; $\mu = 0.0012$ pascal-sec) from the bottom of a supply tank to the bottom of a hold tank, Liquid level in the hold tank is 60 m above that in the supply tank. The pipeline (0.15-m diameter; 210-m length) connecting the tanks contains two gate valves and four elbows. If the flow rate is to be 0.051 m/sec, what is the cost to run the pump for one day? Energy cost is one dollar per HP-day.

3-42. A fire truck moves water from a river through a hose (0.1-m diameter; roughness of 4.57×10^{-5} m) to a height 30 meters above the river level. The water, which moves at a rate of 0.032 m^3/sec, exits at a velocity of 30 m/sec. What power would be needed by the truck's pump? Assume that the hose's overall equivalent length is 100 m.

REFERENCES

1. A. L. Prasuhn, *Fundamentals of Fluid Mechanics*, Prentice-Hall, Englewood Cliffs, NJ (1980).
2. L. W. Moody, *Trans. ASME* **66**, 672 (1944).
3. W. L. McCabe, J. C. Smith, and P. Harriott, *Unit Operations of Chemical Engineering*, McGraw-Hill, New York (1985).
4. J. M. Coulson and J. F. Richardson, *Chemical Engineering*, Vols. 1 and 2, Pergamon, London (1978).
5. Crane Technical Paper No. 410, Crane Company, Chicago, IL.
6. W. L. McCabe and J. C. Smith, *Unit Operations of Chemical Engineering*, McGraw-Hill, New York (1967).
7. W. H. McAdams, *Heat Transmission*, McGraw-Hill, New York (1954), Chapter 5.
8. R. G. Griskey, *Chemical Engineering for Chemists*, American Chemical Society, Washington, D.C. (1997).

4

COMPLEX FLOWS

INTRODUCTION

The flow of Newtonian fluids in conduits is a very important aspect of chemical and petroleum processes. While important, however, such flows do not constitute the entirety of fluid mechanics used in the process industries.

In this chapter we will explore some other significant areas of fluid dynamics relevant to processes. The topics that will be dealt with include flow around objects, motion of particles, flow through packed beds, non-Newtonian fluids, and agitation and mixing.

FLOW AROUND OBJECTS

Flow around objects is a somewhat more complicated situation than flow in conduits. The basis for describing flow around a submerged object is the drag force, which is the force in the flow direction exerted by the fluid on the solid surface.

The earliest treatment of such behavior was that of Newton, who proposed that the drag force on a sphere for flowing air was

$$F_D = \pi R^2 \rho_{\text{air}} \frac{V^2}{2} \tag{4-1}$$

where πR^2 was the projected cross-sectional area of the sphere, and V is the velocity of the air.

More correctly the equation is

$$\frac{F_D}{A_P} = C_D \rho \frac{V^2}{2} \tag{4-2}$$

The C_D term is the drag coefficient and, in essence, plays the same role as f does for conduit flow. A_P is the projected area.

Dimensional analysis and experiment lead us to the conclusion that

$$C_D = \phi(\text{Re}_p) \tag{4-3}$$

where Re_p is a particle Reynolds number defined as

$$\text{Re}_p = \frac{D_p V \rho}{\mu} \tag{4-4}$$

where μ and ρ are the viscosity and density of the fluid, V is the average velocity, and D_P is the average particle diameter.

The behavior of C_D with particle Reynolds number for spheres, disks, and cylinders is shown in Figure 4-1. Note that the curves appear to pass through regions of behavior. Also, note the obvious "bend" in the curves for spheres and cylinders in the vicinity of $\text{Re}_p = 10^6$. These "kinks" are due to a phenomenon called *boundary layer separation*, which takes place when the fluid's velocity change is so large that the fluid no longer adheres to the solid surface.

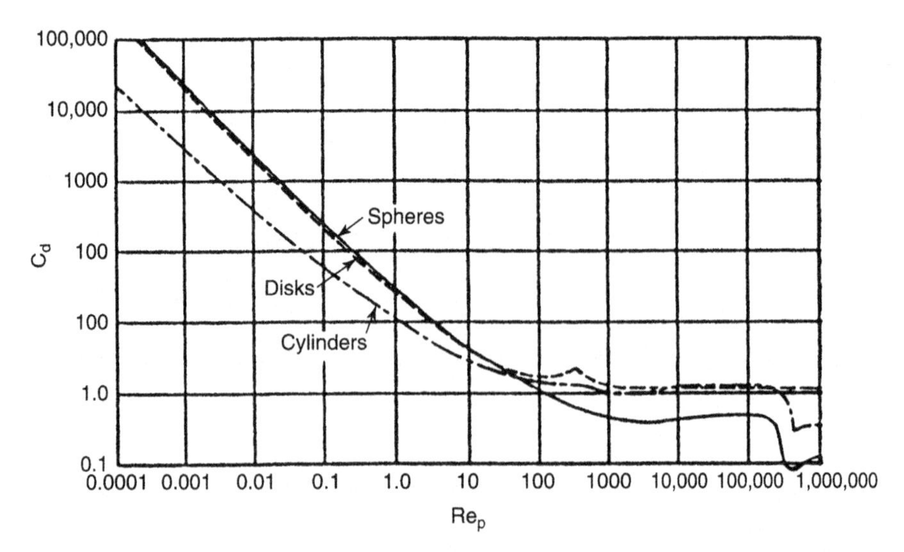

Figure 4-1. Drag coefficients versus Reynolds number. (Reproduced with permission from reference 1. Copyright 1940, American Chemical Society.)

Table 4-1 Ranges, Values, and Flow Types for Spheres

Range	b_1	m	Flow Regime
$Re_p < 2$	24	1.0	Laminar
$2 < Re_p < 500$	18.5	0.6	Transition
$500 < Re_p < 200{,}000$	0.44	0.0	Turbulent

It is possible to relate the C_D and Re_p empirically:

$$C_D = \frac{b_1}{Re_p^m} \tag{4-5}$$

where b_1 is a constant and m is a power. Table 4-1 gives the appropriate values for Re_p ranges. Also given are the types of flow that are encountered. Note that laminar, transition, and turbulent flow occur for flow around objects, albeit not in the same form as for flow in conduits.

In the region where $Re_p < 2$ we encounter the flow type known as *Stokes' Law flow*. For such situations

$$V = \frac{2R^2 g(\rho_{\text{particle}} - \rho_{\text{fluid}})}{9\mu} \tag{4-6}$$

where R is the particle radius and μ is the fluid viscosity.

Another way of identifying the flow regime is to substitute the V from Stokes' Law into the Reynolds number and see that

$$Re_p = \frac{D_p V \rho}{\mu} = \frac{D_p^3 g \rho_{\text{fluid}}(\rho_{\text{particle}} - \rho_{\text{fluid}})}{18\mu^2} \tag{4-7}$$

If A is taken to be

$$A = D_p \left[\frac{A_e \rho_{\text{fluid}}(\rho_{\text{particle}} - \rho_{\text{fluid}})}{\mu^2}\right]^{1/3} \tag{4-8}$$

then

$$Re_p = \frac{A^3}{18} \tag{4-9}$$

This enables the following A values to be determined that correspond to the Re_p value (Table 4-2).

At high Reynolds numbers (i.e., turbulent flow) the drag coefficient value is 0.44. For this case the velocity, V, is

$$V = 1.75\left(\frac{g(\rho_p - \rho)D_p}{\rho}\right)^{1/2} \tag{4-10}$$

Table 4-2 *A* Related to *Re*$_p$

Re_p	A
2.0	3.3
500.0	43.6
200,000.0	2360.0

Table 4-3 Relation of *n* to *Re*$_p$ (2)

Re_p	n
0.1	4.6
1	4.3
10	3.7
10^2	3.0
10^3	2.5

A special case that can occur is hindered settling. In this situation the fluid surrounding the particles has velocity gradients which interfere with particle motion. The terminal settling velocity for this case is then (2)

$$V_{ts} = V(\epsilon)^n \tag{4-11}$$

The ϵ is the porosity, defined as the volume of voids to total volume. Exponent n is given as a function of Re_p in Table 4-3.

FLOW THROUGH PACKED BEDS

Flow through a packed bed represents a highly complex process. In essence, we can visualize the packed bed to be a collection of intermeshed tubes of varying cross section. This gives a tortuous path for the fluid and thus makes it necessary to take a semiempirical approach to such flows.

Basically, the hydraulic radius approach is used to derive appropriate flow equation. For a packed bed we have

$$\text{R}_\text{h} = \left[\frac{\text{Volume available for flow}}{\text{Total wetted surface}}\right] \tag{4-12}$$

$$R_h = \frac{\left[\dfrac{\text{Volume of voids}}{\text{Bed volume}}\right]}{\left[\dfrac{\text{Wetted surface}}{\text{Bed volume}}\right]} = \frac{\epsilon}{a} \tag{4-13}$$

where ϵ is porosity (defined previously). Also,

$$a = a_v(1 - \epsilon)$$

where a_v is the total particle surface divided by the volume of the particles.

Then, by using the superficial velocity ∇_0, the bed height L, and particle diameter D_p, an equation is obtained by fitting to experimental data.

The result is the Ergun equation (3) :

$$\frac{\Delta p g_c}{L} \frac{\phi_s D_p}{\rho V_0^2} \frac{\epsilon^3}{1 - \epsilon} = \frac{150(1 - \epsilon)}{\phi_s D_p V_0 \rho / \mu} + 1.75 \qquad (4\text{-}14)$$

which covers both laminar and turbulent flow.

The laminar and turbulent regions are determined by a Reynolds number defined as

$$\text{Re} = \frac{D_p \overline{V}_0 \rho}{\mu} \qquad (4\text{-}15)$$

The ϕ_s term is the sphericity and is obtained by the relation

$$\phi_s = \frac{6}{D_p} \left[\frac{\text{Particle surface area}}{\text{Particle volume}} \right] \qquad (4\text{-}16)$$

Some typical sphericities are given in Table 4-4.

For the laminar range (Re < 1.0) the form of equation (4-14) that is used is

$$\frac{\Delta p g_c \phi_s^2 D_p^2 \epsilon^3}{L \overline{V}_0 \mu (1 - \epsilon)^2} = 150 \qquad (4\text{-}17)$$

Table 4-4 Values for Various Objects

Item	ϕ_s
Sphere, cube, or cylinder ($L = D$)	1.0
Raschig ring	
L = outside; inside = 1 outside diameter diameter 2 diameter	0.58
L = outside; inside = 3 outside diameter diameter 4 diameter	0.33
Berl saddles	0.3
Sharp-pointed sand particles	0.95
Rounded sand	0.83
Coal dust	0.73
Crushed glass	0.65
Mica flakes	0.28

In the case of turbulent flow (Re > 1000) the form is

$$\frac{\Delta p}{\rho L} \frac{g_c}{\overline{V}_0^2} \frac{\phi_s D_p \epsilon^3}{1 - \epsilon} = 1.75 \tag{4-18}$$

Transition flow (1.0 < Re < 1000) requires equation (4-19).

For cases where there are a mixture of different particle sizes a surface-mean diameter $\overline{D}_s$ in used. These are obtained by the number of particles N_i in each size range or the mass fraction in each size range x_i:

$$\overline{D}_s = \frac{\sum_{i=1}^{n} N_i D_{pi}^3}{\sum_{i=1}^{n} N_i D_{pi}^2} \tag{4-19}$$

$$\overline{D}_s = \frac{1}{\sum_{i=1}^{n} \frac{x_i}{D_{pi}}} \tag{4-20}$$

NON-NEWTONIAN FLUIDS

Many important fluids do not obey Newton's Law of Viscosity. These include polymer solutions, polymer melts, foods, paints, inks, and various slurries.

Because of this a different approach than that used in Chapter 3 must be taken. This approach utilizes the format of the power law [equation (2-11)]:

$$\tau_{Yx} = K \left(\frac{d\gamma}{dt} \right)^n \tag{2-11}$$

Equation (2-11) indicates a linear relationship on a logarithmic plot. Actually, the data have some curvature, but linearity applies to fairly long ranges.

If $(D\Delta P/4L)$, the wall shear stress is plotted against the function $(8V/D)$; a relationship similar to that indicated by equation (2-11) is found (Figure 4-2):

$$\left(\frac{D\Delta P}{4L} \right) = \tau_{\text{wall}} = K' \left(\frac{8V}{D} \right)^{n'} \tag{4-21}$$

Frequently the n of equation (2-11) and the n' of equation (4-21) are close in value:

$$n \cong n' \tag{4-22}$$

Then K and K' are also related:

$$K' = K \left(\frac{3n+1}{4n} \right)^n \tag{4-23}$$

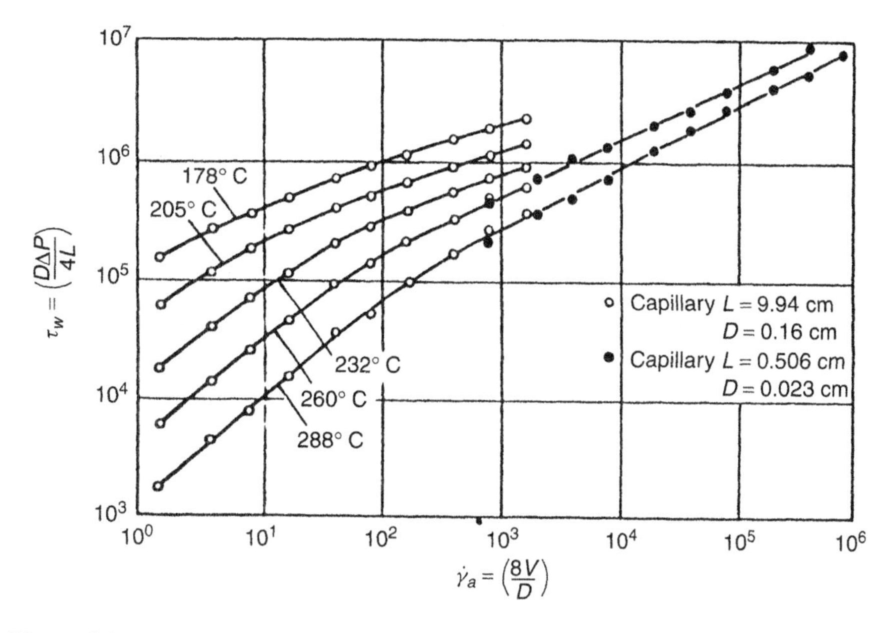

Figure 4-2. Logarithmic plot of $(D\Delta P/4L)$ versus $(8V/D)$. (Adapted from reference 4.)

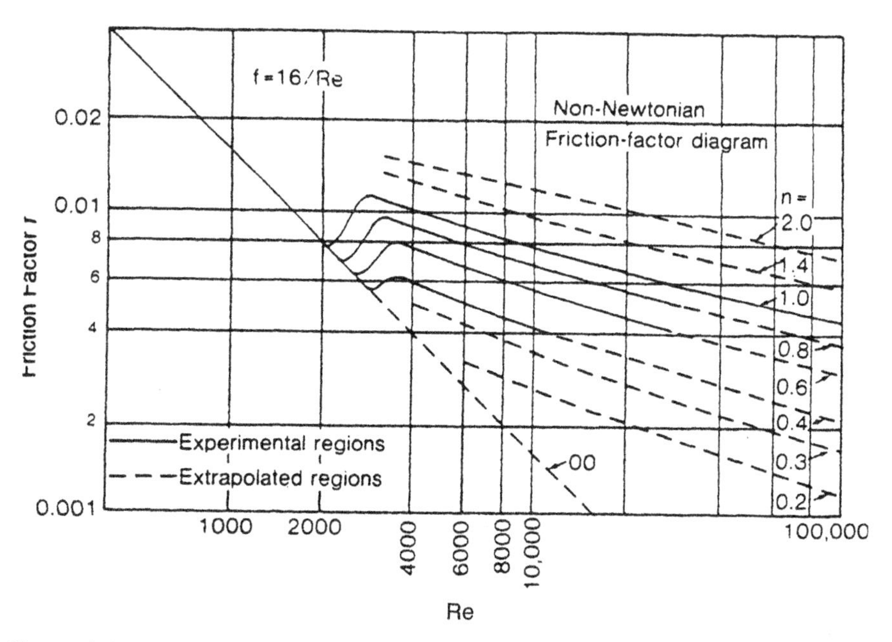

Figure 4-3. Friction factor versus modified Reynolds number for non-Newtonians. (Reproduced with permission from reference 5. Copyright 1959, American Institute of Chemical Engineers.)

It is possible to derive a modified Reynolds number for the non-Newtonian fluid such that

$$\mathrm{Re}' = \frac{D^{n'} V^{2-n'} \rho}{g_c K' 8^{n'-1}} \tag{4-24}$$

Note that when $n = 1.0$ (Newtonian fluid), equation (4-24) gives the Reynolds number defined earlier.

By using the foregoing approach, a friction-factor-modified Reynolds number correlation (Figure 4-3) can be developed. Note that a number of lines exist in the turbulent region (i.e., one for each value of n').

AGITATION AND MIXING

One of the most ubiquitous operations in the chemical process industries is the agitation of fluid systems. Such agitation is undertaken to bring about the mixing (i.e., blending, homogenizing) of such systems.

While agitation and mixing are obviously interrelated, technical understanding of each differs greatly. It turns out that we are able to do a reasonably good job with agitation but less so with mixing. This occurs because we rely on power consumption to define agitation while "goodness of mixing" is much more nebulous.

By using dimensionless analysis we can define a group P_0 (the power number)

$$P_0 = \frac{Pgc}{N^3 \rho (D^1)^5} \tag{4-25}$$

that is related to the Reynolds number Re* for an agitator system:

$$\mathrm{Re}^* = \frac{N\rho (D^1)^2}{\mu} \tag{4-26}$$

The terms in the above are D' (agitator or impeller diameter), N (the agitator speed revolution per unit time), and P (power consumed).

Note that equation (4-26) came from the originally defined Reynolds number by using D' for D and ND' (tip speed) for V.

The relation between P_0 (which is actually the drag coefficient, C_d, for the agitation system) and Re* for an agitator system is dependent on the concept of geometrical similarity. If properly handled, this means that a unique relation between P_0 and Re* exists for any agitation system.

For a system with no appreciable vortex (i.e., baffled as per Figure 4-4) the significant geometric factors are individual ratios of height of liquid (Z_1), height of impeller (Z_i), tank diameter (D_t), and baffle width (W) to the agitator diameter (D').

Incidentally, baffles may have various configurations and particular geometries — for example, placed at an angle, varied in number, or actually perforated. The design of these elements is based on process and/or experiential factors.

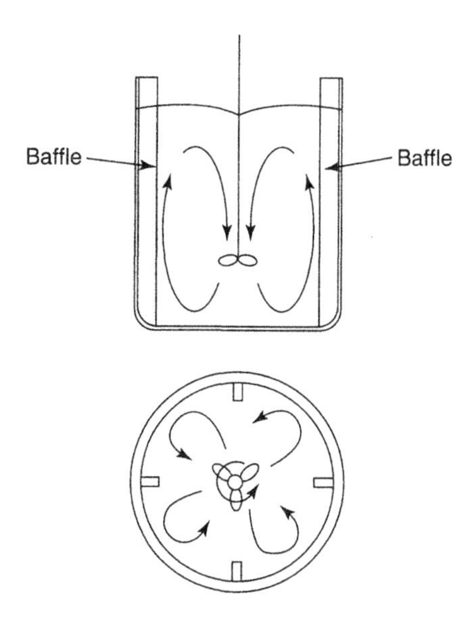

Figure 4-4. Baffled agitator system (8).

Figure 4-5 gives the power number versus Reynolds number correlation for different types of agitators. The pitch referred to is the axial distance that a free propeller would move in a nonyielding liquid in one revolution.

As Figure 4-5 shows, each of the agitator systems has similar behavior. For example, at high Reynolds numbers (above 1000 to 1500) the power number is essentially constant. This corresponds to the turbulent flow behavior for $C_d =$ 0.44 (i.e., a constant value). At lower Reynolds numbers the power number has behavior similar to that of equation (4-5).

In an unbaffled tank system the vortex can play a role. Here, the power number will be affected by the behavior of the liquid surface (i.e., the effect of gravity).

This means that the power number now becomes a function of two dimensionless groups, namely, the Reynolds number (Re) and the Froude number (Fr):

$$P_0 = \phi(\text{Re}, \text{Fr}) \tag{4-27}$$

The Froude number, which is the ratio of inertial to gravity forces, is defined as

$$\text{Fr} = \frac{\overline{V}^2}{gD} \tag{4-28}$$

or for the agitation system ($D = D'$; $\overline{V} = ND'$)

$$\text{Fr} = \frac{D^1 N^2}{g} \tag{4-29}$$

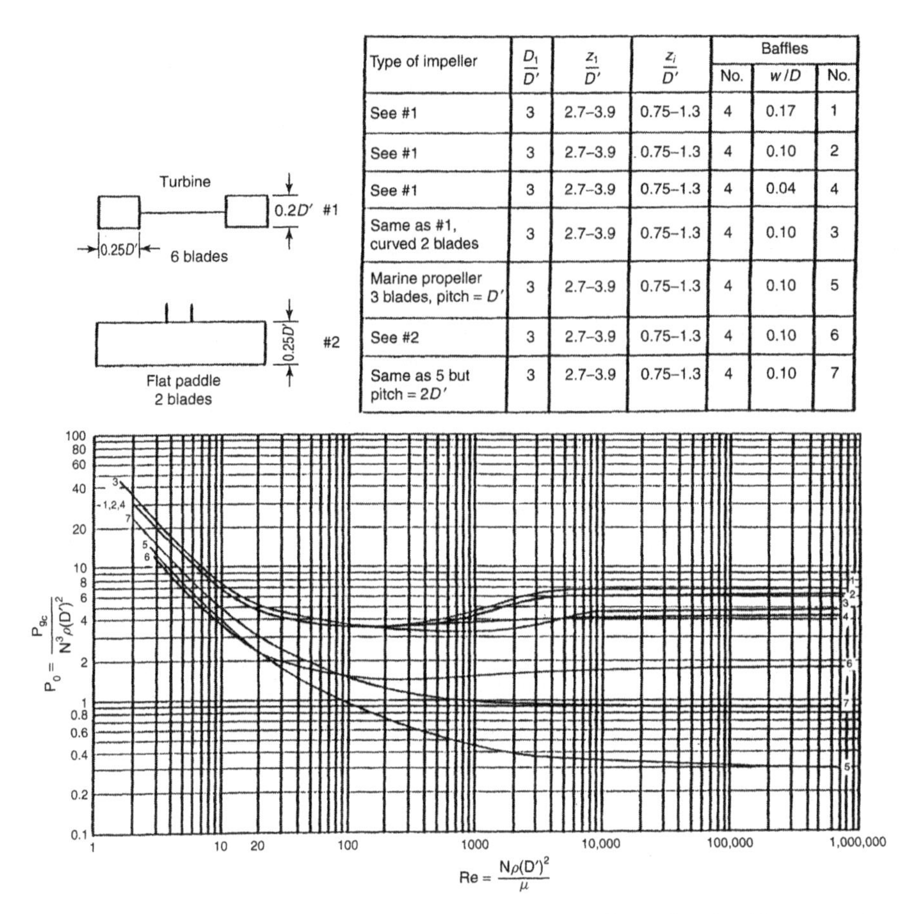

Type of impeller	$\frac{D_1}{D'}$	$\frac{z_1}{D'}$	$\frac{z_i}{D'}$	Baffles No.	Baffles w/D	No.
See #1	3	2.7–3.9	0.75–1.3	4	0.17	1
See #1	3	2.7–3.9	0.75–1.3	4	0.10	2
See #1	3	2.7–3.9	0.75–1.3	4	0.04	4
Same as #1, curved 2 blades	3	2.7–3.9	0.75–1.3	4	0.10	3
Marine propeller 3 blades, pitch = D'	3	2.7–3.9	0.75–1.3	4	0.10	5
See #2	3	2.7–3.9	0.75–1.3	4	0.10	6
Same as 5 but pitch = $2D'$	3	2.7–3.9	0.75–1.3	4	0.10	7

Figure 4-5. Power curves for baffled agitator systems. (Reproduced with permission from reference 6. Copyright 1950, American Institute of Chemical Engineers.)

Data for unbaffled agitators are shown in Figure 4-6. See that the effect of Froude number is negligible below Reynolds numbers of 300. This is because the surface effect is small in this region (i.e., laminar flow).

A special case that we can encounter is the preparation of a suspension. Empirical equations (4-30) and (4-31) can be used to calculate the power needed to suspend a solid. These are

$$\frac{Pg_c}{g\rho_m(\text{Volume})_m\overline{V}} = (1-\epsilon_m)\left(\frac{D_t}{D_i}\right)^{1/2} e^{4.35\beta} \tag{4-30}$$

and

$$\beta = \frac{Z_S - E}{D_t} - 0.1 \tag{4-31}$$

Type of Impeller	$\frac{D_t}{D'}$	$\frac{z_1}{D'}$	$\frac{z_i}{D'}$	a	b	No.
Marine propellers, 3 blades, pitch = 2D'	3.3	2.7–3.9	0.75–1.3	1.7	18	1
Same as #1 but pitch = 1.05D'	2.7	2.7–3.9	0.75–1.3	2.3	18	2
Same as #1 but pitch = 1.04D'	4.5	2.7–3.9	0.75–1.3	0	18	3
Same as #1 but pitch = D'	3	2.7–3.9	0.75–1.3	2.1	18	4

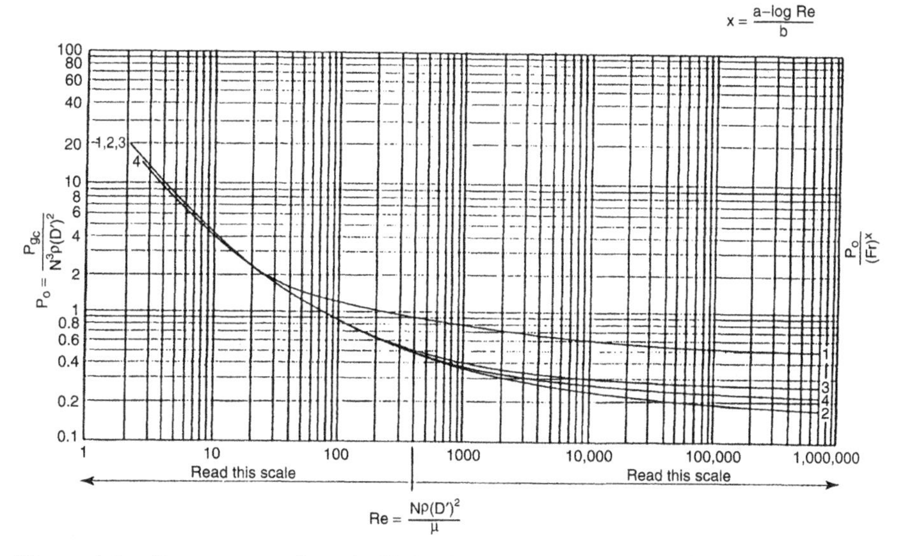

Figure 4-6. Power curves for unbaffled agitator systems. (Reproduced with permission from reference 6. Copyright 1950, American Institute of Chemical Engineers.)

where ρ_m is the density and $(\text{Volume})_m$ is the volume of the solid–liquid suspension; not including the clean zone above Z_S (the top of the suspension), $\overline{V}$ is the terminal settling velocity calculated from Stokes' law, ϵ_m is the volume fraction of liquid in the suspension region, and E is the clearance between the impeller and the tank bottom.

Mixing, as pointed out earlier, is more difficult to characterize in a quantitative sense. Correlations in this area are highly empirical and for the most part limited. One that is available is shown in Figure 4-7 (for turbine agitators). The ordinate f_t is a complicated function of agitation system parameters.

$$f_t = \frac{t_t(N)^{2/3}(D')^{4/3}(g)^{1/6}(D')^{1/2}}{(Z_1)^{1/2}(D_T)^{3/2}} \tag{4-32}$$

where t_t is the blending time, N is the agitation speed, D' is the agitation diameter, Z_1 is the height of liquid, and D_T is the tank diameter.

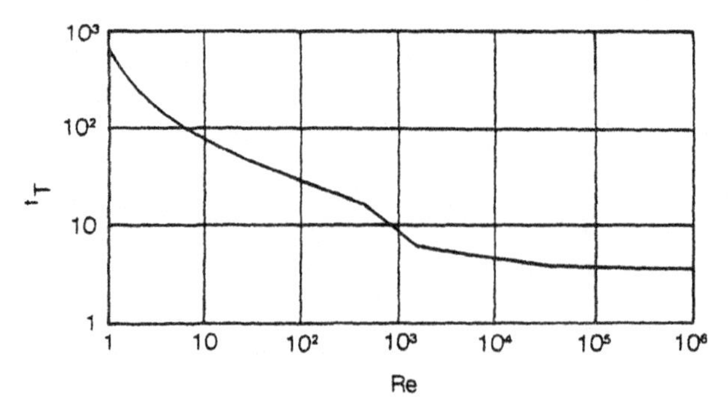

Figure 4-7. Mixing times as a function of Reynolds numbers for miscible fluids in turbine agitated vessels. (Reproduced with permission from reference 7. Copyright 1960, American Institute of Engineers.)

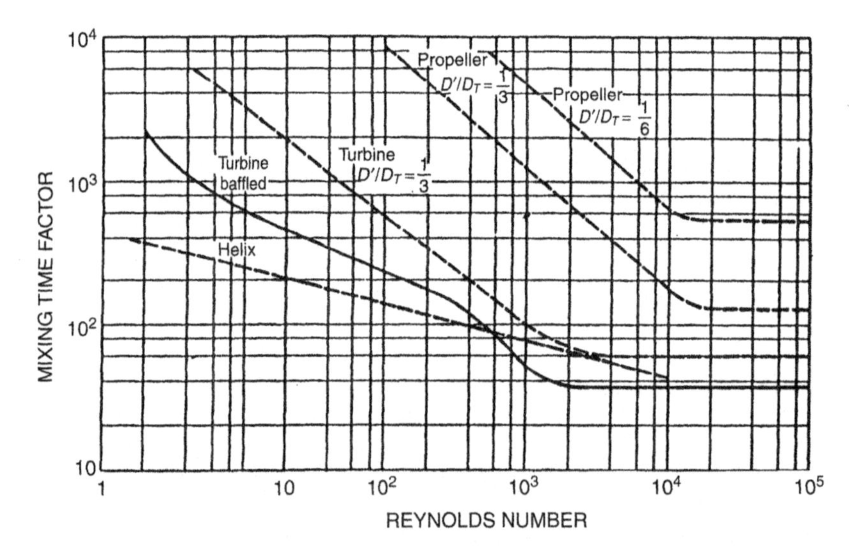

Figure 4-8. Mixing time factor (N times mixing time) as a function of Reynolds number. (Reproduced with permission from references 9 and 10. Copyright 1966 and 1972, American Institute of Chemical Engineers.)

Another mixing time correlation is given in Figure 4-8 for the case of $D_T = Z_1$. All curves except the one marked "Turbine baffled" are for unbaffled tanks. Note that the product of N with the mixing time becomes a constant for large Reynolds numbers.

In many cases, mixing can only be characterized by indirect methods (dispersal of dyes; use of tracers; temperature profiles; conductivity measurements of "salted" systems, etc.). Frequently, the only method available is the process result itself, which unfortunately is not always precise.

WORKED EXAMPLE PROBLEMS

Example 4-1 Drops of oil (15 microns in diameter) are to be settled from an air mixture (atmospheric pressure, 21°C). Oil specific gravity is 0.9. How high should the chamber be for a settling time of 1 minute?

We assume that the system can be treated as a case of flow around a sphere even though the drops are moving and the air is still. See that the important factor is the relative velocity between the gas and spheres (drops).

In order to proceed we must determine the character of the flow. This can be done by using the A factor defined in equation (4-8).

The quantities needed are

$$Ae = 9.8 \text{ m/sec}^2, \qquad \mu_{\text{air}} = 1.8 \times 10^{-5} \text{ kg/m sec}$$

$$D_p = 15 \times 10^{-6} \text{ m}, \qquad \rho_{\text{particle}} = 898.6 \text{ kg/m}^3$$

Next,

$$A = (15 \times 10^{-6} \text{ m}) \left[\frac{9.8 \text{ m/sec}^2 \times 1.2 \text{ kg/m}^3 \times 898.6 \text{ kg/m}^3}{(1.8 \times 10^{-5} \text{ kg/m sec})^2} \right]^{1/3}$$

$$A = 0.479$$

This is well below the limit for Stokes' law flow ($A = 3.3$). Hence, we use equation 4-6

$$\overline{V} = \frac{2r^2 g(\rho_{\text{particle}} - \rho_{\text{fluid}})}{9\mu}$$

$$\overline{V} = \frac{2(7.5 \times 10^{-6} \text{ m})^2 \; 9.8 \text{ m/sec}^2 (898.6 \text{ kg/m}^3)}{9(1.8 \times 10^{-5} \text{ kg/m sec})}$$

$$\overline{V} = 0.0061 \text{ m/sec}$$

In one minute the particles will settle (0.0061×60 sec) or 0.366 m. Hence, the chamber should be 0.366 m high.

Example 4-2 Solid particles (average density of 2800 kg/m^3) are settling in water (30°C). What is the terminal velocity for the particles? Also, what would be the velocity of the system in a centrifugal separator with an acceleration of 390 m/sec^2?

The water properties are

$$\mu = 8.0 \times 10^{-4} \text{ kg/m sec}$$

$$\rho = 996 \text{ kg/m}^3$$

We find A in order to determine the flow regime:

$$A = (1.5 \times 10^{-4}\ \text{m}) \left[\frac{9.8\ \text{m/sec}^2 \times 996\ \text{kg/m}^3\ (2800 - 996)\ \text{kg/m}^3}{(8 \times 10^{-4}\ \text{kg/m sec})^2} \right]^{\frac{1}{3}}$$

$$A = 4.53$$

This exceeds the Stokes' law range. Using the expression below for gravity settling

$$V = \left[\frac{4\ g(\rho_p - \rho) D_p}{3 C_d \rho} \right]^{1/2}$$

and assuming Re_p is 3.75 and obtaining a C_d value (9.0) from Figure 4-1, we obtain

$$V = \left[\frac{4 \times 9.8\ \text{m/sec}^2 (2800 - 996)\ \text{kg/m}^3 \times 1.5 \times 10^{-4}\ \text{m}}{(3)\ (9.0)(996\ \text{kg/m}^3)} \right]^{1/2}$$

$$V = 0.0198\ \text{m/sec}$$

Checking for Re_p, we have

$$\text{Re}_p = \frac{(1.5 \times 10^{-4}\ \text{m})(0.0198\ \text{m/sec})\ [996\ \text{kg/m}^3]}{[8 \times 10^{-4}\ \text{kg/m sec}]}$$

$$\text{Re}_p = 3.70$$

This result is satisfactory; hence V is 0.0198 m/sec.

In the second case we use the acceleration of 390 m/sec^2:

$$A = (1.5 \times 10^{4}\ \text{m}) \left[\frac{390\ \text{m/sec}^2 \times 996\ \text{kg/m}^3 (2800 - 996)\ \text{kg/m}^3}{(8 \times 10^{-4}\ \text{kg/m sec})^2} \right]^{1/3}$$

$$A = 15.46$$

The A value places this case in the transition region. Once again we assume a Reynolds number (55) and obtain a c_d.

$$V = \left[\frac{4 \times 390\ \text{m/sec}^2 (2800 - 996)\ \text{kg/m}^3 \times 1.5 \times 10^{-4}\ \text{m}}{(3)\ (55)\ (996\ \text{kg/m}^3)} \right]^{1/2}$$

$$V = 0.306\ \text{m/sec}$$

The calculated Reynolds number is then

$$\text{Re}_p = \frac{(1.5 \times 10^{-4}\ \text{m})(0.306\ \text{m/sec})\ [996\ \text{kg/m}^3]}{[8 \times 10^{-4}\ \text{kg/m sec}]}$$

$$\text{Re}_p = 57.1$$

Hence velocity is 0.306 m/sec.

Example 4-3 Particles of a mineral ($D_p = 1 \times 10^{-4}$ m; $\rho = 4000$ kg/m^3) are settling in an organic liquid ($\rho = 1600$ kg/m^3; $\mu = 1 \times 10^{-3}$ kg/m sec) The volume fraction of solids is 0.3.

This will be a case of hindered settling. Our first step is to get A:

$$A = (1 \times 10^{-4}\ \text{m}) \left[\frac{9.8\ \text{m/sec}^2 \times 1600\ \text{kg/m}^3 (4000 - 1600)\ \text{kg/m}^3}{(1 \times 10^{-3}\ \text{kg/m sec})^2} \right]^{1/3}$$

$$A = 3.35$$

The A value places the system essentially in the Stokes' law region. Hence,

$$V = \frac{(9.8\ \text{m/sec}^2)(1 \times 10^{-4}\ \text{m})^2 (4000 - 1600)\ \text{kg/m}^3}{18(1 \times 10^{-3}\ \text{kg/m sec})}$$

$$V = 0.0131\ \text{m/sec}$$

$$\text{Re}_p = \frac{(1 \times 10^{-4}\ \text{m})(0.0131\ \text{m/sec})(1600\ \text{kg/m}^3)}{(1 \times 10^{-3}\ \text{kg/m sec})}$$

$$\text{Re}_p = 2.10$$

The exponent value n for equation (4-11) is obtained from Table 4-3. By interpolation, n is found to be 4.2. Porosity, ϵ, is 0.7. Then

$$V_{ts} = (0.013\ \text{m/sec})(0.7)^{4.2}$$

$$V_{ts} = 0.0029\ \text{m/sec}$$

Example 4-4 A catalytic packed bed reactor uses cube-shaped pellets (0.005 m on a side). Pellet density is 1600 kg/m^3. The bed density is 960 kg/m^3. Dimensions of the bed are 0.1-m^2 cross section and 2.0-m length. Superficial velocity of the vapor flowing through the bed is 1.0 m/sec. Vapor properties are a density of 0.65 kg/m^3 and a viscosity of 1.5×10^{-5} kg/m sec.

What is the pressure drop through the bed?

$$\text{Bed porosity} = \epsilon = \frac{\rho_{\text{catalyst}} - \rho_{\text{bed}}}{\rho_{\text{catalyst}}}$$

$$\epsilon = \frac{1600 - 960}{1600} = 0.4$$

The value of ϕ_s is 1.0 (from Table 4-4).

Next, using equation (4-14) we obtain

$$\Delta P = \left[\frac{150(1 - \epsilon)\mu}{\phi_s D_p V_0 \rho} + 1.75 \right] \frac{L(1 - \epsilon)\rho V_0^2}{g_c \phi_s D_p \epsilon^3}$$

$$\Delta P = \left[\frac{(150)\ (1 - 0.4)(1.5 \times 10^{-5}\ \text{kg/m sec}}{(1.0)(0.005\ \text{m})(1.0\ \text{m/sec})(0.65\ \text{kg/m}^3)} + 1.75 \right]$$

$$\times \frac{2\ \text{m}(0.6)(0.65\ \text{kg/m}^3)(1\ \text{m/sec})}{(\text{kg m/N sec}^2)(1.0)(0.005\ \text{m})(0.4)^3}$$

$$\Delta P = 2437.5\ \text{N/m}^2$$

Example 4-5 What would the pressure drop in a ion exchange bed (0.5 m in depth, particle diameters of 0.001 m, bed porosity of 0.25) be for turbulent flow of water through the bed?

At turbulent flow, Re is 1000 so that

$$1000 = \frac{D_p V_0 \rho}{\mu}$$

$$V_0 = \frac{(1000)\ (\mu)}{D_p \rho}$$

Assuming water at room temperature gives $\rho = 1000\ \text{kg/m}^3$ and $\mu = 1 \times 10^{-3}$ kg/m sec:

$$V_0 = \frac{(1000)\ (1 \times 10^{-3}\ \text{kg/m sec})}{(0.001\ \text{m})(1000\ \text{kg/m}^3)}$$

$$V_0 = 1.0\ \text{m/sec}$$

Then for turbulent flow [equation (4-18)] we have

$$\Delta P = \frac{1.75 \rho L V_0^2 (1 - \epsilon)}{gc \phi_s D_p \epsilon^3}$$

$$\Delta P = \frac{(1.75)(1000\ \text{kg/m}^3)(0.5\ \text{m})(1.0\ \text{m/sec})^2(0.75)}{(\text{kg m/N sec})(1.0)(0.001\ \text{m})(0.25)^3}$$

$$\Delta P = 2362.5\ \text{N/m}^2$$

Example 4-6 A screw extruder is to process a polymer (n of 0.46; K of $2.07 \times 10^4\ \text{N/m}^2\ \text{sec}^n$ density of 760 kg/m^3) through a circular die (diameter of 0.0127 m). Desired output is 9.581 cm^3/sec with developed pressure of $2.414 \times 10^6\ \text{N/m}^2$ (at the end of the extruder and the die entrance). What length should the die be?

First find the average velocity in the die:

$$\overline{V} = \frac{9.581\ \text{m}^3/\text{sec})(4)}{(\pi)(1.27\ \text{cm})^2}$$

$$\overline{V} = 7.56\ \text{cm/sec} = 0.0756\ \text{m/sec}$$

$$n' \cong n = 0.46$$

$$K' = K\left[\frac{3n+1}{4n}\right]^n = 2.33 \times 10^4 \frac{\text{N}}{\text{m}^2}\ \text{sec}^n$$

$$\text{Re}' = \frac{D^{n'}V^{2-n'}}{g_c K' 8^{n'-1}}$$

$$\text{Re}' = \frac{(0.0127\ \text{m})^{0.46}(0.0756\ \text{m/sec})^{1.54}\ 760\ \text{kg/m}^3}{\left[\dfrac{\text{kg m}}{\text{N sec}^2}\right]\left[23{,}300\dfrac{\text{N}}{\text{m}^2}\ \text{sec}^{0.46}\right](8)^{-0.54}}$$

$$\text{Re}' = 2.83 \times 10^{-4}$$

From Figure 4-3 the f value is 56,600. The length L is

$$L = \frac{\Delta P g_c D}{2 f \rho V^2}$$

$$L = \frac{\left[2.414 \times 10^6\ \text{N/m}^2\right]\left[\text{kg m/N sec}^2\right][0.0127\ \text{m}]}{[2][56{,}600]\left[760\ \text{kg/m}^3\right]\left[0.0756\ \text{m/sec}\right]^2}$$

$$L = 0.0624\ \text{m}$$

Example 4-7 A non-Newtonian slurry ($n' = 0.40$; $K' = 2.1\ \text{N/m}^2\ \text{sec}^{n'}$; $\rho = 950\ \text{kg/m}^3$) flows through a tube (diameter of 0.06 m; length of 20 m) with an average velocity of 7 m/sec. What is the pressure drop?

$$\text{Re}' = \frac{(0.06\ \text{m})^{0.4}(7\ \text{m/sec})^{1.6}\left[950\ \text{kg/m}^3\right]}{\left[\dfrac{\text{kg m}}{\text{N sec}^2}\right]\left[2.1\dfrac{\text{N}}{\text{m}^2}\ \text{sec}^{0.4}\right](8)^{-0.6}}$$

$$\text{Re}' = 11{,}495$$

Then from Figure 4-3 using the turbulent flow line for $n' = 0.4$ the f value is found to be 0.0031.

$$\Delta P = \frac{2 f L \rho V^2}{g_c D}$$

$$\Delta P = \frac{2(0.0031)(20\ \text{m})\left[950\ \text{kg/m}^3\right]\left[7\ \text{m/sec}\right]^2}{\left[\dfrac{\text{kg m}}{\text{N sec}^2}\right](0.06\ \text{m})}$$

$$\Delta P = 9.62 \times 10^4 \frac{\text{N}}{\text{m}^2}$$

Example 4-8 Calculate the power required for agitation for a three-blade marine impeller of 0.610-m pitch and 0.610-m diameter operating at 100 rpm in an unbaffled tank containing water at 21.1°C. The tank diameter is 1.83 m, the depth of liquid is 1.83 m, and the impeller is located 0.610 m from the tank bottom.

Data : $\mu = 0.001$ kg/m sec, $\rho = 998$ kg/m^3

For the above case (since $D_T/D' = 3.0$, $Z_l/D' = 3.0$, $Z_i/D' = 1.0$) the proper curve is no. 4 in Figure 4-6.

Hence

$$\text{Re} = \frac{(D')^2 N}{\mu} = \frac{(0.6\text{ m})^2 \left(\dfrac{100}{60\text{ sec}}\right)(998\text{ kg/m}^3)}{0.001\text{ kg/m sec}}$$

$$\text{Re} = 5.99 \times 10^5$$

Since Re > 300, we must read the right-hand scale, $P_0/(\text{Fr})^{x'}$ where $x = (\text{a} - \log\text{Re})/b$, where a and b are constants and Fr is the Froude number $(N^2 D'/g)$.

Then, from the table with Figure 4-6, the values of a and b are, respectively, 2.1 and 18, and from the chart we have

$$\frac{P_o}{(\text{Fr})^x} = 0.23$$

Also,

$$x = \frac{a - \log\text{Re}}{b} = \frac{2.1 - 5.82}{18} = -0.206$$

so that

$$\frac{P_o}{(\text{Fr})^{-0.206}} = 0.23$$

$$\text{Fr} = \frac{\left(\dfrac{100}{60\text{ sec}}\right)^2 (0.6\text{ m})}{9.81\text{ m/sec}^2} = 0.170$$

$$P_o = \frac{0.23}{(0.173)^{0.206}} = 0.33$$

Then,

$$P = P_o N^3 (D')^5 \rho = 0.33 \left(\frac{100}{60\text{ sec}}\right)^3 (998\text{ kg/m}^3)(0.6\text{ m})^5 = 118.6\text{ watts}$$

$$P = 118.6\text{ watts} \times \frac{\text{H. P.}}{746\text{ watts}} = 0.16\text{ H.P.}$$

Example 4-9 Two miscible liquids are to be mixed in a baffled turbine agitated vessel ($D = 0.61$ m; $D_t = 1.83$ m; $Z_1 = 1.0$ m) with a speed of 80 revolutions per minute. How long will it take for the two liquids to be completely mixed? The viscosity and density are, respectively, 9.8×10^{-4} pascal-sec and 990 kg/m^3.

First, determine the Reynolds number

$$\text{Re}^* = \frac{N\rho(D')^2}{\mu} = \frac{(80\ \text{min}^{-1})(\text{min}/60\ \text{sec})(990\ \text{kg/m}^3)(0.61\ \text{m})^2}{(9.8 \times 10^{-4}\ \text{pascal-sec})}$$

$$\text{Re}^* = 501{,}070$$

Then, from Figure 4-7 the f_t value is 3.98. Hence the blending time is

$$t_t = \frac{f_t(z_1)^{1/2}(D_T)^{3/2}}{(N)^{2/3}(D')^{4/3}(g)^{1/6}(D')^{1/2}}$$

$$t_t = \frac{(3.98)(1.00\ \text{m})^{1/2}(1.83\ \text{m})^{3/2}}{(1.333\ \text{sec})^{2/3}(0.61\ \text{m})^{11/6}(9.80\ \text{m/sec}^2)^{1/6}}$$

$$t_t = 12.44\ \text{sec}$$

PROBLEMS

4-1. In the processing of instant coffee, extract particles (diameter of 4×10^{-5} m) drop through air at a temperature of 150°C. If the extract density is 1030 kg/m^3, what is the terminal velocity of the particles?

4-2. What is the terminal settling velocity for a range of solid particles (D_p ranges from 1.5×10^{-4} to 1.75×10^{-4} m)? The solid has a density of 2800 kg/m^3.

4-3. A steel ball ($\rho = 7850$ kg/m^3) of 0.0254-m diameter is falling through a thick viscous material ($\rho = 1258$ kg/m^3 $\mu = 0.8$ pascal-sec). What is its velocity?

4-4. A tall cylindrical chimney is 1.5 meters in diameter. Wind is blowing horizontally at a velocity of 8.94 m/sec. For this case what is the wind's force per unit height of chimney?

4-5. If a raindrop has a diameter of 2.54×10^{-5} m and has been falling for a reasonable elapsed time, what is its velocity?

4-6. A skydiver falls in free fall before using the parachute. What is his velocity? Assume $C_d = 0.6$ and his projected area in the direction of fall is 0.1 m^2.

4-7. Pellets of a material are made by spraying molten drops into cold air (20°C) in a tower. The pellets solidify as they fall. Desired diameter is 6×10^{-3} m for the material ($\rho = 1350$ kg/m^3). What is the pellet's terminal velocity?

4-8. Calculate the external surface area and average particle size for a crushed ore. The solid has a density of 4000 kg/m^3. In a bed with a porosity of

0.47 the pressure drop is 1.9×10^6 pascals/m for a superficial velocity of 4.57×10 m/sec.

4-9. Propane gas flows through a packed bed in a column ($L = 15.24$ m; $D = 6.10$ m). The packing consists of 0.0254-m-diameter spheres with a porosity of 0.40. For isothermal conditions (260°C) and an exit pressure of 2.07×10^5 pascals, what is the entrance pressure?

4-10. Air at 37.8°C and atmospheric pressure flows through a packed bed made up of 0.0127-m-diameter spheres. The bed is 1.22 m in diameter and 2.44 m high. Mass flow rate of air is 0.95 kg/sec. Bed porosity is 0.36. Find the pressure drop.

4-11. Laminar flow in a porous media can be represented by Darcy's law:

$$V_0 = \frac{-k}{\mu}\frac{dP}{dl}$$

The parameter k, the permeability, is frequently used to characterize the medium. A cylindrically shaped porous core sample from an oil field is 0.02 m in diameter and 0.15 m long. The volumetric flow rate through the sample with a pressure drop of 101,000 N/m^2 is 10^{-6} m^3/sec. What is the permeability of the sample?

4-12. Water flows through a bed made up of spherical particles (half of these have a specific surface of 787 m^{-1} and the others 1181 m^{-1}). The bed is 0.32 m in diameter and 1.5 m high with a porosity of 0.4. A height of 0.3 m of water exists above the bed. What is the volumetric flow throughput if the water is at 21°C.

4-13. A bed of cubes (6.5×10^{-3} m on a side) are used in a heater for air. The container is a cylinder 2 m deep. Air enters at 27°C and 6.89×10^5 pascals and then leaves at 205°C. Mass flow rate is 1.26 kg/sec for 1 m^2 of free cross section. What is the pressure drop across the bed?

4-14. A gas (M.W. of 30) is passed through a 0.04-m-diameter tube packed with cylindrical catalyst pellets ($L = D = 3.2 \times 10^{-3}$ m). The gas enters at 350°C and 2.03×10^5 pascals with a superficial velocity of 1 m/sec. Bed porosity is 0.38. What is the pressure drop through the bed? How does pressure drop change if cylinder size is increased to 4.8×10^{-3} m?

4-15. A polyethylene melt is flowing through a 3-m-long pipe (diameter 0.05 m) with a volumetric flow rate of 9.4×10^{-4} m^3/s. What is the maximum point velocity in the tube? Indicate all assumptions.

4-16. A small-scale piping system (diameter 0.025 m, length 0.25 m) gives the following results for a non-Newtonian fluid:

Mass Flow Rate (kg/sec)	Pressure Drop (N/m^2)
1.92×10^{-4}	17,243
4.08×10^{-4}	34,486
8.8×10^{-4}	68,975
1.95×10^{-3}	1.38×10^5
4.58×10^{-3}	2.76×10^5

If the pressure drop in a 0.03-m-diameter pipe is limited to 13.8×10^5 N/m^2 at a volumetric flow rate of 6.26×10^{-5} m^3/s, what length of pipe is needed?

4-17. The rheological data for polymer melt are as follows:

τ (dynes/cm^2)	$\dot{\gamma}$ (s^{-1})
14.1	0.9
58.5	10
228	100
344	200
435	300
502	400
780	800
820	1000

What diameter of circular pipe will give a pressure drop of 109.1×10^5 N/m^2 for a volumetric flow of 1.9×10^{-3} m^3/sec?

4-18. A power law fluid (n' of 0.29; k' of 2.6 N-sec$^{n'}$/m^2) flows in a 0.04-m tube with an average velocity of 6.4 m/sec. What would the pressure drop be if the tube were 28 m long?

4-19. In Figure 4-3 the transition Reynolds number for turbulent flow increases in value as n' decreases. The line for n' at 0.0 does not show a transition. Explain why this occurs.

4-20. A molten polymer is to be extruded through a heptagonal die with an average velocity of 0.04 m/sec. The available pressure drop through the 0.1-m-long die is 55.18×10^5 N/m^2. The power law parameters for the polymer are $n = 0.53$ and $k = 3 \times 10^4$ N-sec^n/m^2. What is the characteristic dimension for the die (i.e., side of heptagon)?

4-21. Develop a correlation between the critical Reynolds number (point at which laminar flow ends) and the n' value (i.e., see Figure 3-14) from 1.0 to 0.0.

4-22. The residence time ratio (RTR), which is an important scale-up parameter for flow reactors, can be determined either by dividing the maximum velocity by the average velocity or by using the ratio of average residence time by minimum residence time. Typical values in Newtonian flow systems are 1.0 (plug flow), 1.25 (turbulent flow), and 2.0 (laminar flow). Can these cases be matched by flow situations for pseudoplastic fluids? Prove your answers.

4-23. A molten polymer (n' of 0.71 and a K' of 2100 N-sec^n/m^2) flows through a tube (diameter 0.01 m, length 0.5 m). If the pressure drop is 87.7 $\times$ 10^5 N/m^2, what is the fluid's average velocity? Also, what is its maximum velocity?

4-24. A fluid is to be agitated in a baffled tank (four baffles) at a speed of 110 rpm. The system that uses a marine propeller has the same dimensions as the system of Example 4-8. The fluid's viscosity and density are, respectively, 0.01 pascal-sec and 960 kg/m^3. What is the required power?

4-25. A highly viscous fluid (viscous fluid (viscosity of 120 pascal-sec, density of 1200 kg/m^3) is to be agitated with a six-bladed turbine. The system parameters are: $D_T = 1.83$ m, $D' = 0.61$ m, pitch $= D'$, $Z_1 =$ 1.83 m, $Z_i = 0.61$ m. What would the power requirements be for unbaffled and baffled tanks (four baffles)?

4-26. An agitation system is studied in a pilot plant. The unit (a six-bladed turbine impeller) had the following dimensions: $D' = 0.1$ m, $D_T = 0.3$ m. The two miscible liquids in the system are mixed in 15 sec at a Reynolds number of 10,000 with a power input of 0.4 kW/m^3. What power input would be required for a 15-sec blending time in a 1.83-m-diameter tank? If power input is kept constant, what would the blending time be in the 1.83-m-diameter tank?

4-27. A six-bladed turbine gives a mixing time of 29 sec in a baffled tank. System dimensions are: $D_T = 1.44$ m, $D' = 0.48$ m, $Z_1 = 1.44$ m. The speed was 75 rpm for the fluid ($\mu = 0.003$ pascal-sec $\rho = 1080$ kg/m^3). What values of mixing time would be required if the same power per unit volume was used for impeller diameters one-fourth or six-tenth of the tank diameter?

4-28. A viscous fluid ($\mu = 1$ pascal-sec, $\rho = 750$ kg/m^3) is to be mixed in an unbaffled tank equipped with a three-bladed marine propeller. The agitator is equipped with a 7.46-kW motor. Can this motor move the agitator at a speed of 900 rpm? System dimensions are: $D' = 0.3$ m, $D_T =$ 1.2 m, $Z_1 = 1.2$ m, $Z_i = 0.3$ m pitch $= D'$.

4-29. Repeat problem 4-28 for a baffled tank (four baffles of W of 0.127 m) used with an agitator whose pitch is $2D$'.

4-30. A model of a process unit (a reactor) uses 0.001 kg of feed compared to 0.5 kg for the production unit. The larger unit that uses a six-bladed turbine has the following dimensions: $D' = 0.6$ m, $D_T = 2$ m, $Z_1 = 2$ m. If the optimum speed in the smaller unit is 350 rpm, find the speed at which the large unit should operate if

(a) Reynolds numbers are constant

(b) Power per unit volumes are the same

(c) Blending time is constant

REFERENCES

1. C. E. Lapple and C. B. Shepherd, *Ind. Eng Chem.* **32**, 605 (1940).
2. A. D. Maude and R. L. Whitmore, *Br. J. Appl. Phys.* **9**, 477 (1958).
3. S. Ergun, *Chem. Eng. Prog.* **48**, 49 (1952).
4. F. Rodriguez, *Principles of Polymer Systems*, McGraw-Hill, New York (1970).
5. D. W. Dodge and A. B. Metzner, *AIChE J.* **5**, 198 (1959).
6. J. H. Rushton, E. Costich, and H. J. Everett, *Chem. Eng. Prog.* **46**, 395; 467 (1950).
7. K. W. Norwood and A. B. Metzner, *AIChE J.* **6**, 432 (1960).
8. W. L. McCabe, J. C. Smith, and P. Harriott, *Unit Operations of Chemical Engineering*, McGraw-Hill, New York (1985).
9. L. A. Cutter, *AIChE J.* **12**, 35 (1966).
10. M. Moo-Young, K. Tichar, and F. A. L. Dullien, *AIChE J.* **18**, 178 (1972).

5

HEAT TRANSFER; CONDUCTION

INTRODUCTION

Practically all processes in the chemical, petroleum, and related industries require the transfer of energy. Typical examples are the heating and cooling of process streams, phase changes, evaporations, separations (distillations, etc.), solutions, crystallizations, and so on.

The basic underlying principle governing such systems is the First Law of Thermodynamics for a control volume or open system. In Chapter 2, this approach was used to develop the Bernoulli balance used with macroscopic fluid mechanics systems. Here we will use a different form but one that nonetheless emanates from the First Law.

After this form is developed, we will apply it to various flow and nonflow situations.

THE EQUATION OF ENERGY

The First Law of Thermodynamics is in essence a statement of the conservation of energy. For a flowing system, we can write such a balance in word form as

$$\begin{pmatrix}\text{Energy rate}\\ \text{of accumulation}\\ \text{in control volume}\end{pmatrix} = \begin{pmatrix}\text{Rate of heat flow}\\ \text{across surface}\end{pmatrix} + \begin{pmatrix}\text{Power (work per}\\ \text{unit time) out}\\ \text{at surface}\end{pmatrix} + \begin{pmatrix}\text{Energy rate of}\\ \text{flow at inlets}\end{pmatrix} - \begin{pmatrix}\text{Energy rate of}\\ \text{flow at outlets}\end{pmatrix} \tag{5-1}$$

If we consider a pure fluid with no internal heating sources other than the fluid's viscous dissipation, we can show (for rectangular coordinates) that

$$\rho \hat{C}_v \left(\frac{\partial T}{\partial t} + v_x \frac{\partial T}{\partial x} + v_y \frac{\partial T}{\partial y} + v_z \frac{\partial T}{\partial z} \right) = - \left[\frac{\partial q_x}{\partial x} + \frac{\partial q_y}{\partial y} + \frac{\partial q_z}{\partial z} \right]$$

Convection or temperature change of moving fluid — Heat conduction

$$- T \left(\frac{\partial p}{\partial T} \right)_\rho \left(\frac{\partial v_x}{\partial x} + \frac{\partial v_y}{\partial y} + \frac{\partial v_z}{\partial z} \right) - \left\{ \tau_{xx} \frac{\partial v_x}{\partial x} + \tau_{yy} \frac{\partial v_y}{\partial y} + \tau_{zz} \frac{\partial v_z}{\partial z} \right\}$$

Expansion effects — Viscous dissipation

$$- \left\{ \tau_{xy} \left(\frac{\partial v_x}{\partial y} + \frac{\partial v_y}{\partial x} \right) + \tau_{xz} \left(\frac{\partial v_x}{\partial z} + \frac{\partial v_z}{\partial x} \right) + \tau_{yz} \left(\frac{\partial v_y}{\partial z} + \frac{\partial v_z}{\partial y} \right) \right\} \quad (5\text{-}2)$$

Viscous dissipation

Similarly, equations (5-3) and (5-4) give the forms for cylindrical and spherical coordinates:

$$\begin{aligned}
\rho \hat{C}_v \left(\frac{\partial T}{\partial t} + v_r \frac{\partial T}{\partial r} + \frac{v_\theta}{r} \frac{\partial T}{\partial \theta} + v_z \frac{\partial T}{\partial z} \right) &= - \left[\frac{1}{r} \frac{\partial}{\partial r}(r q_r) + \frac{1}{r} \frac{\partial q_\theta}{\partial \theta} + \frac{\partial q_z}{\partial z} \right] \\
& - T \left(\frac{\partial p}{\partial T} \right)_\rho \left(\frac{1}{r} \frac{\partial}{\partial r}(r v_r) + \frac{1}{r} \frac{\partial v_\theta}{\partial \theta} + \frac{\partial v_z}{\partial z} \right) - \left\{ \tau_{rr} \frac{\partial v_r}{\partial r} + \tau_{\theta\theta} \frac{1}{r} \left(\frac{\partial v_\theta}{\partial \theta} + v_r \right) \right. \\
& \left. + \tau_{zz} \frac{\partial v_z}{\partial z} \right\} - \left\{ \tau_{r\theta} \left[r \frac{\partial}{\partial r} \left(\frac{v_\theta}{r} \right) + \frac{1}{r} \frac{\partial v_r}{\partial \theta} \right] + \tau_{rz} \left(\frac{\partial v_z}{\partial r} + \frac{\partial v_r}{\partial z} \right) \right. \\
& \left. + \tau_{\theta z} \left(\frac{1}{r} \frac{\partial v_z}{\partial \theta} + \frac{\partial v_\theta}{\partial z} \right) \right\}
\end{aligned} \quad (5\text{-}3)$$

$$\begin{aligned}
\rho \hat{C}_v \left(\frac{\partial T}{\partial t} + v_r \frac{\partial T}{\partial r} + \frac{v_\theta}{r} \frac{\partial T}{\partial \theta} + \frac{v_\phi}{r \sin\theta} \frac{\partial T}{\partial \phi} \right) &= - \left[\frac{1}{r^2} \frac{\partial}{\partial r}(r^2 q_r) \right. \\
& \left. + \frac{1}{r \sin\theta} \frac{\partial}{\partial \theta}(q_\theta \sin\theta) + \frac{1}{r \sin\theta} \frac{\partial q_\phi}{\partial \phi} \right] - T \left(\frac{\partial p}{\partial T} \right)_\rho \left(\frac{1}{r^2} \frac{\partial}{\partial r}(r^2 v_r) \right. \\
& \left. + \frac{1}{r \sin\theta} \frac{\partial}{\partial \theta}(v_\theta \sin\theta) + \frac{1}{r \sin\theta} \frac{\partial v_\phi}{\partial \phi} \right) - \left\{ \tau_{rr} \frac{\partial v_r}{\partial r} + \tau_{\theta\theta} \left(\frac{1}{r} \frac{\partial v_\theta}{\partial \theta} + \frac{v_r}{r} \right) \right. \\
& \left. + \tau_{\phi\phi} \left(\frac{1}{r \sin\theta} \frac{\partial v_\phi}{\partial \phi} + \frac{v_r}{r} + \frac{v_\theta \cot\theta}{r} \right) \right\} - \left\{ \tau_{r\theta} \left(\frac{\partial v_\theta}{\partial r} + \frac{1}{r} \frac{\partial v_r}{\partial \theta} - \frac{v_\theta}{r} \right) \right. \\
& \left. + \tau_{r\phi} \left(\frac{\partial v_\phi}{\partial r} + \frac{1}{r \sin\theta} \frac{\partial v_r}{\partial \phi} - \frac{v_\phi}{r} \right) + \tau_{\theta\phi} \left(\frac{1}{r} \frac{\partial v_\phi}{\partial \theta} + \frac{1}{r \sin\theta} \frac{\partial v_\theta}{\partial \phi} - \frac{\cot\theta}{r} v_\phi \right) \right\}
\end{aligned} \quad (5\text{-}4)$$

The foregoing forms of the Energy Equation can be rewritten for Newtonian fluids (constant density and thermal conductivity) together with the appropriate terms for the q energy fluxes (i.e., q_x, q_y, q_z, etc.). These expressions are given below for the rectangular case:

$$q_x = -k\frac{\partial T}{\partial x}, \qquad q_v = -k\frac{\partial T}{\partial y}, \qquad q_z = -k\frac{\partial T}{\partial z} \tag{5-5}$$

the cylindrical case:

$$q_r = -k\frac{\partial T}{\partial_r}, \qquad q_\theta = -k\frac{1}{r}\frac{\partial T}{\partial \theta}, \qquad q_z = -k\frac{\partial T}{\partial z} \tag{5-6}$$

and the spherical case:

$$q_r = -k\frac{\partial T}{\partial r}, \qquad q_\theta = -k\frac{1}{r}\frac{\partial T}{\partial \theta}, \qquad q_\phi = -k\frac{1}{r\sin\theta}\frac{\partial T}{\partial \phi} \tag{5-7}$$

Equations (5-8), (5-9), and (5-10) give the revised forms of equations (5-2) through (5-4). The rectangular coordinate equation is (5-8), while (5-9) and (5-10) are, respectively, for the cylindrical and spherical cases.

$$\underbrace{\rho\hat{C}_p\left(\frac{\partial T}{\partial t} + v_x\frac{\partial T}{\partial x} + v_y\frac{\partial T}{\partial y} + v_z\frac{\partial T}{\partial z}\right)}_{\text{Convection}} = \underbrace{k\left[\frac{\partial^2 T}{\partial x^2} + \frac{\partial^2 T}{\partial y^2} + \frac{\partial^2 T}{\partial z^2}\right]}_{\text{Heat conduction}}$$

$$\underbrace{+ 2\mu\left\{\left(\frac{\partial v_x}{\partial x}\right)^2 + \left(\frac{\partial v_y}{\partial y}\right)^2 + \left(\frac{\partial v_z}{\partial z}\right)^2\right\} + \mu\left\{\left(\frac{\partial v_x}{\partial y} + \frac{\partial v_y}{\partial x}\right)^2 + \left(\frac{\partial v_x}{\partial z} + \frac{\partial v_z}{\partial x}\right)^2 + \left(\frac{\partial v_y}{\partial z} + \frac{\partial v_z}{\partial y}\right)^2\right\}}_{\text{Viscous dissipation}} \tag{5-8}$$

Then, for cylindrical coordinates we have

$$\rho\hat{C}_p\left(\frac{\partial T}{\partial t} + v_r\frac{\partial T}{\partial r} + \frac{v_\theta}{r}\frac{\partial T}{\partial \theta} + v_z\frac{\partial T}{\partial z}\right) = k\left[\frac{1}{r}\frac{\partial}{\partial r}\left(r\frac{\partial T}{\partial r}\right) + \frac{1}{r^2}\frac{\partial^2 T}{\partial \theta^2} + \frac{\partial^2 T}{\partial z^2}\right]$$

$$+ 2\mu\left\{\left(\frac{\partial v_r}{\partial r}\right)^2 + \left[\frac{1}{r}\left(\frac{\partial v_\theta}{\partial \theta} + v_r\right)\right]^2 + \left(\frac{\partial v_z}{\partial z}\right)^2\right\} + \mu\left\{\left(\frac{\partial v_\theta}{\partial z} + \frac{1}{r}\frac{\partial v_z}{\partial \theta}\right)^2\right.$$

$$\left. + \left(\frac{\partial v_z}{\partial r} + \frac{\partial v_r}{\partial z}\right)^2 + \left[\frac{1}{r}\frac{\partial v_r}{\partial \theta} + r\frac{\partial}{\partial r}\left(\frac{v_\theta}{r}\right)\right]^2\right\} \tag{5-9}$$

and for spherical coordinates we have

$$\rho\hat{C}_p\left(\frac{\partial T}{\partial t} + v_r\frac{\partial T}{\partial r} + \frac{v_\theta}{r}\frac{\partial T}{\partial \theta} + \frac{v_\phi}{r\sin\theta}\frac{\partial T}{\partial \phi}\right) = k\left[\frac{1}{r^2}\frac{\partial}{\partial r}\left(r^2\frac{\partial T}{\partial r}\right)\right.$$

$$\left. + \frac{1}{r^2\sin\theta}\frac{\partial}{\partial \theta}\left(\sin\theta\frac{\partial T}{\partial \theta}\right) + \frac{1}{r^2\sin^2\theta}\frac{\partial^2 T}{\partial \phi^2}\right]$$

$$+ 2\mu \left\{ \left(\frac{\partial v_r}{\partial r} \right)^2 + \left(\frac{1}{r}\frac{\partial v_\theta}{\partial \theta} + \frac{v_r}{r} \right)^2 + \left(\frac{1}{r \sin\theta}\frac{\partial v_\phi}{\partial \phi} + \frac{v_r}{r} + \frac{v_\theta \cot\theta}{r} \right)^2 \right\}$$

$$+ \mu \left\{ \left[r\frac{\partial}{\partial r}\left(\frac{v_\theta}{r}\right) + \frac{1}{r}\frac{\partial v_r}{\partial \theta} \right]^2 + \left[\frac{1}{r\sin\theta}\frac{\partial v_r}{\partial \phi} + r\frac{\partial}{\partial r}\left(\frac{v_\phi}{r}\right) \right]^2 \right.$$

$$\left. + \left[\frac{\sin\theta}{r}\frac{\partial}{\partial \theta}\left(\frac{v_\phi}{\sin\theta}\right) + \frac{1}{r\sin\theta}\frac{\partial v_\theta}{\partial \phi} \right]^2 \right\} \tag{5-10}$$

Equations (5-2), (5-3), and (5-4) can also be applied to non-Newtonians by using the appropriate expression for the stresses. Also, all of the equations (5-2) through (5-5) and (5-8) through (5-10) can be used for other heat generation cases by adding an appropriate generation term to the right-hand side of the foregoing equations. This term would reflect the type of generation that takes place. Some examples are the energy associated with a chemical reaction (heat of reaction, heat of combustion) or a phase change (latent heat of vaporization, heat of fusion, heat of sublimation) or from another source (electrical, nuclear, etc).

STEADY-STATE CONDUCTION IN STATIC SYSTEMS

All of the velocity terms in the Energy Equation will disappear for a static system. If there is no variation of temperature with time and no internal heat sources, equations (5-5) or (5-6) or (5-7) will be applicable.

Suppose we have a one-dimensional system through which thermal energy (heat) is flowing. For a constant cross section and A (see Figure 5-1) we then have

$$\frac{q^1}{A} = -k\frac{dT}{dx} \tag{5-11}$$

where q^1 is the energy flow per unit time. If the q^1 value is constant, we have

$$q^1 = \frac{kA\Delta T}{\Delta x} \tag{5-12}$$

This is the steady-state conduction solution in one dimension for rectangular coordinates which shows that the temperature profile is linear in a slab of a given material.

If a number of slabs of different materials are put together (see Figure 5-2), we have an analogous situation to a set of electrical resistance in series.

For such a system the heat flow per unit time q^1 corresponds to current; the temperature change corresponds to voltage; and the thickness divided by thermal conductivity times area corresponds to the electrical resistance.

Hence, for a set of slabs as in Figure 5-2 the q^1 value is

$$q^1 = \frac{T_1 - T_4}{\dfrac{\Delta x_a}{k_a A} + \dfrac{\Delta x_b}{k_b A} + \dfrac{\Delta x_c}{k_c A}} \tag{5-13}$$

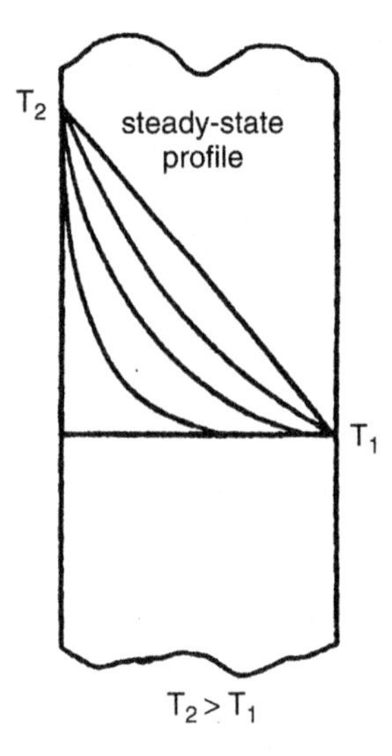

Figure 5-1. Unsteady (curves) and steady-state (straight line) temperature profiles in a slab. (Reproduced with permission from reference 5. Copyright 1997, American Chemical Society.)

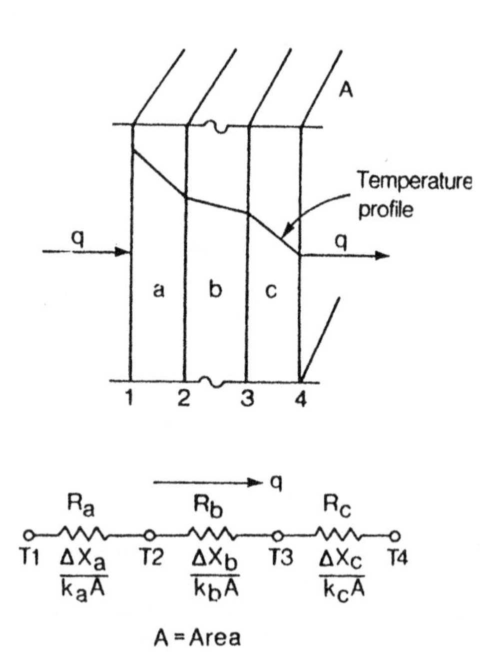

Figure 5-2. Steady-state conduction in multiple slabs with electrical analog. (Reproduced with permission from reference 5. Copyright 1997, American Chemical Society.)

If we have a cylinder that has only radial conduction, then

$$q^1 = -kA\frac{dT}{dr} \tag{5-14}$$

Now because $A = 2\pi rL$ for the cylinder surface, we obtain for a hollow cylinder (inner radius r_i and outer radius r_o) a q^1 as shown:

$$q^1 = \frac{2\pi kL(T_i - T_o)}{\ln(r_o/r_i)} \tag{5-15}$$

For multiple cylindrical sections we have

$$q^1 = \frac{2\pi L(\Delta T)_{\text{overall}}}{\dfrac{\ln(r_2/r_1)}{k_a} + \dfrac{\ln(r_3/r_2)}{k_b} + \dfrac{\ln(r_4/r_3)}{k_c}} \tag{5-16}$$

GENERAL CASE FOR STATIC SYSTEMS

The most general case for the static system is one in which there is the possibility of heat generation and unsteady-state heat transfer. For such a case where all velocities are zero by combining equations (5-2) and (5-5) we obtain

$$\rho C_P\frac{\partial T}{\partial t} = \frac{\partial}{\partial x}\left(\frac{k\partial T}{\partial x}\right) + \frac{\partial}{\partial y}\left(\frac{k\partial T}{\partial y}\right) + \frac{\partial}{\partial z}\left(\frac{k\partial T}{\partial z}\right) + \dot{q} \tag{5-17}$$

The $\dot{q}$ term is for heat generation (i.e., heat of reaction, latent heat, etc.).

For a steady-state ($\partial T/\partial t = 0$) one-dimensional slab with a k independent of position, equation (5-17) becomes

$$\frac{\partial^2 T}{\partial X^2} + \frac{\dot{q}}{k} = 0 \tag{5-18}$$

which is the basic equation for one-dimensional slab conduction with heat generation.

The corresponding cylindrical coordinate equation to (5-17) for only radial conduction is

$$\frac{\partial^2 T}{\partial r^2} + \frac{1}{r}\frac{\partial T}{\partial r} + \underbrace{\frac{1}{r^2}\frac{\partial^2 T}{\partial \theta^2}}_{\substack{\text{No } \theta \\ \text{con-} \\ \text{duct-} \\ \text{ion}}} + \underbrace{\frac{\partial^2 T}{\partial z^2}}_{\substack{\text{No } z \\ \text{con-} \\ \text{duct-} \\ \text{ion}}} + \frac{\dot{q}}{k} = \frac{\rho C\rho}{k}\frac{\partial T}{\partial T} \tag{5-19}$$

$$\frac{\partial^2 T}{\partial r^2} + \frac{1}{r}\frac{\partial T}{\partial r} + \frac{\dot{q}}{k} \tag{5-20}$$

Again this represents the basic equation for a cylinder with heat generation and radial conduction.

In unsteady-state cases, temperature change with time must be considered. Hence, for a system in rectangular coordinates without heat generation we obtain

$$\rho C_P \frac{\partial T}{\partial t} = \frac{\partial}{\partial x}\left(\frac{k\partial T}{\partial x}\right) + \frac{\partial}{\partial y}\left(\frac{k\partial T}{\partial y}\right) + \frac{\partial}{\partial z}\left(\frac{k\partial T}{\partial z}\right) \tag{5-21}$$

Note that the solution of equation (5-21) is complex even for the one-dimensional case because two partial differentials are involved. Fortunately, a great deal of work has been done in this area. The results of these efforts are tabulated in textbooks such as those authored by Carslaw and Jaeger (1) and Crank (2).

In addition, sets of charts have been developed that related temperature to position and time for slabs, cylinders, and spheres (1, 3, 4); these are shown in Figures 5-3 through 5-5.

The terms in these charts are: T_1, surface temperature; T_0, temperature at a given point at zero time; T, the temperature at that point when a time, t, has elapsed; α, the materials' thermal diffusivity; x_1, a characteristic dimension (a radius, or half thickness); x, the position for T_0 and T; n, a dimensionless position (x/x); m, a dimensionless function. $\dfrac{k}{hx_1}$ (the ratio of k to h (a film coefficient) times x_1).

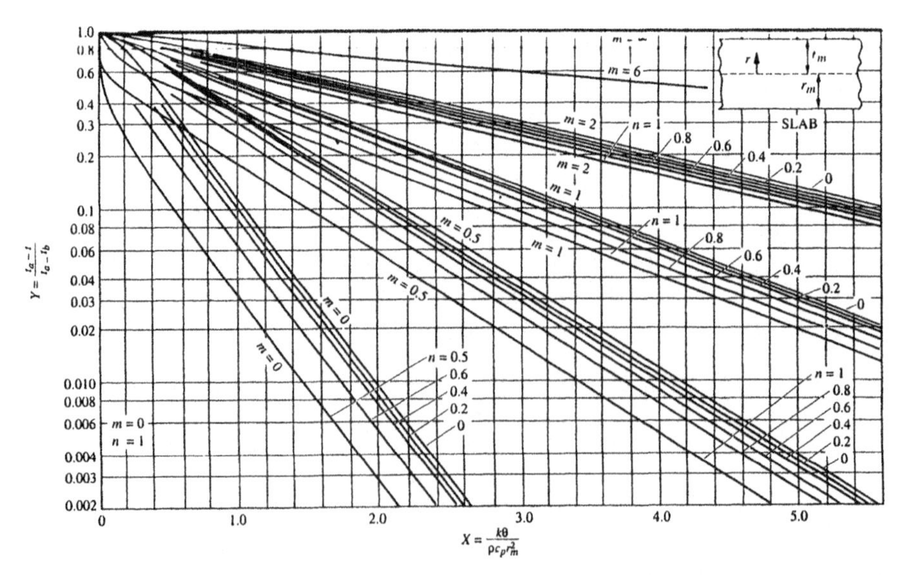

Figure 5-3. Temperature as a function of time and position in a slab. (Reproduced with permission from reference 3. Copyright 1923, American Chemical Society.)

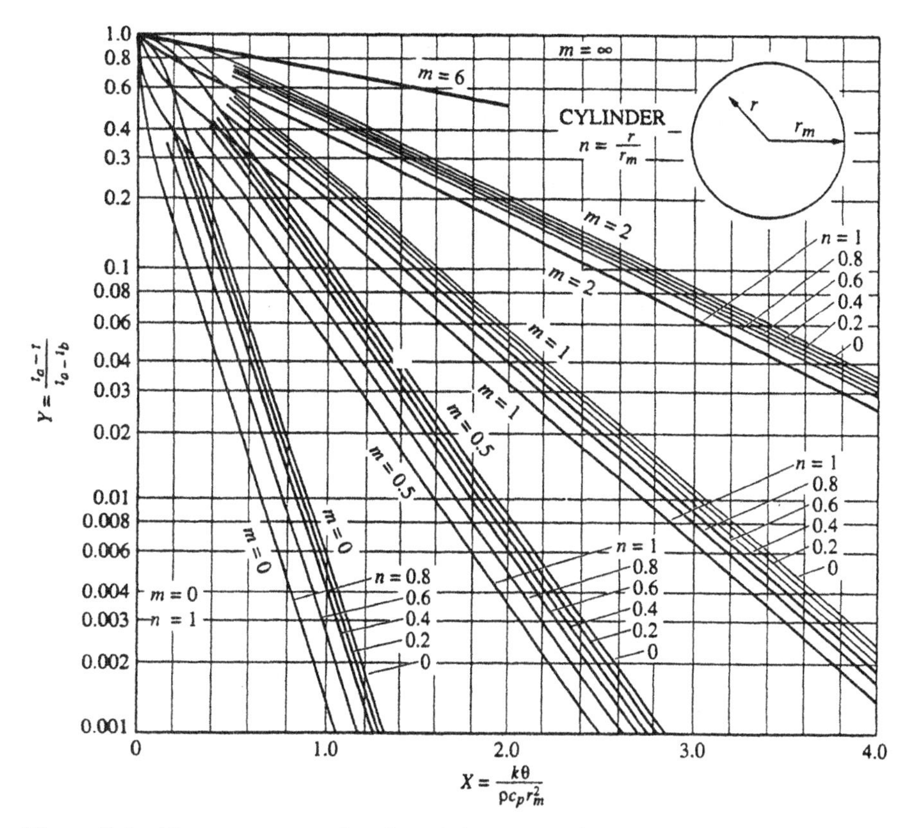

Figure 5-4. Temperature as a function of time and position in a long cylinder. (Reproduced with permission from reference 3. Copyright 1923, American Chemical Society.)

WORKED EXAMPLES

Example 5-1 A cold storage room has walls constructed of 0.102-m corkboard contained between double wooden walls each 0.0127 m thick. What is the rate of heat loss if wall surface temperature is −12.2°C inside and 21.1°C outside? Also, what is the temperature at the interface between the outer wall and the corkboard?

Values of thermal conductivity for wood and corkboard are 0.1073 W/m °C and 0.0415 W/m °C, respectively.

Because this is a case of multiple slab resistances, we use equation (5-13).

$$q^1 = \frac{(\Delta T)_{\text{overall}}}{\dfrac{\Delta x_a}{k_a A} + \dfrac{\Delta x_B}{k_B A} + \dfrac{\Delta x_c}{k_c A}}$$

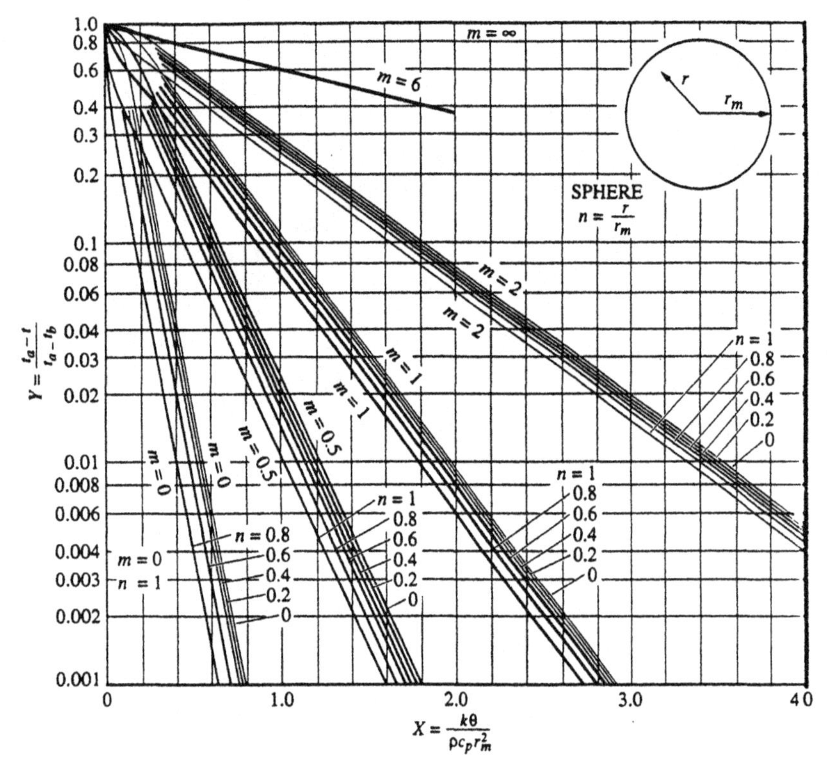

Figure 5-5. Temperature as a function of time and position in sphere. (Reproduced with permission from reference 3. Copyright 1923, American Chemical Society.)

Taking 1 m^2 of surface for the area A and substituting the appropriate values gives

$$q^1 = \frac{(21.1 + 12.2)^\circ\text{C}}{\dfrac{(0.0127\ \text{m})}{(0.1073\ \text{W/m}\ ^\circ\text{C})(1\ \text{m}^2)} + \dfrac{(0.102\ \text{m})}{(0.0415\ \text{W/m}\ ^\circ\text{C})(1\ \text{m}^2)} + \dfrac{(0.0127\ \text{m})}{(0.1073\ \text{W/m}\ ^\circ\text{C})(1\ \text{m}^2)}}$$

$$q^1 = 11.51\ \text{W}$$

The above represents the rate of heat loss. Because this applies to all segments of the system (wood–corkboard–wood), we can use the value for the outside wooden wall.

$$\frac{q^1}{A} = 11.51\ \text{W/m}^2 = (0.1073\ \text{W/m}\ ^\circ\text{C})(21.1 - T_{\text{interface}})\left(\frac{1}{0.0127\ \text{m}}\right)$$

$$T_{\text{interface}} = 19.7^\circ\text{C}$$

Example 5-2 A steel pipe of outside diameter of 0.051 m is insulated with a 0.0064-m thickness of asbestos followed by a 0.0254-m layer of fiber glass. If the pipe wall is 315.6°C and the outside insulation is 37.8°C, what is the temperature between the asbestos and fiber glass?

The thermal conductivities for asbestos and fiber glass are 0.166 W/m °C and 0.0485 W/m °C.

Because this is a system of cylindrical resistances, we use equation (5-16), where a stands for asbestos and b stands for fiber glass.

$$q^{1} = \frac{2\pi L(\Delta T)}{\dfrac{\ln(r_2/r_1)}{k_a} + \dfrac{\ln(r_3/r_2)}{k_b}}$$

Substituting the various values gives

$$q^{1} = \frac{2\pi L(315.6 - 37.8)^{\circ}\mathrm{C}}{\dfrac{\ln(0.0318/0.0254)}{0.166\ \mathrm{W/m\ ^{\circ}C}} + \dfrac{\ln(0.057/0.0318)}{0.0485\ \mathrm{W/m\ ^{\circ}C}}}$$

$$\frac{q^{1}}{L} = 129.5\ \mathrm{W/m}$$

However, this heat flow is the same for all parts of the system. Hence, if we consider only the asbestos ring, we have

$$\frac{q^{1}}{L} = 129.5\ \mathrm{W/m} = \frac{2\pi(315.6 - T_{\mathrm{interface}})}{\dfrac{\ln 1.25}{0.166\ \mathrm{W/m\ ^{\circ}C}}}$$

$$T_{\mathrm{interface}} = 288.3^{\circ}\mathrm{C}$$

Example 5-3 A 0.0032-m-diameter wire 0.3048 m long has a voltage of 10 volts imposed on it. Outer surface temperature is maintained at 93.3°C. What is the center wire temperature? (k of wire is 22.5 W/m °C; resistivity is 70 ohm-cm).

Because the wire's length is many times its diameter, we consider only radial conduction and heat generation. This means that we use equation (5-20).

$$\frac{\partial^2 T}{\partial r^2} + \frac{1}{r}\frac{\partial T}{\partial r} + \frac{\dot{q}}{k} = 0$$

Before we can solve it we must find $\dot{q}$, the heat generation per unit volume. In this case, the heating is due to the electrical heating. Therefore, for the volume

$$\dot{q}\pi r^2 L = \frac{(\mathrm{Voltage})^2}{\mathrm{Resistance}}$$

$$\dot{q} = \frac{(\mathrm{Voltage})^2}{\mathrm{Resistance}\ \pi r^2 L} = \frac{(10)^{2}}{(\mathrm{Resistance})\ \pi(0.016\ \mathrm{cm})^2(30.48\ \mathrm{cm})}$$

The resistance is determined from the resistivity.

$$\text{Resistance} = 70 \text{ ohm-cm } \frac{30.48 \text{ cm}}{\pi(0.16 \text{ cm})^2}$$

$$\text{Resistance} = 0.0268 \text{ ohm}$$

Thus the heat generation per unit volume is

$$\dot{q} = 1.539 \times 10^9 \text{ W/m}^3$$

The boundary conditions in this case are

$$r = 0, \qquad T = T_0$$

$$r = R, \qquad T = T_W = 93.3^\circ\text{C}$$

Solving the differential equation gives

$$T_0 - T_W = \frac{\dot{q}r^2}{4k}$$

Substituting and determining T_0 yields

$$T_0 = \frac{(1.539 \times 10^9 \text{ W/m}^3)(0.0016 \text{ m})^2}{4(22.5 \text{ W/m }^\circ\text{C})} + 93.3^\circ\text{C}$$

$$T_0 = 137.1^\circ\text{C}$$

Example 5-4 How long a time will be needed to raise the centerline temperature of a slab (hard rubbery material) to 132.2°C? The slab originally is at 26.7°C and is 0.0127 m thick. Thermal diffusivity ($k/\rho C\rho$) of the slab is 2.978×10^{-7} m^2/sec.

In this case we can use Figure 5-3 for the slab. The parameters n and m must first be obtained.

$$n = \frac{x}{x_1} = \frac{0}{(0.0064)} = 0$$

$$m = \frac{0.159 \text{ W/m }^\circ\text{C}}{h(0.0064)}$$

If the h from the metal to the rubber slab is assumed to be 5678 W/m^2 °C (a reasonable value), m is 0.00442. Even with a less likely h value of 567.8 w/m^2 °C, the m is 0.0442. Hence, we assume m is 0.

Next, calculating the dimensionless temperature (the ordinate of Figure (5-3)), we obtain

$$Y = \frac{141.7 - 132.2}{141.7 - 26.7} = 0.0821$$

Using this ordinate and reading to the line for $m = 0, n = 0$ gives an abscissa of 1.13. Then because

$$1.13 = \frac{kt}{\rho C\rho x_1^2}$$

We solve for t and obtain a value of 612 sec.

Example 5-5 Find the temperature profile for the laminar flow of a Newtonian fluid (constant density and thermal conductivity) if there is a constant energy flux at the wall.

Flow is along the tube axis (z direction). This means that the only velocity is V_z. Also, viscous dissipation is neglected. The proper equation in this case is (5-9), which upon evaluation

$$\rho\hat{C}_\rho\left(\frac{\partial T}{\partial t} + v_r\frac{\partial T}{\partial r} + \frac{v_\theta}{r}\frac{\partial T}{\partial \theta} + v_z\frac{\partial T}{\partial z}\right) = k\left[\frac{1}{r}\frac{\partial}{\partial r}\left(r\frac{\partial T}{\partial r}\right) + \frac{1}{r^2}\frac{\partial^2 T}{\partial \theta^2} + \frac{\partial^2 T}{\partial z^2}\right]$$

(Steady state) (No V_r, no $V\theta$) (No θ change) (z convection larger than z conduction)

becomes

$$\rho\hat{C}_\rho\left(v_z\frac{\partial T}{\partial z}\right) = k\left[\frac{1}{r}\frac{\partial}{\partial r}\left(r\frac{\partial T}{\partial r}\right)\right]$$

For the developed flow case (see Example 2-2) we have

$$V_z = (V_z)_{\text{max}}\left[1 - \left(\frac{r}{R}\right)^2\right]$$

Substituting for V_z and using the following boundary conditions, we obtain

$$r = 0, \qquad \frac{\partial T}{\partial r} = 0$$

$$r = R, \qquad T = T_W$$

$$0 \leqq r \leqq R, \qquad \frac{\partial T}{\partial z} = \text{constant}$$

The last condition results because the wall heat flux is constant and the average fluid temperature increases linearly with z.

Solving the energy equation gives

$$T_W - T = \frac{(V_z)\max\rho C\rho}{16R^2 k}\left(\frac{\partial T}{\partial z}\right)(3R^2 - 4r^2R^2 + r^4)$$

Alternatively, using the centerline temperature T_c, we obtain

$$T - T_c = \frac{\rho C\rho (V_z) \max R^2}{k4} \left(\frac{\partial T}{\partial z}\right) \left[\left(\frac{r}{R}\right)^2 - \frac{1}{4}\left(\frac{r}{R}\right)^4\right]$$

Example 5-6 What is the maximum temperature attained in a lubricant between two rotating cylinders? The outer cylinder (radius of 6 cm) rotates at an angular velocity of 8000 revolutions per minute. Clearance between the cylinders (both at 60°C) is 0.03 cm.

Lubricant viscosity, density, and thermal conductivity are, respectively, 0.1 kg/m sec, 1200 kg/m^3, and 0.13 J/sec m °C.

In this system, clearance is a small distance. Hence, the rotating cylinders can be represented by a system of parallel plates (Figure 5-6) with the top plate (outer cylinder moving with a velocity (the angular velocity times the radius) of 8 m/sec.

For the system of Figure 5-6, we can take the Equation of Energy and reduce it to a solvable form by the following.

$$\rho \hat{C}_p \left(\frac{\partial T}{\partial t} + v_x \frac{\partial T}{\partial x} + v_y \frac{\partial T}{\partial y} + v_z \frac{\partial T}{\partial z} \right) = k \left[\frac{\partial^2 T}{\partial x^2} + \frac{\partial^2 T}{\partial y^2} + \frac{\partial^2 T}{\partial z^2} \right]$$

(Steady state) (No convection) (Only x conduction)

$$+ 2\mu \left\{ \left(\frac{\partial v_x}{\partial x}\right)^2 + \left(\frac{\partial v_y}{\partial y}\right)^2 + \left(\frac{\partial v_z}{\partial z}\right)^2 \right\} + \mu \left\{ \left(\frac{\partial v_x}{\partial y} + \frac{\partial v_y}{\partial x}\right)^2 \right.$$

$$\left. + \left(\frac{\partial v_x}{\partial z} + \frac{\partial v_z}{\partial x}\right)^2 + \left(\frac{\partial v_y}{\partial z} + \frac{\partial v_z}{\partial y}\right)^2 \right\}$$

(V_z only a function of x)

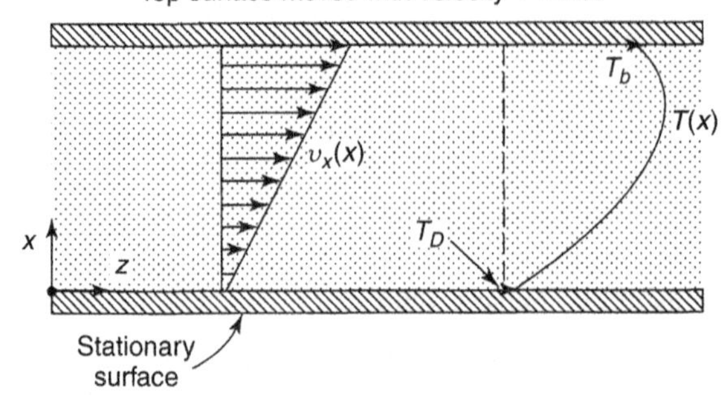

Figure 5-6. Physical analog to rotating cylinders with viscous dissipation. (Reproduced with permission from reference 6. Copyright 1960, John Wiley and Sons.)

so that

$$-k\frac{\partial^2 T}{\partial x^2} = \mu\left(\frac{\partial V_z}{\partial x}\right)^2$$

The velocity term can be obtained from the Equation of Motion z component (2-27). From this equation (with pressure and gravitational effects both zero), we obtain

$$\frac{\partial^2 V_z}{\partial x^2} = 0$$

Integrating with the boundary conditions

$$V_z = V \text{ (i.e., } \Omega R), \qquad x = \mathrm{B}$$

$$V_z = 0, \qquad x = 0$$

the relation

$$\frac{V_z}{V} = \frac{x}{B}$$

is determined.

Then

$$-k\frac{d^2 T}{dx^2} = \mu\frac{V^2}{B^2}$$

Solving with boundary conditions

$$T = T_0, \qquad x = 0$$

$$T = T_1, \qquad x = B$$

we obtain

$$T = T_0 + (T_1 - T_0)\frac{x}{B} + \frac{\mu V^2}{2\,k}\frac{x}{B}\left(1 - \frac{x}{B}\right)$$

or

$$\frac{T - T_0}{T_1 - T_0} = \frac{x}{B} + \frac{\mu V^2}{2k(T_1 - T_0)}\frac{x}{B}\left(1 - \frac{x}{B}\right)$$

The dimensionless grouping in the above expression is known as the Brinkman number (Br).

$$\mathrm{Br} = \frac{\mu V^2}{k(T_1 - T_0)} = \frac{\text{Heat generated by viscous dissipation}}{\text{Conduction heat transfer}}$$

This group indicates the impact of viscous dissipation effects.

For the case at hand, $T_1 = T_0$ and

$$T - T_0 = \frac{\mu V^2}{2k}\frac{x}{B}\left(1 - \frac{x}{B}\right)$$

The maximum temperature will occur when $x = 0.5B$—that is, giving the largest value of

$$\frac{x}{B}\left(1 - \frac{x}{B}\right)$$

$$T = 60°\text{C} + \frac{(0.1 \text{ kg/m/sec})(8 \text{ m/sec})^2}{2(0.13 \text{ J/sec m °C})}\frac{1}{2}\left(\frac{1}{2}\right)$$

$$T = 66.15°\text{C}$$

PROBLEMS

5-1. Find the expression for heat flow through a pipe wall (inside radius r_i, outside radius r_o). The temperatures of the inside and outside walls are T_i and T_o ($T_i > T_o$), respectively. Thermal conductivity varies linearly from k_o at T_o to k_i at T_i.

5-2. A non-Newtonian fluid described by the Ostwald–De Waele power law is contained between two concentric cylinders [outer of radius R moves with an angular velocity of Ω; inner cylinder of radius $(R - b)$ is stationary]. Find the temperature profile in the annular space between the two cylinders if the inner cylinder is at T_1 and the outer one is at T_2.

5-3. Vented well-insulated spherical containers are used to store liquified gases. The inner and outer radii of the insulation of such a device are r_1 and r_2. Temperatures are T_1 (at r_1) and T_2 (at r_2).

The insulation's thermal conductivity is a function of temperature such that

$$k = k_1 + (k_2 - k_1)\frac{T - T_1}{T_2 - T_1}$$

where k_1 and k_2 are thermal conductivities at T_1 and T_2.

How much liquid oxygen (enthalpy of vaporization of 6.85×10^6 joules/kg mole) would evaporate from a container whose outside radius is 2.15 m (T of 273 K; k of 0.155 W/m °C) and whose inner radius is 1.85 m (T of 90 K; k of 0.121 W/m °C)?

5-4. A set of cylindrical samples (all 0.03 m in diameter are put together end to end. The outside radius of the overall system is heavily insulated. The five cylinders (A, B, C, D, E) comprising the system have lengths and thermal conductivities as follows (A: 0.071 m, 238 W/m °K; B: 0.04 m, 73 W/m °K; C: 0.04 m, 15 W/m °K; D: 0.04 m, 1126 W/m °K; E: 0.07 m, 395 W/m °K).

The ends of the system (A and E) are at 0°C and 100°C, respectively. Calculate the temperatures at each interface in the system.

5-5. A pipe of 0.10-m diameter is insulated with a 0.02-m thickness of a material. The inner and outer temperature of the insulation are at 800°K and

490°K, respectively. If the rate of heat loss per unit length is 603 W/m, what is the insulation's thermal conductivity?

5-6. A furnace wall is made up of refractory brick, brick, and steel. The refractory brick's face to the fire is at 1150°C, and the outside of the steel is 30°C. The pertinent data for the materials in terms of thickness and thermal conductivities are as follows: refractory, 0.2 m and 1.5 W/m °C; brick, 0.1 m and 0.14 W/m °C; steel, 0.006 m and 45 W/m °C. There may be thin layers of air between the brick and the steel. Estimate the equivalent thickness of these layers in terms of the brick if heat loss from the furnace is 300 W/m^2.

5-7. A jacketed polymerization batch reactor uses water at 27°C to cool the reaction at 50°C. The reactor is made of stainless steel with a wall thickness of 0.0125 m. A thin layer of polymer is left on the reactor surface ($k =$ 0.156 W/m °C). If the energy transferred from the reactor is 7413 W/m^2, what is the thickness of the polymer residue and the temperature drop through the metal.

5-8. A solid (0.025 m thick) has a cross-sectional area of 0.1 m^2. One face is at 38°C, and the other is at 94°C. The temperature at the center is 60°C when the energy flow through the solid is 1 KW. Find the thermal conductivity as a function of temperature.

5-9. The outer surface temperature of a 0.05-m-diameter pipe is 177°C. Insulation (0.05 m thick) covers the pipe. Temperature on the outside of the insulation is 37.8°C. The thermal conductivity of the insulation is given by $k = 0.86 + (0.0180)T$, where k is in W/m °C and T is in °C. Calculate heat loss per meter length of pipe.

5-10. A furnace is built with firebrick (0.225 m thick; k of 1.4 W/m °K), insulating brick (0.120 m thick; k of 0.2 W/m °K), and building brick (0.225 m thick; k of 0.7 W/m °K). Inside and outside temperatures are 1200 and 330 K, respectively. Calculate heat loss per unit area and the temperature at the interface of the firebrick and insulating brick.

5-11. A liquid is falling over an inclined plane (angle of B with the vertical). The outer edge of the film ($x = 0$) is kept at T_0 and the solid surface ($x - \delta$) at T. Viscosities at T_0 and T_δ are μ_0 and μ_δ.

If viscosity is a function of temperature as

$$\mu = C_1 \exp(C_2/T)$$

find the velocity distribution in the film.

5-12. Find the temperature distribution for a viscous fluid in steady flow between two broad parallel plates both of which are at T_0. Consider viscous dissipation but neglect property temperature dependence.

5-13. Two concentric porous spherical shells (radii R_1 and R_2) are used in a cooling system. Air is charged to the inner sphere where it is cooled and then moves through the inner shell to the outer shell and then to the atmosphere (i.e. transpires). The inner surface of the outer sphere is at T_2 and the outer surface of the inner sphere is at a lower temperature T_0. Relate heat removal to mass flow rate of the gas.

5-14. What would the temperature distribution be between the two shells of the preceding problem: $R_1 = 5 \times 10^{-4}$ m; $R_2 = 1 \times 10^{-4}$ m; $T_1 = 300°\text{C}$. $T_2 = 100°\text{C}$; $k = 0.0257$ W/mK; $C_p = 1047$ J/kg °K.

5-15. An oven's wall consists of three materials A, B, and C. The thermal conductivities of A and C are 20 and 50 W/m °K, respectively. Thicknesses of A, B, and C are 0.30, 0.15, and 0.15 m. Temperature at the inside surface (i.e., A) is 600°C. Outside surface temperature (C) is 20°C. If the heat flux is 5000 W/m, what is k for material B?

5-16. A steel pipe (I.D. of 0.10 m; O.D. of 0.112 m) is covered with 0.10 m of insulation A ($k = 0.40$ W/m °K) and 0.05 m of insulation B ($k = 0.20$ W/m °K). The inside of the pipe wall is at 300°C, while the outside of insulation B is at 25°C. What is the heat flow rate and the temperatures at both interfaces (i.e., steel–insulation A; insulation A–insulation B)?

5-17. A ceramic truncated cone has a circular cross section given by $D = 0.0025(h + 0.05)$, where h is the distance measured from the truncated section. The distance from the truncated section to the base is 0.20 m. The temperatures at the base and truncated section are 600 and 400°K. Lateral surface of the cone is heavily insulated, and the ceramic's k value is 3.46 W/m °K. Find the temperature distribution in the cone (assume one dimensional behavior). Also calculate the heat flow rate.

5-18. Liquid oxygen retention is believed to be enhanced by use of transpiration cooling (see Problem 5-13). In this case, oxygen at −183°C is contained in a spherical unit. Immediately adjacent to this sphere is a very thin gas space also spherically shaped. On the outside of the gas space there is porous insulation. Oxygen that evaporates, enters the gas space, and then enters through the porous insulation (outside surface temperature of 0°C). The properties of the system are for oxygen ($C_p = 0.921$ kW/kg °K enthalpy of vaporization 213.3 kJ/kg) and a k of 0.035 W/mK for the insulation. Dimensions of the unit are 0.3048 m diameter for the oxygen container, and the insulation thickness is 0.1524 m. Calculate the rate of heat gain and evaporation loss with and without transpiration.

5-19. Show that as the viscosities at the solid surface and the film edge approach each other (i.e., $\mu_\delta \rightarrow \mu_0$) in a falling film, then

$$v_z = \frac{\rho g \delta^2 \cos\beta}{2\mu_0}\left[1 - \left(\frac{x}{\delta}\right)^2\right]$$

Refer to Problem 5-11.

5-20. Two large, flat, porous horizontal plates are separated by a small distance B. The upper plate at $y = B$ is at T_B, while the lower plate at $y = 0$ is to be kept at T_0. In order to reduce the heat to be removed from the lower plate, a cooling gas is passed upward through both plates. Find the temperature distribution in the space B and determine the conduction heat flow to the lower plate.

5-21. Find the temperature profile for a fluid in laminar flow between two parallel plates. This flow is brought about because the upper plate (separation of b) moves with a velocity V_0.

5-22. Fins are used to enhance heat transfer. Consider a fin (length L, width W, thickness $2B$) attached to a large vertical surface. Assume no heat is lost from the end or edges of the fin. Use a coordinate system such that x is 0 at midthickness (i.e., B and $-$B for edges), Z is 0 at the wall, and L is at the end of the fin; y ranges from 0 at one edge to L at the other edge. Find the temperature profile in the fin. Indicate all assumptions.

5-23. A spherical nuclear fuel element is surrounded by a spherical shell of metal that shields the system. The heat generated within the fuel element is a function of position. This behavior can be approximated by the expression

$$\dot{q} = \dot{q}_o \left[1 + b \left(\frac{r}{R^{(F)}} \right)^2 \right]$$

where $\dot{q}_o$ is the volumetric rate of heat production at the element's center. Find the equation for the temperature distribution in the system. What is the system's maximum temperature?

5-24. A catalytic reactor unit uses a packed bed in the annular section or torus of a system. The inner wall of the system is at a constant temperature T, and the heat generation from the reaction is q_R. The bed's effective thermal conductivity is k_{eff}. If the inside and outside radii are 0.020 and 0.0125 m, respectively, $k_{eff} = 0.52$ W/mK, $q_R = 5.58 \times 10^5$ J/sec m^3, and $T_0 = 480°$C, calculate the outer wall temperature.

5-25. A shielding wall (thickness of B) for a nuclear reactor is exposed to gamma rays that result in heat generation given by

$$\dot{q} = \dot{q}_0 e^{-ax}$$

where a and $\dot{q}_o$ are constants and x is distance measured from the inside to the outside of the shielding wall. Inside and outside wall temperatures are T_1 and T_2. What is the temperature distribution in the wall?

5-26. A wall of thickness M has its left face ($x = 0$) heavily insulated and its right face ($x = M$) at temperature T_1. What is the temperature distribution if internal heat generation is given by

$$\dot{q}(x) = \dot{q}_o(1 - x/M)$$

where $\dot{q}_o$ is a constant?

5-27. A semiconductor material ($k = 2$ W/mK, resistivity $= 2 \times 10^{-5}$ ohm m) is used in the form of a cylindrical rod (0.01-m diameter, 0.04 m long). The rod's surface is well-insulated, and the ends are kept at temperatures of 0°C and 100°C. If 10 amperes of current pass through the rod, what is its midpoint temperature? What is the heat flow rate at each end?

5-28. A viewing port in a furnace is made of a quartz window (L thick). The inner window surface ($x = 0$) is irradiated with a uniform heat flux $\dot{q}_o$ from the furnace's hot gases. A fraction of this energy is absorbed at the inner surface. The remaining energy is absorbed as it passes through the window. Heat generation in the quartz is given by

$$\dot{q}(x) = (1 - K)\dot{q}_0 \alpha e^{-\alpha x}$$

where K is the fraction absorbed at the inner surface and α is the quartz's absorption coefficient. If the heat loss at the windows outer surface is given by $h\ (T_L - T_\infty)$, find the temperature distribution in the quartz.

5-29. A cylindrical system (0.02 m diameter) reacts chemically to uniformly generate 24,000 W/m throughout its volume. The chemically reacting material (k of 0.5 W/m °K) is encapsulated within a second cylinder (outside radius of 0.02 m, k of 4 W/m °K). If the interface temperature between the cylinders is 151°C, find the temperatures at the center of the reacting mass and the outside surface.

5-30. A carbon heating element is made into a rectangular shape (0.075 m wide, 0.0125 m thick, and 0.91 m long). When 12 volts are applied to the ends of the element, the surface reaches a temperature of 760°C. What is the temperature at the center of the bar? Properties are $k = 4.94$ W/m °K, and resistivity $= 4 \times 10^{-5}$ ohm m.

5-31. Water pipes in a household have been frozen. In order to open up the system, it is decided to melt the ice by passing an electric current I through the pipe wall (R_1 and R_2 are the inner, and outer pipe radii; electrical resistance per unit length is R_c' (Ω/m). The pipe is well-insulated on the outside, and during the melting process the pipe remains at a constant temperature T_0. Find temperature as a function of r. How long will it take to melt the ice if current is 100 amperes, $R_1 = 0.05$ m, and R_c' is 0.30 Ω/m?

5-32. Find the temperature distribution for a plane wall ($2L$ thick) with a uniformly distributed heat source $\dot{q}$. The left hand face is at T_1 and the right-hand face a T_2.

5-33. Cans of vegetables are to be heated in a steam unit by stacking them vertically. The cans originally at 30°C are heated for 2700 seconds by steam at 115°C. The heat transfer coefficient from the steam is 4542 W/m^2 °K. Properties of the can contents are $k = 0.83$ W/m °K and thermal diffusivity of 2×10^{-7} m^2/sec. What is the temperature in the center of the can?

5-34. A steel ball (0.05-m diameter; k of 43.3 W/mK at a temperature of 700°K) is immersed in a liquid at 395 K. What is the temperature at the center after one hour if the h is 11.4 W/m^2 K? Thermal diffusivity of the ball is 1.2×10^{-5} m^2/sec.

5-35. A five-pound roast at room temperature is placed in an oven at 350°F. It was found that the center of the roast reached 200°F after 3.32 hours. What is heat transfer coefficient between the oven and the roast?

5-36. Suppose that a cylinder, sphere, and slab all had the same value of the abscissa in Figures 5-3, 5-4, and 5-5. Based on this situation, how would they compare in terms of temperature change?

5-37. A cylindrical rod (0.10 m in diameter; 0.2 m long) is heated to a temperature of 595°C and then immersed into a hydrocarbon bath that keeps the rod's surface temperature at 149°C. What is the radial temperature profile after 120 seconds and after 180 seconds? The thermal diffusivities for the rod are 1.14×10^{-5} m^2/sec at 149°C and 6.45×10^{-6} m^2/sec at 595°C.

5-38. A slab of metal (0.3 m by 0.3 m with a thickness of 0.024 m) at 594°C is cooled at a temperature of 37.8°C. Find the centerline temperature and at a point 0.003 m from the surface. The h value is 817.67 W/m^2 °K, α for the metal is 1.55×10^{-5} m^2/sec, and k is 51.1 W/m °C.

5-39. A metal sphere (0.025-m diameter) at 600°C is immersed for 30 sec in an oil bath at 40°C. What is the temperature at the sphere's center? The α for the sphere is 6.84×10^{-6} m^2/sec; k is 24.2 W/m °C; h is 1500 W/m^2.

5-40. A sphere made of an experimental material is to be processed in a two-step operation. The sphere (diameter of 0.01 m) at 400°C is cooled in air until its center temperature reaches 335°C (h is 10 W/m^2 °K). Next, the sphere is placed in a water bath at 20°C (h is 6000 W/m^2 °K). What times would be required for each step? The α and k for the sphere are 6.66×10^{-6} m^2/sec and 20 W/m °K, respectively.

5-41. A researcher is attempting to find the heat transfer coefficient for air flowing over a sphere (0.125 m in diameter; $k = 398$ w/m °K; α of 1.15×10^{-4} m^2/sec; originally at 66°C. After 20 seconds in an air stream at 27°C, the sphere's centerline temperature is 33°C. What is h?

5-42. Ball bearings (0.2-m diameter) are to be cooled from their initial temperature of 400°C by air at −15°C in a cooling chamber equipped with a conveyer belt. The h is 1000 W/m^2 °K and the ball-bearing properties are $k = 50$ W/m °K, $\alpha = 2 \times 10^{-5}$ m^2/sec, and $C_p = 450$ J/kg K. Operating conditions are to be such that 70 percent of the initial thermal energy of the ball bearings above −15°C be removed. What is the residence time required? What should the drive velocity of the conveyer belt be?

5-43. A long cylinder (0.04-m diameter; C_p of 1068 J/kg °K; ρ of 3970 kg/m^3; α of 5.26×10^{-6} m^2/sec at 800°K) is cooled in a fluid at 300 K (h of

1600 W/m^2 °K) for 35 sec. It is then wrapped in heavy insulation. What will the temperature of the rod's centerline be after a very long time?

5-44. A hailstone (0.005-m diameter formed at −30°C falls through air at 5°C. If h is 250 W/m^2 °K, how long will it take for the outer surface of the stone to melt?

5-45. A heating unit is to be designed for metal rods (diameter of 0.05 m) to be drawn at a velocity of 0.0073 m/sec. The air in the heating unit is at 750°C (h is 125 W/m^2 °K). The original temperature of the rod is 50°C. In order to reach a centerline temperature of 600°C, what should be the length of the heating chamber? Rod properties are $\alpha = 1.11 \times 10^{-5}$ m^2/sec, $\rho =$ 7832 kg/m^3, and $C_p = 559$ J/kg °K.

REFERENCES

1. H. S. Carslaw and J. C. Jaeger, *Conduction of Heat in Solids*, Oxford University Press, London (1959).
2. J. Crank, *The Mathematics of Diffusion*, Oxford University Press, London (1956).
3. H. P. Gurney and J. Lurie, *Ind. Eng. Chem. 15*, 1170 (1923).
4. H. C. Hottel, Personal communication quoted in W. H. Rosenhow and H. Y. Choi, *Heat Mass and Momentum Transfer*, Prentice-Hall, Englewood Cliffs, NJ (1961).
5. R. G. Griskey, *Chemical Engineering for Chemists*, American Chemical Society, Washington, D.C. (1997).
6. R. B. Bird, W. E. Stewart, and E. N. Lightfoot, *Transport Phenomena*, John Wiley & Sons, New York (1960).

6

FREE AND FORCED CONVECTIVE HEAT TRANSFER

INTRODUCTION

In Chapter 5 we derived the Equation of Energy and then applied it to various cases as, for example, in static systems (no flow). These situations were essentially cases of conduction or conduction together with heat generation. The flow cases that were treated were restricted to laminar flow and simplified geometries and boundary conditions.

Situations that involve either turbulent flow, complex geometries, or difficult boundary conditions make it extremely difficult to obtain solutions of the Equation of Energy. For these cases, we must instead take a semiempirical approach which uses the concept of the heat transfer coefficient h. The defining equation for h is given as

$$q^1 = hA\Delta T \tag{6-1}$$

where A is the surface area, ΔT is the temperature driving force, and q^1 is the heat flow per unit time (i.e., in watts, etc.).

Using first principles and dimensionless forms, we will derive the basic format describing the heat transfer coefficient. Next, we will use experimental data or combination of analytical solutions of the Energy Equation and experimental data to obtain equation for h.

BASIC RELATIONSHIP FOR THE HEAT TRANSFER COEFFICIENT, *h*

Consider (Figure 6-1) the transfer of heat with a fluid flowing in a cylindrical tube. The inner wall temperature is T_0, and the fluid's average temperature is T_b.

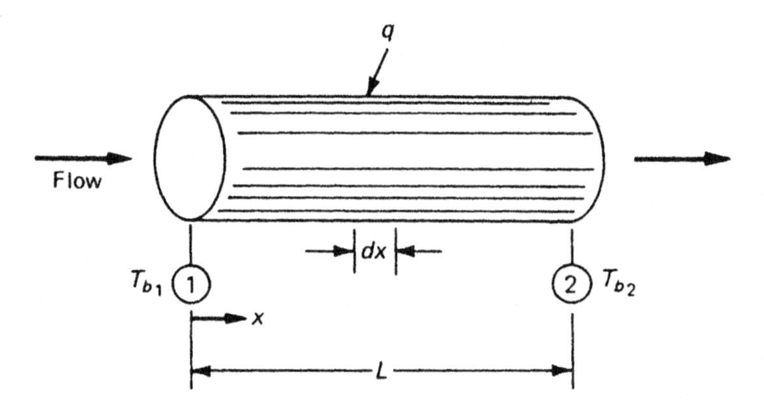

Figure 6-1. Fluid Flow in a cylindrical tube. (Adapted from reference 18.)

Equation (6-1) for the system is

$$q^1 = h(\pi DL)(T_0 - T_b) \tag{6-2}$$

with $\Delta T = (T_0 - T_b)$ and $A = \pi DL$,

Furthermore, because the heat flow can be defined by radial conduction at the wall, the total heat flow is given as

$$q^1 = \int_0^L \int_0^{2\pi} \left(k\frac{\partial T}{\partial r} \right) \bigg|_{r=R} R\, d\theta\, dz \tag{6-3}$$

The above assumes heat is added in the $(-r)$ direction.

Substituting for q^1 gives

$$h = \frac{1}{\pi DL(T_0 - T_{b1})} \int_0^L \int_0^{2\pi} \left(+k\frac{\partial T}{\partial r} \right) \bigg|_{r=R} R\, d\theta\, dz \tag{6-4}$$

Now if we make equation (6-4) dimensionless by defining $r^* = r/D$, $z^* = z/D$, and $T^* = (T - T_0)/(T_b - T_0)$, we obtain

$$\frac{hD}{k} = \frac{1}{2\pi L/D} \int_0^{L/D} \int_0^{2\pi} \left(-\frac{\partial T^*}{\partial r^*} \right) \bigg|_{r^*=\frac{1}{2}} d\theta\, dz^* \tag{6-5}$$

The hD/k term is a dimensionless number defined as the Nusselt number.

Further analysis shows that the dimensionless temperature is a function of various groups including r^*, θ, z^*, the Reynolds number, Re, the Brinkman number, Br (Example 5-6), and another dimensionless group the Prandtl number, Pr.

$$\mathrm{Pr} = \frac{C_p \mu}{k} \tag{6-6}$$

The result, when applied to equation (6-5), shows that

$$\mathrm{Nu} = \phi\left(\mathrm{Re},\ \mathrm{Pr},\ \mathrm{Br},\ \frac{L}{D}\right) \tag{6-7}$$

An alternative approach is to consider the situation wholly from an empirical viewpoint. For convection heat transfer without sizeable viscous dissipation (i.e., a low value of Brinkman number) the Nusselt number must be related to flow (i.e., Reynolds number) and energy. In the latter case, we use the ratio approach pointed out earlier for the Reynolds and Brinkman numbers. Hence,

$$\text{Prandtl} = \mathrm{Pr} = \frac{\text{Momentum diffusivity}}{\text{Thermal diffusivity}} \tag{6-8}$$

and

$$\mathrm{Pr} = (\mu/\rho)/(k/\rho C_p) = \frac{C_p \mu}{k} \tag{6-9}$$

Note that if the Prandtl and Reynolds numbers are combined and multiplied by and divided by $(T_0 - T_b)$, we obtain

$$\begin{aligned}\text{(Prandtl number) (Reynolds number)} = \mathrm{Pr\ Re} &= \frac{\rho C_p V (T_0 - T_b)/D}{k(T_0 - T_b)/D^2} \\ &= \frac{\text{Heat transport by convection}}{\text{Heat transport by conduction}}\end{aligned} \tag{6-10}$$

The combination of Pr Re (a dimensionless number) is known as the *Peclet number*.

Continuing in an empirical sense, we note in laminar flow that successive layers of fluid introduce the dimensionless ratio L/D. Furthermore, this ratio also becomes important for turbulent flow in the entrance region.

Finally, the question of temperature dependence of the fluid μ, ρ, C_p, and k must be addressed because large changes of temperature caused alter the Reynolds and/or Prandtl number. Generally, the property most readily affected by temperature (and not always so) is the viscosity. With this in mind, the Nusselt number dependence also includes a viscosity ratio μ_b/μ_w, where μ_b is the viscosity at the fluids' bulk average temperature and μ_w is the viscosity at the fluids' average wall temperature.

The resultant Nusselt number function is then

$$\frac{hD}{k} = \phi\left(\mathrm{Re},\ \mathrm{Pr},\ L/D,\ \frac{\mu_b}{\mu_w}\right) \tag{6-11}$$

If viscous dissipation is important, then

$$\frac{hD}{k} = \phi\left(\mathrm{Re},\ \mathrm{Pr},\ \mathrm{Br},\ L/D,\ \frac{\mu_b}{\mu_w}\right) \tag{6-12}$$

If sizeable fluid density changes occur as a function of temperature, the preceding approaches must be altered. Basically in such situations, fluid motion near a heated or cooled surface occurs because of buoyancy effects. This phenomenon in which velocity and temperature distributions are intimately connected is called *natural* or *free convection*. The mode of convection where some external force accounts for the motion is termed *forced convection*.

Figure 6-2 illustrates the situations for a cake of ice. In the free convection case, fluid motion is caused by the chilled air near the ice surface while the fan moves the fluid for the forced convection case. Some cases of free convection that are commonly observed are motion near convectors ("radiators") in domestic heating and "heat" rising from pavements.

When free convection occurs, the pressure and gravitational terms of the Equation of Motion $[-\nabla P + \rho g]$ [written as

$$-\frac{\partial P}{\partial x} + \rho g_x, \quad -\frac{\partial P}{\partial y} + \rho g_Y, \quad \text{and} \quad -\frac{\partial P}{\partial z} + \rho g_z$$

in equation (2-22) through (2-27)] are replaced by $-\rho\beta g(T - T_0)$. The β is the coefficient of thermal expansion.

The result of this change is that dimensional analysis yields the following form for the Nusselt number:

$$\text{Nu} = \frac{hD}{k} = \phi(\text{Gr}, \text{Pr}) \tag{6-13}$$

The Grashof number, Gr, is defined as

$$\text{Grashof number} = \text{Gr} = \left(\frac{L^3 \rho g \beta \Delta T}{\mu^2}\right) \tag{6-14}$$

Figure 6-2. Examples of free and forced convection. (Reproduced with permission from reference 22. Copyright 1997, American Chemical Society.)

Also,

$$\text{Gr} = \frac{\text{Buoyancy forces}}{\text{Viscous forces}} \frac{\text{Inertial forces}}{\text{Viscous forces}} \tag{6-15}$$

Furthermore,

$$\frac{\text{Gr}}{(\text{Re})^2} = \frac{\text{Buoyancy forces}}{\text{Inertial forces}} \tag{6-16}$$

FORCED CONVECTION HEAT TRANSFER COEFFICIENT EXPRESSIONS FOR CONDUIT FLOW

As mentioned earlier, equation for heat transfer coefficients can be obtained by fitting the relations of equations (6-11) or (6-12) to experimental data. In some cases, existing analytical solutions of the Equation of Energy are modified with experimental data to provide working equations.

The latter situation is used in the case of laminar flow in circular tubes. An analytical solution of the Equation of Energy for fully developed laminar flow gives the following expression for the Nusselt number (where h_a is the average heat transfer coefficient):

$$\text{Nu}_a = \frac{h_a D}{k} = 1.62\ \text{Re}^{1/3}\text{Pr}^{1/3}\left(\frac{D}{L}\right)^{1/3} \tag{6-17}$$

This was modified by Sieder and Tate (1) using experimental data to the form

$$\text{Nu}_a = \frac{h_a D}{k} = 1.86\ \text{Re}^{1/3}\text{Pr}^{1/3}\left(\frac{D}{L}\right)^{1/3}\left(\frac{\mu_b}{\mu_w}\right)^{0.14} \tag{6-18}$$

The last term represents the correction for possible viscosity variation with temperature as discussed earlier. Equation (6-18) is valid for

$$\text{Re Pr}\left(\frac{D}{L}\right) > 10 \tag{6-19}$$

Also, see that the heat transfer coefficient in equation (6-18) would approach zero for long tubes. In such situations with constant wall temperature the following relation (2) can be used for circular tubes

$$\frac{h_a D}{k} = 3.66 + \frac{0.0668\left(\frac{D}{L}\right)\text{Re Pr}}{1 + 0.04\left[\frac{D}{L}\ \text{Re Pr}\right]^{2/3}} \tag{6-20}$$

For long lengths in equation (6-20) (i.e., D/L approaching zero) the Nusselt number will have the asymptotic value of 3.66.

Expressions for turbulent flow in circular tubes were determined from experimental data. One such equation (using Nu as a function only of Re and Pr) is

$$\frac{hD}{k} = 0.023 \text{ Re}^{0.8}\text{Pr}^{1/3} \tag{6-21}$$

If we compare equation (6-17) and (6-21), we see that the differences occur in the multiplier term (1.62 versus 0.023), Reynolds number power ($\frac{1}{3}$ versus 0.8), and D/L power ($\frac{1}{3}$ versus 0). The Prandtl number power ($\frac{1}{3}$) is the same for both. A phenomenological approach to these effects indicate that the change in flow from laminar to turbulent is reflected both in the larger power on the Reynolds number and the elimination of the D/L effect (transformation from layers to vortices and eddies). Also, the change in flow does not alter the Prandtl number power.

The turbulent flow (circular tubes) counterpart to equation (6-18) is given below (1):

$$\text{Nu}_a = \frac{h_a D}{k} = 0.026 \text{ Re}^{0.8}\text{Pr}^{1/3}\left(\frac{\mu_b}{\mu_w}\right)^{0.14} \tag{6-22}$$

Figure 6-3 presents laminar, transition and turbulent behavior. For turbulent flow cases where $10 < L/D < 400$ (3) we have

$$\text{Nu}_a = \frac{h_a D}{k} = 0.036 \text{ Re}^{0.8}\text{Pr}^{1/3}\left(\frac{D}{L}\right)^{0.055} \tag{6-23}$$

Expressions for noncircular tubes are presented in a variety of sources (4–6). In many cases, it is possible to use the relation between equivalent diameter and

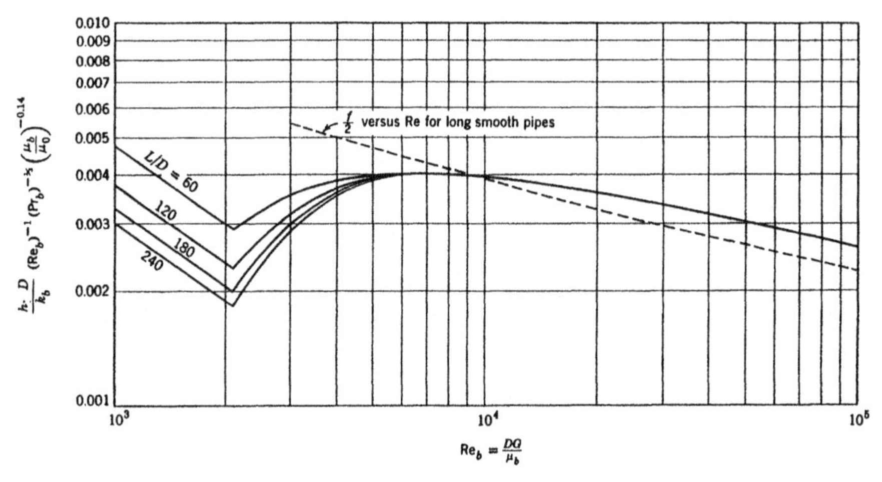

Figure 6-3. Nusselt number behavior for tube flow. (Reproduced with permission from reference 1. Copyright 1936, American Chemical Society.)

hydraulic radius (see Chapter 2):

$$D_{eq.} = 4R_h \tag{6-24}$$

Exceptions are discussed in reference 6.

One additional point should be made about the foregoing heat transfer equations. As can be seen, the average heat transfer coefficient was obtained in each case. The same equation can also be used to find h values based on temperatures other than the average bulk value.

FORCED CONVECTION HEAT TRANSFER COEFFICIENT EXPRESSION FOR FLOW OVER SURFACES

Flow over objects such as cylinders constitute a complex case of heat transfer. This complexity principally results because of flow situations. For this reason a semiempirical approach is used to determine heat transfer coefficients. The relation used for cylinders (and objects as well) is

$$\frac{hD}{k_f} = c\left(\frac{VD\rho_f}{\mu_f}\right)^n (\mathrm{Pr}_f)^{1/3} \tag{6-25}$$

The f subscript refers to the temperature at which the properties are determined. This temperature (the film temperature) is

$$T_f = \frac{T_0 + T_\infty}{2} \tag{6-26}$$

where T_0 is the temperature of the surface and T_∞ is the approach temperature of the flowing fluid.

Values of C and n for normal flow (fluid flows perpendicular to the cylinder) are given in Table 6-1 (6, 7) as a function of Reynolds number.

If the cylinder is not circular (i.e., a square or a hexagon), the values of Table 6-2 (8) can be used.

Heat transfer coefficients for flows at various orientations (other angles) are given elsewhere (9).

Table 6-1 Constants for Equation (6-25)

$\mathrm{Re}_{d_f}^a$	C	n
0.4–4	0.989	0.330
4–40	0.911	0.385
40–4,000	0.683	0.466
4,000–40,000	0.193	0.618
40,000–400,000	0.0266	0.805

[a] The Reynolds number is based on the cylinder diameter and the film temperature.

Table 6-2 Constants for Equation (6-25) for Noncircular Cylinders

Geometry	Re	C	n
→ ◇ (diamond), D	$5 \times 10^3 - 10^5$	0.246	0.588
→ □ (square), D	$5 \times 10^3 - 10^5$	0.102	0.675
→ ⬡ (hexagon, vertex forward), D	$5 \times 10^3 - 1.93 \times 10^4$ $1.95 \times 10^4 - 10^5$	0.160 0.0385	0.638 0.782
→ ⬡ (hexagon, flat side forward), D	$5 \times 10^3 - 10^5$	0.153	0.638
→ | (vertical plate), D	$4 \times 10^3 - 1.5 \times 10^4$	0.228	0.731

In many process situations, flow is over a bank of tubes rather than a single tube. There are two basic possible configurations (in-line and staggered; see Figure 6-4). Geometry is an important consideration; in particular, the S_p and S_n spacings as shown in Figure 6-4.

Experiments show that the values of the average heat transfer coefficient vary with the number of vertical rows until 10 or more rows are used. The h value in this latter case remains constant. Also, the velocity used in calculating the Reynolds number is not the fluid's approach velocity but rather the maximum velocity found by using the minimum flow area:

$$V_{\text{max}} = V_{\infty}\left(\frac{S_n}{S_n - D}\right) \tag{6-27}$$

Table 6-3 (10) gives the values of C and n for equation (6-25) as functions of geometry for 10 or more rows. Cases involving fewer than 10 rows can be handled by using the correction factors of Table 6-4.

Pressure drops for flow of gases over tube banks are given by

$$\Delta P = \frac{2f'G^2N}{\rho}\left(\frac{\mu_{\text{wall}}}{\mu_{\text{bulk}}}\right)^{0.14} \tag{6-28}$$

where G is the mass velocity (kg/m^2 sec) at minimum flow area and N is the number of transverse rows.

The f' is a modified friction factor given by either equation (6-29) or (6-30). For staggered tubes

$$f' = \left[0.25 + 0.118 \Bigg/ \left|\left(\frac{S_n - D}{D}\right)\right|^{1.08}\right]\text{Re}^{-0.16} \tag{6-29}$$

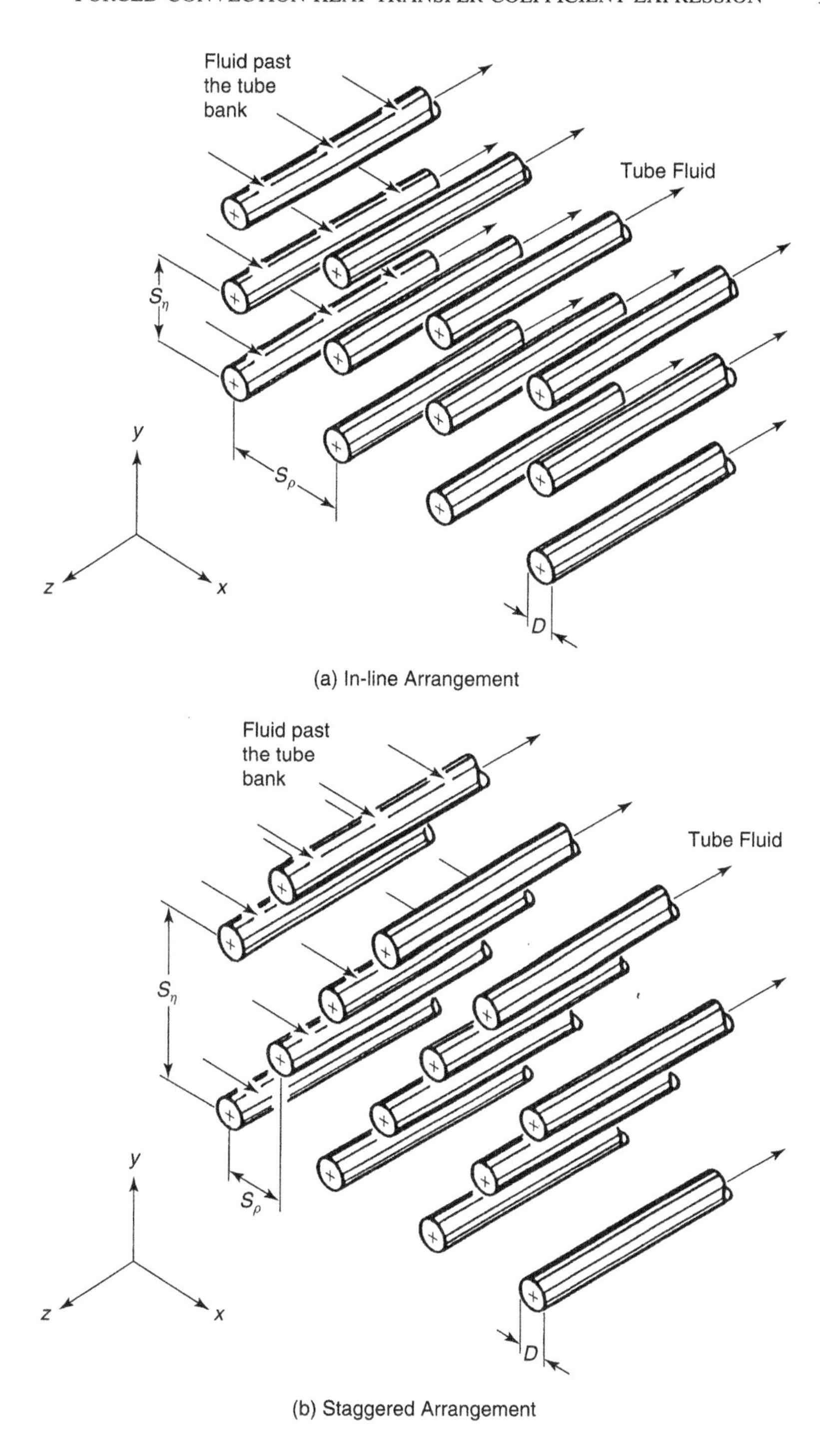

Figure 6-4. In-line and staggered tube arrangements. (Adapted from reference 21.)

Table 6-3 Constants for Equation (6-25); Ten or More Rows of Tubes[a]

		S_n/D							
		1.25		1.5		2.0		3.0	
	S_p/D	C	n	C	n	C	n	C	n
In-line	1.25	0.386	0.592	0.305	0.608	0.111	0.704	.0703	0.752
	1.5	0.407	0.586	0.278	0.620	0.112	0.702	.0753	0.744
	2.0	0.464	0.570	0.332	0.602	0.254	0.632	0.220	0.648
	3.0	0.322	0.601	0.396	0.584	0.415	0.581	0.317	0.608
Staggered	0.6							0.236	0.636
	0.9					0.495	0.571	0.445	0.581
	1.0			0.552	0.558				
	1.125					0.531	0.563	0.575	0.560
	1.25	0.575	0.556	0.561	0.554	0.576	0.556	0.579	0.562
	1.5	0.501	0.568	0.511	0.562	0.502	0.568	0.542	0.568
	2.0	0.448	0.572	0.462	0.568	0.535	0.556	0.498	0.570
	3.0	0.344	0.592	0.395	0.580	0.488	0.562	0.467	0.574

[a] Blank sections indicate lack of data.

Table 6-4 Ratio of *h* for *N* Rows Deep to That for 10 Rows Deep

Form	1	2	3	4	5	6	7	8	9	10
Staggered	0.68	0.75	0.83	0.89	0.92	0.95	0.97	0.98	0.99	1.0
In-line	0.64	0.80	0.87	0.90	0.92	0.94	0.96	0.98	0.99	1.0

For in-line tubes

$$f' = \left(0.044 + \frac{0.08 S_p/D}{\left|\left(\dfrac{S_n - D}{D}\right)\right|^{0.43+1.13D/S_p}} \right) \mathrm{Re}^{-0.15} \tag{6-30}$$

Heat transfer correlations also exist for other shapes. For spheres (11)

$$\mathrm{Nu} = \frac{hD}{k} = 2 + 0.60\ \mathrm{Re}^{1/2}\mathrm{Pr}^{1/3} \tag{6-31}$$

whereas for flat plates (with parallel flow)

$$\mathrm{Nu} = \frac{hD}{k} = 0.664\ \mathrm{Re}_L^{1/2}\mathrm{Pr}^{1/3} \tag{6-32}$$

where Re_L is LV_ρ/μ (L is plate length). Equation (6-32) is for laminar flow ($\mathrm{Re}_L < 300{,}000$).

In turbulent flow (6) the flat plate equation is

$$\mathrm{Nu} = \frac{hD}{k} = 0.0366\ \mathrm{Re}_L^{0.8}\mathrm{Pr}^{1/3} \tag{6-33}$$

FREE CONVECTION HEAT TRANSFER COEFFICIENT EXPRESSIONS

As we have seen, the critical dimensionless groups for free or natural convection are the Grashof and Prandtl numbers. In general (the h will be a mean value),

$$\mathrm{Nu}_{\mathrm{mean}} = \phi(\mathrm{Gr}, \mathrm{Pr}) \tag{6-34}$$

The complexities of free convection heat transfer make it necessary to mainly use empirical relations based on experimental data.

Correlations of Nu with the Gr Pr product are shown in Figures 6-5 and 6-6 for long horizontal cylinders and vertical plates. The vertical plate data can also

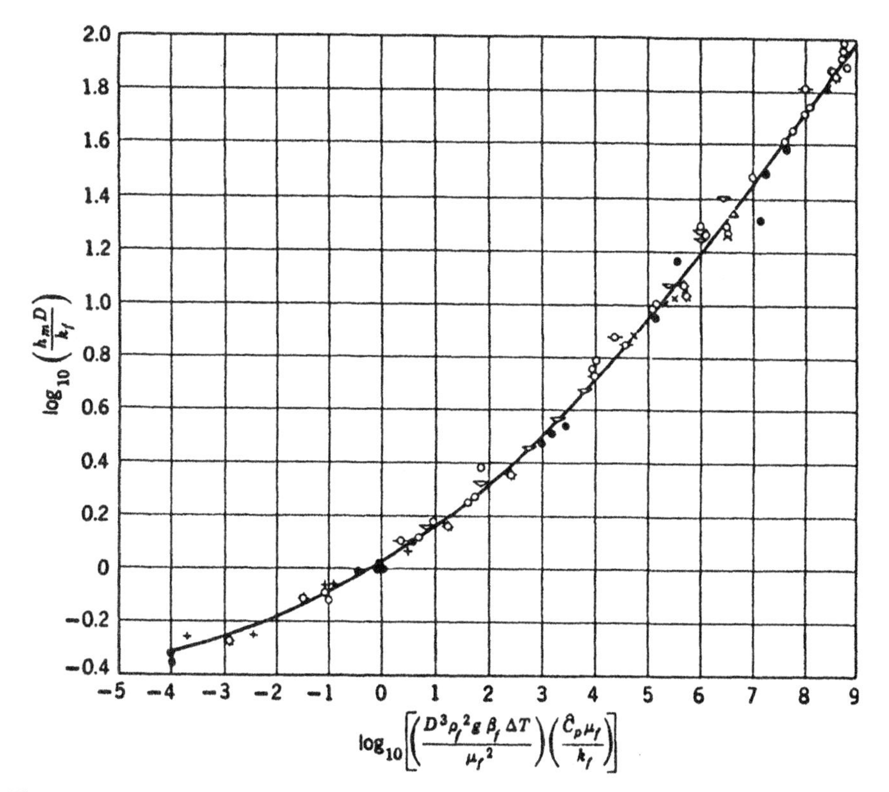

Figure 6-5. Nusselt numbers for free convection from long horizontal cylinders. (Reproduced with permission from reference 23. Copyright 1960, John Wiley & Sons.)

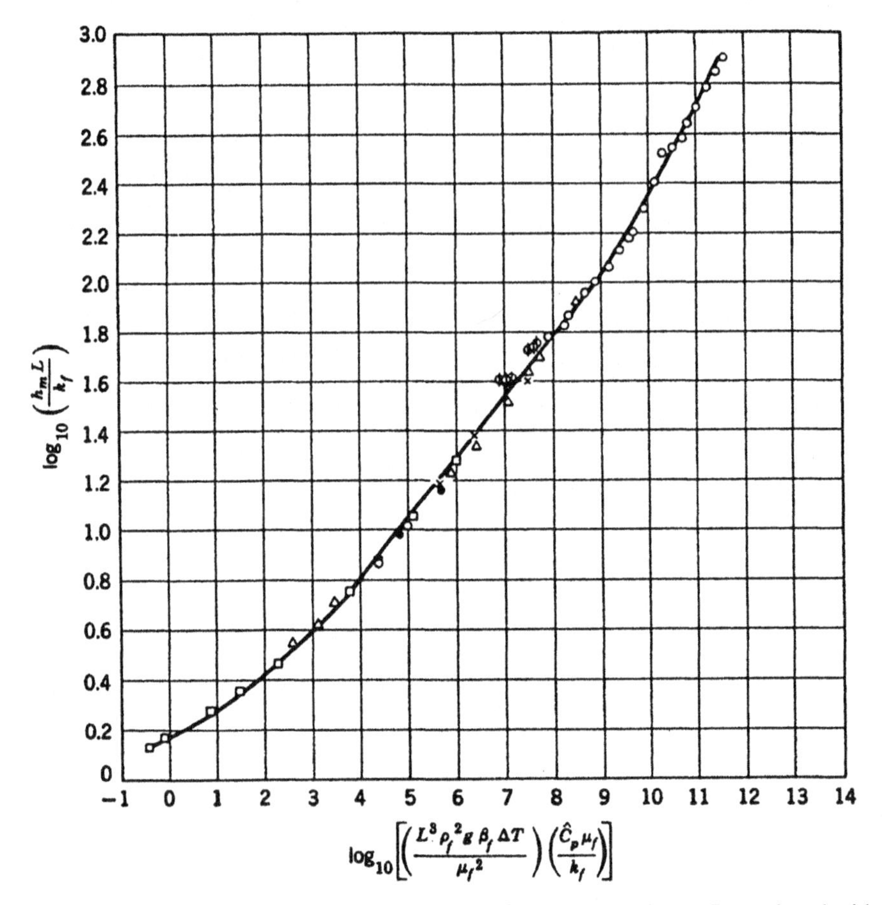

Figure 6-6. Nusselt numbers for free convection from vertical plates. (Reproduced with permission from reference 23. Copyright 1960, John Wiley & Sons.)

be used for vertical cylinders if the following criterion is met

$$\frac{D}{L} \geqq \frac{35}{\mathrm{Gr}_L^{1/4}} \tag{6-35}$$

In Figure 6-5 it can be seen that for a considerable range the Nusselt relation is linear with Gr Pr (10^4 to 10^9). The applicable relation (5) is

$$\mathrm{Nu}_m = \frac{h_m D}{k_f} = 0.518(\mathrm{Gr}_f \mathrm{Pr}_f)^{1/4} \tag{6-36}$$

The appropriate relation (5) for high Gr Pr values (10^9 to 10^{12}) is

$$\mathrm{Nu}_m = \frac{h_m D}{k_f} = 0.13(\mathrm{Gr}_f \mathrm{Pr}_f)^{1/3} \tag{6-37}$$

Likewise, in Figure 6-6 the linear portion (Gr Pr from 10^4 to 10^9) is given by

$$\mathrm{Nu}_m = \frac{h_m L}{k_f} = 0.59(\mathrm{Gr}_f \mathrm{Pr}_f)^{1/4} \tag{6-38}$$

for plates (5) and

$$\mathrm{Nu}_m = \frac{h_m D}{k_f} = 0.59(\mathrm{Gr}_f \mathrm{Pr}_f)^{1/4} \tag{6-39}$$

for cylinders (5).

The high Gr Pr range (10^9 to 10^{13}) is given by (12, 13)

$$\mathrm{Nu}_m = 0.10(\mathrm{Gr}_f \mathrm{Pr}_f)^{1/3} \tag{6-40}$$

Table 6-5 gives the values of A, C, and n of equation (6-41) for various cases:

$$\mathrm{Nu}_m = A + C(\mathrm{Gr}_f \mathrm{Pr}_f)^n \tag{6-41}$$

Cases involving free convection in enclosed spaces are less well defined. The usual method is to determine an "effective" thermal conductivity k_e. Discussion of this approach is given elsewhere (6, 8, 18).

Another situation that can result is combined forced and free convection. A discussion of such situations is given in reference 19.

A typical relation is the following equation (20):

$$\mathrm{Nu} = 1.75\left(\frac{\mu_b}{\mu_w}\right)^{0.14} [\mathrm{Gz} + 0.012(\mathrm{Gz}\ \mathrm{Gr}^{1/3})^{4/3}]^{1/3} \tag{6-42}$$

for mixed forced and free convection in laminar flow in a horizontal tube. Gz is the Graetz number [Re Pr(D/L)].

Table 6-5 Constants for Equation (6-41)

Case	$\mathrm{Gr}_f \mathrm{Pr}_f$ Range	A	C	n	Reference
Upper surface heated plates or Lower surface cooled plates	2×10^4 to 8×10^6	0	0.54	$\frac{1}{4}$	14, 15
Upper surface heated plates or Lower surface cooled plates	8×10^6 to 1×10^{11}	0	0.15	$\frac{1}{3}$	14, 15
Lower surface heated plates or Upper surface cooled plates	10^5 to 10^{11}	0	0.58	$\frac{1}{5}$	14, 16
Spheres	1 to 10^5	2	0.43	$\frac{1}{4}$	17

WORKED EXAMPLES

Before considering the solution of convection heat transfer examples, we will review the steps followed in such cases. While the discussion applies to heat transfer coefficients based on average temperature, it can also be applied to other heat transfer coefficients.

For conduit flow:

1. Find the fluids' bulk average temperature. This is done by averaging the average bulk inlet and outlet temperatures.
2. Determine the physical properties (μ, ρ, C_p, and k) needed to evaluate the Reynolds and Prandtl numbers. In some cases a wall average temperature is also needed.
3. Calculate the Reynolds and Prandtl numbers.
4. Select the appropriate convection heat transfer equation applicable to the system (i.e., laminar, turbulent, length, etc.).
5. Carry out the solution.

In the case of flow over an object:

1. Find the fluids' film temperature (i.e., average of object surface temperature and fluid approach temperature).
2. Determine physical properties at film temperature.
3. Calculate Reynolds and Prandtl numbers.
4. Select the appropriate equation or the C and n values needed.
5. Solve the equation.

Let us now consider some pertinent examples.

Example 6-1 In a heat exchanger, water flows through a long copper tube (inside diameter 2.2 cm) with an average velocity of 2.13 m/sec. The water is heated by steam condensing at 150°C on the outside of the tube. Water enters at 15°C and leaves at 60°C. What is the heat transfer coefficient, h, for the water?

First, we evaluate the average bulk temperature of the water, which is $(15 + 60)/2$ or 37.5°C.

Evaluating water properties (from the Appendix) at this temperature yields

$$\rho = 993 \text{ kg/m}^3, \qquad C_p = 4.17 \times 10^3 \text{ J/kg °C}$$
$$\mu = 0.000683 \text{ kg/m sec}, \qquad k = 0.630 \text{ W/m °C}$$

Now we calculate the Reynolds and Prandtl numbers.

$$\text{Re} = \frac{(0.022 \text{ m})(2.13 \text{ m/sec})(993 \text{ kg/m}^3)}{(0.000683 \text{ kg/m sec})} = 68{,}100$$

$$\text{Pr} = \frac{(4.17 \times 10^3 \text{ J/kg °C})(0.000683 \text{ kg/m sec})}{(0.630 \text{ W/m °C})}$$

$$\text{Pr} = 4.53$$

The flow is turbulent, and because the tube is a long one (i.e., no L/D effect) we use the Nusselt number relation:

$$\frac{hD}{k} = 0.026 \text{ Re}^{0.8}\text{Pr}^{0.33}\left(\frac{\mu_b}{\mu_w}\right)^{0.14}$$

All of the quantities in the above equation are known ($\mu_b =$ 0.000683 kg/m sec) except μ_w. To get this value, we have to know the fluids' average wall temperature. This temperature is between the fluids' bulk average temperature of 37.5°C and the outside wall temperature of 150°C. We use a value of 93.75°C (the average of the two). At this temperature

$$\mu_w = 0.000306 \text{ kg/m sec}$$

Then

$$h = \frac{(0.026)(0.0630 \text{ W/m °C})}{(0.022 \text{ m})}(68{,}100)^{0.80}(4.53)^{0.33}\left(\frac{0.000683}{0.000306}\right)^{0.14}$$

and

$$h = 10{,}086.75 \text{ W/m}^2 \text{ °C}$$

Example 6-2 Oil enters a 1.25-cm diameter, 3-m-long tube at 38°C. The tube wall is maintained at 66°C, and flow velocity is 0.3 m/sec. What is the total heat transfer to the oil and oil exit temperature?

We immediately encounter a problem with this case. The difficulty is that the oil exit temperature is unknown and, as such, we cannot find the systems' bulk average temperature.

However, by making use of thermodynamics and, in particular, the Second Law of Thermodynamics we can conclude that the oil exit temperature cannot exceed 66°C. We can, therefore, commence the solution by using an assumed exit temperature (in this case 50°C).

This implies that average temperature of the fluid is $(50 + 38)/2$ or 44°C. For this temperature we obtain the appropriate properties from the table below:

Temperature (°C)	ρ(kg/m^3)	μ(kg/m sec)	C_p J/hg °C)	k(W/m °C)
40	867.05	0.210	1964	0.144
60	864.04	0.0725	2047	0.140
80	852.02	0.0320	2131	0.138

and calculate Reynolds and Prandtl numbers:

$$\text{Re} = \frac{DV\rho}{\mu} = \frac{(0.0125\ \text{m})(0.3\ \text{m/sec})(873.7\ \text{kg/m}^3)}{(0.1825\ \text{kg/m sec})}$$

$$\text{Re} = 18.0$$

$$\text{Pr} = \frac{C_p\mu}{k} = \frac{(1981\ \text{J/kg}\ ^\circ\text{C})(0.1825\ \text{kg/m sec})}{(0.143\ \text{W/m}\ ^\circ\text{C})}$$

$$\text{Pr} = 2528$$

Because flow is laminar and length is not excessive, we use the Nusselt relation below:

$$\frac{hD}{k} = 1.86(\text{Re})^{0.333}(\text{Pr})^{0.333}\left(\frac{D}{L}\right)^{0.333}\left(\frac{\mu_b}{\mu_w}\right)^{0.14}$$

substituting the values including the μ_w of 0.0605 kg/m sec at 66°C.

$$h = \frac{(0.143\ \text{w/m}\ ^\circ\text{C})}{0.0125\ \text{m}}(1.86)(18.0)^{0.333}(2528)^{0.333}\left(\frac{1}{240}\right)^{0.333}\left(\frac{0.1825}{0.0605}\right)^{0.14}$$

$$h = 142.7\ \text{w/m}^2\ ^\circ\text{C}$$

In order to check the result, we can use the First Law of Thermodynamics (i.e., heat transferred equals the oil enthalpy change).

From the First Law of Thermodynamics we have

$$Q = \Delta H$$

and

$$hA[T_w - (T_b)_{\text{average}}] = (mC_p)_{\text{oil}}(T_{\text{exit}} - T_{\text{in}})$$

and solving for T exit [where $(T_b)_{\text{average}} = (T_{\text{exit}} - T_{\text{in}})/2$] we obtain

$$T_{\text{exit}} = 44.6^\circ\text{C}$$

This value will not alter the properties appreciably hence the h value is correct so that

$$q^1 = hA\Delta T$$

$$q^1 = (142.7\ \text{W/m}^2\ ^\circ\text{C})\pi(0.0125\ \text{m})(3\ \text{m})(66^\circ\text{C} - 413^\circ\text{C})$$

$$q^1 = 415.24\ \text{W}$$

Example 6-3 Which system (a 2.54-cm-diameter pipe with wall temperature at 21.11°C or a 1.27-cm-diameter pipe with wall temperature at 4.44°C) will give the least pressure drop for water flowing at 0.454 kg/sec which is cooled from 65.56°C to 26.67°C?

For both cases the bulk average temperature is given by

$$T_{\text{bulk}} \text{ for fluid} = \frac{65.6 + 26.67}{2} = 46.14°\text{C}$$

The fluid properties (from the Appendix) are

$$\mu = 5.829 \times 10^{-4} \text{ kg/m sec}, \qquad k = 0.6542 \text{ W/m °C}$$
$$C_p = 4186.8 \text{ J/kg °C}, \qquad \rho = 988.8 \text{ kg/m}^3$$

The Prandtl number for both cases is the same:

$$\text{Pr} = \frac{C_p\mu}{k}\,\frac{(4186.8 \text{ J/kg °C})(5.829 \times 10^{-4} \text{ kg/m sec})}{0.6542 \text{ w/m °C}}$$

$$\text{Pr} = 3.73$$

Now, consider case 1:

$$\text{Re} = \frac{DV\rho}{\mu} = \frac{DV\rho\dfrac{\pi D^2}{4}}{\mu\dfrac{\pi D^2}{4}} = \frac{Dw}{\mu\dfrac{\pi D^2}{4}} \qquad (w \text{ is mass flow rate})$$

$$\text{Re} = \frac{(0.0254 \text{ m})(0.454 \text{ kg/sec})}{(5.829 \times 10^{-4} \text{ kg/m sec}(0.0254 \text{ m})^2\pi)} = 39{,}040$$

$$\mu_b = 5.829 \times 10^{-4} \text{ kg/m sec} \quad \text{if } T_w = 37.74°\text{C},$$

$$\mu_w = 6.821 \times 10^{-4} \text{ kg/m sec}$$

Note that $(\mu_b/\mu_w)^{0.14} \cong 1.0$, so that

$$h = \frac{k}{D}0.023 \text{ Re}^{0.8}\text{Pr}^{1/3} = \frac{0.6542 \text{ w/m °C}}{0.0254 \text{ m}}(39{,}040)^{0.8}(3.73)^{1/3}(0.023)$$

and

$$h = 4328 \text{ w/m}^2 \text{ °C}$$

Now we again make use of the First Law of Thermodynamics:

$$Q = \Delta H$$

and $hA\Delta T = wC_p(T_{\text{exit}} - T_{\text{inlet}})$

$$A = (L\pi D)$$

$$L = \frac{wC_p(T_{\text{exit}} - T_{\text{inlet}})}{h\pi D\Delta T}$$

$$L = \frac{(0.454 \text{ kg/sec})(4186.8 \text{ J/kg °C})(26.67\text{°C} - 65.56\text{°C})}{(4328 \text{ w/m}^2 \text{ °C})(0.0254 \text{ m})(2.11\text{°C} - 46.14\text{°C})\pi}$$

$$L = 4.86 \text{ m}$$

Next, we calculate ΔP [using equation (3-17)]

$$\frac{\Delta P}{L} = \frac{2f_\rho V^2}{g_c D}$$

The f value (0.0055) is found from Figure 3-3 for the Re of 39,040. Substituting and calculating ΔP gives

$$\Delta P = \frac{2f_\rho V^2 L}{g_c D} = \frac{(2)(0.0055)(4.86 \text{ m})(988.8 \text{ kg/m}^3)(0.906 \text{ m/sec})^2}{1 \text{ kg m/N sec}^2 \text{ } (0.0254 \text{ m})}$$

$$\Delta P = 1708 \text{ N/m}^2$$

We now repeat for the 1.27-cm-diameter pipe at 4.44°C.

$$\text{Re} = \frac{Dw}{\mu\left(\frac{\pi D^2}{4}\right)} = \frac{(0.0127 \text{ m})(0.454 \text{ kg/sec})}{(5.829 \times 10^{-4} \text{ kg/m sec})\pi/4(0.0127 \text{ m})^2}$$

$$\text{Re} = 78{,}080$$

Again,

$$\frac{hD}{k} = 0.023 \text{ Re}^{0.8}\text{Pr}^{1/3}$$

because $(\mu_b/\mu_w)^{0.14} \cong 1.0$

$$h = \frac{0.6542 \text{ w/m °C}}{0.0127 \text{ m}}(78{,}080)^{0.8}(3.73)^{1/3}(0.023)$$

$$h = 15{,}074 \text{ W/m}^2 \text{ °C}$$

Then, from the enthalpy balance ($Q = \Delta H$) we have

$$L = \frac{wC_p(T_{\text{exit}} - T_{\text{inlet}})}{h\pi D\Delta T}$$

$$L = \frac{(0.454 \text{ kg/sec})(26.67 - 65.56)\text{°C}(4186.8 \text{ J/kg °C})}{(15{,}074 \text{ w/m}^2 \text{ °C})(\pi)(0.0127 \text{ m})(4.44\text{°C} - 46.14\text{°C})}$$

$$L = 2.95 \text{ m}$$

Repeating the ΔP calculation, we find (Figure 3-3) that $f = 0.0047$ at the Re of 78,080.

$$\Delta P = \frac{2fL_\rho V^2}{gcD} = \frac{2(0.0047)(2.95 \text{ m})(988.8 \text{ kg/m}^3)(3.625 \text{ m/sec})^2}{1 \text{ kg m/N sec}^2(0.0127 \text{ m})}$$

$$\Delta P = 28{,}370 \text{ N/m}^2$$

Therefore, the 0.0127-m pipe gives the higher pressure drop.

Example 6-4 The double-pipe heat exchanger is essentially a set of concentric pipes. One fluid flows within the smaller pipe and the other in the torus or annulus.

For such an exchanger (inside pipes outer diameter is 2 cm; outside pipes' inner diameter is 4 cm), water flows in the annular space at an average velocity of 1.5 m/sec. The water, which cools an organic flowing in the central pipe, has a temperature change from 16°C to 28°C.

In this case calculate the heat transfer coefficient for the water assuming that the exchanger is heavily insulated and that the wall temperature on the inside of the annulus (i.e., outside of central pipe) is 30°C.

The bulk average temperature of the water is

$$T_b = \frac{16^\circ\text{C} + 28^\circ\text{C}}{2} = 22^\circ\text{C}$$

For this temperature

$$\mu_b = 9.67 \times 10^{-4} \text{ kg/m sec}, \qquad C_p = 4186 \text{ J/kg °C}$$

$$\rho = 998 \text{ kg/m}^3, \qquad k = 0.599 \text{ W/m °C}$$

Also, $\mu_w = 8.516 \times 10^{-4}$ kg/m sec (at 30°C).

We next determine the Reynolds and Prandtl numbers. However, before doing so, we must determine the D_{eq} for the annulus or torus:

$$D_{eq} = 4R_h$$

The R_H value is given by

$$R_H = \frac{\pi(D_0^2 - D_1^2)}{4(\pi D_0 + \pi D_1)}$$

Note that the annulus has *two* wetted perimeters:

$$R_H = \frac{1}{4}\frac{(D_0^2 - D_1^2)}{(D_0 + D_1)} = \frac{1}{4}(D_0 - D_1)$$

and

$$D_{eq} = (D_0 - D_1) = (0.04 - 0.02) \text{ m}$$

$$D_{eq} = 0.02 \text{ m}$$

Then,

$$\text{Re} = \frac{D_{eq} V_\rho}{\mu} = \frac{(0.02 \text{ m})(1.5 \text{ m/sec})(998 \text{ kg/m}^3)}{(9.67 \times 10^{-4} \text{ kg/m sec})}$$

$$\text{Re} = 30{,}962$$

Also,

$$\text{Pr} = \frac{C_p \mu}{k} = \frac{(4186 \text{ J/kg °C})(9.67 \times 10^{-4} \text{ kg/m sec})}{(0.599 \text{ J sec/m °C})}$$

$$\text{Pr} = 6.75$$

For turbulent flow we use the relation

$$\frac{h D_{eq}}{k} = 0.023 \text{ Re}^{0.8} \text{Pr}^{1/3} \left(\frac{\mu_b}{\mu_w}\right)^{0.14}$$

$$h = \frac{(0.599 \text{ W/m °C})}{(0.02 \text{ m})} (0.023)(30{,}962)^{0.8}(6.76)^{1/3} \frac{(9.67)}{(8.51)}^{0.14}$$

$$h = 5191 \text{ W/m}^2 \text{ °C}$$

Example 6-5 Six rows of tubes 15.24 m are set up as an in-line arrangement. Tubes are 0.0064 m in diameter and $S_n = S_p = 0.0192$ m. Tube wall temperature is 93.3°C. Atmospheric air is forced across at an inlet velocity of 4.57 m/sec. What is the total heat transfer per unit length of the tube bank?

We first calculate T_f assuming that the air is at 21.1°C.

$$\text{T}_f = \frac{93.3 + 21.1}{2} = 57.2\text{°C}$$

For this temperature and atmospheric pressure the air properties are:

$$\mu_f = 0.0000197 \text{ kg/m sec}, \qquad \rho_f = 1.0684 \text{ kg/m}^3$$

$$k_f = 0.0284 \text{ W/m °C}, \qquad C_{\rho f} = 1.047 \text{ kJ/kg}$$

Next, we find the Reynolds and Prandtl numbers. In order to find the former, we need the velocity, V_{max}, in the bank itself.

$$v_{max} = 4.57 \text{ m/sec} \left(\frac{0.0192}{0.0192 - 0.0064}\right) = 6.86 \text{ m/sec}$$

$$\text{Re} = \frac{(0.0064\ \text{m})(6.86\ \text{m/sec})(1.0684\ \text{kg/m}^3)}{0.0000197\ \text{kg/m sec}}$$

$$\text{Re} = 2381$$

$$\text{Pr} = \frac{(1.047\ \text{kJ/ky})(0.0000197\ \text{kg/m sec})}{(0.0284\ \text{W/m °C})}$$

$$\text{Pr} = 0.726$$

We now use equation (6-25):

$$\frac{hD}{k} = C\text{Re}^n\text{Pr}^{1/3}$$

The C and n values are obtained from Table 6-3. We use the in-line arrangement with geometric factors.

$$\frac{S_n}{D} = \frac{S_p}{D} = \frac{0.0192\ \text{m}}{0.0064\ \text{m}} = 3.0$$

The values of C and n are, respectively, 0.317 and 0.608. Hence,

$$h = \frac{(0.0284\ \text{W/m °C})}{(0.0064\ \text{m})}(0.317)(2381)^{0.608}(0.726)^{1/3}$$

$$h = 142.9\ \text{W/m}^2\ \text{°C}$$

This h is, however, for ten rows. We must now correct to six rows. The correction factor from Table 6-4 is (0.94).

$$\text{h} = 142.9\ \text{W/m}^2\ \text{°C}(0.94) = 133.3\ \text{W/m}^2\ \text{°C}$$

Then, $q^1 = hA\Delta T$, where A is the *total surface* per unit length.

$$q^1 = 133.3\ \text{W/m}^2\ \text{°C}(5.974\ \text{m}^2/\text{m})(72.2\text{°C})$$

$$q^1 = 57{,}495\ \text{W/m}$$

Example 6-6 Air at atmospheric pressure and 27°C is blown across a long 4.0-cm-diameter tube at a velocity of 20 m/sec. What dimension (i.e., side) of a square duct would be needed to give the same heat transfer to the duct? Wall temperature is 50°C in both cases.

We first determine the film temperature, T_f:

$$T_f = T_{\text{film}} = \frac{(50 + 27)}{2} = 38.5\text{°C}$$

For this T_f (from the Appendix)

$$\frac{\mu}{\rho} = 17.74 \times 10^{-6}\ \text{m}^2/\text{sec}$$

$$k = 0.02711\ \text{w/m}\ ^\circ\text{C}$$

$$\text{Pr} = 0.70$$

Determining the Reynolds number for the circular tube.

$$\text{Re} = \frac{DV\rho}{\mu} = \frac{DV}{\mu/\rho} = \frac{(0.04\ \text{m})(20\ \text{m/sec})}{(17.74 \times 10^{-6}\ \text{m}^2/\text{sec})}$$

$$\text{Re} = 45,100\ \text{(for the circular tube)}$$

Now we again use equation (6-25):

$$h = \frac{k}{D} C(\text{Re})^n (P_r)^{1/3}$$

For the Re of 45,100 (Table 6-1) we have

$$C = 0.0266, \qquad n = 0.805$$

$$h = \frac{(0.02711\ \text{w}/^\circ\text{C})}{(0.04\ \text{m})}(0.0266)(45,100)^{0.805}(0.7)^{1/3}$$

$$h = 89.32\ \text{w/m}^2\ ^\circ\text{C}$$

$$\frac{q^1}{L} = (89.32\ \text{w/m}^2\ ^\circ\text{C})\pi(0.04\ \text{m})(50.27)^\circ\text{C}$$

$$\frac{q^1}{L} = 258.1\ \text{w/m}$$

The square duct is essentially a noncircular cylinder. As such its values of C and n are given in Table 6-2. Note that the side of the duct is taken as D.

In order to solve for the correct side, we use equation (6-25). Here the side D will appear as shown below:

$$q^1 = hA\Delta T$$

and

$$q^1 = \frac{Ck}{D}\left(\frac{DV\rho}{\mu}\right)^n \text{Pr}^{1/3}(4D)\Delta T$$

Essentially we can solve directly for D. However, in order to do so, we must know C and n. As a starting point, assume that the side of the square is 0.04 m.

This means that the value of Re is still 45,100, and C and n (from Table 6-2) are, respectively, 0.102 and 0.675. Then,

$$D^n = \frac{q^1 n}{Ck\text{Pr}^{1/3}4(\Delta T)(V\rho)^n}$$

Substituting the values of each quantity including the q^1 of 258.1 W/m yields

$$D = 0.0297 \text{ m}$$

Hence, the square duct must have a dimension of 0.0297 m for each side.
The new Reynolds number will be 33,490, but this will still give the same C and n values. Hence, the 0.0297-m dimension is correct.

Example 6-7 Consider two cases:

Case 1: Vertical plate (0.9 m high; 0.5 m wide at 32°C with air at 22°C).

Case 2: Vertical plate (0.5 m high; 0.9 m wide at 22°C with air at 32°C).

Which case gives the higher rate of heat transfer?
This is a free convection situation. The film temperature T_f is

$$T_f = \frac{22 + 32}{2} = 27°\text{C}$$

Properties of air for this temperature are

$$\mu = 1.983 \times 10^{-5} \text{ kg/m sec}, \qquad C_\rho = 1.0057 \text{ kJ/kg °C}$$

$$\rho = 1.177 \text{ kg/m}^3, \qquad k = 0.02624 \text{ W/m °C}$$

If the case is assumed to be laminar (i.e., Gr Pr $\leqq 10^9$) then

$$\frac{h_m L}{k_f} = 0.59(\text{Gr}_f \text{Pr}_f)^{1/4}$$

Note that the film temperature and properties are the same in both cases. The only difference will be length, L, so that

$$\frac{(h_m)_1}{(h_m)_2} = \frac{L_2}{L_1}\left(\frac{L_1^3 \Delta T}{L_2^3 \Delta T}\right)^{1/4} = \left(\frac{L_2}{L_1}\right)^{1/4}$$

also

$$\frac{q_1^1}{q_2^1} = \frac{(h_m)_1 A_1 \Delta T_1}{(h_m)_2 A_2 \Delta T_2} = \frac{(h_m)_1}{(h_m)_2}$$

$$\frac{q_1^1}{q_2^1} = \frac{L_2}{L_1}^{1/4} = \left(\frac{0.5}{0.9}\right)^{1/4} = 0.863$$

$$q_1^1 = 0.863 q_2^1$$

Hence, case 2 has a higher heat transfer rate.

Next, we check the assumption of laminar conditions for both cases:

$$\text{Gr Pr} = \frac{g\beta \Delta T L^3 C_\rho \rho^2}{k}$$

For case 1

$$\text{Gr Pr} = \frac{(9.8 \text{ m m/sec}^2)(10°\text{K})(0.9)^3(1005.7 \text{ J/kg °K})(1.177 \text{ kg/m}^3)^2}{(300.16°\text{K})(1.983 \times 10^{-5} \text{ kg/m sec})(0.02624 \text{ J/sec °C m})}$$

$$\text{Gr Pr} = 6.38 \times 10^8$$

For case 2

$$\text{Gr Pr} = (6.38 \times 10^8)\left(\frac{0.5}{0.9}\right)^3 = 1.09 \times 10^8$$

As can be seen, both cases are for laminar conditions.

Example 6-8 A long duct (square cross section) is at 15°C, and the surrounding air is at 39°C. Find the rate of heat transfer per unit length. All sides of the duct are surrounded by air. Duct sides are 0.32 m.

This is a case of free convection. The combined heat transfer will involve two vertical plates (the two sides of the duct) and a colder horizontal plate (upward) with warmer air (top of duct) as well as a cooler horizontal plate (downward) with warmer air.

The film temperature is the same for all cases.

$$T_f = \frac{15°\text{C} + 39°\text{C}}{2} = 27°\text{C}$$

Air properties for this film temperature are

$$\mu = 1.98 \times 10^{-5}/\text{m sec}, \qquad C_p = 1005.7 \text{ J/kg °C}$$
$$\rho = 1.177 \text{ kg/m}^3, \qquad k = 0.02624 \text{ W/m °C}$$

In order to select appropriate empirical equation, we must first find Gr Pr for each case.

For duct sides (vertical plates) we have

$$\text{Gr Pr} = \frac{g\beta \Delta T L^3 C_p \rho^2}{k}$$

Using air properties and $\beta = 1/300.16°\text{K}$, we have

$$\text{Gr Pr} = \frac{(9.8 \text{ m/sec}^2)(24°\text{K})(0.32 \text{ m})^3(1005.7 \text{ J/kg °K})(1.177 \text{ kg/m}^3)^2}{(300.16°\text{K})(1.983 \times 10^{-5} \text{ kg/m sec})(0.02624 \text{ J/sec m °K})}$$

$$\text{Gr Pr} = 6.88 \times 10^7$$

Hence, the appropriate equation is

$$\frac{h_m L}{k_f} = 0.59(\text{Gr}_f \text{Pr}_f)^{1/4}$$

and, taking $L = 0.32$ m, we have

$$h_m = (0.59)(0.02624 \text{ W/m °C})\left(\frac{1}{0.32 \text{ m}}\right)(6.88 \times 10^7)^{1/4}$$

$$h_m = 4.41 \text{ W/m}^2 \text{ °C}$$

Next, consider the top of the duct (heated plate facing upward; i.e., upper surface of heated plate).

The Gr_fPr_f is the same value as above ($L = 0.32$ m and $T_f = 300.16°\text{K}$):

$$\frac{h_m L}{k_f} = 0.15(\text{Gr}_f \text{Pr}_f)^{1/3}$$

$$h_m = (0.15)(0.02624 \text{ W/m °C})\left(\frac{1}{0.32 \text{ m}}\right)(6.88 \times 10^7)^{1/3}$$

$$h_m = 5.04 \text{ W/m}^2 \text{ °C}$$

Finally, for the bottom of the duct (lower surface of heated plate) we have

$$\frac{h_m L}{k_f} = 0.58(\text{Gr}_f \text{Pr}_f)^{1/5}$$

Then

$$h_m = (0.58)(0.02624 \text{ W/m °C})\left(\frac{1}{0.32 \text{ m}}\right)(6.88 \times 10^7)^{1/5}$$

$$h_m = 1.76 \text{ W/m}^2 \text{ °C}$$

The overall heat transferred per unit length of duct is then

$$q^{11} = \{(2)\ (4.41\ \text{w/m}^2\ ^\circ\text{C})(0.32\ \text{m}) + (5.04\ \text{W/m}^2\ ^\circ\text{C})(0.32\ \text{m}) + (1.76\ \text{W/m}^2\ ^\circ\text{C})(0.32\ \text{m})\}\{24^\circ\}$$

$$q^{11} = 120\ \text{W/m}$$

PROBLEMS

6-1. Water at a mass flow rate of 1 kg/sec is heated from 30°C to 70°C in a tube whose outer surface is at 100°C. How long must a 0.025-m tube be to carry out the results?

6-2. Air (atmospheric conditions) flows through a 0.15-m-diameter tube (10 m in length). If the air enters at 60°C and the tube surface temperature is 15°C, what will the exit temperature and heat loss be?

6-3. Engine oil flows through a 0.003-m-diameter tube that is 30 m long. The oil enters at 60°C and the wall temperature is kept at 100°C. Find the oil outlet temperature and the average heat transfer coefficient.

6-4. Air (285 K, atmospheric pressure) enters a 2-m-long rectangular duct (0.075 by 0.150 m) whose surface is at 400 K. Air mass flow rate is 0.10 kg/sec. Find the heat transfer rate and the outlet temperature.

6-5. What would the average heat transfer coefficient (fluid properties at 37.8°C) be respectively for air, water, and engine oil if the average Nusselt number is 4.6?

6-6. Liquid ammonia flows through a 0.025-m-diameter tube (2.5 m long) at a mass flow rate of 0.454 kg/sec. If the ammonia enters at 10°C and leaves at 38°C, what must the average wall temperature be?

6-7. Water at 25°C (2 kg/sec) flows through a 4-m-long tube (0.04-m diameter). If the tube surface is kept at 90°C, what is the water exit temperature and rate of heat transfer.

6-8. Water (mass flow rate of 2 kg/sec) enters a long pipe at 25°C and 1000 bars. The pipe is heated such that 10^5 watts are transferred. If the water leaves at 2 bars, what is its outlet temperature?

6-9. Atmospheric air enters a 3-m-long, 0.05-m-diameter tube at 0.005 kg/sec. The h is 25 W/m^2 °K, there is a uniform heat flux at the surface of 1000 W/m^2. Find the outlet and inlet air temperatures and sketch the axial variation of surface temperature.

6-10. Engine oil flows through a 0.003-m-diameter, 30-m-long tube at a mass flow rate of 0.02 kg/sec. The tube wall is kept at 100°C, which makes the outlet temperature 60°C. Find the oil inlet temperature.

6-11. Steam (at 120°C) condenses on the outside of a horizontal pipe at 30 kg/hr. Water flows through the pipe (0.025-m diameter; 0.8 m long) at an average velocity of 1 m/sec. The inlet water temperature is 16°C. Assuming that the only important thermal resistance is the water convection, find the average heat transfer coefficient. Latent heat of the steam is 2202 kJ/kg.

6-12. Water (at 8°C) enters a 0.01-m-diameter tube at a volumetric flow rate of 400 cm/sec. If the tube wall is kept at 250°C and the outlet temperature is 50°C, how long is the tube?

6-13. Water (mass flow rate of 1 kg/sec) flows into a 0.025-m-diameter tube at 15°C and leaves at 50°C. Tube wall temperature is kept 14°C higher than the water temperature. What is the tube length?

6-14. Water at 26.7°C flows into a 0.0032-m-diameter pipe (1.83 m long) with an average velocity of 0.038 m/sec. If the tube wall is kept at 82.2°C, what is the heat transfer rate?

6-15. Water (mass flow rate of 3 kg/sec) is heated from 5°C to 15°C in a 0.05-m-diameter tube (wall at 90°C). What is the tube length?

6-16. Water at 10°C flows through a 0.025-m-diameter tube (15 m long) at a rate of 05 kg/sec. If the tube wall is 15°C higher than the water temperature, what is the outlet temperature?

6-17. Oil flows through a 0.05-m-diameter pipe at 1 m/sec. At a given point the oil is at 50°C ($\mu = 2.1$ centipoise; $\rho = 880$ kg/m^3). If the outside steam is at 130°C, what is the heat transfer coefficient at that point?

6-18. Air flows through a steam-heated tubular heater. What would be the effect on heat transferred divided by ΔT for the following cases (assuming that the air heat transfer coefficient controls): (a) double gas pressure with fixed mass flow rate; (b) double mass flow rate; (c) double number of heater tubes; (d) halve tube diameter.

6-19. If equation (6-22) is divided by Re Pr, we obtain the Colburn equation. In this equation both the Pr and Cp for air are very slowly varying functions of temperature. Thus, h increases with $\mu^{0.2}$. Explain this indicated behavior and find the dependence of h on temperature.

6-20. A heavily insulated, electrically heated pipe (heat generation 10^6 W/m^3) has inner and outer diameters of 0.02 and 0.04 m, respectively. If water enters (0.1 kg/sec) at 20°C and leaves at 40°C, what is the tube length?

6-21. A 0.025-m-diameter cylinder whose temperature is 150°C is placed in an air stream (1 atmosphere, 38°C) whose velocity is 30 m/sec. What is the heat loss per meter of length for the cylinder?

6-22. If a person (surface temperature of 24°C) can be approximated by a cylinder 1.83 m high and 0.3048 m in diameter, what will be the heat loss if the wind (−1.11°C) flows at a rate of 13.41 m/sec?

6-23. A sphere (0.003-m diameter, 93°C) is placed in a water stream (38°C, 6 m/sec). What is the heat transfer rate?

6-24. Compare the heat transfer rates per unit length for a stream of air and a stream of water (velocities of 6 m/sec; 20°C) flowing over a 65°C cylinder (diameter of 2.5×10^{-5} m).

6-25. A staggered tube arrangement ($S_n = S_p = 0.02$ m, tube diameter of 6.33×10^{-3} m) at a surface temperature of 90°C is used to heat atmospheric air at 20°C with an inlet velocity of 4.5 m/sec. Six rows of tubes (50 tubes high) are used. What is the rate of heat transfer (per unit tube length)?

6-26. An in-line tube bank ($S_n = S_p = 0.002$ m, $D = 0.001$ m, 10 rows of 50 tubes each) is to cool a flue gas stream at 427°C (inlet velocity of 5 m/sec). Cold water flows in the tubes (surface temperature of 27°C). What is the rate of heat transfer?

6-27. Compare the rate of heat transfer to cross-flow air (25°C; velocity of 15 m/sec) for the following surfaces at 75°C (circular cylinder of 0.01-m diameter; square duct 0.01-m sides; vertical plate 0.01 m high).

6-28. A tube bank (12 rows high, 6 rows deep) is arranged in a staggered manner (tube centers form an equilateral triangle of 0.045-m sides), and air (1 atm, 20°C) flows across the bank at 10 m/sec approach velocity. Tubes have a diameter of 0.026 m and length of 4 m. Determine the heat transfer rate if the tubes are at 100°C.

6-29. Air at a velocity of 25 m/sec flows across a duct with a film temperature of 80°C. Compare the heat transfer with a 0.05-m-diameter circular duct and a square duct (0.05 m).

6-30. Air (3.5 mega N/m^2, 38°C) flows across a staggered tube bank (400 tubes, 0.0125-m diameter, 20 rows high, $S_p = 3.75$ cm, and $S_n = 2.5$ cm) approach velocity is 9 m/sec. For a 1.5-m tube length and a surface temperature of 200°C, find the air exit temperature.

6-31. Thin metal strips (0.006 m wide) normal to air flow (from a fan) are used to dissipate heat in an electric heater. Seven 0.35-m strips are used with the air velocity of 2 m/sec and temperature of 20°C. If the strips are heated to 870°C, estimate the total convection heat transfer (in actuality, radiation will supply a large amount of the transfer).

6-32. A wire (1.3×10^{-4}-m diameter, 0.0125 m long) electrically heated encounters a cross-flow air stream (230 m/sec, −30°C, 54×10^3 N/m^2). What electric power is needed to keep the wire's surface temperature at 175°C?

6-33. Platinum wire (0.004-m diameter, 0.10 m long) electrically heated is placed horizontally in a 38°C container of water. If the wire is kept at 93°C, calculate the heat lost.

6-34. An oven door (0.5 m high, 0.7 m wide) has an average surface temperature of 32°C. What is the heat loss if the surrounding air is at 22°C.

6-35. Air (1 atm) is contained between two 0.5-m sided "square" vertical plates (0.015 m apart) at temperatures of 100°C and 40°C. Compute the free convection heat transfer.

6-36. A 0.025-m-diameter sphere with an imbedded electrical heater is immersed in various quiescent mediums at 20°C (atmospheric air, water, ethylene glycol). Calculate the power needed to keep the sphere surface at 94°C.

6-37. A rectangular cavity is formed from two parallel 0.5-m sided square plates 0.05 m apart (insulated lateral boundaries). If the heated plate is kept at 325 K and cooled at 275 K, find the heat flux for the cases of Table 6-5.

6-38. A food processing vat contains oil at 205°C. A shell at 60°C surrounds the tank on the vertical sides. The air space separating the vat and shell is 0.35 m high and 0.03 m thick. Estimate (per square meter of surface) the free convection loss.

6-39. A heater for engine oil in a large vessel is a 0.30 by 0.30-m sided "square" plate at 100°C. What is the heat transfer rate for oil at 20°C?

6-40. A 0.075-m-diameter tube using 120°C steam is to heat an area with an ambient air temperature of 17°C. If the total heating required is 29,308 W, what pipe length is needed?

6-41. An average radiant heat flux of 1100 W/m^2 impinges the outside wall (6 m high) of a building. If 95 W/m^2 is conducted through the wall, estimate the outside wall temperature. Ambient air is at 20°C.

6-42. Air at 1 atm and 27°C passes through a horizontal 0.025-m tube (0.4 m long) with an average velocity of 30 cm/sec. The tube wall is maintained at 140°C. This system is a mixed free-forced convection case. Calculate the h and compare to an assumption of only forced convection.

REFERENCES

1. E. N. Sieder and C. E. Tate, *Ind. Eng. Chem.* **28**, 1429 (1936).
2. H. Hausen, *V.D.I.Z.* **4**, 91 (1943).
3. W. Nusselt, *Forsch. Geb. Ing.* **2**, 309 (1931).
4. T. R. Irvine, in *Modern Developments in Heat Transfer*, W. Ibele, editor, Academic Press, New York (1963).
5. W. H. McAdams, *Heat Transmission*, McGraw-Hill, New York (1954).
6. J. D. Knudsen and D. L. Katz, *Fluid Dynamics and Heat Transfer*, McGraw-Hill, New York (1958).
7. R. Hilpert, *Forsch. Geb. Ing.* **4**, 220 (1933).
8. M. Jakob, *Heat Transfer*, Vol. 1, John Wiley and Sons, New York (1949).

9. R. G. Griskey and R. E. Willins, *Can. J. Chem. Eng.* **53**, 500 (1975).
10. E. D. Grimson, *Trans ASME* **59**, 583 (1937).
11. W. E. Ranz and W. R. Marshall, *Chem. Eng. Prog.* **48**, 141 (1952).
12. F. J. Bayley, *Proc. Inst. Mech. Eng.* **169**(20), 361 (1955).
13. C. Y. Warner and V. S. Arpaci, *Int. J. Heat Mass Transfer* **11**, 397 (1968).
14. T. Fujii and H. Imura, *Int. J. Heat Mass Transfer* **15**, 755 (1972).
15. J. R. Lloyd and W. R. Moran, ASME Paper 74-WA/HT-66.
16. S. N. Singh, R. C. Birkebak, and R. M. Drake, *Prog. Heat Mass Transfer* **2**, 87 (1969).
17. T. Yuge, *J. Heat Transfer* **82**, 214 (1960).
18. J. P. Holman, *Heat Transfer*, fourth edition, McGraw-Hill, New York (1976).
19. B. Metais and E. R. G. Eckert, *J. Heat Transfer* **86**, 295 (1964).
20. C. K. Brown and W. H. Gauvin, *Can. J. Chem. Eng.* **43**, 306 (1965).
21. B. V. Karlekar and R. M. Desmond, *Heat Transfer*, second edition, West Publishing, St. Paul, MN (1982).
22. R. G. Griskey, *Chemical Engineering for Chemists*, American Chemical Society, Washington, D.C. (1997).
23. R. B. Bird, W. E. Stewart, and E. N. Lightfoot, *Transport Phenomena*, John Wiley and Sons, New York (1960).

7

COMPLEX HEAT TRANSFER

INTRODUCTION

The semiempirical approaches for the heat transfer coefficient can also be applied to more complex situations than those described earlier. Such cases include packed beds, agitated systems, non-Newtonian fluids, and heat transfer with phase change.

In this chapter we will consider these situations and present the recommended approaches for such processes. Attention will also be directed to the phenomenological bases for the relations.

HEAT TRANSFER IN PACKED BEDS

In Chapter 6 we derived the relationship for the dimensionless groups [see equations (6-2) through (6-5)]. A similar approach can be used for flows through packed beds. This is done by writing the relation

$$dq^1 = h_L(a_v S dz)(T_0 - T_b) \tag{7-1}$$

This equation in differential form is the analog to equation (6-2). The h_L is a local heat transfer coefficient, S is the bed cross section, dz is the differential height, and a_v is the surface per unit volume. The combination of $a_v S dz$ is the heat transfer surface.

It can again be shown that the Nusselt number is a function of the Reynolds and Prandtl numbers. The resultant semiempirical equations (fitted from experimental data) are (1)

$$\frac{h_L}{(C_p)_b G_0} = 0.91\psi(\text{Re}_f)^{-0.51}(\text{Pr}_f)^{-2/3} \tag{7-2}$$

for Re_f less than 50. The $(C_p)_b$ is the fluid C_p at bulk temperature.

G_0 is the mass velocity of the fluid (the fluid mass flow rate divided by the bed cross section). The ψ term (2) is a shape factor (spheres 1.0; cylinders 0.91; flakes 0.86; Raschig rings 0.79; partition rings 0.67; Berl saddles 0.80).

Re_f and Pr_f are defined as follows:

$$\text{Re}_f = \frac{G_0}{a_v \mu_f \psi} \tag{7-3}$$

$$\text{Pr}_f = \left(\frac{C_p \mu}{k}\right)_f \tag{7-4}$$

When Re_f is greater than 50 we have

$$\frac{h_L}{(C_p)_b G_0} = 0.61\psi \ \text{Re}^{-0.41}\text{Pr}^{-2/3} \tag{7-5}$$

Heat transfer of this type occurs in such processes as fixed bed chemical reactors and as packed bed mass transfer units.

HEAT TRANSFER IN NON-NEWTONIAN SYSTEMS

Non-Newtonian systems are implicit in many important industrial processes. This includes not only the polymer industries (plastics, resins, fibers, elastomers, coatings, etc.) but also such important areas as food processing.

The complex nature of the flows encountered as well as such factors as compressibility and viscous dissipation make it necessary to use special relations to describe such heat transfer. One aspect that is fortunate is that the vast majority of process situations involving non-Newtonians are in laminar flow.

For polymer solutions flowing in a circular tube (3–5) we have

$$\frac{h_a D}{k} = 1.75\left(\frac{3n^1+1}{4n^1}\right)^{1/3}\left(\frac{WC_p}{k_L}\right)^{1/3}\left(\frac{K_b^1}{K_w^1}\right)^{0.14} \tag{7-6}$$

The n^1 and K^1 terms are from the relation

$$\left(\frac{D\Delta P}{4L}\right) = K^1\left(\frac{8V}{D}\right)^{n^1} \tag{7-7}$$

a modified form of the Ostwald–De Waele (see Chapter 2) Power Law. The b and w subscripts refer to bulk and average wall temperature. W is mass flow rate and L is the length. Also,

$$\text{Gz} = \text{Graetz number} = \frac{WC_p}{kL} \tag{7-8}$$

Equation (7-6) holds if Gz is greater than 20 and n^1 exceeds 0.10. Lower values of Gz and n' are handled by equation (7-9).

$$\frac{h_a D}{k} = 1.75\Delta^{1/3}\left(\frac{WC_p}{kL}\right)^{1/3}\left(\frac{K_b^1}{K_w^1}\right)^{0.14} \tag{7-9}$$

The Δ factor is obtained from Figures 7-1 and 7-2.

The preceding equations are not applicable for highly viscous non-Newtonians such as molten polymers because of factors such as the materials' compressibility or severe viscous heating effects. As an example of the latter, consider the processing of certain polymers in screw extruders where the shear heating (viscous dissipation) causes the fluid to rise significantly in temperature.

In the case of highly viscous systems, the use of Figure 7-3 is recommended. This figure is a plot of Nusselt number versus Graetz number for the highly viscous systems. The B^* terms are Brinkman numbers (negative when the fluid is being heated, positive if the fluid is cooled). The solid line for $B^* = 0.0$ is the classical Graetz–Nusselt relationship. Theoretical predictions of the effect of viscous dissipation are depicted for B^* values of 1.0, 0.5, −0.5, and −1.0.

Lines fitted to symbols represent actual experimental data. As can be seen, the effect of viscous heating is much greater than that predicted theoretically. The experimental data also include the severe thermal expansion effects encountered when processing such materials as molten polymers.

Use of this relation for process cases will be covered in the Worked Examples section.

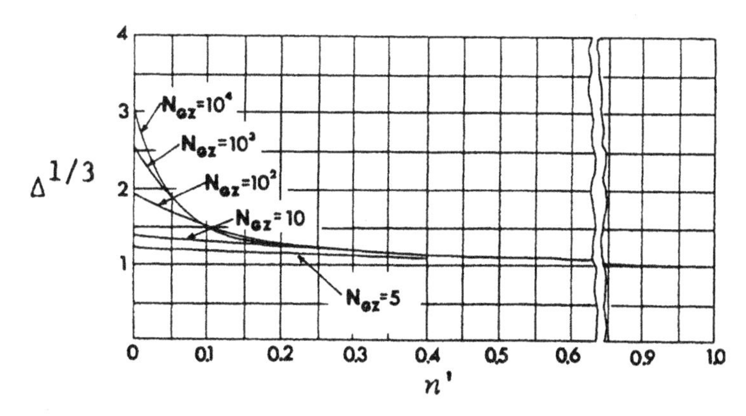

Figure 7-1. Factor $\Delta^{1/3}$ for laminar flow. (Reproduced with permission from references 3–5. Copyright 1957, American Institute of Chemical Engineers. Copyrights 1960 and 1964, *Chemical Engineering Science*.)

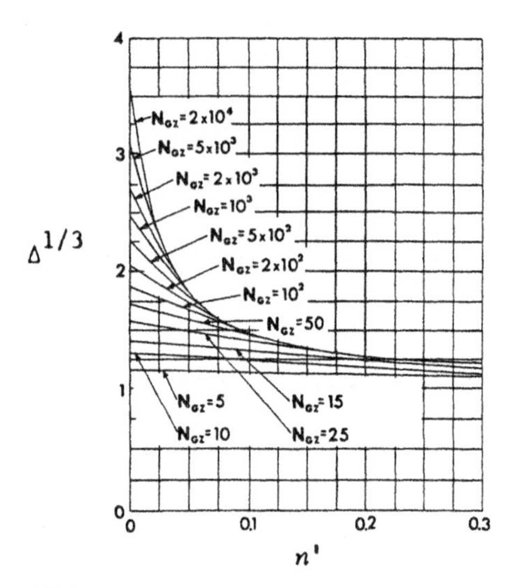

Figure 7-2. Factor $\Delta^{1/3}$ for laminar flow and low n values. (Reproduced with permission from references 3–5. Copyright 1957, American Institute of Chemical Engineers. Copyrights 1960 and 1964, *Chemical Engineering Science*.)

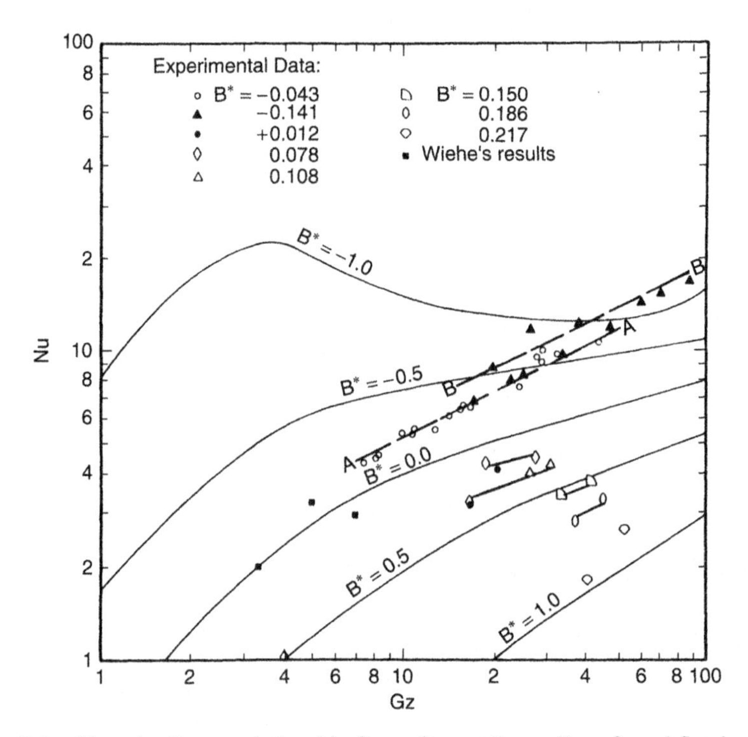

Figure 7-3. Nusselt–Graetz relationship for molten or thermally softened flowing polymers with viscous dissipation and thermal expansion effects. (Reproduced with permission from references 6 and 7. Copyrights 1972 and 1975, Society of Plastics Engineers.)

HEAT TRANSFER IN AGITATED SYSTEMS

Many industrial processes are carried out in agitated tanks with accompanying heat transfer. In such cases, the vessel is jacketed. The heat transfer fluid used for heating or cooling circulates through the jacket. Heating or cooling coils are also sometimes immersed in the vessel as well.

The semiempirical heat transfer correlations are of the form

$$\frac{hD_T}{k} = C(\mathrm{Re}^1)^a \mathrm{Pr}^{1/3}\left(\frac{\mu}{\mu_w}\right)^b \tag{7-10}$$

Table 7-1 gives the various C, a, and b values for different agitator systems. The Re^1 is as defined by equation (4-26):

$$\mathrm{Re}^1 = \frac{\mathrm{N}\rho(D^1)^2}{\mu} \tag{7-11}$$

HEAT TRANSFER WITH PHASE CHANGE

Heat transfer with phase change is widely used in the process industries. Basically, the association of a latent heat (say for example of vaporization) gives such heat transfer a greater impact than ordinary cases. This can be realized by simply seeing that a latent heat change can be orders of magnitude greater than sensible heat ($C_p\,dT$) changes.

While all forms of phase changes (melting, solidification, vaporization, condensation, sublimation, etc.) can occur in industrial processes, there are essentially only two that are well-described in a technical sense. These are vaporization (boiling) and condensation. The latter's behavior is the better known of these two forms.

Table 7-1 Constants for Equation (7-10)

Type[a]	c	a	b	Re^1 range	Reference
Paddle, no baffle	0.36	2/3	0.21	300 to 3×10^5	8,9
Flat blade turbine, no baffle	0.54	2/3	0.14	30 to 3×10^5	10
Flat blade turbine, with baffle	0.74	2/3	0.14	500 to 3×10^5	10, 11
Anchor, no	1.0	1/2	0.18	10 to 300	9
baffles	0.36	2/3	0.18	300 to 4×10^4	
Helical ribbon, no baffles	0.633	0.5	0.18	8 to 10^5	12
Paddle, no baffles, coil	0.87	0.62	0.14	300 to 4×10^5	8

[a]Heat transfer coefficients with jacket except for last case which is for the coil.

Condensation heat transfer is an intricate process because it essentially involves two phases (vapor and liquid condensate). Furthermore, the latent heat associated with the phase change complicates the heat transfer situation.

Let us first consider condensation of a vapor on a vertical surface. In this case the condensate in the form of a liquid laminar film will flow down the surface (see Figure 7-4). As can be seen, maximum velocity occurs on the outside of the film and minimum velocity (zero) at the solid surface. An analytical treatment for the mean heat transfer coefficient for the entire surface yields

$$h_m = 0.925 \left(\frac{k_f^3 \rho_f^2 g \lambda}{L \Delta T \mu_f} \right)^{1/4} \tag{7-12}$$

Actually, when fit to experimental data, this relationship gives equation (7-13), which is the recommended form.

$$h_m = 1.13 \left(\frac{k_f^3 \rho_f^2 g \lambda}{\mu_f L \Delta T} \right)^{1/4} \tag{7-13}$$

Note that the only difference between equations (7-13) and (7-12) is the multiplier (i.e., 1.13 instead of 0.925).

Equation (7-13) can be rewritten in the form of a Reynolds number relation as

$$h_m \left(\frac{\mu_f^2}{k_f^3 \rho_f^2 g} \right)^{1/3} = 1.80 \left(\frac{4\Gamma}{\mu_f} \right)^{-1/3} \tag{7-14}$$

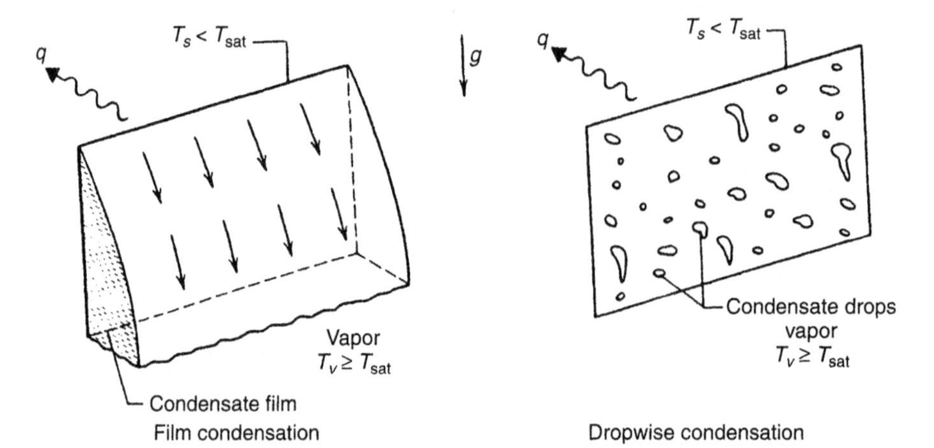

Figure 7-4. Laminar condensate film on vertical surface (24). Film and dropwise condensation on a vertical surface (Adapted with permission from reference 25. Copyright 1981, John Wiley and Sons Inc.)

where

$$\mathrm{Re} = \frac{4\Gamma}{\mu_f} \tag{7-15}$$

The Γ term is the mass flow rate of condensate per unit perimeter (wall width, tube circumference, etc.).

Film temperature, T_f, is defined as shown below:

$$\text{Film temperature} = T_f = T_{\text{condensation}} - 3/4(T_{\text{condensation}} - T_{\text{wall}}) \tag{7-16}$$

When the Reynolds number exceeds 1800, flow becomes turbulent. In such cases (for vertical surfaces) we have

$$h_m \left(\frac{\mu_f^2}{k_f^3 \rho_f^2 g}\right)^{1/3} = 0.0077 \left(\frac{4\Gamma}{\mu_f}\right)^{0.4} \tag{7-17}$$

If we have a case of horizontal tubes, then for a vertical tier of N such horizontal tubes we obtain

$$h_m = 0.725 \left(\frac{k_f^3 \rho_f^2 g \lambda}{N D_{\text{tube}} \mu_f \Delta T}\right)^{1/4} \tag{7-18}$$

or

$$h_m = 0.95 \left(\frac{k_f^3 \rho_f^2 g L}{\mu_f W}\right)^{1/3} \tag{7-19}$$

Usually, the condensate flow for horizontal tubes is laminar. If, however, $2\Gamma'/\mu_f$ exceeds 2100 (where Γ' is W/L), then equation (7-17) should be modified to determine h_m as per

$$h_m \left(\frac{\mu_f^2}{k_f^3 \rho_f^2 g}\right)^{1/3} = 0.0077 \left(\frac{2\Gamma'}{\mu_f}\right)^{0.4} \tag{7-20}$$

All of the preceding has been based on the assumption of film condensation. There is another condensation mode, namely, dropwise condensation (see Figure 7-4). In this case, droplets coalesce and flow in rivulets over the surface. The result is that parts of the surface are not covered and hence can contact the vapor directly (i.e., without the resistance of the liquid film). Because of this we obtain extraordinarily high heat transfer rates with dropwise condensation which can have heat transfer coefficients four to ten times those of film condensation heat transfer coefficients.

The question might be fairly asked as to why dropwise condensation is not used exclusively in process work. Very simply put, the answer is that it cannot be maintained because it is a result of surface behavior (even with coated surfaces, dropwise behavior takes place only for a time). Because of this, designs are almost always based on film condensation.

An interesting process happening, however, is that frequently surfaces become contaminated and dropwise condensation then causes the heat transfer coefficient to markedly increase for a period of time (i.e., until the surface contamination disappears). Ultimately, film condensation again becomes the controlling mechanism, and the coefficient drops to its regular value.

Boiling heat transfer is more complicated than condensation heat transfer. This can be seen in Figure 7-5, which plots heat flux q^1/A or h versus the difference of the temperature and the fluids' saturation temperature (boiling point). In boiling there are a number of regions encountered. The heat flux or heat transfer coefficient changes from point to point as shown.

In order to get a better feel for this situation, let's consider a simple laboratory experiment in which we will place an electric immersion heater into a beaker of water. At the beginning of the experiment, we will first notice movement of the water. This free convection takes place at low temperature differences. As the heater's surface temperature difference increases, so does the heat flux (Figure 7-5). Ultimately, bubbles begin to form at the heating surface. This corresponds to the nucleate boiling region of Figure 7-5. Here again the flux or heat transfer coefficient increases until it reaches a peak. As the bubbles begin to coalesece at the heat transfer surface, a thermal resistance (vapor film) is formed which reduces the flux and coefficient. Complete film formation reduces the flux and coefficient to the low point shown in Figure 7-5. See that flux increases past this point by radiation.

As Figure 7-5 clearly demonstrates, boiling is much more complicated than condensation. This is underscored if we compare the two aspects of condensation (film and dropwise) to Figure 7-5. The boiling analogs are nucleate boiling

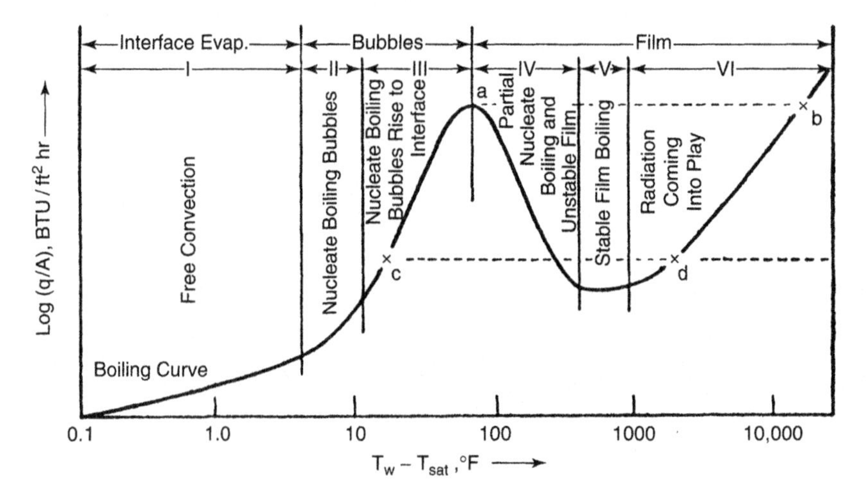

Figure 7-5. Typical boiling curve. (Reproduced with permission from reference 24. Copyright 1948, American Society of Mechanical Engineers.)

(dropwise condensation) and film boiling (film condensation). In essence, then condensation would correspond only to a portion of the boiling curve.

It becomes apparent that each region of the boiling curve requires its own correlation. For example, in the free convection region

$$\mathrm{Nu} = 0.61(\mathrm{Gr\ Pr})^{1/4} \tag{7-21}$$

In the nucleate boiling region there are a number of empirical correlations. One (13) is

$$C_\ell(T_w - T_{\text{sat}}) = C' \left[\frac{(q/A)}{\mu\ell\lambda} \sqrt{\frac{g_c\sigma}{g(\rho_\ell - \rho_g)}} \right]^{1/3} (\mathrm{Pr})^{1.7} \tag{7-22}$$

The l and g subscripts refer to liquid and gaseous phases, C_l is the liquid $C_{p'}\sigma$ is the surface tension, and C' is an empirical constant dependent on the surface. A list of C' values for various fluid–surface combinations is given in Table 7-2.

Equation (7-22) and Table 7-2 give us some interesting insights into boiling heat transfer. First of all we see that fluid-surface interaction is extremely important (the presence of the surface tension in equation (7-22) and the C' factor). Furthermore, from Table 7-2 there does not appear to be any possible correlation of C' values for various systems.

If there is convection as well as nucleate boiling, then the total flux is given by

$$\left(\frac{q}{A}\right)_{\text{total}} = \left(\frac{q}{A}\right)_{\text{moderate boiling}} + \left(\frac{q}{A}\right)_{\text{convection}} \tag{7-23}$$

This means that we can then have enhanced heat transfer for such situations.

The peak heat flux (the maximum of Figure 7-5) can be estimated with the relation (18)

$$\frac{(q/A)_{\max}}{\rho_g\lambda} = 143(g)^{1/4}\left(\frac{\rho_1 - \rho_g}{\rho_g}\right)^{0.6} \tag{7-24}$$

where g is the acceleration of gravity in G's.

Table 7-2 Values of the Coefficient C' for Various Liquid–Surface Combinations

Fluid–heating-surface Combination	Reference	C'
Water–copper	14	0.013
Water–platinum	15	0.013
Water–brass	16	0.0060
n-Butyl alcohol–copper	14	0.00300
Isopropyl alcohol–copper	14	0.00250
n-Pentane–chromium	17	0.015
Benzene–chromium	17	0.010
Ethyl alcohol–chromium	17	0.027

Finally, for a film boiling on a tube, the heat transfer coefficient is taken to be a combination of a conduction h and a radiative h:

$$h = h_{\text{conduction}} + h_{\text{radiative}} \tag{7-25}$$

and (19)

$$h_{\text{conduction}} = 0.62\left[\frac{k_g^3\rho_g(\rho_l - \rho_g)g(\lambda + 0.4C_g\Delta T)}{D_{\text{Tube}}\mu_g\Delta T}\right]^{1/4} \tag{7-26}$$

and

$$h_{\text{radiative}} = \frac{\sigma\varepsilon(T_w^4 - T_{\text{sat}}^4)}{T_w - T_{\text{sat}}} \tag{7-27}$$

where σ is the Stefan–Boltzmann constant ($\sigma = 5.669 \times 10^{-8}$ w/m^2 °K^4) and σ is the emissivity of the surface.

WORKED EXAMPLES

Example 7-1 It is desired to carry out a process in a packed column with heat transfer. The column can be packed either with 1.27-cm Raschig rings or 1.27-cm Berl saddles. Porosities (ε) for both are 0.63. The a_v values for the Raschig rings and Berl saddles are, respectively, 33.83 and 43.28 m^2/m^3. Which packing would require the lower superficial mass velocity (G_0)?

Let the Raschig ring be case 1 and the Berl saddle case 2. The heat transfer expression is

$$\frac{h_L}{Cp_b G_0}\text{Pr}_f^{2/3} = 0.91\psi\ \text{Re}^{-0.51}$$

for Re < 50.

If we write the above for both cases $(h_L)_1$ and $(h_L)_2$ and then obtain the ratio of $(h_L)_1/(h_L)_2$, we obtain

$$\frac{(h_L)_1}{(h_L)_2} = \frac{\psi_1(G_0)_1\ \text{Re}_1^{-0.51}}{\psi_2(G_0)_2\ \text{Re}_2^{-0.51}}$$

Substituting the Re relation gives

$$\frac{(h_L)_1}{(h_L)_2} = \frac{(G_0)_1^{0.49}\psi_1^{1.51}a_1^{0.51}}{(G_0)_2^{0.49}\psi_2^{1.51}a_2^{0.51}}$$

Now for equivalent process result assume $(h_L)_1a_1 = (h_L)_2a_2$ because

$$dq^1 = h_L(aSdz)(\Delta T)$$

Then,

$$1.0 = \frac{(G_0)_1^{0.49}\psi_1^{1.51}a_1^{1.51}}{(G_0)_2^{0.49}\psi_2^{1.51}a_2^{1.51}}$$

Also,

$$a = a_v(1 - \varepsilon)$$

This gives values of $a_1 = 12.52\ \text{m}^2/\text{m}^3$ and $a_2 = 16.01\ \text{m}^2/\text{m}^3$. Values of ψ_1 and ψ_2 are 0.79 and 0.80. Substituting and solving, we obtain

$$(G_0)_1 = 2.34(G_0)_2$$

when Re < 50.

In the second case (Re > 50) we find the h_L ratio as

$$\frac{(h_L)_1}{(h_L)_2} = \frac{(G_0)_1^{0.59}\psi_1^{1.41}a_1^{0.41}}{(G_0)_2^{0.59}\psi_2^{0.59}a_2^{0.41}}$$

With equal heat transfer (i.e., $h_L a$ values the same)

$$1.0 = \frac{(G_0)_1^{0.59}(\psi_1 a_1)^{1.41}}{(G_0)_2^{0.59}(\psi_2 a_2)^{1.41}}$$

$$(G_0)_1 = 1.86(G_0)_2$$

Hence, in both cases the Raschig rings would require a higher superficial mass velocity.

Example 7-2 A polymer solution (n of 0.5; K at 90°F of 51-lb mass sec^{n-2} ft^{-1}; viscosity activation energy of 14,900 Btu/lb mole) is fed into a 1-inch inside-diameter stainless steel tube (10 feet long) at a mass flow rate of 750 lb mass/hour and a temperature of 90°F. The velocity profile is fully developed prior to entering the heated tube. Heat is supplied by steam condensing at 20 psia.

Remaining fluid properties are: density = 58 lb mass/ft^3, specific heat = 0.6 Btu/lb mass °F, thermal conductivity = 0.5 Btu/ft °F hr.

Calculate exit temperature of the fluid.

The temperature of the condensing steam is 227.96°F. For purposes of calculation we will take this as 228°F.

Next we use equation (7-6) where

$$\text{Nu} = 1.75\left(\frac{3n+1}{4n}\right)^{1/3}(\text{Gz})^{1/3}\left(\frac{K_B}{K_w}\right)^{0.14}$$

From the given data

$$\left(\frac{3n+1}{4n}\right) = \left(\frac{2.5}{2.0}\right) = 1.25$$

$$\text{Gz} = \frac{WC_p}{kL} = \frac{(750\ \text{lb mass/hr})(0.6\ \text{Btu/lb mass °F})}{(0.5\ \text{Btu/ft °F hr})(10\ \text{ft})} = 90$$

The value of K_B is taken at the average of the entering and wall temperatures:

$$\left(\frac{90+228}{2}\right)\text{°F or 159°F } K_w \text{ is taken at 228°F.}$$

Hence,

$$K_B = (K)159\text{°F} = (\text{K})90\text{°F } \exp\left[\frac{E}{R}\left(\frac{1}{619} - \frac{1}{550}\right)\right]$$

$$K_w = (K)228\text{°F} = (\text{K})90\text{°F } \exp\left[\frac{E}{R}\left(\frac{1}{688} - \frac{1}{550}\right)\right]$$

solving these equations yields

$$K_B = 11.182 \text{ lb mass sec}^{2-n} \text{ ft}^{-1} \text{ and } K_w = 3.315 \text{ lb mass sec}^{2-n} \text{ ft}^{-1}$$

Then,

$$h = \frac{k}{D}1.75\left(\frac{3n+1}{4n}\right)^{1/3}(\text{Gz})^{1/3}\left(\frac{K}{K_w}\right)^{0.14}$$

$$h = \frac{0.5 \text{ Btu/ft °F hr}}{1/12 \text{ ft}}(1.75)(1.25)^{1/3}(90)^{1/3}\left(\frac{11.182}{3.315}\right)^{0.14}$$

$$h = 69.73\frac{\text{Btu}}{\text{ft}^2 \text{ °F hr}}$$

Next by enthalpy balance,

$$wC_p\Delta T \text{ fluid} = hA(T_{\text{wall}} - T_{\text{fluid}})_{\text{average}}$$

$$wC_p(T_2 - T_1) = hA\left[T_{\text{wall}} - \left(\frac{T_2+T_1}{2}\right)\right]$$

$$\left(750\frac{\text{lb mass}}{\text{hr}}\right)\left(\frac{0.6 \text{ Btu}}{\text{lb mass °F}}\right)(T_2 - 90\text{°F})$$

$$= \left(69.73\frac{\pi}{12} \times 10\right) \text{ ft}^2\left[288 - \left(\frac{T_2+90}{2}\right)\right]$$

Solving for T_2 gives a value of 137°C.

Example 7-3 Thermally softened polymethylmethacrylate entering at 426°F is being cooled in a 9.6-foot-long tube (inside diameter of 1 inch). The flow rate of the polymer is 317.9 g/min. Wall temperature is 395°F. Compare the exit temperature calculated using the appropriate B^* value to that for $B^* = 0.0$.

The B^* value is given by

$$B^* = \frac{T_{\text{wall}} w n}{(3n+1)k\rho\pi R(T_1 - T_{\text{wall}})}$$

For polymethylmethacrylate the T_{wall} and n values can be obtained from the rheological data presented by Westover (20). Density and thermal conductivity are taken from the work of Heydemann and Guicking (21) and Griskey, Luba, and Pelt (22), respectively. The computation gives a value of 0.150 for B^*.

The Graetz number for this case is 39.7. For this Gz the corresponding Nusselt numbers are 5.8 (at $B^* = 0.0$) and 3.5 (at $B^* = 0.150$), respectively (Figure 7-3). Then, $q = hA\Delta T$.

By enthalpy balance we have

$$wC_p \Delta T_{\text{fluid}} = hA(T_{\text{wall}} - T_{\text{fluid}})$$

and

$$wC_p(T_2 - T_1) = hAT_{\text{wall}} - \frac{(T_1 + T_2)}{2}$$

Then

$$\frac{WC_p}{hA}(T_2 - 426^\circ\text{F}) = \left[410^\circ\text{F} - \left(\frac{426^\circ\text{F}}{2} - \frac{T_2}{2}\right)\right]$$

The w is 317.9 g/min, C_p is 0.9 Btu/lb mass °F (from reference 23 and A is given by πDL. If the h values for ($B^* - 0.150$ and $B^* = 0.0$) 4.02 and 6.67 Btu/hr ft^2 °F are substituted the results are

$$T_2 = 425.6 \text{ (for } B^* = 0.150)$$

$$T_2 = 415.9 \text{ (for } B^* = 0.0)$$

Example 7-4 Molten polyethylene at a flow rate of 3.575 lbm/min is heated in a 1-inch inside-diameter tube 10 feet long. The polymer enters the tube (wall temperature of 430°F) at an inlet temperature of 390°F.

Conservatively, what would be the effect of neglecting viscous dissipation on the polymer's exit temperature?

Calculating the Graetz number we obtain

$$\text{Gz} = \frac{wC_p}{kL} \frac{(3.575 \text{ lbm/min})(0.606 \text{ Btu/lbm-F})}{(0.002167 \text{ Btu/min-ft-F})(10 \text{ ft})}$$

$$\text{Gz} = 100$$

Without viscous dissipation ($B^* = 0.0$ in Figure 7-3) we have

$$\text{Nu} = 7.8$$

With viscous dissipation (conservative approach is to use the data points):

$$\text{Nu} = 18.5$$

Then, for the polymer we obtain

$$wC_p(T_2 - T_1) = hA\,\{T_{\text{wall}} - [(T_2 + T_2)/2)]\}$$

$$\frac{wC_p}{hA}(T_2 - T_1) = T_{\text{wall}} - T_1/2 - T_2/2$$

$$T_2\left(\frac{wC_p}{hA} + 1/2\right) = T_{\text{wall}} + T_1\left(\frac{wC_p}{hA} - 1/2\right)$$

$$h(\text{for no dissipation}) = \frac{\text{Nu}\,k}{D}$$

$$h = \frac{(7.8)(0.002167\ \text{Btu/min-°F-ft})}{(1/12\ \text{ft})}$$

$$h = 0.4811\ \text{Btu/min-°F}$$

Likewise, with viscous heating we obtain

$$h = \frac{(18.50)(0.002167\ \text{Btu/min-°F-ft})}{(1/12\ \text{ft})}$$

$$h = 0.2028\ \text{Btu/min-°F}$$

Then by solving we obtain

$$T_2 = 396°\text{F (no dissipation)}$$

$$T_2 = 403°\text{F (conservative calculation)}$$

Example 7-5 What length of 0.0508-m outside-diameter vertical tube is required to condense 0.1827 kg/sec of saturated stream at 127.8°C if tube wall temperature is 72.2°C?

$$\Gamma = \frac{W}{\pi D} = \frac{0.1827\ \text{kg/sec}}{(0.0508)\ \text{m}} = 1.141\ \text{kg/sec m}$$

$$T_f = 127.8°\text{C} - 3/4(127.8 - 72.2)°\text{C} = 85.1°\text{C}$$

$$\mu_f = 0.00033\ \text{kg/m sec}$$

Then

$$\text{Re} = \frac{4\Gamma}{\mu_f} = \frac{(1.141\ \text{kg/sec})}{0.00033\ \text{kg/sec}} = 13{,}800$$

Thus, condensate flow is turbulent and

$$h_m = 0.0077(\text{Re})^{0.4}\left(\frac{k_f^3\rho_f^2 g}{\mu_f^2}\right)^{1/3}$$

$$h_m = 0.007(13{,}800)^{0.4}\left(\frac{(0.673\ \text{W/m °C})^3(869.78\ \text{kg/m}^3)^2(9.8\ \text{m/sec}^2)}{(0.00033\ \text{kg/m sec})^2}\right)^{1/3}$$

$$h_m = 9581\ \text{W/m}^2\ \text{°C}$$

and by enthalpy balance (heat transferred condenses vapor) we have

$$q = \text{w}\lambda = (0.183\ \text{kg/sec})(2179.3\ \text{kJ/kg})$$

$$q = 398{,}575\ \text{W}$$

$$A = \frac{398{,}575\ \text{W}}{9581\ \text{W/m}^2\ \text{°C}\ (55.6\text{°C})} = 0.748\ \text{m}^2$$

$$L = \frac{0.748\ \text{m}^2}{\pi(0.051)} = 4.67\ \text{m}$$

Example 7-6 What is the value of h_m for dry saturated steam (100°C) condensing, outside a bank of horizontal tubes, 16 tubes high. Average temperature of the outer tube surface is 93.33°C (tube O.D. is 0.0254 m).

$$T_f - 100 - \tfrac{3}{4}(6.67) = 95\text{°C}$$

Then assuming laminar case (because turbulent flow in horizontal tubes is relatively rare), we obtain

$$h_m = 0.725\left(\frac{k_f^3\rho_f^2 g\lambda}{ND_0\mu_f\Delta T}\right)^{1/4}$$

$$h_m = 0.725\left(\frac{(0.680\ \text{W/m °C})^3(858.57\ \text{k/m}^3)^2(9.8\ \text{m/sec}^2)^2(2271\ \text{kJ/kg})}{16(0.0254\ \text{m})(0.00030\ \text{kg/m sec})(6.67\text{°C})}\right)^{1/4}$$

$$h_m = 6507\ \text{W/m}^2\ \text{°C}$$

Checking the assumption of laminar condensate flow, we obtain

$$A = 16\pi D_0 L = 16\pi(0.0254\ \text{m})L$$

$$A = 1.277L\ \text{m}^2$$

and

$$w = \frac{h_m A(\Delta T)}{\lambda} = \frac{(6507\ \text{W/m}^2\ \text{°C})(1.277L\ \text{m}^2)(6.67\text{°C})}{2271\ \text{kJ/kg}}$$

$$w = 0.0244L\ \text{kg/sec}$$

$$\Gamma' = \frac{w}{L} = \frac{0.0244L}{L} = 0.0244\frac{\text{kg}}{\text{sec m}}$$

so that

$$\frac{2\Gamma'}{\mu_f} = \frac{(2)\ (0.0244\ \text{kg/m sec})}{0.00030\ \text{kg/m sec}} = 162.7$$

This is less than 2100 and hence flow is laminar.

Example 7-7 Compare the relative heat fluxes for the following nucleate boiling cases:

Case 1: Water respectively with surfaces of copper, platinum, brass, and stainless steel for the same $(T_w - T_{sat})$

Case 2: n-Pentane with chromium and copper surfaces for the same $(T_w - T_{sat})$

Note that in equation (7-22) the only changes from surface to surface will be the C^1 values.

Then for case 1 with water we have

$$(q/A)_a/(q/A)_b/(q/A)_c/(q/A)_d = (C^1)_a^3/(C^1)_b^3/(C^1)_c^3/(C^1)_d^3$$

where a is copper, b is platinum, c is stainless steel, and d is brass.

Then, from Table 7-2 and a C^1 value of stainless steel (polished) of 0.008 we obtain

$$(q/A)_a/(q/A)_b/(q/A)_c/(q/A)_d = (0.013)^3/(0.013)^3/(0.008)^3/(0.006)^3$$

$$(q/A)_a/(q/A)_b/(q/A)_c/(q/A)_d = 2.2/2.2/0.51/0.22$$

$$(q/A)_a/(q/A)_b/(q/A)_c/(q/A)_d = 10/10/2.32/1.0$$

Hence, copper and platinum would have a flux ten times that of brass or (10/2.32) 4.3 times that of stainless steel.

For the n-pentane case using Table 7-2 and aC^1 value of 0.0154 for copper, we obtain

$$(q/A)_{\text{chromium}}/(q/A)_{\text{copper}} = (0.015)^3/(0.0154)^3$$

$$(q/A)_{\text{chromium}}/(q/A)_{\text{copper}} = 0.924$$

Hence, chromium would have a flux 92.4 percent of copper's.

Example 7-8 A tank equipped with a paddle agitator (unbaffled) is both jacketed and outfitted with a heating coil. What is the rate of heat transfer in watts if the agitator (0.5 m) is rotating at 200 rpm.? The tank surface area is 10.2 m^2. Coil surface area is 0.5 m^2. Tank diameter is 1.5 m.

The fluid is originally at 298°K and the wall temperatures (both jacket and coil) are at 350°K. Fluid properties are

$$\rho = 970 \text{ kg/m}^3, \qquad C_p = 2000 \text{ J/kg °K}$$
$$\mu(\text{at } 298°\text{K}) = 1.1 \text{ kg/m sec}, \qquad \mu_w(\text{at } 350°\text{K}) = 0.1 \text{ kg/m sec}$$
$$k = 0.18 \text{ W/m °K}$$

The Reynolds number is

$$\text{Re}^1 = \frac{(D^1)^2 N\rho}{\mu} = \frac{(0.5 \text{ m})^2(200)\ (970 \text{ kg/m}^3)}{(\text{min})(60 \text{ sec/min})(1.1 \text{ kg/m sec})}$$
$$\text{Re}^1 = 735$$

Likewise, the Prandtl number is

$$\text{Pr} = \frac{(2000 \text{ J/kg °K})(1.1 \text{ kg/m sec})}{(0.18 \text{ W/m °K})}$$
$$\text{Pr} = 12{,}220$$

The h values are

$$(h)_{\text{jacket}} = \frac{k}{D_T}(0.36)(\text{Re}^1)^{2/3}(\text{Pr})^{1/3}\left(\frac{\mu}{\mu_w}\right)^{0.21}$$

for the jacket, and

$$(h)_{\text{coil}} = \frac{k}{D_T}(0.74)(\text{Re}^1)^{2/3}(\text{Pr})^{1/3}\left(\frac{\mu}{\mu_w}\right)^{0.14}$$

for the coil.

Substituting in each case gives

$$(h)_{\text{jacket}} = \frac{0.18 \text{ W/m °K}}{(1.5 \text{ m})}(0.36)(735)^{2/3}(12{,}220)^{1/3}\left(\frac{1.1}{0.1}\right)^{0.21}$$
$$(h)_{\text{jacket}} = 134.1 \text{ W/m}^2 \text{ °K}$$
$$(h)_{\text{coil}} = \frac{0.18 \text{ W/m °K}}{(1.5 \text{ m})}(0.74)(735)^{2/3}(12{,}220)^{1/3}\left(\frac{1.1}{0.1}\right)^{0.14}$$
$$(h)_{\text{coil}} = 233.1 \text{ W/m}^2 \text{ °K}$$

Then, the total heat flux is

$$q^1 = [(134.1\ \text{W/m}^2\ ^\circ\text{K})(10.2\ \text{m}^2) + (233.1\ \text{W/m}^2\ ^\circ\text{K})(0.5\ \text{m}^2)](350^\circ\text{K} - 298^\circ\text{K})$$

$$q^1 = 77{,}187\ \text{W}$$

PROBLEMS

7-1. A pilot plant reactor uses air that passes through a 0.051-m-diameter tube (flow gives a Re of 12,000). What would be the effect of filling the pipe with 0.0127-diameter particles of alumina on heat transfer?

7-2. A multitube reactor is used to carry out an exothermic gas reaction. Catalyst is packed in 0.025-m-diameter tubes and boiling water is used in the reactor jacket. Feed and jacket temperatures are 116°C. During operation the average reactor temperature rises to 121°C a short distance into the reactor and slowly goes down to 116.2°C at the end of the reactor. Heat transfer resistance is about the same for the bed and the wall. If the tube diameter is increased to 0.038 m (same catalyst), what jacket temperature should be used to keep the peak of 121°C? Sketch both cases' temperature profiles. Estimate steam pressures.

7-3. Rock piles have been considered for thermal energy storage systems. Consider the axial flow of air (1 kg/sec; 90°C) through such a system in which 25°C spherical rocks (0.03-m diameter) are placed in a cylinder (1-m diameter, 2 m high). The bed has a void space of 0.42. Rock density and C_p are 2300 kg/m^3 and 879 J/kg °K. Calculate the total heat transfer rate.

7-4. A fixed bed is prepared by pouring alumina powder of 1.17×10^{-4} m-diameter size into a 0.140 m diameter cylinder to a depth of 0.254 m. Bed density is 1026 kg/m^3 and the temperature is 24°C. Calculate the h for the bed if atmospheric air flows through the system at G of 2.4×10^4 kg/sec m^2. The specific heat and particle density are 245 J/kg °C and 2676 kg/m^3. Calculate the heat transfer coefficient.

7-5. A pebble heater is a fixed-bed device used to heat steam and/or other gases to temperatures higher than could be obtained in units solely fabricated of metal. For gases heated to 1037°C, changing pebble size from 0.0079 m to 0.0127 m increased the maximum allowable gas flow (kg/m sec) and the overall heat transfer coefficient (J/sec m K) but decreased the bed pressure drop. Comment on this occurrence.

7-6. A polymer solution (n of 0.45; K of 130 Newtons-sec^n/m^2 at 306°K) is flowing in a 0.025-m-diameter tube. It enters at 311°K and leaves at 327°K. Wall temperature is constant at 367°K. The activation energy for K is 13.66 kJ/g-mole. Properties are

$\rho = 1050\ \text{kg/m}^3$
$C_p = 2.09\ \text{kJ/kg-°K}$
$k = 1.21\ \text{W/m °K}$

What is the flow rate of the solution if the heat exchanger is 1.52 m long?

7-7. The polymer solution in the preceding problem can be processed in one of two different systems. For each case, the solution's mass flow rate (0.5 kg/sec) and inlet, and outlet temperatures (65.6 and 26.7°C) are the same. Which pipe diameter (0.02-m, wall at 21°C or 0.01 m, wall at 5°C) will give the lower pressure drop?

7-8. Thermally softened polystyrene is heated in a circular tube (diameter of 0.03 m) that is 3 m long. The mass flow rate of the polymer is 400 g/min. What wall temperature will be needed if the polymer's average inlet and exit temperature are 490 and 513°K? Rheological data for the polymer are $n = 0.22$ and $K = 2.2 \times 10^4\ \text{N sec}^n/\text{m}^2$.

7-9. Molten polypropylene is to be cooled from 260°C to 220°C. If the heat exchanger of the previous problem is available (with the same wall temperature), what mass flow of polypropylene can be accommodated? The n and K values for the polymer are, respectively, 0.4 and $4.1 \times 10^3\ \text{N sec}^n/\text{m}^2$.

7-10. A polyethylene at 175°C (ρ_0 of 0.920 g/cm^3) is pumped into a 0.02-m-diameter pipe whose wall temperature is 120°C. At what length of pipe will the polyethylene solidify? Flow data for the polyethylene are $n = 0.48$ and $k = 6.5 \times 103\ \text{N sec}^n/\text{m}^3$. The polymer mass flow rate is 200 g/min.

7-11. Nylon 6 is flowing at an average velocity of 0.02 m/sec in a 0.03-m-diameter tube 4 m long. The tube wall temperature is 565°K and the polymer exits at 510°K. What is the polymer's inlet temperature? The n and K values for the nylon are 0.65 and $1.85 \times 10^3\ \text{N sec}^n/\text{m}^2$.

7-12. A polymer solution (n of 0.42 and k of 110 N sec^n/m^2) is flowing through a rectangular duct (width, 1 m; height, 0.03 m) for a length of 3 m. The duct's wall temperature is 358°K, and the solution's inlet and outlet temperatures are 310 and 323°K, respectively. What is the solution's average velocity in the tube? Solution properties are:

$\rho = 1050\ \text{kg/m}^3$
$C_p = 1.19\ \text{kJ/kg °K}$
$k = 1.20\ \text{W/m °K}$

7-13. A non-Newtonian fluid flows at 4.55 kg/min inside a 0.0254-m-diameter tube that is 1.52 m long. The wall is kept at 93.3°C. If the fluid enters at 37.8°C, what is its exit temperature?

The rheological properties are n of 0.40 and K of 140.18 (37.8°C) and 62.63 (93.3°C). Physical properties are density of 1042 kg/m^3, k of 0.050 cal/sec cm °C, and C_p of 0.5 cal/gm °C.

7-14. In certain types of agitated process units, scrapers are attached to the agitator. Empirical equations describing such a case are given below:

$$h_i = 2\sqrt{\frac{k\rho c_p n B}{\pi}}$$

$$\frac{h_j D_a}{k} = 4.9\left(\frac{D_a \overline{V} \rho}{\mu}\right)^{0.57}\left(\frac{c_p \mu}{k}\right)^{0.47}\left(\frac{D_a n}{\overline{V}}\right)^{0.17}\left(\frac{D_a}{L}\right)^{0.37}$$

where B is number of blades, n is the rotational speed, V is the bulk average longitudinal velocity, D_a is the scraper diameter, and L is the exchanger length. What would be the effect of changing properties (density, thermal conductivity, specific heat) or process conditions (agitator speed, agitator diameter, longitudinal velocity) as predicted by each of the above equations?

References for above are as follows: P. Harriott, *Chem. Eng. Prog. Symp. Ser. I* **166**(29), 137 (1959); and A. H. P. Skelland, *Chem. Eng. Sci. I*, 166 (1958).

7-15. Liquid styrene (60°C) is heated in a 1.83-m steam jacketed unit. The agitator is a six-bladed standard turbine. For a stirrer speed of 2.67 r/sec, calculate the heat transfer coefficient for the inner wall. How much would this value change if a pitched blade turbine was used.

7-16. A liquid (density of 961 kg/m^3, C_p of 2500 J/kg °K, k of 0.173 W/m °K) in a 1.83-cm-diameter tank at 300 K is heated by hot water in a jacket (constant wall surface temperature of 355.4°K). The agitator (flat blade turbine of 0.61-m diameter) rotates at 1.67 r/sec. What is the heat transfer coefficient? Viscosity values are 1 kg/m sec at 300 K and 0.084 kg/m sec at 355.4°K.

7-17. A 2-m-diameter agitated tank contains 6200 kg (dilute aqueous solution). The agitator (turbine of diameter 0.67 m) operates at 2.33 r/sec. If the jacket uses steam condensing at 110°C and the fluid is at 40°C, what is the heat transfer coefficient?

7-18. If in the preceding problem the heat transfer area is 14 m^2 and the steam heat transfer coefficient is 10 kW/m^2 °C, what would the rate of heat transfer be? Wall thickness is 0.01 m.

7-19. A 0.032-m-diameter tube is used to condense n-propyl alcohol at atmospheric pressure. If the cooling water inside the tube keeps the pipe outside at 25°C, contrast the amounts of condensation using vertical or horizontal tubes. Condensation temperature of the alcohol is 97.8°C. Alcohol properties at 43.3°C are: density of 779 kg/m^3, k of 0.17 W/m °K, λ of 687 kJ/kg, and viscosity of 1.4×10^{-3} kg/m sec.

7-20. Four hundred tubes (0.0064-m diameter) in a square configuration are used to condense steam at atmospheric pressure. Calculate the steam condensed per hour per unit length if the tube walls are kept at 88°C.

7-21. A vertical plate, 1.2 m high and 0.30 m wide, is kept at 70°C and exposed to atmospheric pressure steam. What is the heat transfer and total steam mass condensed per hour.

7-22. A heat exchanger is to condense 636 kg/hr using a square array of 400 tubes (0.0127-m diameter). Estimate the length of tubes needed, assuming that the tube wall temperature is maintained at 97°C.

7-23. Saturated steam (at 6.8 atm) condenses on a 0.0254-m-diameter vertical tube maintained at 138°C. Calculate the heat transfer coefficient.

7-24. What is the ratio of horizontal condensation to vertical condensation for a tube of diameter D and length L? Assume laminar flow.

7-25. What is the condensation rate of saturated steam at 1.5×10^5 pascals on a vertical tube (1 m long, 0.1-m diameter) whose surface is at 94°C?

7-26. Repeat Problem 7-25 for a horizontal tube case.

7-27. A condenser is made up of a horizontal tube bank (0.0305 m diameter; 3.048 m long). The unit condenses 11,364 kg/hr of steam at 0.136 atm pressure. If the tube walls are at 34.4°C, how many tubes are needed?

7-28. If the steam in the preceding problem condensed in dropwise fashion, how many tubes would be needed?

7-29. An electrical current passed through a long wire (0.001-m diameter) dissipates 4085 W/m. The wire is contained in water at atmospheric pressure and has a surface temperature of 128°C. Determine the boiling heat transfer coefficient and estimate the liquid–surface interaction coefficient.

7-30. A steel bar (0.02-m diameter; 0.20 m long) is raised to a temperature of 455°C and then submerged in liquid water at one atmosphere pressure. If the bar's emissivity is 0.9, estimate the initial rate of heat transfer.

REFERENCES

1. F. Yoshida, D. Ramaswami, and O. A. Hougen, *AIChE J.* **8**, 5 (1962).
2. B. Gamson, *Chem. Eng. Prog.* **47**, 19 (1951).
3. A. B. Metzner, R. D. Vaughn, and G. L. Houghton, *AIChE J.* **3**, 92 (1957).
4. A. B. Metzner and D. F. Gluck, *Chem. Eng. Sci.* **12**, 185 (1960).
5. D. R. Oliver and V. G. Jensen, *Chem. Eng. Sci.* **19**, 115 (1964).
6. I. Saltuk, N. Siskovic, and R. G. Griskey, *Polymer Eng. Sci.* **12**, 402 (1972).
7. R. G. Griskey, P. Notheis, W. Fedoriw, and S. Victor, *Proc. 33rd ANTEC Soc Plastics Eng.* **XXI**, 459 (1975).

8. T. H. Chilton, T. B. Drew, and R. H. Jebens, *Ind. Eng. Chem.* **36**, 510 (1944).
9. V. W. Uhl, *Chem. Eng. Prog. Symp.*, **51**, 93 (1953).
10. R. A. Bowman, A. C. Mueller, and W. N. Nagle, *Trans. A.S.M.E.* **62**, 283 (1940).
11. G. Brooks and G. Su, *Chem. Eng. Prog.* **55**, 54 (1959).
12. M. D. Gluz and L. S. Pavlushenko, *J. Appl. Chem. U.S.S.R.* **39**, 2323 (1966).
13. W. M. Rosenhow, *Trans ASME* **74**, 969 (1952).
14. E. L. Piret and H. S. Isbin, *AIChE Heat Transfer Symposium*, St. Louis, MO, December 1953.
15. J. N. Addams, D.Sc. dissertation, Massachusetts Institute of Technology (1948).
16. D. S. Cryder and A. C. Finalborgo, *Trans. Am. Inst. Chem. Engs.* **33**, 346 (1937).
17. M. T. Cichelli and C. F. Bonilla, *Trans. Am. Inst. Chem. Eng.* **41**, 755 (1945).
18. W. M. Rosenhow and P. Griffith, *AIChE–ASME Joint Heat Transfer Symposium*, Louisville, KY, March 1955.
19. L. A. Bromley, *Chem. Eng. Prog.* **46**, 221 (1950).
20. R. F. Westover, in *Processing of Thermoplastic Plastics*, E. C. Bernhardt, editor, Reinhold, New York (1965), p. 554.
21. P. Heydemann and H. D. Guicking, *Kolloid Z.* **193**, 16 (1963).
22. R. G. Griskey, M. Luba, and T. Pelt, *J. Appl. Polymer Sci.* **23**, 55 (1979).
23. R. G. Griskey and D. O. Hubbell, *J. Polymer Sci.* **12**, 853 (1968).
24. E. F. Fauber and R. L. Scorah, *Trans. ASME* **70**, 369 (1948).
25. F. B. Incropera and D. P. DeWitt, *Fundamentals of Heat Transfer*, John Wiley and Sons, New York (1981).

8

HEAT EXCHANGERS

INTRODUCTION

In the preceding chapters, we have considered the transfer of energy in both the conduction and convection modes. We will now develop methods of combining these different aspects so that they can be applied to process situations. In particular, we call the devices *heat exchangers*.

Heat exchangers allow thermal interchange (but not necessarily mass interchange) between flowing streams. The streams are usually separated from each other by a solid wall. Hence, we have a combination of convection heat transfer (the flowing streams) together with conduction heat transfer (through the solid surfaces).

A wide variety of heat exchanger units can be used in process work. Very often, complex flow patterns are used to enhance the heat transfer. These patterns make it necessary to use special semiempirical techniques to design such units.

The evaporator, a unit with widespread industrial use, is another case where heat transfer plays a prominent role. Again, the complex nature of many of the fluids processed in industry frequently makes it necessary to use special design techniques.

COMBINED CONVECTION AND CONDUCTION

Figure 8-1 depicts a combined conduction–convection system. Basically, the inner cylinder could be a pipe and the outer cylinder a layer of insulation. One fluid flows inside the pipe and another is outside the insulation. The temperature profiles in the solid objects and between the fluids and solid surfaces are shown.

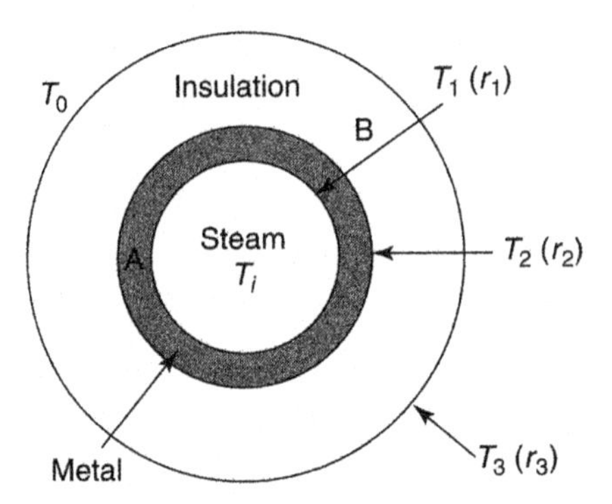

Figure 8-1. Combined conduction and convection (1).

In order to develop a combined conduction–convection system, we make use of the approach developed earlier—namely, that of the electrical analog. For such a system, the energy or heat transferred is analogous to electrical current. Furthermore, we know the heat transferred is directly related to the temperature driving force (ΔT). Finally, we also know that the heat transferred will also be directly related to the available surface area (A). On this basis, we can write an empirical equation of the form

$$q^1 = AU(T_i - T_0) \tag{8-1}$$

where U is the *overall heat transfer coefficient.*

For the system of Figure 8-1, we can write convective heat transfer expressions for the fluid inside the inner cylinder:

$$q^1 = h_i 2\pi r_1 L(T_i - T_l) \tag{8-2}$$

where h_i is the convective heat transfer coefficient and $2\pi r_1 L$ is the inner surface-area.

Likewise for the outside of the cylindrical insulation, the convective heat transfer is

$$q^1 = h_0 2\pi r_3 L(T_3 - T_0) \tag{8-3}$$

Then, for the overall system

$$\frac{1}{AU} = \frac{1}{2\pi L r_1 h_i} + \frac{\ln(r_2/r_1)}{2\pi L k_a} + \frac{\ln(r_3/r_2)}{2\pi L k_b} + \frac{1}{2\pi L h_0 r_3} \tag{8-4}$$

where the two logarithmic terms represent conduction in the appropriate cylindrical sections.

Table 8-1 Typical U Values

System	U_1(W/m^2 °C)
Air heater (molten salt to air)	34
Oil preheater	613
Reboiler (condensing steam to boiling water)	2839–4543
Steam-jacketed vessel evaporating milk	2839

Table 8-2 Typical Fouling Factors

Fluid	R_f, m^2 °C/W
Seawater (T less than)	8.81×10^{-5}
Seawater (T more than)	1.76×10^{-4}
Oil	7.10^{-4} to 8.8×10^{-4}
Refrigerant	1.76×10^{-4}

A specific designated U value can be determined if based on a given surface area. For example, if we choose A_1, the inner surface area ($A_1 = 2\pi r_1 L$), then we can transform equation (8-4) into equation (8-5):

$$U_1 = \frac{1}{\dfrac{1}{h_1} + \dfrac{r_1}{k_a} \ln\left(\dfrac{r_2}{r_1}\right) + \dfrac{r_1}{k_b} \ln\left(\dfrac{r_3}{r_2}\right) + \dfrac{r_1}{r_3 h_0}} \tag{8-5}$$

This definition could as well be based on $A_2(2\pi r_2 L)$ or $A_3(2\pi r_3 L)$. In all cases, however, see that $A_1 U_1 = A_2 U_2 = A_3 U_3$.

Some typical U values are given in Table 8-1.

Deposits can be formed on solid surfaces in heat exchange systems. In effect, such deposits add another conductive resistance to the heat transfer system. The change caused by such a deposit is accounted for by a fouling factor R_f.

$$R_f = \frac{1}{U_{\text{fouled}}} - \frac{1}{U_{\text{clean}}} \tag{8-6}$$

Some typical fouling factors are given in Table 8-2.

In practice, such factors are found by plotting are found by plotting $1/U$ versus flow rate. Clean and fouled curves will be displaced for the same flow rates.

COMPENSATING FOR TEMPERATURE AND GEOMETRY

The design of heat exchangers requires that certain approaches must be taken because of the system's geometry and flow patterns. For example, the fluids interchanging energy can be made to flow either cocurrently or counter-currently. Furthermore, the changes in temperature are such that the temperature driving force for heat transfer is changing as illustrated in Figure 8-2. In order to compensate for a changing temperature difference, we use an averaging technique based on a logarithmic approach.

In order to do this, we must use a differential approach for parallel flow heat exchangers (double pipe as in Figure 8-3).

$$dq^1 = U(T_h - T_c)\,dA \tag{8-7}$$

and

$$dq^1 = (WC_p)_{\text{hot}} dT_h = (WC_p)_{\text{cold}} dT_c \tag{8-8}$$

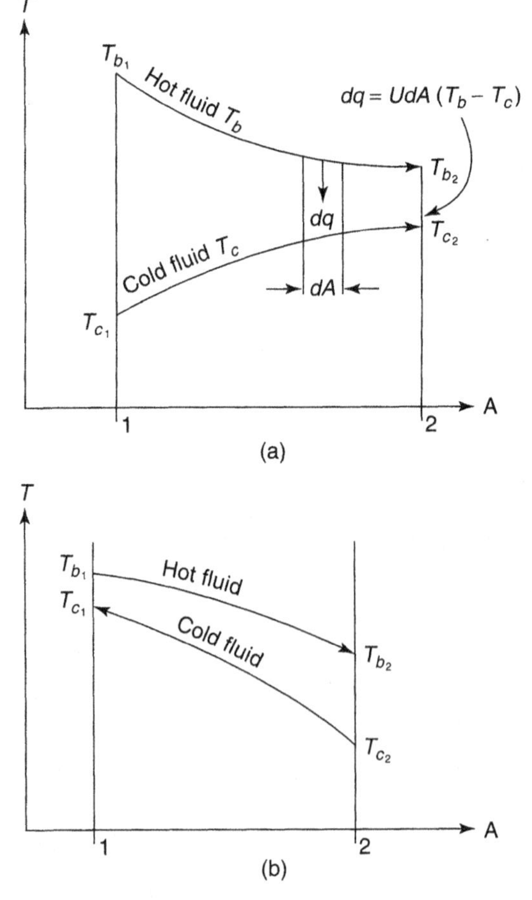

Figure 8-2. Temperature behavior in parallel (a) and counterflow (b) systems. (Reproduced with permission from reference 2. Copyright 1997, American Chemical Society.)

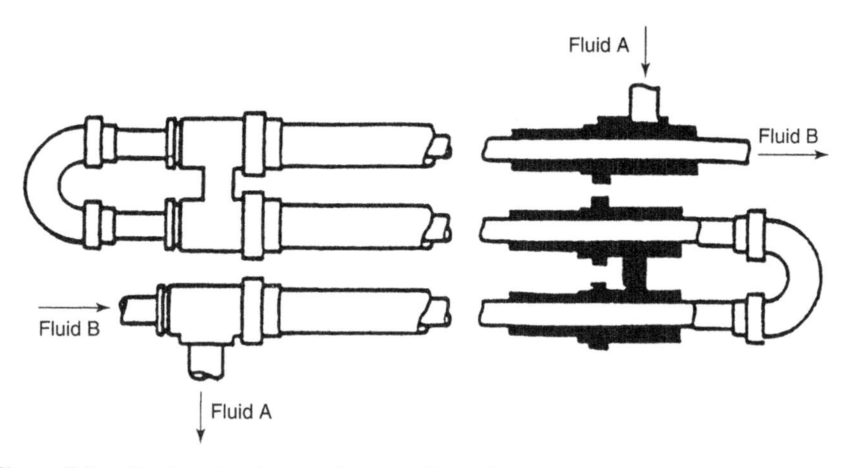

Figure 8-3. Double-pipe heat exchanger. (Reproduced with permission from reference 2. Copyright 1997, American Chemical Society.)

where the W is mass flow rate, and the h and c subscripts are hot and cold, respectively. If we equate the dq^1 values and solve

$$\frac{d(T_h - T_c)}{T_h - T_c} = -U\left(\frac{1}{(WC_P)_{\text{hot}}} + \frac{1}{(WC_p)_{\text{cold}}}\right)dA \tag{8-9}$$

then

$$\ln\left(\frac{T_{h2} - T_{c2}}{T_{h1} - T_{c1}}\right) = -UA\left(\frac{1}{(WC_p)_{\text{hot}}} + \frac{1}{(WC_p)_{\text{cold}}}\right) \tag{8-10}$$

Finally,

$$(WC_p)_{\text{hot}} = \frac{q}{T_{h1} - T_{h2}} \tag{8-11}$$

$$(WC_p)_{\text{cold}} = \frac{q}{T_{c2} - T_{c1}} \tag{8-12}$$

and

$$q = UA\left[\frac{(T_{h2} - T_{c2}) - (T_{h1} - T_{c1})}{\ln\left(\dfrac{T_{h2} - T_{c2}}{T_{h1} - T_{c1}}\right)}\right] \tag{8-13}$$

The temperature terms in the bracket are known as the log mean temperature difference. Thus

$$q^1 = UA\Delta T_{lm} \tag{8-14}$$

In cases, other than parallel flow (i.e., cross-flow, multiple pass shell, and tube exchangers, etc.) we use the form

$$q^1 = UAF\Delta T_{lm} \tag{8-15}$$

where F is a geometrical correction factor.

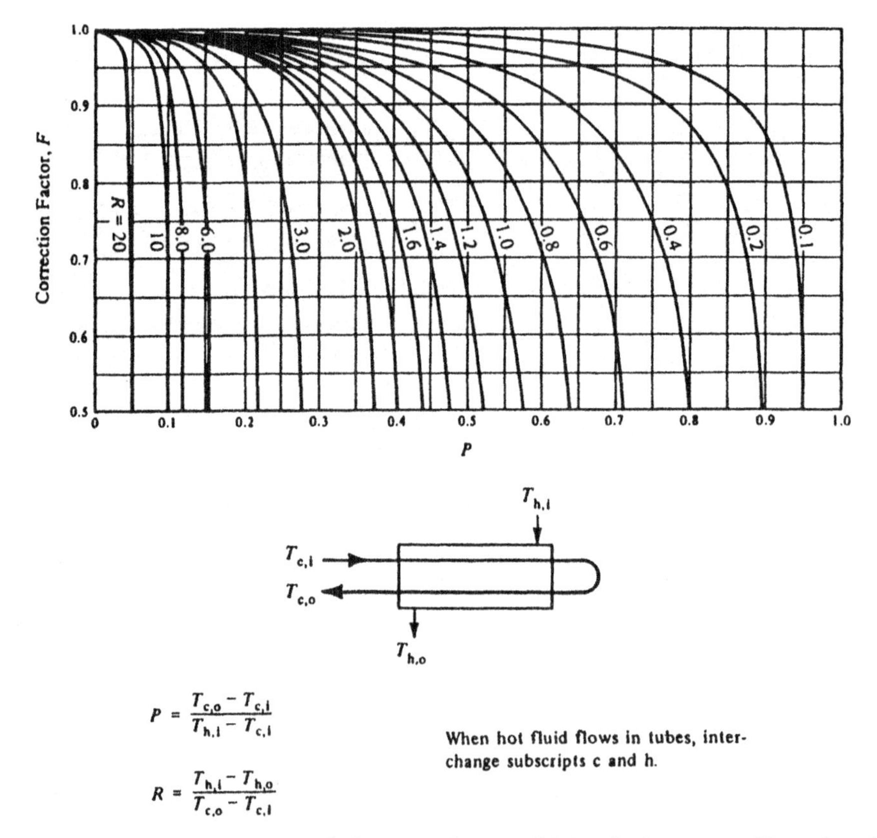

Figure 8-4. F factor for one shell pass and any multiple of tube passes. (Reproduced with permission from reference 3. Copyright 1940, American Society of Mechanical Engineers.)

Values of F are given for various exchangers (see Figures 8-4, 8-5, 8-6, and 8-7) related to pertinent system temperatures.

HEAT EXCHANGER DESIGN; THE EFFECTIVENESS—NTU METHODS

Equation (8-15), together with charts of the form of Figures 8-4, 8-5, 8-6, and 8-7, can be used to design a heat exchanger. However, in such instances we must either know $(\Delta T)_{lm}$ or be able to easily find it.

If we do not know $(\Delta T)_{lm}$, the process becomes quite difficult, requiring an extensive trial and error. In order to get around this problem, we use a technique based on effectiveness defined as

$$\varepsilon = \text{Effectiveness} = \frac{\text{Actual heat transfer}}{\text{Maximum possible heat transfer}} \tag{8-16}$$

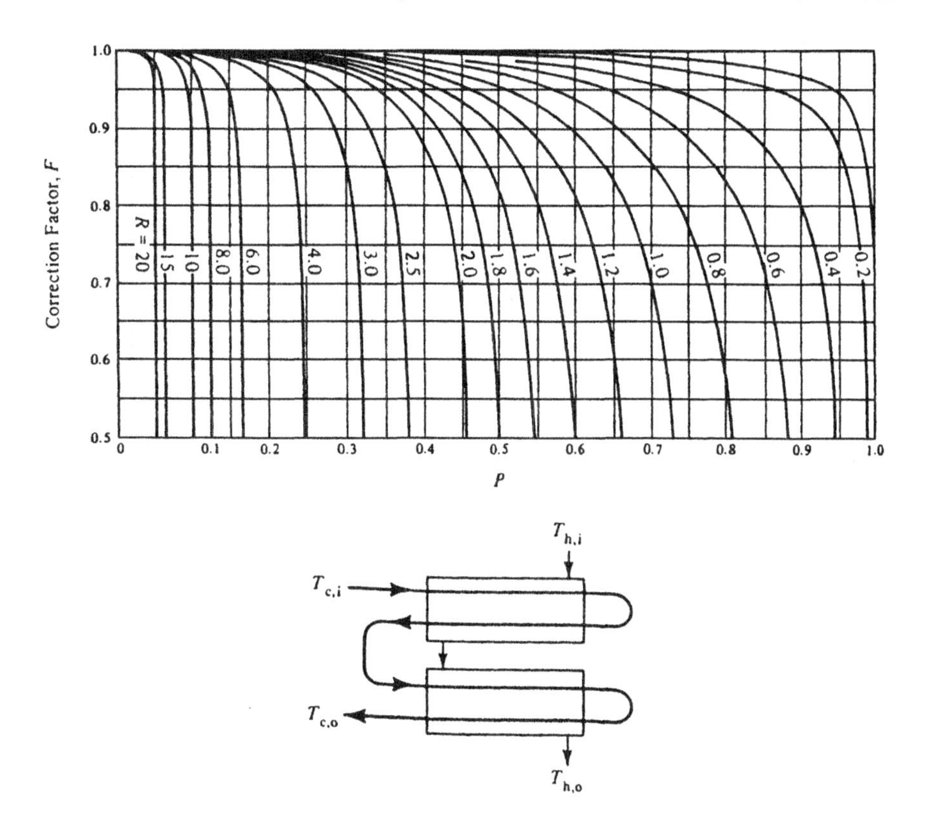

$P = \frac{T_{c,o} - T_{c,i}}{T_{h,i} - T_{c,i}}$ $R = \frac{T_{h,i} - T_{h,o}}{T_{c,o} - T_{c,i}}$ When hot fluid flows in tubes, interchange subscripts c and h.

Figure 8-5. *F* factor for two shell passes and any multiple of tube passes. (Reproduced with permission from reference 3. Copyright 1940, American Society of Mechanical Engineers.)

For parallel flow heat exchangers we have

$$q = (WC_p)_h(T_{h1} - T_{h2}) = (WC_p)_c(T_{c1} - T_{c2}) \tag{8-17}$$

and for counterflow heat exchangers we obtain

$$q = (WC_p)_h(T_{h1} - T_{h2}) = (WC_p)_c(T_{c1} - T_{c2}) \tag{8-18}$$

We can define one fluid as having a *maximum temperature change*. This fluid will then have a *maximum value* of (WC_p) because of the energy balance. Then

$$q_{\text{maximum}} = (WC_p)_{\text{minimum}}(T_h \text{ inlet} - T_c \text{ inlet}) \tag{8-19}$$

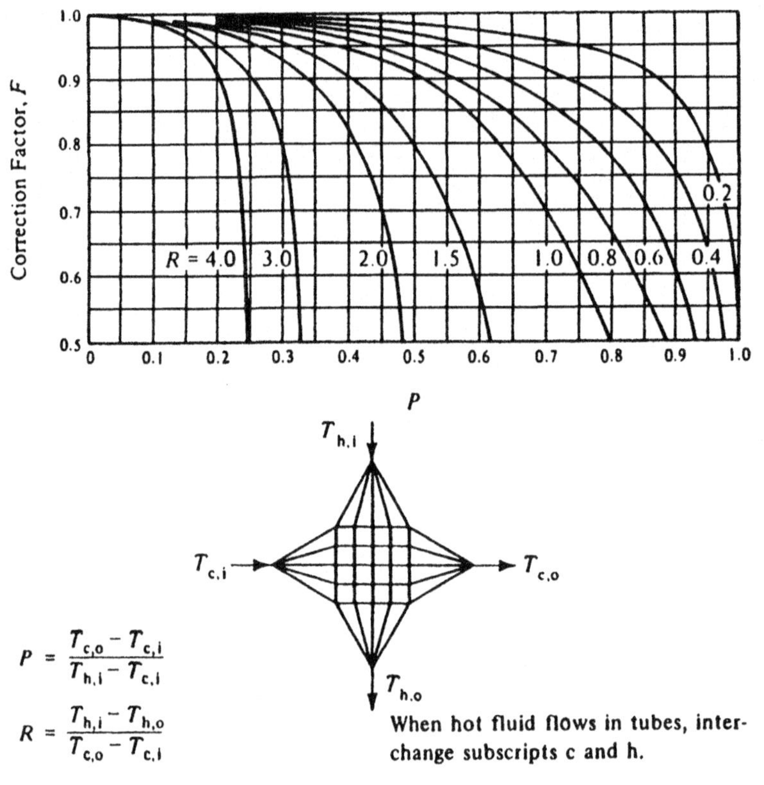

Figure 8-6. F factor for single pass cross-flow exchanger (both fluids unmixed). (Reproduced with permission from reference 3. Copyright 1940, American Society of Mechanical Engineers.)

Based on this the effectiveness, ε, is then (for parallel flow)

$$\varepsilon_h = \frac{T_{h1} - T_{h2}}{T_{h1} - T_{h1}} \tag{8-20}$$

$$\varepsilon_c = \frac{T_{c2} - T_{c1}}{T_{h1} - T_{c1}} \tag{8-21}$$

Manipulation of equation (8-10), together with the effectiveness approach, gives us the following type of solution:

$$\varepsilon = \frac{1 - \exp\left[\dfrac{-UA}{\dfrac{1}{(WC_p)_{\min}} + \dfrac{1}{(WC_p)_{\max}}}\right]}{1 + \dfrac{(WC_p)_{\min}}{(WC_p)_{\max}}} \tag{8-22}$$

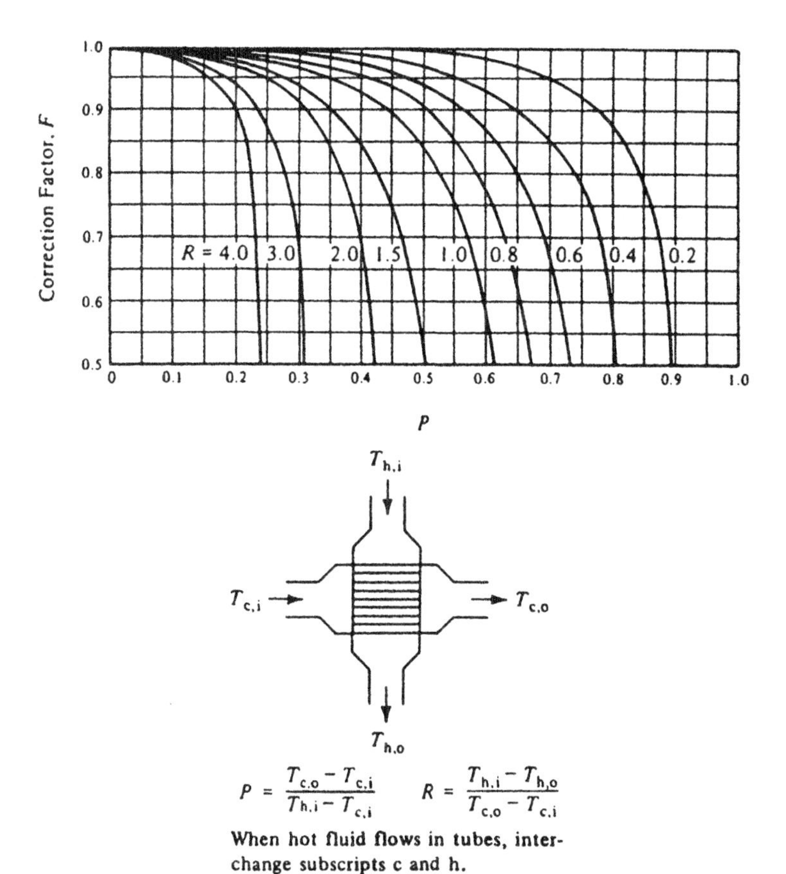

Figure 8-7. *F* factor for single pass cross-flow exchanger (one fluid mixed). (Reproduced with permission from reference 3. Copyright 1940, American Society of Mechanical Engineers.)

The overall result is a series of solutions for ε in terms of the ratios of

$$\frac{(WC_p)_{\min}}{(WC_p)_{\max}} \quad \text{and} \quad \frac{AU}{(WC_p)_{\min}}$$

The first term is also known as $C_{\min}/C_{\max}$ and R, while the second term is referred to as NTU. Solutions are shown for various systems in Figures 8-8 through 8-13 as well as in Table 8-3.

The heat transfer surfaces and inlet/outlet temperatures are important inputs to heat exchanger design. There are, however, additional factors that must also be considered. There include:

1. Materials of construction
2. Friction flow losses

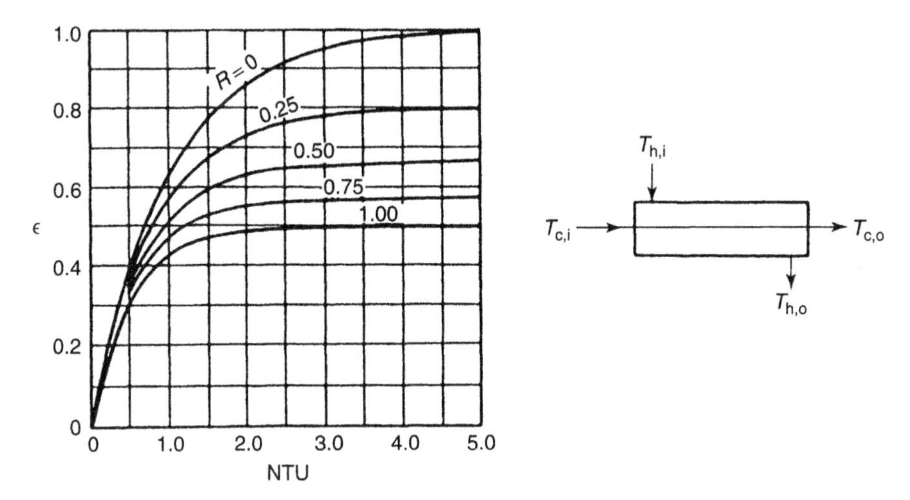

Figure 8-8. Parallel flow heat exchanger effectiveness relation (4). (With permission of W. M. Kays.)

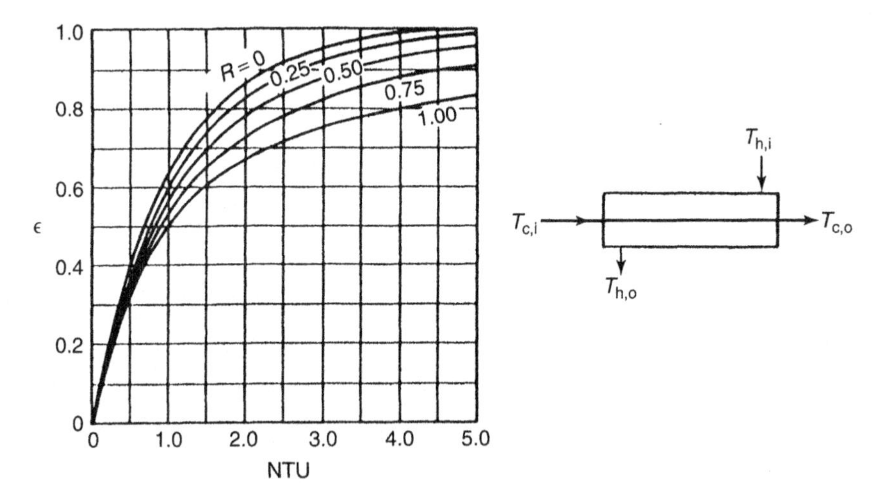

Figure 8-9. Counterflow heat exchanger effectiveness relation (4). (With permission of W. M. Kays.)

3. Deposits; fouling
4. Ease of maintenance
5. Economic aspects

The question of fouling and deposits has been discussed earlier in this chapter where the concept of reduced heat transfer (i.e., additional thermal resistance) was discussed. Dealing with such situations involves, where possible, preventing

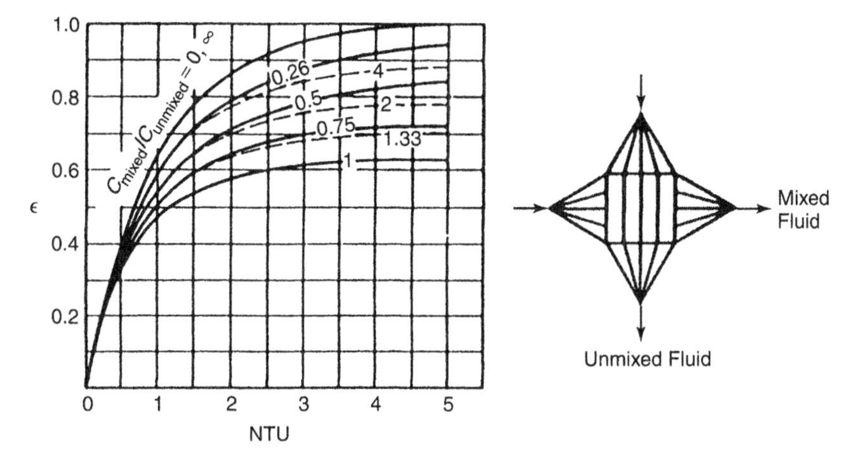

Figure 8-10. Cross-flow exchanger effectiveness relation (one fluid mixed) (4). (With permission of W. M. Kays.)

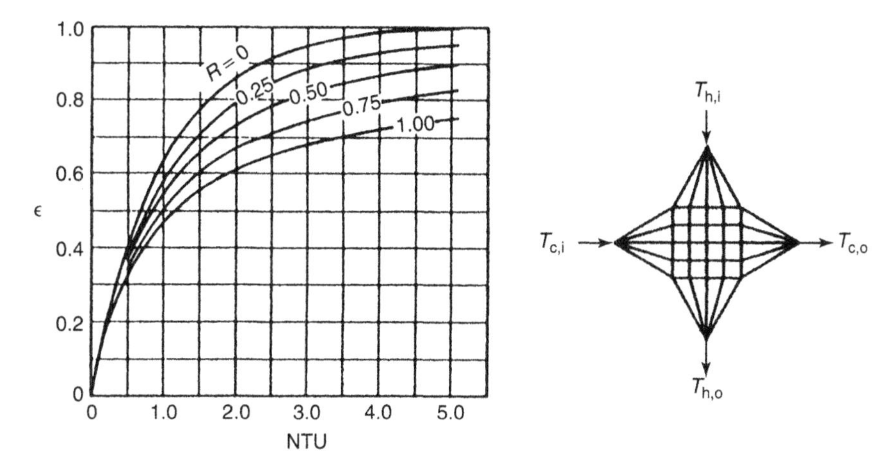

Figure 8-11. Cross-flow exchanger (both fluids unmixed) effectiveness relation (4). (With permission of W. M. Kays.)

the occurrence of fouling and deposits on heat transfer surface. This general area is one of great importance to heat transfer practitioners.

The potential for attack by corrosive fluids is something that must always be taken into consideration for design. Solutions to the problem include the uses of alloys, coated surfaces, or nonmetallics. In another view, the physical configuration of the unit becomes important. Here the flow pattern and the frictional losses become important aspects. For example, if an exchanger does the job thermally but incurs inordinately high pressure drops, it then becomes an inappropriate design.

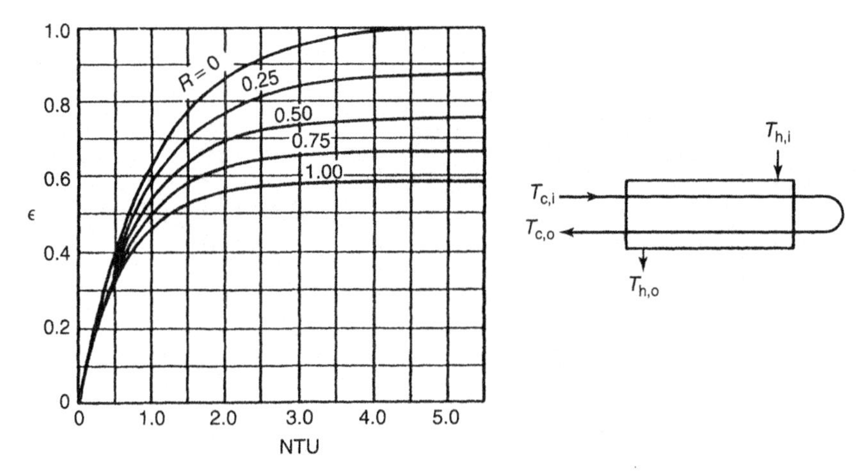

Figure 8-12. Relation for effectiveness (1–2 parallel counterflow exchangers) (4). (With permission of W. M. Kays.)

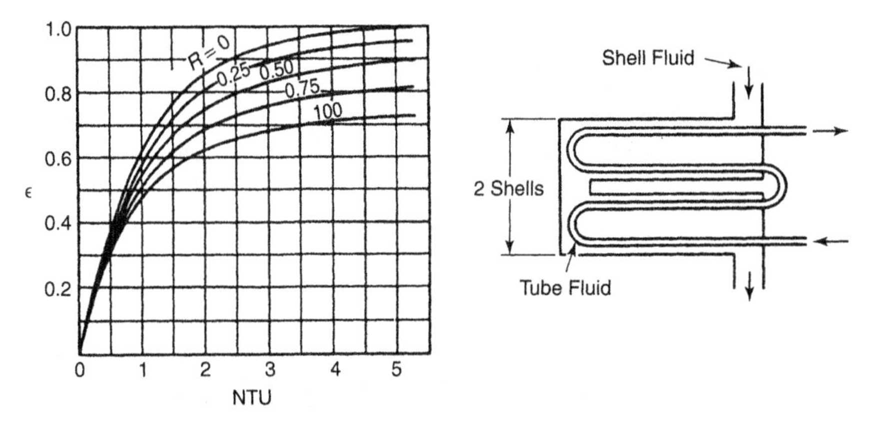

Figure 8-13. Relation for effectiveness of 2–4 multipass counterflow exchanger (4). (With permission of W. M. Kays.)

An area related to flow itself is the requirement of ease of maintenance. If a complicated unit is designed that does not allow appropriate maintenance, then serious problems may be the result. All of the design factors must then be put together in such a way as to optimize the capital and operational costs and hence satisfy the economic aspects.

The augmentation or enhancement of heat transfer is a specialized topic that is frequently utilized in design. Here, physical changes in the system are used to increase the rate of heat transfer. A widely used technique, the finned tube, is shown in Figure 8-14. The fins, which act both as conducting "fingers" and enhancers of fluid mixing, bring about improved heat transfer.

Table 8-3 Heat Exchanger Effectiveness Relations (2, 3)

System	Effectiveness Relation
Double pipe	
Parallel flow	$\varepsilon = \dfrac{1-\exp[-N(1+C)]}{1+C}$
Counter flow	$\varepsilon = \dfrac{1-\exp[-N(1-C)]}{1-C\exp[-N(1-C)]}$
Cross-flow	
Both fluids unmixed	$\varepsilon = 1-\exp\left[\dfrac{\exp(-NCn)-1}{Cn}\right]$ where $n = N^{-0.22}$
Both fluids mixed	$\varepsilon = \left[\dfrac{1}{1-\exp(-N)} + \dfrac{C}{1-\exp(-NC)} - \dfrac{1}{N}\right]^{-1}$
C_{max} mixed, C_{min} unmixed	$\varepsilon = (1/C)\{1-\exp[C(1-e^{-N})]\}$
C_{max} unmixed, C_{min} mixed	$\varepsilon = 1-\exp\{(1/C)[1-\exp(-NC)]\}$
Shell and tube	
One shell pass; 2, 4, 6 tube passes	$\varepsilon = 2\left\{1+C+(1+C^2)^{1/2}\dfrac{1+\exp[-N(1+C^2)^{1/2}]}{1-\exp[-N(1+C^2)^{1/2}]}\right\}^{-1}$

where N is NTU or $UA/(WC_p)_{min}$ and $C = (WC_p)_{min}/(WC_p)_{max}$.

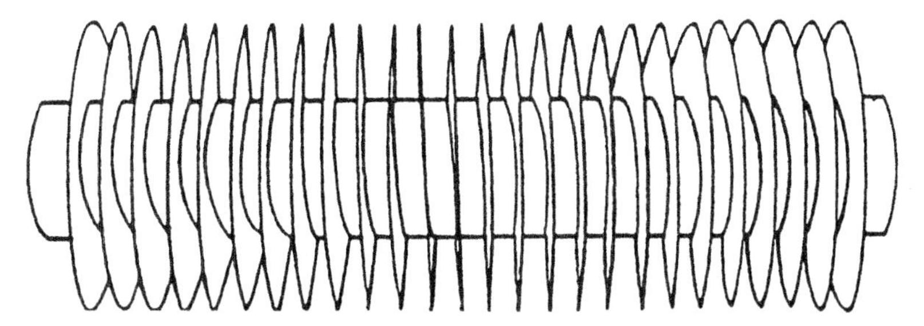

Figure 8-14. Finned tube heat exchanger. (Reproduced with permission from reference 2. Copyright 1997, American Chemical Society.)

Compact heat exchangers (Figure 8-15) represent another method of augmenting heat transfer.

A variety of other ways have been used to enhance heat transfer. For example, grooves (or, in some cases, rifled grooves) have been cut into tube surfaces. In other cases, vibration has been used as an aid to heat transfer.

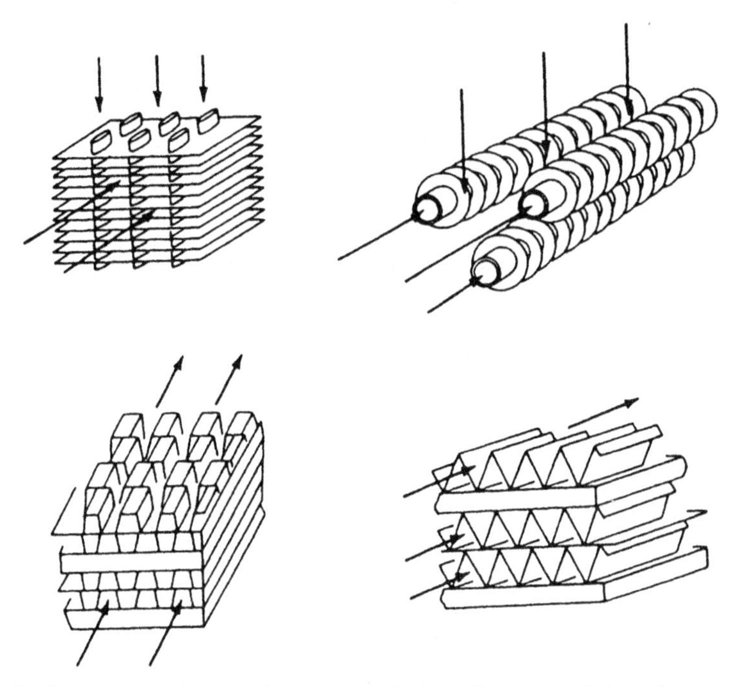

Figure 8-15. Compact heat exchangers are devices that use a high surface-to-volume ratio to augment heat transfer (4). (With permission of W. M. Kays.)

WORKED EXAMPLES

Example 8-1 Water at 98°C flows through a steel pipe (inside diameter of 0.05 m, outside diameter of 0.06 m) at a velocity of 0.25 m/sec. The horizontal pipe is in contact with atmospheric air at 20°C. What is the overall heat transfer coefficient for the system based on the pipe's outside surface area?

At 98°C the properties of water are

$$\mu = 2.82 \times 10^{-4} \text{ kg/m sec}, \qquad \rho = 960 \text{ kg/m}^3$$

$$K = 0.68 \text{ W/m °C}, \qquad C_p = 4244 \text{ J/kg °C}$$

The Reynolds number for the flowing water is

$$\text{Re} = \frac{\text{DV}_\rho}{\mu} = \frac{(0.05 \text{ m})(0.25 \text{ m/sec})(960 \text{ kg/m}^3)}{2.82 \times 10^{-4} \text{ kg/m sec}}$$

$$\text{Re} = 42{,}550$$

Let us assume that the inside wall temperature (T_i) is close to 98°C. If this is the case, then the viscosity correction term can be ignored and we can use the relation

$$\frac{h_i D}{k} = 0.023 \text{ Re}^{0.8} \text{ Pr}^{0.33}$$

for the water.

We solve as follows:

$$h_i = \frac{(0.023)(0.68 \text{ W/m °C})}{(0.05 \text{ m})}(42{,}550)^{0.8}(1.76)^{1/3}$$

$$h_i = 1906 \text{ W/m}^2 \text{ °C}$$

On the outside, we have free convection transfer to a horizontal tube. In the laminar flow case we have

$$h_m = 0.518\frac{k_f}{D}(\text{Gr}_f \text{Pr}_f)^{1/4}$$

For air at atmospheric pressure (i.e., ideal gas behavior) it can be shown that

$$h_m = 1.32\frac{(\Delta T)^{1/4}}{D}$$

where $\Delta T = (T_0 - T_{\text{air}})$, so that

$$h_m = 1.32\frac{(T_0 - T_{\text{air}})^{1/4}}{D}$$

Furthermore, all of heat fluxes are the same (i.e., for water, steel wall, and air). Hence, for a 1-m length of pipe we obtain

$$h_i \pi D_i (T_{\text{water}} - T_i) = \frac{\pi k (T_i - T_0)}{\ln(r_0/r_i)} = h_m \pi D_0 (T_0 - T_{\text{air}})$$

If the simplified h_m equation is combined with the h_m term above, then

$$h_m \pi D_0 (T_0 - T_{\text{air}}) = \pi D_0^{3/4} 1.32 (T_0 - T_{\text{air}})^{5/4}$$

Then we can obtain two equations in T_i and T_0

$$(T_{\text{water}} - T_i) = 7.55(T_i - T_0)$$

$$(T_i - T_0) = 1227(T_0 - T_{\text{air}})^{5/4}$$

If we solve by trial and error (since $T_{\text{water}} = 98$°C and $T_{\text{air}} = 20$°C).

We find that $T_i = 97.65$°C and $T_0 = 97.6$°C. This means that the original assumption for h_i is correct. Furthermore, h_m will be given by

$$h_m = \frac{(1.32)(97.6 - 20)^{1/4}}{(0.06)^{1/4}} = 7.91 \text{ W/m}^2 \text{ °C}$$

Finally, U_0 will be

$$U_0 = \frac{1}{\dfrac{D_0}{D_i h_i} + \dfrac{D_0 \ln(D_0/D_i)}{2k} + \dfrac{1}{h_0}}$$

$$U_0 = \frac{1}{6.30 \times 10^{-4} + 1.22 \times 10^{-4} + 0.126}$$

$$U_0 = 7.86 \text{ W/m}^2 \text{ °C}$$

Note that the controlling portion of the system is the free convection segment. In other words, the water and the metal wall have excellent heat transfer while the air side is poor.

Example 8-2 A shell and tube heat exchanger is used to heat water (in the tube side) from 30°C to 45°C at a mass flow rate of 4 kg/sec. The fluid used for heating (shell side) is water (entering temperature of 90°C) with a mass flow rate of 2.0 kg/sec. A single shell pass is utilized. The overall heat transfer coefficient (based on inside tube area) is 1390 W/m^2 °C.

Tubes are 1.875 cm (inside diameter) and require an average water velocity of 0.375 m/sec. Available unit floor space limits the tube length to 1.75 m.

For this heat exchanger find the number of passes, tubes per pass, and tube length.

As a start, we find the heating fluid's exit temperature from the First Law of Thermodynamics:

$$W_c (C_p)_c \Delta T_c = W_h (C_p)_h \Delta T_h$$

$$\Delta T_h = \frac{(4.0 \text{ kg/sec})(4174 \text{ J/kg °C})(15\text{°C})}{(2.0 \text{ kg/sec})(4179 \text{ J/kg °C})}$$

Note that the C_p for the hot fluid was taken to be for an average temperature using an estimated ΔT of 30°C.

$$\Delta T_h = 30\text{°C}$$

Actually, C_p could have been taken to be the same for both cases.

The $(T_h)_{\text{exit}}$ is then (90 – 30)°C, or 60°C.

Heat transferred for the system is

$$q^1 = (W C_p \Delta T)_c$$

$$q^1 = (4.0 \text{ kg/sec})(4174 \text{ J/kg °C})(15\text{°C})$$

$$q^1 = 250.4 \text{ kW}$$

Next we determine the logarithmic mean temperature for a counterflow system

$$(\Delta T)_{lm} = \frac{(90 - 45) - (60 - 30)}{\ln[(90 - 45)/(60 - 30)]}$$

$$(\Delta T)_{lm} = 37^{\circ}\mathrm{C}$$

The total required surface area for the exchanger will be

$$A = \frac{q^1}{U(\Delta T)_{lm}} = \frac{250.4 \text{ kW}}{1390\dfrac{\text{W}}{\text{m}^2\ {}^{\circ}\text{C}}(37^{\circ}\text{C})}$$

$$A = 4.87 \text{ m}^2$$

Next, using the relation between mass flow rate and average velocity, we can find the cross-sectional area:

$$w = \rho(\text{C.S.A.})\ \text{V}$$

$$\text{C.S.A.} = \frac{4.0 \text{ kg/sec}}{(993 \text{ kg/m}^3)(0.37 \text{ m/sec})}$$

$$\text{C.S.A.} = 0.0107 \text{ m}^2$$

Then, the number of tubes required is

$$n = \frac{0.0107 \text{ m}^2}{\dfrac{\pi}{4}(0.01875 \text{ m})^2}$$

$$n = 38.75 \text{ tubes}$$

which we round off to 39 tubes.

The tube length can be calculated by

$$L = \frac{\text{Total surface area}}{n\pi D}$$

$$L = \frac{4.87 \text{ m}^2}{(39)\ (\pi)(0.01875)} = 2.12 \text{ m}$$

This exceeds the length limit of 1.875 m. In order to get a workable system, we go to a two-pass system. Using Figure 8-4 with the parameters of

$$P = \frac{t_2 - t_1}{T_1 - t_2} = \frac{45 - 30}{90 - 30} = 0.25$$

$$R = \frac{90 - 60}{45 - 30} = 2.0$$

we obtain an F value of 0.945.

Then, the new required heat transfer surface is

$$A = \frac{250.4 \text{ kW}}{(1390 \text{ W/m}^2\ ^\circ\text{C})(0.945)(37^\circ\text{C})}$$

$$A = 5.15 \text{ m}^2$$

and the length L is

$$L = \frac{5.15 \text{ m}^2}{(2)(39)(\pi)(0.01875)} = 1.12 \text{ m}$$

The length is acceptable. The final system, therefore, involves two passes of 39 tubes with a length of 1.12 m.

Example 8-3 A given heat exchanger can be designed to have a U value of 2270 W/m^2 °C with a variety of different configurations (double pipe parallel; double pipe counterflow; cross-flow with both fluids unmixed and cross-flow with one fluid mixed). Find the heat transfer surfaces for each case if equal flow rates of water (1.26 kg/sec) are used and one fluid is cooled from 94°C to 72°C while the other fluid is initially at 38°C.

The final temperature of the 38°C fluid can be found from the First Law of Thermodynamics:

$$W_C(C_p)_C(T_{\text{final}} - 38)^\circ\text{C} = W_H(C_p)_H(94 - 72)^\circ\text{C}$$

But, $W_C = W_H$; also the C_p's are essentially the same. Hence, T_{final} for the cold fluid should be about 60°C. Using the value to determine the C_p and then calculating the T_{final}, we get

$$T_{\text{final}} = 38^\circ\text{C} + \frac{(4196 \text{ J/kg }^\circ\text{C})}{(4176 \text{ J/kg }^\circ\text{C})}(22^\circ\text{C})$$

$$T_{\text{final}} = 60.1^\circ\text{C}$$

Next, we calculate $(\Delta T)_{lm}$ for the system

$$(\Delta T)_{lm} = \frac{(94 - 38)^\circ\text{C} - (72 - 60.1)^\circ\text{C}}{\ln[(94 - 38)/(72 - 60.1)]}$$

$$\Delta T_{lm} = 28.5^\circ\text{C}$$

Then heat transferred is given by

$$q^1 = (1.26 \text{ kg/sec})(4196 \text{ J/kg }^\circ\text{C})(22^\circ\text{C})$$

$$q^1 = 116.3 \text{ kW}$$

Now consider each case in order.

For a *parallel-flow double-pipe system* we have

$$A = \frac{q}{U(\Delta T)_{lm}} = \frac{116{,}300 \text{ W}}{(2270 \text{ W/m}^2 \text{ °C})(28.5\text{°C})}$$

$$A = 1.80 \text{ m}^2$$

In the double-pipe counterflow case, ΔT is 34°C and

$$A = \frac{116{,}300 \text{ W}}{(2270 \text{ W/m}^2 \text{ °C})(34\text{°C})}$$

$$A = 1.51 \text{ m}^2$$

Next for the *crossflow exchanger with both fluids unmixed*, we use Figure 8-6 to get F:

$$P = \frac{(98 - 72)}{(60.1 - 38)} = 0.996$$

$$R = \frac{(60.1 - 38)}{(98 - 72)} = 0.37$$

The F value is 0.953. Hence A is given by

$$A = \frac{116{,}300 \text{ W}}{(2270 \text{ W/m}^2 \text{ °C})(0.953)(28.5\text{°C})}$$

$$A = 1.89 \text{ m}^2$$

Finally for the *cross-flow exchanger with one fluid mixed* the temperature parameters are the same. The F value (from Figure 8-7) is 0.938.

$$A = \frac{116{,}300 \text{ W}}{(2270 \text{ W/m}^2 \text{ °C})(0.938)(28.5\text{°C})}$$

$$A = 1.92 \text{ m}^2$$

Example 8-4 A cross-flow heat exchanger (unmixed fluids) is used as an air heater. Hot water at 80°C flowing in tubes is used to heat the air. The exchanger's U value is 220 W/m^2 °C. Available surface area is 10 m^2.

For an air flow rate of 2.25 m^3/sec and a temperature change from 15°C to 30°C, determine the water's exit temperature.

The density of the inlet air is 1.223 kg/m^3 (at 15°C). Hence, the mass flow rate of the cold fluid (air) is

$$W_c = (1.223 \text{ kg/m}^3)(2.25 \text{ m}^3\text{/sec})$$

$$W_c = 2.75 \text{ kg/sec}$$

Also, the heat transferred is then

$$q^1 = (W_c)(C_p)_c \Delta T_c$$
$$q^1 = (2.75\text{ kg/sec})(1006\text{ J/kg }^\circ\text{C})(15^\circ\text{C})$$
$$q^1 = 41520\text{ W}$$

Now we must choose a minimum fluid. If we select water, a trial-and-error solution will be required. Hence, we begin with the choice of air.

$$(WC_p)_{\min} = (2.75\text{ kg/sec})(1006\text{ J/kg }^\circ\text{C})$$
$$(WC_p)_{\min} = 2767\text{ W/}^\circ\text{C}$$

Also,

$$\frac{UA}{(WC_p)_{\min}} = \frac{(220\text{ W/m}^2\ ^\circ\text{C})(10\text{ m}^2)}{2767\text{ W/}^\circ\text{C}}$$

and the effectiveness ε is

$$\varepsilon = \frac{(30-15)^\circ\text{C}}{(80-15)^\circ\text{C}} = 0.230$$

Next we consult Figure 8-11. The point of $\varepsilon = 0.230$ and $AU/(WC_p)_{\min} = 0.795$ does not fall on any of the curves. This indicates that water is the minimum fluid. The following relation holds:

$$(WC_p)_{\max} = 2767\text{ W/}^\circ\text{C}$$
$$(\Delta T)_h = \frac{41{,}250}{(WC_p)_{\min}}$$
$$\varepsilon = \frac{\Delta T_h}{80-15} = \frac{\Delta T_h}{65}$$

Our procedure is to assume a value of $(WC_p)_{\min}/(WC_p)_{\max}$ and then calculate $(WC_p)_{\min}$. This can be used to find $\text{AU}/(WC_p)_{\min}$. This value with the $(WC_p)_{\min}/(WC_p)_{\max}$ will (from Figure 8-11) give an ε value that can be compounded to a value calculated from the above ε equation.

Then, we obtain the following:

$\frac{(WC_p)_{\min}}{(WC_p)_{\max}}$	$(WC_p)_{\min}$	$\Delta T_h\,^\circ$C	ε From Chart	ε Calculated
0.4	1107	37.26	0.74	0.57
0.3	830.1	49.69	0.84	0.76
0.25	691.5	59.65	0.90	0.92
0.26	719.4	57.34	0.885	0.88

Therefore take the WC_p ratio as 0.258, and the WC_p value for water is 713.9. From the First Law of Thermodynamics the T_{exit} for the water is

$$T_{exit} = 80^\circ C - \frac{41{,}250 \text{ W}}{713.9 \text{ W}/^\circ C}$$

$$T_{exit} = 22.2^\circ C$$

Example 8-5 A shell-and-tube heat exchanger (two shell passes and four tube passes) uses ethylene glycol (C_p of 2742 J/kg °C), which goes from 130°C to 70°C and has a mass rate of 1.25 kg/sec. The water enters at 32°C and leaves at 82°C and U for the system is 830 W/m^2 °C. What is the heat transfer surface needed? Also, find the mass flow rate of the water.

The water mass flow rate can be found from the First Law of Thermodynamics:

$$W_W(4175 \text{ J/kg }^\circ C)(82 - 32)^\circ C = (1.25 \text{ kg/sec})(2742 \text{ J/kg }^\circ C)(130 - 70)^\circ C$$

$$W_W = 0.985 \text{ kg/sec}$$

The WC_p values are then

$$(WC_p)_w = (0.985 \text{ kg/sec})(4175 \text{ j/kg }^\circ C)$$

$$(WC_p)_w = 4112 \text{ J}/^\circ C \text{ sec}$$

$$(WC_p)_G = (1.25 \text{ kg/sec})(2742 \text{ J/kg }^\circ C)$$

$$(WC_p)_G = 3428 \text{ J}/^\circ C \text{ sec}$$

Then, the WC_p ratio is

$$\frac{(WC_p)_{min}}{(WC_p)_{max}} = \frac{3428}{4112} = 0.834$$

while ε is

$$\varepsilon = \frac{130 - 70}{130 - 32} = 0.612$$

Then from Figure 8-13 we have

$$\frac{UA}{(WC_p)_{min}} = 1.57$$

$$A = \frac{(1.57)(3428 \text{ J}/^\circ C \text{ sec})}{830 \text{ W/m}^2 \text{ }^\circ C}$$

$$A = 6.48 \text{ m}^2$$

Example 8-6 In a gas turbine system the regenerator is a heat exchanger that preheats combustion air by using high-temperature combustion exhaust gases.

The actual mass of fuel burned is small compared to the air mass used. Furthermore, the specific heats of both the exhaust gases and air are close in value. Using these assumption, derive effectiveness relations for the regenerator using either parallel or counterflow conditions.

We begin for parallel flow. According to equation (8-10), we obtain

$$\left(\frac{T_{h2} - T_{c2}}{T_{h1} - T_{c1}}\right) = \exp\left[-UA\left(\frac{1}{(WC_p)_{\text{hot}}}\right) + \left(\frac{1}{(WC_p)_{\text{cold}}}\right)\right]$$

but

$$(WC_p)_{\text{hot}} = (WC_p)_{\text{cold}}$$

Hence,

$$\left(\frac{T_{h2} - T_{c2}}{T_{h1} - T_{c1}}\right) = \exp\left[-\frac{2UA}{(WC_p)_{\text{cold}}}\right]$$

Then for the cold fluid the efficiency ε is

$$\varepsilon = \frac{T_{c2} - T_{c1}}{T_{h1} - T_{c1}}$$

From the First Law of Thermodynamics (because the WC_p's are the same) we have

$$T_{h2} - T_{h1} = T_{c1} - T_{c2}$$

Also,

$$\frac{T_{h2} - T_{c2}}{T_{h1} - T_{c1}} = \frac{T_{h1} - 2T_{c2} + T_{c1}}{T_{h1} - T_{c1}}$$

In turn,

$$\frac{T_{h1} - 2T_{c2} + T_{c1}}{T_{h1} - T_{c1}} = \frac{(T_{h1} - T_{c1}) + (T_{c1} - T_{c2}) + (T_{c1} - T_{c2})}{(T_{h1} - T_{c2})}$$

The right-hand side is also $1 - 2\varepsilon$, so that

$$1 - 2\varepsilon = \exp\left(-\frac{2\text{UA}}{WC_p}\right)$$

$$\varepsilon = \frac{1 - \exp\left(\frac{2UA}{WC_p}\right)}{2}$$

For the counterflow case, we use the efficiency expression together with the heat transferred:

$$\varepsilon = \frac{q^1}{q^1_{\max}} = \frac{UA(T_h - T_c)}{WC_p(T_{h1} - T_{c1})}$$

$$\varepsilon = \frac{AU}{WC_p}\frac{T_{h1} - T_{c1}}{T_{h1} - T_{c1}}$$

$$\varepsilon = \frac{AU}{WC_p}\frac{[1 + T_{c2} - T_{c1}]}{T_{h1} - T_{c2}}$$

$$\varepsilon = \frac{AU}{WC_p}[1 - \varepsilon]$$

$$\varepsilon = \frac{\dfrac{AU}{WC_p}}{1 + \dfrac{AU}{WC_p}}$$

PROBLEMS

8-1. Water at 65.6°C flows through a stainless steel tube ($k = 52$ W/m °K) of 0.0254-m inside diameter. The heat transfer coefficient for the water and the inside surface is 752.56 W/m^2 °K. Likewise, the heat transfer coefficient between the outer tube surface and the air surrounding it is 39.75 W/m^2 °K. What is the overall heat transfer coefficient, U, based on the inside diameter? Estimate U after the tube has been in use for several years.

8-2. A double-pipe heat exchanger (inside tube diameter of 0.5 m cools engine oil from 160°C to 60°C. Water at 25°C is used as a coolant. Mass flow rates are 2 kg/sec for both fluids. Estimate the length of the exchanger if the overall heat transfer coefficient is 250 W/m^2 °K.

8-3. An oil is heated in a two-pass vertical tube (0.0221-m inside diameter; 0.0254-m outside diameter; tube material is mild steel) condensing steam on the outside is used to heat the oil from 15.6 to 65.6°C. Oil properties are (at 15.6°C): A specific gravity of 0.840; thermal conductivity of 0.135 W/m °K; specific heat of 2 kJ/kg °K; and a viscosity of 0.005 Pa-sec. Oil viscosity at 65.6°C is 0.0018 Pa-sec. Steam is saturated at 3.4 atmospheres. Oil volumetric flow is 5040 gallons/hour. Velocity of oil in tubes should be 0.91 m/sec. How many and what length of tubes are required?

8-4. Benzene (C_p of 1.9 kJ/kg °K and specific gravity of 0.88) is to be cooled from 350°K to 300°K with 290°K water. The heat exchanger uses water passing through tubes (0.022-m inside and 0.025-m outside diameter). For a mass flow rate of benzene of 1.25 kg/sec calculate the tube length and minimum water quantity (water temperature not to rise above 320°K). The heat transfer coefficients for the water and benzene are 0.85 and 1.70 kW/m^2 °K, respectively.

8-5. A vertical shell and tube (0.025-m outside diameter, 0.0016-m wall thickness; 2.5-m length) is used to condense benzene at a rate of 1.25 kg/sec on the outside of the tubes. Cooling water enters the tubes at 295°K and has a velocity of 1.05 m/sec. What is the number of tubes needed for a water single pass? Benzene condensation temperature is 353°K (latent heat is 394 kJ/kg).

8-6. Water at a mass flow rate of 2.5 kg/sec is heated from 25°C to 65°C in a counterflow double-pipe exchanger by oil cooling from 138°C to 93°C (C_p of 2.1 kJ/kg °C). The exchanger is to be replaced by two smaller exchangers (equal areas) by bleeding off 0.62 kg/sec of water at 50°C. Overall heat transfer coefficients are 450 W/m^2 °C. Overall oil flow is the same but split for the two exchanger system. Find the areas of the smaller exchangers and respective oil flow rates.

8-7. A four-tube pass shell and tube heat exchanger heats 2.5 kg/sec of water from 25 to 70°C in the tubes. The heating fluid is water at 93°C (flow rate of 5 kg/sec). Overall heat transfer coefficient is 800 W/m^2 °C. If both the overall heat transfer coefficient and the hot fluid rate are constant, find the percent heat transfer reduction as a function of cold fluid mass flow rate.

8-8. An automobile air conditioner has a condenser that removes 60,000 BTU/hr from a refrigerant (auto speed 40 mph; ambient temperature of 95°F). Refrigerant temperature is 150°F. U for the finned tube exchanger is 35 BTU/hr ft^2 with an air temperature rise of 10°F. If the U varies as a 0.7 power of velocity and air mass flow directly as velocity, find the percent performance redirection of the condenser a function of velocity (40 mph to 10 mph). Refrigerant (Freon 12) temperature is constant at 150°F.

8-9. The hot fluid in a double-pipe exchanger enters at 65°C and leaves at 40°C, while the cold fluid enters at 15°C and leaves at 30°C. Is the exchanger operating at counter or parallel flow? What is the exchanger effectiveness if the cold fluid is the minimum fluid?

8-10. A shell and tube exchanger (one shell pass, two tube passes) condenses steam (0.14 bar). The 130 brass tubes (inner and outer diameters of 0.0134 and 0.0159 m) have a length of 2 m. Coolant (water) enters the tubes at 20°C with an average velocity of 11.25 m/sec. Condensation heat transfer coefficient is 13,500 W/m^2 °K. What are the water outlet temperature, the condensation steam rate and the overall heat transfer coefficient?

8-11. Water at different temperatures flows through a single pass cross-flow heat exchanger (fluids unmixed). Hot water (10,000 kg/hr) enters at 90°C and cold water (20,000 kg/hr) at 10°C. What is the cold water exit temperature if the exchanger effectiveness is 60 percent.

8-12. A tube and shell exchanger uses oil as both the hot and cold fluids. The cold fluid is heated from 37.8°C to 148.9°C with oil at 204.4°C. Both streams have equal flow rates, and they have the same viscosity at a given temperature. What streams should go to the tube or shell? If the cooler oil is at a greater pressure, where should the streams flow? Assume h shell $\sim C_p^{0.3} k^{0.7} \mu^{-0.3} (\mu/\mu_w)^{0.14}$.

8-13. An oil is to be heated in horizontal multipass heater by steam at 4.4 atmospheres. The steel tubes have inside and outside diameters of 0.016 and 0.019 m with a maximum length of 4.57 m. The oil enters at 0.91 m/sec with a volumetric flow rate of 150 gallons/min and a temperature of 37.8°C. If the oil is completely mixed after each pass, how many passes are needed to reach to exit temperature of 82.2°C?

8-14. Water at 18°C enters the shell of a two-tube pass; one-shell pass exchanger at a mass flow rate of 4000 kg/hr. Engine oil flows into the tubes at 2000 kg/hr and 150°C. The surface area of the exchanger is 14 m^2. Assume U based on the outside area is 200 W/m^2 °K. What are the fluids exit temperatures? C_p of the oil is 2600 J/kg °K.

8-15. A cross-flow (both fluids unmixed) exchanger cools air entering at 40°C (4000 kg/hr). Water entering at 5°C (4600 kg/hr) is the coolant. If the U is 150 W/m^2 °C and the surface area is 25 m^2, find both exit temperatures.

8-16. Water (7.5 kg/sec) is to be heated from 85°C to 99°C in a shell and tube exchanger. Condensing steam at 345 kN/m^2 supplies the necessary energy. The exchanger consists of one shell pass (two tube passes of 30 2.5-cm outside-diameter tubes). If U is 2800 W/m^2 °C, find the tubes length. If the tubes become fouled, what would the water's exit temperature be?

8-17. Air at 207 kN/m^2 and 200°C at 6 m/sec flows through a copper tube (2.5-cm inside diameter, length of 3 m, wall thickness of 0.8 mm. Atmospheric air (20°C) flows perpendicular to the tube (velocity of 12 m/sec). What is the air exit temperature from the tube? What would be the effect of cutting the hot air flow in half?

8-18. Hot water at 90°C and 4 m/sec flows in the inner tube of a double-pipe steel heat exchanger (inner-tube inside diameter of 2.5 cm and wall thickness of 0.8 mm; outside tubes inside diameter is 3.75 cm). Oil at 20°C flows in the annulus at 7 m/sec. If the exchanger length is 6 m calculate U.

8-19. Oil is cooled by water that flows at the rate of 0.1 kg/sec per tube through 2-m-long tubes with outside diameter of 19 mm and 1.3-mm wall thickness. The oil flows on the outside and in the opposite direction at a rate of 0.075 kg/sec with an inlet temperature of 370°K. The water's inlet temperature is 280°K. Oil and water side heat transfer coefficients are, respectively, 1.7 and 2.5 kW/m^2 °K. Oil specific heat is 1.9 kJ/kg °K.

8-20. A liquid boils at 340°K on the inside of a metal surface heated by condensing steam on the outside (constant steam to metal heat transfer coefficient of 11 kW/m^2 °K). The metal's thickness is 3 mm, and its thermal conductivity is 42 W/m °K. The inner heat transfer coefficients (i.e., metal surface to boiling liquid) are functions of the temperature differences (metal to boiling liquid) as shown below:

T(K)	h(kW/m^2 °K)
22.2	4.43
27.8	5.91
33.3	7.38
36.1	7.30
38.9	6.81
41.7	6.36
44.4	5.73
50.0	4.54

What value of steam temperature will give a maximum rate of evaporation.

8-21. A shell-and-tube exchanger (120 tubes of 22-mm inside diameter and 2.5-m length) condenses benzene at 35°K outside with 290°K inlet water temperature. The condensation rate is 4 kg/sec using a water velocity of 0.7 m/sec.

Condensing vapor coefficient (based on inside area) is 2.25 kW/m^2 °K. Heat of vaporization is 400 kJ/kg. After a period of time a resistance scale (0.0002 m^2 K/W) develops. What water velocity has to be used to maintain the same condensation rate? Assume that the water side coefficient varies with velocity raised to 0.81 power.

8-22. A shell-and-tube exchanger (25-mm outside diameter, 22-mm inside diameter) condenses benzene in the shell. Water, the coolant, flows at 0.03 m/sec and goes from 290°K to 300°K. Water side heat transfer coefficient is 850 W/m^2 °K. What will the tubing length be?

8-23. A cross-flow heat exchanger uses a bundle of 132 tubes perpendicular to the flow in a 0.6-m square duct. Water at 150°C, 0.5 m/sec velocity enters the tubes (inside and outside diameters of 10.2 and 12.5 mm). Air enters at 10°C with a volumetric flow rate of 1 m^3/sec. The heat transfer coefficient on the outside of the tubes is 400 W/m^2 °K. What are the fluid outlet temperatures?

8-24. A double-pipe exchanger has a uniform overall heat transfer coefficient. The cold fluid (0.125 kg/sec) enters at 40°C and leaves at 95°C. The hot fluid (0.125 kg/sec) enters at 210°C. Cold and hot fluids C_p's are 4200 and 2100 J/kg °K, respectively. For this system, what is the maximum heat transfer rate and the effectiveness, and should flow be co- or counter-current? Also find the ratio of required areas for both flow cases.

8-25. A shell-and-tube heat exchanger (one shell pass, two tube passes) has a heat transfer surface of 15 m and an overall heat transfer coefficient of 800 W/m^2 °K. Ethylene glycol (2 kg/sec, T of 60°C) and water (5 kg/sec, T of 10°C) are the respective fluids. Find the rate of heat transfer and the fluid exit temperatures.

8-26. Hot flue gases (200°C) heat water (2.5 kg/sec) from 35°C to 85°C. and leave at 93°C. If U is 180 W/m^2 °K, what is the exchanger's area?

8-27. A cross-flow heat exchanger uses air (mixed) at 25°C to cool water (unmixed; 0.067 kg/sec) from 99°C to 60°C. The air flow rate is 0.233 kg/sec. Find the heat transfer area if U is 80 W/m^2 °K.

8-28. Air at 30°C is used to cool air from 100°C to 55°C in a cross-flow exchanger. The cold air is unmixed and the hot air is mixed. The U is 70 W/m^2 °K. Mass flow rates for the cold and hot air are 1.94 and 0.833 kg/sec. What is the cold air exit temperature and the heat exchanger surface?

8-29. A parallel flow exchanger placed in operation has a U of 800 W/m^2 °K and a capacity of 2500 W/°K. The respective temperatures are 30°C and 200°C (cold fluid) and 360°C and 300°C (hot fluid). After a long period of operation the cold fluid only reaches 120°C and the hot fluid leaves at a temperature above 300°C. If the capacity rates are the same, what is the reason for the change?

8-30. A recuperator cools a turbine exhaust (450°C, 9 kg/sec) with pressurized hot water (150°C) in 2.5-cm tubes (i.d.) and 5-m length. The water exit temperature is limited to 210°C. If the gases flow across the tubes and U is 60 W/m^2 °K, find the number of tubes if the exchanger efficiency is 70 percent.

8-31. If condensing steam at 138°C is substituted for hot oil and a shell and tube exchanger (water two passes on tube side) is used, then repeat Problem 8-6. U for the system is 1700 W/m^2 °C.

8-32. A cross-flow exchanger is used as an air preheater. The air enters at 4.6 kg/sec at 1 atm and 20°C. The hot flue gases have a mass flow rate of 5 kg/sec (375°C). The U and A for the exchanger are 50 W/m^2 °C and 110 m^2. Calculate the rate of heat transfer and exit temperatures if both fluids are unmixed.

8-33. If one fluid of Problem 8-32 is mixed and the other is unmixed, find the heat transfer rate and exit temperatures.

8-34. A process water heater (shell and tube, one shell pass) has condensing steam (150°C) in the shell. Water (2.5 kg/sec, 40°C) enters the tubes (four passes). The water exit temperature is 120°C. Find the exchanger area if $U = 2500$ W/m^2 °C.

8-35. A recuperator uses air to air (both streams unmixed). Both flow rates are 0.5 kg/sec, and hot and cold streams enter at 40°C and 20°C, respectively. What are the exit temperatures if U is 40 W/m^2 C and area is 20 m^2.

8-36. What would the water exit temperature be if the exchanger of Problem 8-34 developed a fouling factor of 0.0002 m^2 °C/W?

8-37. Condensing steam (377°K) in the shell is used to heat oil from 300°K to 344°K (1 m/sec). The tubes have diameters of 41 and 48 mm, respectively. Oil density and specific heat are 900 kg/m^3 and 1.9 kJ/kg °K. After continued use the inside diameter is reduced to 38 mm (fouling; resistance coefficient of 0.0009 m^2 °K/W). Oil side coefficients for a 38-mm inside diameter are:

Oil Side Coefficients (W/m^2 °K)	Oil Temperature (°K)
74	300
80	311
97	322
136	333
244	344

Find the length of the tube bundle.

8-38. Normal hexane (7.5 kg/sec) from the top of a distillation column (356°K) is to be condensed at a pressure of 150 kN/m^2. The heat load is 4.5 MW. If cooling water is available at 289°K and U is 450 W/m^2 °K, specify the type and size of the required exchanger.

8-39. Compute the heat exchanger areas needed for a counterflow double-pipe exchanger and a multipass system (cold fluid two passes through tubes; hot fluid one shell pass; both in same direction). The hot and cold fluids are both water with 20 kg/sec of the warmer fluid going from 360°K to 335°K. The cold fluid (25 kg/sec) enters at 300°K. Find the surface areas for each exchanger if $U = 2$ kW/m^2 °K.

8-40. A 1-m-long co-current flow reactor is used to cool oil with water. Water temperatures are 285°K and 310°K. Oil enters at 420°K and leaves at 370°K. If oil and water mass flow rates, inlet temperatures, and other exchanger dimensions remain the same, how much longer must the exchanger be to reduce the oil exit temperature to 350°K?

8-41. A shell and tube exchanger (one shell pass) uses condensing steam at 200°C to heat 50 kg/sec of water from 60°C to 90°C (U is 4500 W/m^2 °C). What are the values of effectiveness and outlet water temperatures if the inlet stream is changed (180°C, 160°C, 140°C, and 120°C)?

REFERENCES

1. R. Fahien, *Fundamentals of Transport Phenomena*, McGraw-Hill, New York (1983).
2. R. G. Griskey, *Chemical Engineering for Chemists*, American Chemical Society, Washington, D.C. (1997).
3. R. A. Bowman, A. E. Mueller, and W. M. Nagle, *Trans. ASME* **62**, 283 (1940).
4. W. M. Kays and A. L. London, *Compact Heat Exchangers*, 3d ed., Krieger: Malabar, Florida.

9

RADIATION HEAT TRANSFER

INTRODUCTION

We have previously discussed two of the three modes of heat transfer (conduction and convection). This chapter will consider radiation, the third mode of heat transfer.

In nature, energy supplied to a system causes the submacroscopic constituents (molecules, atoms) to be transformed to a higher energy level. The normal tendency of these molecules and atoms to return to low energy levels results in the emission of energy in the form of electromagnetic radiation.

Thermal radiation in itself represents only a small segment of the electromagnetic spectrum (see Figure 9-1), which includes radio waves, infrared, ultraviolet, x-rays, and γ-rays. Thermal radiation comprises the band of wavelengths from 10^{-5} cm to 10^{-2} cm.

Radiant energy (in a vacuum) is related to the speed of light by the relation

$$c = \lambda w \tag{9-1}$$

where c is the speed of light (3×10^8 m/sec), λ is the wavelength, and w is the frequency.

In order to develop an appropriate expression for radiant heat transfer, we use the concept of the photon. This entity (zero mass and charge) has an energy, e, related to the frequency by

$$e = hw \tag{9-2}$$

where h is Planck's constant (6.625×10^{-34} J sec/molecule).

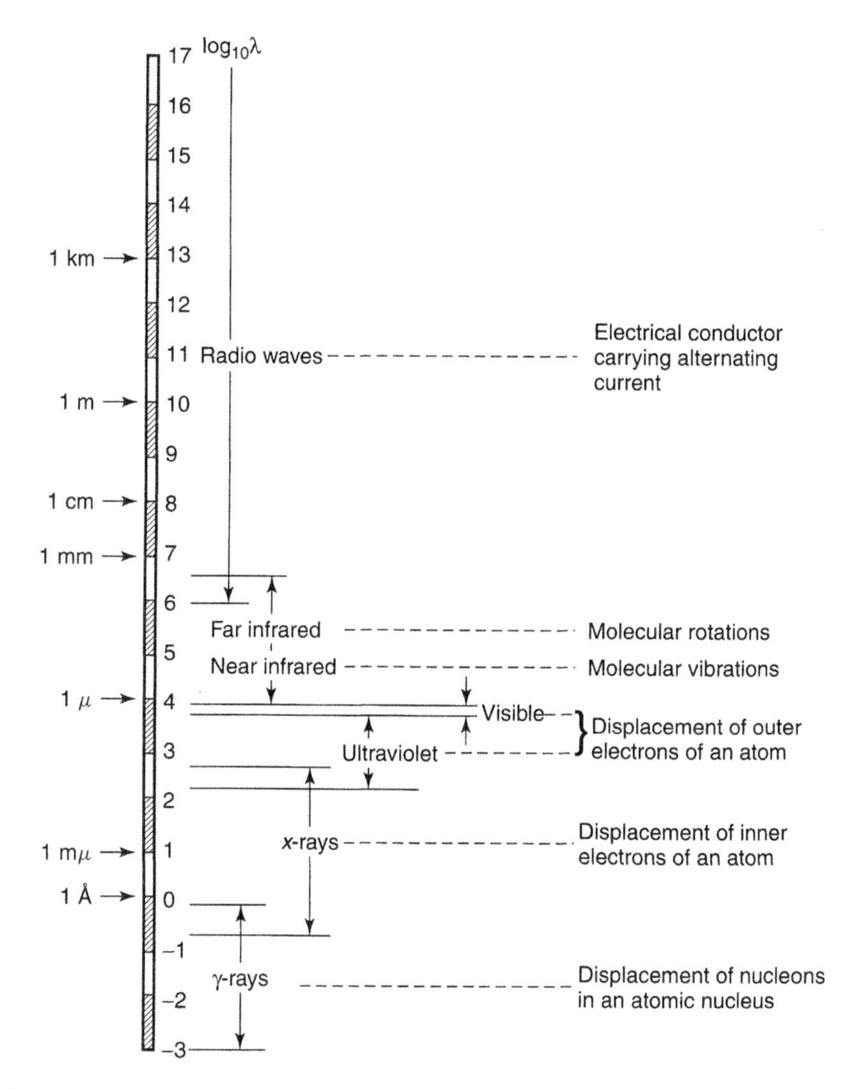

Figure 9-1. Spectrum of electromagnetic energy. (Adapted with permission from reference 9. Copyright 1960, John Wiley and Sons.)

Now if we apply the Laws of Thermodynamics to a "photon gas" (i.e., a gas made up of photons), we ultimately find that radiant energy emitted from an ideal radiator (the blackbody) is

$$E_b = \sigma T^4 \tag{9-3}$$

where E_b is the radiant energy flux (energy per unit area and per unit time, i.e., W/m^2), T is the absolute temperature and σ the Stefan–Boltzmann constant (5.669×10^{-8} W/m^2 °K^4).

PROPERTIES AND ASPECTS OF RADIATION

When incident radiation falls on a body, three results can occur: absorption, reflection, and transmission. If we consider the fraction of each of these, then

$$\alpha + \rho + \tau = 1 \tag{9-4}$$

where α, ρ, and τ represent, respectively, the fractions of absorption, reflection, and transmission.

For many solids and liquids the transmission is small or essentially nil; hence

$$\alpha + \rho = 1 \tag{9-5}$$

As stated earlier, the blackbody is an entity that represents the ideal radiation. Therefore, we can relate the radiant energy for any real body (i.e., non-blackbody) to the blackbody's radiant energy. The ratio of emissive energy fluxes is then the emissivity of the real object:

$$\varepsilon = \frac{E}{E_b} \tag{9-6}$$

where E is W/m^2 for the non-blackbody and E_b is the W/m^2 for the blackbody.

Furthermore, it can be shown that the absorptivity for the non-blackbody is also given by the E/E_b ratio:

$$\alpha = \frac{E}{E_b} \tag{9-7}$$

This results in the equality of emissivity and absorptivity of any body (known as Kirchhoff's identity):

$$\alpha = \varepsilon \tag{9-8}$$

From the preceding, it is obvious that the blackbody's emissivity and absorptivity are both unity. Non-blackbodies (known as gray bodies) have fractional values. Typical values are given in Table 9-1.

Although the blackbody represents an idealization, it can be approximated in nature. This is done by using a very small hole in an enclosure. For such a case the emissivity, ε^1, is then

$$\varepsilon^1 = \frac{\varepsilon}{\varepsilon + f(1-\varepsilon)} \tag{9-9}$$

where ε is the enclosure emissivity and f is the fraction of total internal unity for the hole.

GEOMETRICAL ASPECTS OF RADIATION

A very important aspect of radiative heat transfer is the system geometry. This is accounted for by using radiation shape factors (also called view factor, angle

Table 9-1 Emissivities of Various Surfaces

Surface	Temperature (°K)	Emissivity
Water	273	0.95
	373	0.963
Various oil paints (all colors)	373	0.92 to 0.96
Refractories	872	0.65 to 0.70
Poor radiators	1272	0.75
Refractories	872	0.80 to 0.85
Good radiators	1272	0.85 to 0.90
Abestos paper board	311	0.93
	644	0.945
Red brick	294	0.93
Polished cast iron	473	0.21
Oxidized cast iron	472	0.64
	872	0.78
Polished iron	450	0.052
	500	0.064
Rusted iron	293	0.685
Polished aluminum	500	0.039
	850	0.057
Oxidized aluminum	472	0.11
	872	0.19
Polished copper	353	0.018
Oxidized copper	472	0.57
	872	0.57

Source: References 1 and 2.

factor, configuration factor) defined as follows:

F_{12} = fraction of energy leaving surface 1 which reaches surface 2

F_{21} = fraction of energy leaving surface 2 which reaches surface 1

F_{mn} = fraction of energy leaving surface m which reaches surface n

The shape factors are determined by combining solid geometry with the calculus. Basically, for two objects (see Figure 9-2) we take a differential area dA_1 and connect it to surface 2 by a ray length r_{12}. We then project dA_1 as shown. Normals n_1 and n_2 are erected to the differential areas. The θ_1 and θ_2 angles are between the normals and the ray length r_{12}.

For this system is can be ultimately shown that

$$A_1 F_{12} = F_2 F_{21} = \int_{A_1} \int_{A_2} \cos \theta_1 \theta_2 \frac{dA_1 dA_2}{\pi r_{12}^2} \tag{9-10}$$

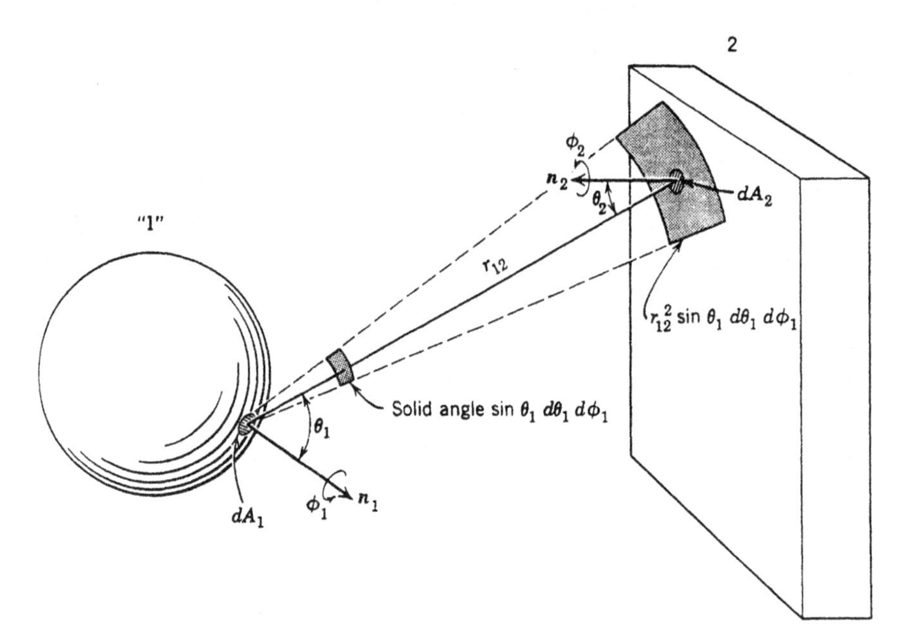

Figure 9-2. Radiation system model. (Adapted with permission from reference 9. Copyright 1960, John Wiley and Sons.)

Determination of the radiation shape factor for a given system is a complex problem. Some results (see Figures 9-3, 9-4, and 9-5) have been published for straightforward cases.

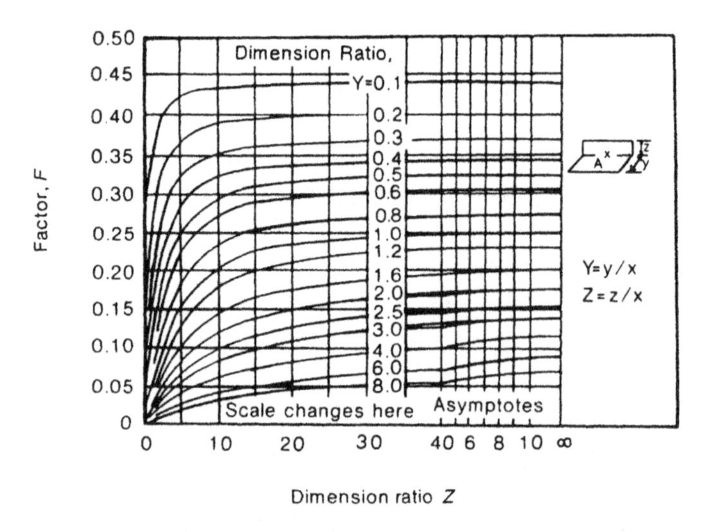

Figure 9-3. Radiation shape factors between adjacent surfaces (1).

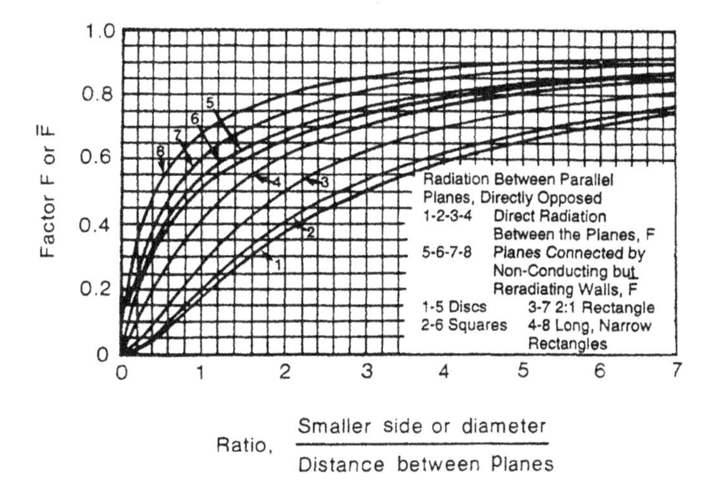

Figure 9-4. Radiation shape factors between parallel planes (1).

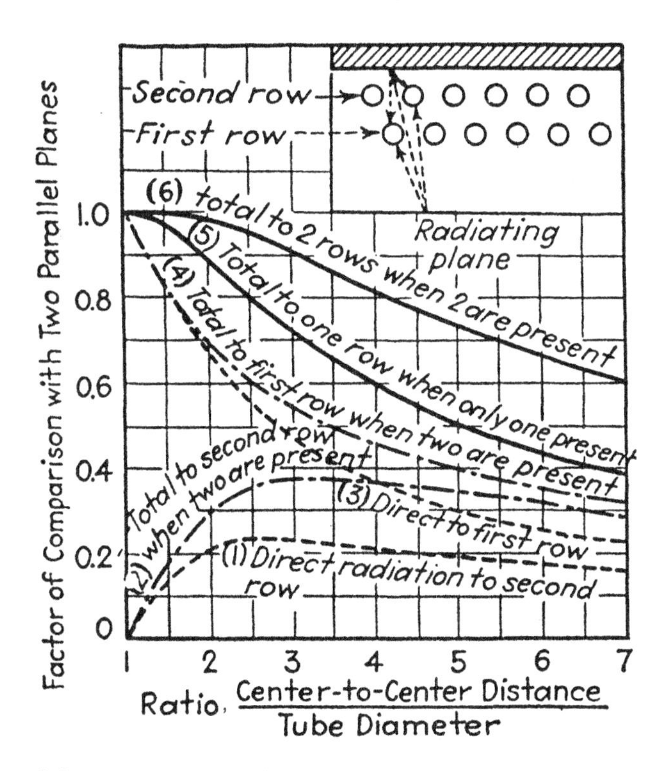

Figure 9-5. Radiation shape factor between a plane and rows of tubes (1).

Radiation heat transfer in a system of blackbodies then becomes

$$q_{\text{net}} = A_1 F_{12}(E_{b1} - E_{b2}) = A_2 F_{21}(E_{b1} - E_{b2}) \tag{9-11}$$

and in turn

$$q_{\text{net}} = A_1 F_{12} \sigma (T_1^4 - T_2^4) = A_2 F_{21} \sigma (T_1^4 - T_2^4) \tag{9-12}$$

If the bodies undergoing radiant heat transfer are not blackbodies and are connected by a third surface that doesn't exchange heat, then

$$q_{\text{net}} = \frac{\sigma A_1 (T_1^4 - T_2^4)}{\dfrac{A_1 + A_2 - 2A_2 F_{12}}{A_2 - A_1 (F_{12})^2} + \left(\dfrac{1}{\varepsilon_1} - 1\right) + \dfrac{A_1}{A_2}\left(\dfrac{1}{\varepsilon_2} - 1\right)} \tag{9-13}$$

where ε_1 and ε_2 are the emissivities of surfaces 1 and 2, respectively.

When one surface or object can completely "see" the other object, then

$$\frac{A_1 + A_2 - 2A_2 F_{12}}{A_2 - A_1 (F_{12})^2} = 1 \tag{9-14}$$

and

$$q = \frac{\sigma A_1 (T_1^4 - T_2^2)}{\dfrac{1}{\varepsilon_1} + \left(\dfrac{A_1}{A_2}\right)\left(\dfrac{1}{\varepsilon_2} - 1\right)} \tag{9-15}$$

A typical case for equation (9-15) is that of infinite parallel planes.

An intermediate case that occurs between the forms of equations (9-13) and (9-15) is the situation when two surfaces in an enclosure exchange radiation only with each other. Then,

$$q_{\text{net}} = \frac{\sigma A_1 (T_1^4 - T_2^4)}{\dfrac{1}{F_{12}} + \left(\dfrac{1}{\varepsilon_1} - 1\right) + \dfrac{A_1}{A_2}\left(\dfrac{1}{\varepsilon_2} - 1\right)} \tag{9-16}$$

In practice, it is possible to find radiation shape factors between parts of a system. Consider, for example, the situation depicted in Figure 9-6 where we need the shape factor between areas 1 and 4.

For such a situation, we first compare the combined areas 1 and 2 to the combined areas 3 and 4. This gives

$$A_{1+2} F_{(1+2)(3+4)} = A_1 F_{1(3+4)} + A_2 F_{2(3+4)} \tag{9-17}$$

Also,

$$A_1 F_{1(3+4)} = A_1 F_{13} + A_1 F_{14} \tag{9-18}$$

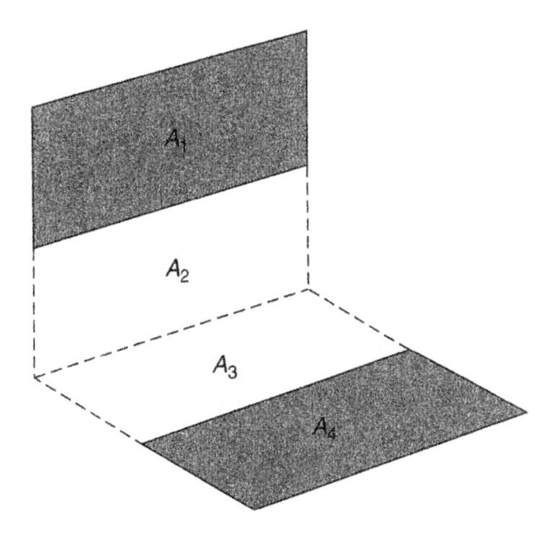

Figure 9-6. Model for determining shape factors between portions of adjacent planes B.

and

$$A_{(1+2)}F_{(1+2)3} = A_1F_{13} + A_2F_{23} \tag{9-19}$$

Combining the foregoing gives

$$A_{(1+2)}F_{(1+2)(3+4)} = A_{(1+2)}F_{(1+2)3} - A_2F_{23} + A_1F_{14} + A_2F_{2(3+4)} \tag{9-20}$$

and

$$F_{14} = \frac{1}{A_1}(A_{(1+2)}F_{(1+2)3} + A_2F_{23} - A_{(1+2)}F_{(1+2)3} - A_2F_{2(3+4)}) \tag{9-21}$$

See that all of the F values in the parentheses can be obtained from Figure 9-3.

More generalized techniques for both adjacent and parallel systems can be found in reference 3.

One additional subject of interest is the effect of radiation shields. Such devices are reflective materials placed between radiating surfaces (see Figure 9-7). In this case if we use a number (n) of such shields all with equal emissivities, we obtain the relation that

$$q_s^1 = \frac{1}{n+1}q^1 \tag{9-22}$$

where q_s^1 is the heat transformed with shields and q^1 is the heat transformed without shields.

EFFECT OF GASES

Simple monatomic and diatomic gases are essentially transparent to radiation. Hence, radiant transfer in such systems can be treated as if a vacuum were

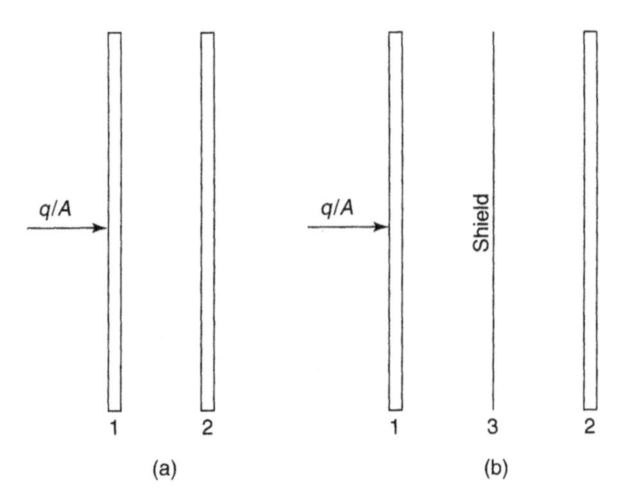

Figure 9-7. Radiation shield. Adapted from (10).

present. On the other hand, the more complex polyatomic gases (water, carbon dioxide, sulfur dioxide) absorb radiation and as such must be considered in the radiant heat transmission.

In treating this situation the base case is to consider a hemisphere of radius L. For this case the net heat transferred:

$$q_{\text{net}} = A_s \sigma (\varepsilon_g T_g^4 - \alpha_g T_s^4) \tag{9-23}$$

where A_s is the surface area, ε_g is the gas emissivity at T_g (the gas temperature), and α_g is the gas absorptivity at T_s (the surface temperature).

If the enclosure is not a blackbody, then

$$\frac{q_{\text{gray}}}{q_{\text{black}}} = \frac{1 + \varepsilon_s}{2} \tag{9-24}$$

for values of $\varepsilon_s \geqq 0.7$ (1) .

The equivalent or mean beam length or a given system at low values of total gas pressure times beam length is four times the enclosure's mean hydraulic radius. Some values of the mean equivalent beam length (L_e) are given in Table 9-2.

Emissivities for gases are given as a function of PL_e and temperature in reference 1. Plots for water vapor are given in Figures 9-8 and 9-9.

Plots for carbon dioxide, sulfur dioxide carbon monoxide, and ammonia are given in reference 1 as adapted from references (5, 6, 7). Also given is a correction for the combined effect of water vapor and carbon dioxide (the principal constituents of flue gases).

Table 9-2 L_e Values

System	L_e
Infinite parallel planes L apart	$1.8L$
Sphere (diameter D) radiation to surface	$0.65D$
Hemisphere (diameter D) radiation of center of base	$0.5D$
Infinite circular cylinder (diameter D) radiation to curved surface	$0.95D$
Cube (side of L) radiation to face	$0.66L$
Circular cylinder (height = diameter D) radiation to entire surface	$0.60D$
Circular cylinder semi-infinite height (diameter D) radiation to entire base	$0.65D$
Any arbitrary shape volume of V with radiation to surface of A	$\frac{3.6(V)}{A}$

Source: References 1 and 4.

WORKED EXAMPLES

Example 9-1 Two parallel blackbodies 1.52 × 3.04 m are spaced 1.52 m apart. One plate is maintained at 537.8°C, and the other is maintained at 260°C. What is the net radiant energy interchange?

In order to solve this problem, we need the radiation shape factor between the surfaces. This involves the use of Figure 9-4. For the situation (no re-radiating walls) we select from curves 1-4. The geometry in this case is a 2 to 1 (i.e., 3.04/1.52) rectangle, which means that curve 3 is the appropriate one to be used. The ratio is

$$\frac{\text{Shorter side}}{\text{Spacing}} = \frac{1.52}{1.52} = 1.0$$

Then, from Figure 9-4 we have

$$F_{12} = 0.285$$

$$q_{\text{net}} = A_1 F_{12}(E_{b1} - E_{b2}) = A_1 F_{12} \sigma (T_1^4 - T_2^4)$$

$$q_{\text{net}} = (4.65 \text{ m}^2)(0.285)(5.67 \times 10^{-8} \frac{\text{W}}{\text{m}^2 \ {}^\circ\text{K}^4})[(1255^\circ\text{K})^4 - (811.1^\circ\text{K})^4]$$

$$q_{\text{net}} = 2432.5 \text{ W}$$

Example 9-2 Find the radiation shape factors between areas 1 and 2 in Figure 9-10.

For case a we have

$$A_1 F_{12} = [A_{(1+4)} F_{(1+4)(3+2)} - A_{(1+4)} F_{(1+4)3} + A_4 F_{43} - A_4 F_{4(3+2)}]$$

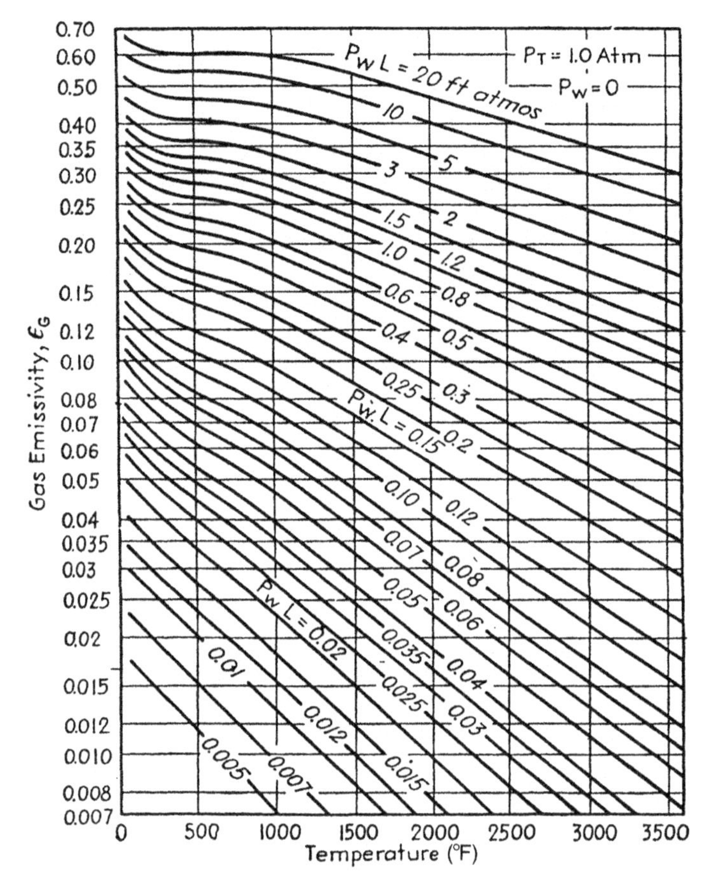

Figure 9-8. Emissivity of water vapor at 1 atmosphere total pressure and low partial pressure (11).

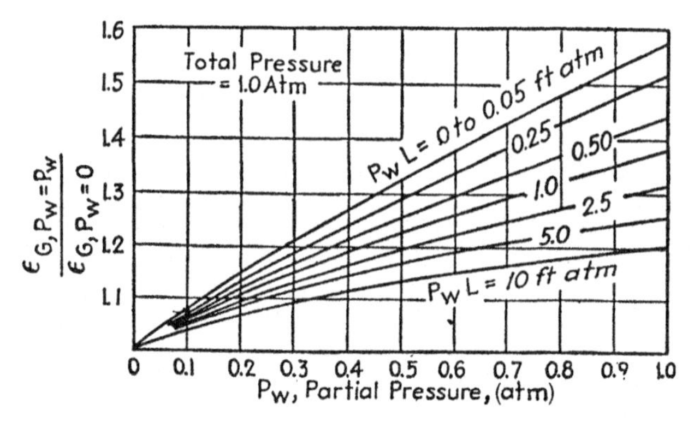

Figure 9-9. Pressure correction factor for water vapor actual emissivity is value from Figure 9-8 times C (11).

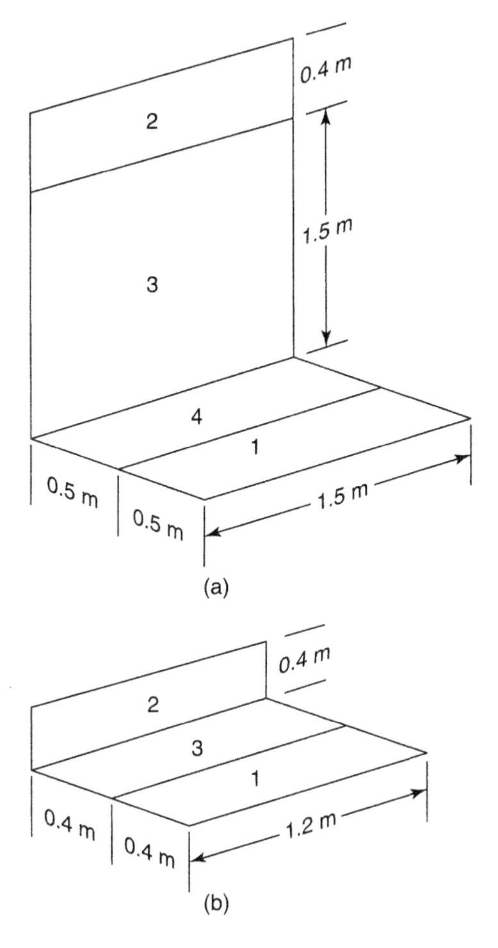

Figure 9-10. Radiation shape factor systems.

Now using Figure 9-3 the appropriate *y/x*, *z/x*, and *F* values are

$$\frac{z_{(3+2)}}{x} = \frac{2}{1.50} = 1.33, \qquad \frac{y_{(1+4)}}{x} = \frac{1}{1.5} = 0.666, \qquad F_{(1+4)(3+2)} = 0.27$$

$$\frac{z_3}{x} = \frac{1.50}{1.50} = 1.0, \qquad \frac{y_{(1+4)}}{x} = \frac{1}{1.5} = 0.666, \qquad F_{(1+4)3} = 0.26$$

$$\frac{z_3}{x} = \frac{1.5}{1.5} = 1.0, \qquad \frac{y_4}{x} = \frac{0.5}{1.5} = 0.33, \qquad F_{43} = 0.35$$

$$\frac{z_{(3+2)}}{x} = \frac{2.0}{1.5} = 1.33, \qquad \frac{y_4}{x} = \frac{0.5}{1.5} = 0.33, \qquad F_{4(3+2)} = 0.36$$

So that,

$$F_{12} = 1/0.75\ \text{m}^2[(1.50\ \text{m}^2)(0.27) - (1.50\ \text{m}^2)(0.26) + (0.75\ \text{m}^2)(0.36) - (0.75\ \text{m}^2)(0.36)]$$

$$F_{12} = 0.03$$

Repeating for case b, we have

$$A_2 F_{2(3+1)} = A_2 F_{23} + A_2 F_{21}$$

$$F_{2(3+1)} = F_{23} + F_{21}$$

Then once again using Figure 9-3 we obtain

$$\frac{z_2}{x} = \frac{0.4}{1.2} = 0.33; \qquad \frac{y_{(3+1)}}{x} = \frac{0.8}{1.2} = 0.66; \ F_{2(3+1)} = 0.32$$

$$\frac{z_2}{x} = \frac{0.4}{1.2} = 0.33; \qquad \frac{y_3}{x} = \frac{0.4}{1.2} = 0.33; \qquad F_{23} = 0.26$$

Solving for F_{21}, we obtain

$$F_{21} = 0.06$$

But,

$$A_2 F_{21} = A_1 F_{12}$$

$$F_{12} = F_{21} = 0.06$$

Example 9-3 A furnace in the form of a cylinder has one end opened to the outside. Dimensions of the furnace are a diameter of 5 cm and a height of 10 cm. Furnace sides and bottom are maintained respectively at temperatures of 1450°K and 1750°K. Surroundings are at 298°K.

What power is needed to keep the furnace operating as described?

We assume that the furnace sides and bottom can be approximated by blackbodies. Radiation heat transfer is presumed to predominate. Furthermore, the only heat loss will be through the opening to the surroundings. Because the surroundings are great in extent, we can also assume that the opening behaves as a blackbody.

Let the temperatures be designated as T_t (top), T_s (side), and T_b (bottom). Likewise, the areas are A_s (side) and A_b (bottom).

$$q_{\text{loss}} = q_s + q_b$$

$$q_{\text{loss}} = A_s F_{st}\sigma(T_s^4 - T_t^4) + A_b F_{bt}\sigma(T_b^4 - T_t^4)$$

We start first by finding F_{bt} using Figure 9-4. The appropriate curve is 1 and the abscissa is 0.5 (i.e., 5/10), which gives

$$F_{bt} = 0.05$$

Next, since

$$F_{bt} + F_{bs} = 1.0$$

we have

$$F_{bs} = 0.95$$

Also, since

$$A_s F_{sb} = A_b F_{bs}$$

we obtain

$$F_{sb} = \frac{\pi(5\ \text{cm}^2)}{4\pi(5\ \text{cm})(10\ \text{cm})}(0.95)$$

$$F_{sb} = 0.119$$

Then by symmetry

$$F_{sb} = F_{st} = 0.119$$

Substituting the values, we obtain

$$q_{\text{loss}} = (\pi)(0.05\ \text{m})(0.1\ \text{m})(0.119)[(1450^\circ\text{K})^4 - (298^\circ\text{K})^4] (5.67 \times 10^{-8}\ \text{W/m}^2\ {}^\circ\text{K}) + \pi/4(0.05\ \text{m})^2(5.67 \times 10^{-8}\ \text{W/m}^2\ {}^\circ\text{K})[1750^\circ\text{K})^4 - (290^\circ\text{K})^4]$$

and

$$q_{\text{loss}} = 463.5\ \text{W}$$

Example 9-4 An oxidized steel tube ($\varepsilon = 0.6$, outside diameter = 0.0762 m) passes through a silica brick furnace ($\varepsilon = 0.8$, inside dimensions 0.152 m × 0.152 m × 0.152 m).

The inside wall furnace temperature is 982.2°C, and the outside of the tube is 537.8°C. What is the rate of heat transfer?

The equation to be used is the one for surfaces enclosed by a nonconducting but re-radiating surface:

$$q_{\text{net}} = \frac{\sigma A_1(T_1^4 - T_2^4)}{\dfrac{A_1 + A_2 - 2A_1F_{12}}{A_2 - A_1(F_{12})^2} + \left(\dfrac{1}{\varepsilon_1} - 1\right) + \dfrac{A_1}{A_2}\left(\dfrac{1}{\varepsilon_2} - 1\right)}$$

In order to use this equation, a value for F_{12} must be found. However, there are no published data (as in Figures 9-3 through 9-5) available for this system. It appears that the F_{12} value will have to be computed as per equation 9-10. Closer inspection of the system, however, shows us that we do not have to follow this route. For the system at hand it is apparent that no part of the enclosure is hidden from the view of the tube. In essence, therefore, every bit of the tube "sees" the furnace. Hence,

$$\frac{A_1 + A_2 - 2A_1F_{12}}{A_2 - A_1(F_{12})^2} = 1/1$$

The areas in the system are

$$A_1 = (\pi)(0.0762)(0.152) = 0.036 \text{ m}^2$$

$$A_2 = [(0152 \times 0.152 \times 0.152 \times 0.152) - 2(\pi/4)(0.0762)^2] \text{ m}^2$$

$$A_2 = 0.130 \text{ m}^2$$

Substituting these values, the emissivities and temperature gives us

$$q_{\text{net}} = \frac{(0.0365 \text{ m}^2)[(1255.5°\text{K})^4 - (811.1)^4](5.669 \times 10^{-8} \text{ W/m}^2\ °\text{K}^4)}{\dfrac{0.0365 \text{ m}^2}{0.130 \text{ m}^2}\left(\dfrac{1}{0.8} - 1\right) + \left(\dfrac{1}{0.6} - 1\right) + \dfrac{1}{1}}$$

$$q_{\text{net}} = 2432.5 \text{ W}$$

Example 9-5 A drying unit moves the material to be dried through the system by a conveyer belt. The width of material on the belt is large enough for the case to give semi-infinite conditions.

Drying is accomplished by a row of electrically heated cylinders (2 cm in diameter and 8 cm apart on a center-to-center basis) placed 16 cm from the belt. The cylinders whose emissivity is 0.9 are at 1700°K. The material ($\varepsilon = 0.6$) is at a temperature of 370°K.

Find the radiant heat transferred.

For this case assume that the unit walls are black.

This situation is that described by equation (9-16). In order to obtain the needed F_{12} value, we use Figure 9-5. The geometric ratio is the center to spacing divided by the tube diameter (8 cm/2 cm or 4). In Figure 9-5, we use the curve for the case when only one row of tubes is present. The F_{12} value is 0.6.

Then by substituting into equation (9-16) we obtain

$$q_{\text{net}} = \frac{(5.67 \times 10^{-8} \text{ W/m}^2\ °\text{K}^4)[(1700°\text{K})^4 - (370°\text{K}^4)]}{\dfrac{1}{0.6} + \left(\dfrac{1}{0.6} - 1\right) + \dfrac{8 \text{ cm}}{\pi(2 \text{ cm})}\left(\dfrac{1}{0.9} - 1\right)}$$

$$\frac{q_{\text{net}}}{A_1} = 190.4 \text{ kW/m}^2$$

Example 9-6 An exhaust gas stream (10 percent water and 90 percent air) flows through a circular chimney (1 m in diameter). The chimney surface is at 1200°K (assume it to be a blackbody).

Temperature and pressure of the gas stream are at 1350°K and 1.5 atmosphere, respectively. The convection heat transfer coefficient between the gas and the chimney wall is 8.5 W/m^2 °K.

We first find the PL_e combination. In this case (from Table 9-2) for the infinite cylinder, $L_e = 0.95$ D. Then,

$$PL_e = (1.5 \text{ atm})(0.10)(0.95)(1 \text{ m})$$

$$PL_e = 0.1425 \text{ atm-m} = 14.44 \text{ kN/m}$$

From Figure 9-8 for the temperature of 1350°K (2430°R) the emissivity value is 0.110. The correction factor (from Figure 9-9) is 1.1, giving a corrected ε of 0.121.

Next we find the absorptivity (at the surface temperature of 1200°K). The value of α from Figure 9-8 is 0.120. The same correction will apply, giving a final absorptivity of 0.132.

Then by substituting into equation (9-23) we obtain

$$\text{Radiant energy flux} = \frac{q_r}{A} = 5.67 \times 10^{-8} \text{ W/m}^2\ {}^\circ\text{K}^4 \times [0.121(1350^\circ\text{K})^4 - 0.132(1200^\circ\text{K})^4]$$

$$\frac{q_r}{A} = 7269 \text{ W/m}^2$$

Also, for the convection portion we have

$$\text{Convective energy flux} = \frac{q_c}{A} = (8.5 \text{ W/m}^2\ {}^\circ\text{K})(150^\circ\text{K})$$

$$\frac{q_c}{A} = 1275 \text{ W/m}^2$$

PROBLEMS

9-1. A square plate (0.3 by 0.3 m) is attached to the side of a spacecraft. The plate whose emissivity is 0.5 is perfectly insulated from the vessel. What would the plate's equilibrium temperature be when the sun's radiant heat flux is 1500 W/m^2? Assume outer space is a blackbody at 0°K.

9-2. Two gray surfaces (one at 482°C with an emissivity of 0.90 and the other at 204°C and an emissivity of 0.25) have a net transfer of radiant energy. Determine the W/m^2 for the following cases: infinite parallel planes 3 m apart; flat squares 2.0 m on a side 1 m apart.

9-3. A block of metal (emissivity of 0.5) has a conical hole machined into it (0.05 m deep with a surface diameter of 0.025 m). Find the radiant energy emitted by the hole if the block is heated to 550°C. Find a value of apparent emissivity for the hole (ratio of actual energy from the hole to that from a black surface having an area the same as the opening and a temperature of 550°C).

9-4. Repeat the calculations of Problem 9-2 for two cases: one where the 482°C object is a spherical shell (3.048 m in diameter) and the 204°C object is a spherical shell (0.3048-m diameter) concentric with the first; also for the instance where the 482°C and 204°C objects are concentric cylinders of diameters 0.254 and 0.229 m, respectively.

9-5. Two steel plates 0.15 m apart are used to heat large lacquered aluminum sheets. If the steel plates are respectively at 300°C and 25°C, what is the heat transferred and the lacquered sheet temperature? Emissivities of the steel and lacquered sheets are 0.56 and 1.0.

9-6. A furnace consists of a carborundum muffle 4.58 by 6.10 m at 1150°C and a row of 0.102-m-diameter tubes on 2.06-m centers 3.048 m above the muffle. The tubes that are at 316°C have an emissivity of 0.8. The muffle's emissivity is 0.7. What is the radiant heat transmission (assume side walls reradiate)?

9-7. What is the distribution of radiant heat to various rows of tubes irradiated from one side? The tubes are 0.102 m in diameter and are set on 0.244-m triangular centers.

9-8. A room ($3 \times 3 \times 3$ m) has one side wall at 260°C and the floor at 90°C and all other surfaces are completely insulated. If all surfaces are black, calculate the net heat transfer between the wall and the floor.

9-9. A building's flat black roof has an emissivity of 0.9, along with an absorptivity of 0.8 for solar radiation. The sun's energy transfer is 946 W/m^2. The temperature of the air and surroundings is 26.7°C. Combined conduction–convection heat transfer is given by $q/A = 0.38(\Delta T)^{1.25}$ where the ΔT is the difference between the roof and the air. Find the roof temperature (assume that the blackbody temperature of space is −70°C).

9-10. A square room 3 by 3 m has its floor at 25°C and its ceiling at 13°C. If the material has an emissivity of 0.8 and the room's height is 2.5 m, find the net energy interchange.

9-11. Repeat Problem 9-9 if the roof is painted with aluminum paint (emissivity of 0.9, absorptivity for solar radiation of 0.5).

9-12. Find the configuration factors F_{12} for the cases shown below:

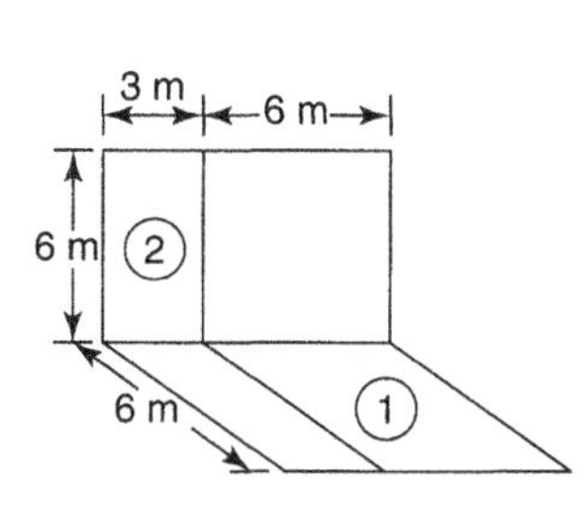

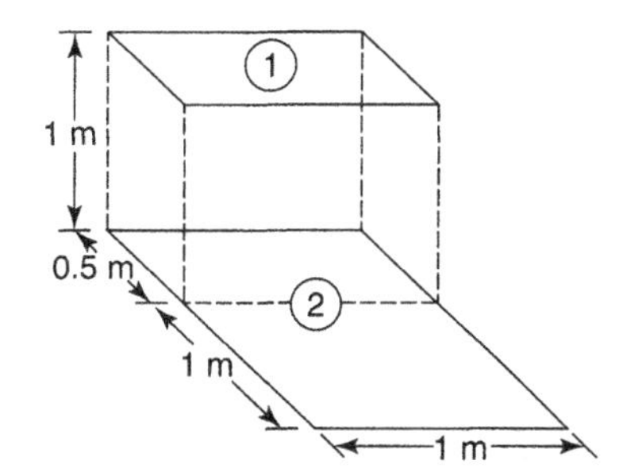

9-13. A long oven's cross-sectional area is a semicircle of diameter 1 m. Materials to be dried are placed on the floor. If the floor is at 325°K and the walls are at 1200°K, what is the drying rate per unit length of the oven in kg/sec m? Assume blackbody behavior.

9-14. Two disk-shaped blackbody objects are set 0.2 m coaxially apart. Both disks have an outer diameter of 0.8 m; however, the upper disk is a torus or ring (inner diameter 0.4 m). If the temperatures of the upper and lower disks are 1000°K and 300°K, respectively, calculate the net radiative exchange.

9-15. A plaster ceiling in a room (4.57 m by 4.57 m, height of 2.44 m) is heated with an installed radiant system. The room's concrete floor and the air in the room are at 23.9°C. The system is to supply 1172 W. If the ceiling and floor emissivities are 0.93 and 0.63, respectively, calculate the required ceiling temperature. Assume that the convective heat transfer coefficient between the ceiling and the air is $h = 0.20(\Delta T)^{1/4}$ Btu/ft^2-h-°F with ΔT in °F.

9-16. The effective blackbody temperature of the outer atmosphere on a clear night is −70°C. Air at 15°C has water at a vapor pressure equal to liquid water or ice at 0°C. If a thin film of water is placed in a sheltered pan with a full view of the sky, will ice form? Assume that the convective h is 2.6 W/m^2 C.

9-17. Two parallel blackbodies (1 m by 1 m) insulated on their reverse sides are at temperatures of 500°K and 750°K. If the objects (separated by 1 m) are located in a large room (walls at 300°K), what is the net radiative heat transfer to the room walls?

9-18. An ice rink (circular) is enclosed by a hemispherical dome. The ice and dome are at 0°C and 15°C, respectively. The rink diameter is 25 m. What is the net radiative transfer (assume blackbodies)?

9-19. A 0.10-m-diameter thin-walled tube is kept at 120°C by steam passing through it. A radiation shield (at 35°C) is installed with a 0.01-m air gap. If the tube and shield have emissivities of 0.8 and 0.10, what is radiant heat transfer per unit length?

9-20. Air circulates at 60°C through a duct. If a thermocouple is inserted into the duct, will there be a sizeable error due to radiation?

9-21. A furnace has a 0.102-m diameter sight hole. If the furnace is at 482°C and the surroundings are at 26.7°C, what is the net radiant heat loss from the hole? Assume blackbody behavior.

9-22. An enclosure in the shape of a tetrahedron has emissivities of 0.4 for all surfaces. One surface is insulated. The other surfaces have temperatures of 4.44°C, 26.7°C, and 149°C. What is the temperature of the fourth surface?

9-23. A container of chilled lemonade has an inside surface area of 648.4 cm^2. If the cap is removed, the resultant hole is 3.23 cm^2. What is the shape factor from the inside of the container to a thumb placed on the hole?

9-24. Two parallel square plates (1.22 by 1.22 m) are maintained at 560°C and 282°C. The emissivities are 0.5 (560°C) and 0.6. Find the heat lost by each plate and determine the net radiant energy to the walls (21.1°C) of a large room.

9-25. Estimate the temperature at which ice will form on a clear night (sky effective radiation temperature is −73.3°C). The convective heat transfer coefficient is 28.4 W/m^2. Neglect water's heat of vaporization. Assume water is a blackbody.

9-26. Three infinite plates are positioned in a parallel setting. The outermost plates are at 1000°K (emissivity of 0.8) and 100°K (emissivity of 0.8). The plate in the center (emissivity of 0.5) is not heated externally. What is the temperature of the center plate?

9-27. An insulated house has a 0.09 m gap between the plaster wall and wooden siding. If the inside wall is at 18.3°C and the outer wall is at −9.4°C, what is the heat loss by radiation and natural convection (assume h is 3.86 W/m^2 K)? How much would the heat loss be reduced by covering the inside wall with aluminum foil?

9-28. A fuel oil ($CH_{1.8}$) is burned with a 20 percent excess air. The combustion product is fed to a 0.152-m-diameter pipe that transfers heat to air blown over the outside of the pipe. At the pipe entry the gases are at 1093°C (surface is at 427°C). Exit conditions are 538°C for the gas and 316°C for the surface. What is the pipe length for (a) blackbody (b) gray (emissivity of 0.8)? The wC_p value for the gas is 52.75 K J/hr °C.

9-29. A furnace has a spherical cavity of 0.5-m diameter. Contents are a gas mixture (CO_2 partial pressure of 0.25 atm; N_2 partial pressure of 0.75 atm) at a total pressure of 1 atm and 1400 K. What cooling rate is needed to maintain the cavity wall (blackbody) at 500°K?

9-30. A furnace consists of two large parallel plates separated by 0.75 m. A gas mixture

Gas	Mole Fraction
O_2	0.20
N_2	0.50
CO_2	0.15
H_2O	0.15

flows between the plates at a total pressure of 2 atm and a temperature of 1300°K. The plates (blackbodies) are kept at 500°K. What is the net radiative heat flux?

9-31. A flue gas (CO_2 and H_2O vapor partial pressures of 0.05 atm and 0.10 atm) is at 1 atm and 1400 K. If the gas flows through a long flue (1-m diameter, surface at 400 K) find the net radiative flux.

9-32. A gas turbine combustion chamber can be simulated as a long tube (0.4-m diameter; surface temperature of 500°C. The flowing combustion gas at 1 atm and 1000°C contains carbon dioxide and water vapor (both mole fractions are 0.15). If the chamber surface is a blackbody, what is the net radiative flux?

REFERENCES

1. H. C. Hottel, in *Heat Transmission*, W. H. McAdams, editor, third edition, editor, McGraw-Hill, New York (1954), Chapter 4.
2. J. H. Perry, *Chemical Engineers Handbook*, third edition, McGraw-Hill, New York (1950), p. 484.
3. D. C. Hamilton and W. R. Morgan, Radiant Interchange Configuration Factors, *NACA Tech Note*, 2836 (1952).
4. E. R. G. Eckert and R. M. Drake, *Analysis of Heat and Mass Transfer*, McGraw-Hill, New York (1972).
5. S. A. Guerrieri, *Research Report*, Massachusetts Institute of Technology (1933).
6. F. J. Port, Sc.D. thesis, Massachusetts Institute of Technology (1940).
7. W. Ullrich, Sc.D. thesis, Massachusetts Institute of Technology (1935).
8. W. M. Rosenhow and H. Y. Choi, *Heat Mass and Momentum Transfer*, Prentice-Hall, Englewood Cliffs, NJ (1961).
9. R. B. Bird, E. N. Lightfoot, and W. E. Stewart, *Transport Phenomena*, John Wiley & Sons, New York (1960).
10. J. P. Holman, *Heat Transfer*, fourth edition, McGraw-Hill, New York (1981).
11. R. B. Egbert, Sc.D. Thesis, Mass. Inst. Tech. (1941).

10

MASS TRANSFER; MOLECULAR DIFFUSION

INTRODUCTION

The transfer of mass is not only an integral but also an essential part of the chemical and process industries. Such transfer is, for example, the basis for the many separation and purification operations used in industry. Furthermore, mass transfer figures largely in both catalyzed and uncatalyzed chemical reaction systems.

We can illustrate the transfer of mass of a very simple experiment by using a long trough or container filled with water. Into this container, we add a few crystals of a solid chemical that will dissolve and color the water (i.e., such as potassium permanganate giving a purple color or a copper compound giving a green color). We would then observe the color moving from the point where the dissolution took place ($c = c_0$) to the rest of the liquid ($c = 0$). This movement (called *ordinary diffusion*) takes place because of a concentration driving force.

Mass transfer, however, can take place because of other driving forces. For example, we can have mass transferred by the action of a pressure gradient (i.e., *pressure diffusion*). Such transport will occur only when such pressure gradients are very large such as with ultracentrifuge (used to separate high-molecular-weight fractions in polymeric or biological systems).

Forced diffusion occurs because of the action of some external force. The most commonly found form of forced diffusion is the action of an electric field on ionic species. Here the value of the force is given by multiplying the electric field strength times the ionic charge. It becomes obvious that such forced diffusion is the basis of electrochemistry.

Finally, the last of the diffusion types is that caused by temperature gradients. Here we have what is known as *thermal diffusion*. Typical applications have

Table 10-1 Flux–Driving Force Relations

Fluxes	Velocity Gradient	Temperature Gradient	Concentration Gradient	Chemical Affinity
Momentum (tensor, 9 components)	$\tau_{yx} = \mu dv/dy$			
Heat (vector, 3 components)		$q_y = k\,dT/dy$ Fourier's Law	Dufour effect	
Mass (vector, 3 components)		Soret effect, thermal diffusion	$J_{A_y} = D_{AB} dC_A/dy$ Fick's Law	
Chemical reaction (scalar)				rate $= k_r C_A^n$

Note: All equations shown are one-dimensional form.

Source: Reference 6, with permission. Copyright 1996, American Chemical Society.

included the separations of gaseous isotopes. A system based on the principle of thermal diffusion was used to separate uranium isotopes.

In order to better grasp the impact of thermal diffusion, consider Table 10-1, which relates one-dimensional fluxes with their driving forces. Note that the fluxes are listed in terms of their mathematical complexity: first momentum (a tensor with nine components) and then heat (vector; three components) and mass (vector; three components) and chemical reaction (a scalar)]. According to Onsager (1) the fluxes are interrelated if their order (second order for a tensor; first order for a vector; zeroth order for a scalar) is the same or differs by two.

Hence, heat and mass transfer are interrelated; and by Onsager's concepts, mass can be transferred by a temperature gradient (*thermal diffusion*, the Soret effect). Note that Onsager also predicts an interaction between momentum transport, chemical reaction, and their driving forces. Although such effects have not been clearly experimentally defined, they are most intriguing because of the possibility of inducing chemical reaction with velocity gradients.

FICK'S FIRST LAW; MASS FLUX; EQUATION OF CONTINUITY OF SPECIES

The basic equation governing mass transfer by ordinary diffusion is Fick's First Law (see Chapter 1):

$$J_{Ay} = -D_{AB}\frac{\partial C_A}{\partial y} \tag{10-1}$$

where J_{Ay} is the mass (molar) flux in the y direction, D_{AB} is the diffusivity, and C_A is the concentration of A.

Furthermore, since mass flux is a vector

$$J = -D_{AB}\left(\frac{\partial C_A}{\partial x}\mathbf{i} + \frac{\partial C_A}{\partial y}\mathbf{j} + \frac{\partial C_A}{\partial z}\mathbf{k}\right) \tag{10-2}$$

where **i**, **j**, and **k** are unit vectors in the x, y, and z directions.

In defining J_A it is stipulated that this flux must be referred to a plane (see Figure 10-1) across which there is no net volume transport (i.e., the plane moves with respect to the fixed apparatus, although the fluid is stagnant). It is more desirable of course to obtain expressions for the flux relative to the apparatus itself instead of to the moving plane.

In order to do this, a new flux N will be introduced. Thus the velocity of the moving plane is given by

$$\mathbf{U}_y = \sum_i \mathbf{N}_i \overline{V}_i \tag{10-3}$$

where $\mathbf{U}_y$ is the velocity, $\mathbf{N}_i$ is the mass flux of the ith component, and $\overline{V}_i$ is the partial molal volume of the ith component.

For a binary system (A and B) we have

$$\mathbf{U}_y = \mathbf{N}_A \overline{V}_A + \mathbf{N}_B \overline{V}_B \tag{10-4}$$

and

$$\mathbf{N}_A = \mathbf{U}_y C_A + J_A \tag{10-5}$$

$$\mathbf{N}_A - C_A(\mathbf{N}_A \overline{V}_A + \mathbf{N}_B \overline{V}_B) = J_A = -D_{AB}\frac{\partial C_A}{\partial y} \tag{10-6}$$

Equation (10-6) is Fick's First Law for the new N flux, as such it represents a starting point for mass transfer. Note that equation (10-6) includes both a molecular diffusion term (i.e., $-D_{AB} dC_A/dy$) and a flux (the term $C_A(N_A V_A + N_B V_b)$ due to fluid bulk motion.

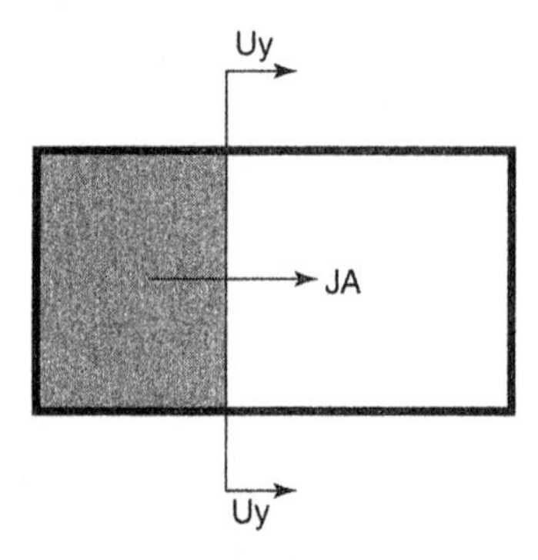

Figure 10-1. Motion of diffusion boundary. (Reproduced with permission from reference 6. Copyright 1997, American Chemical Society.)

Because we track the behavior of particular molecular species in mass transfer, we need to have individual mass balances or equations of continuity for each species. This is especially needed because we can convert species from one form to another by chemical reaction (i.e., A can be changed to B or C, etc.).

In order to develop such an equation, we carry out a mass balance on a space element (see Figure 10-2). Note that the principal terms will include accumulation, the balance of the mass fluxes, and chemical reaction. If we consider this in a differential element that we shrink to an infinitesimal basis, we obtain (in rectangular coordinates)

$$\frac{\partial c_A}{\partial t} + \left(\frac{\partial N_{Ax}}{\partial x} + \frac{\partial N_{Ay}}{\partial y} + \frac{\partial N_{Az}}{\partial z}\right) = R_A \tag{10-7}$$

Likewise, for cylindrical coordinates we obtain

$$\frac{\partial c_A}{\partial t} + \left(\frac{1}{r}\frac{\partial}{\partial r}(rN_{Ar}) + \frac{1}{r}\frac{\partial N_{A\theta}}{\partial \theta} + \frac{\partial N_{Az}}{\partial z}\right) = R_A \tag{10-8}$$

while for spherical coordinates the form is

$$\frac{\partial c_A}{\partial t} + \left[\frac{1}{r^2}\frac{\partial}{\partial r}(r^2 N_{Ar}) + \frac{1}{r\sin\theta}\frac{\partial}{\partial \theta}(N_{A\theta}\sin\theta) + \frac{1}{r\sin\theta}\frac{\partial N_{A\phi}}{\partial \phi}\right] = R_A \tag{10-9}$$

In equations (10-7), (10-8), and (10-9) the $\partial C_A/\partial t$ term represents accumulation, the bracketed N_A terms represent the mass fluxes, and R_A represents a chemical reaction.

Note that similar equations would be written for each of the molecular species present in the system. Also note that if all of the individual equations of continuity of species are summed, the basic overall equation of continuity for flow results.

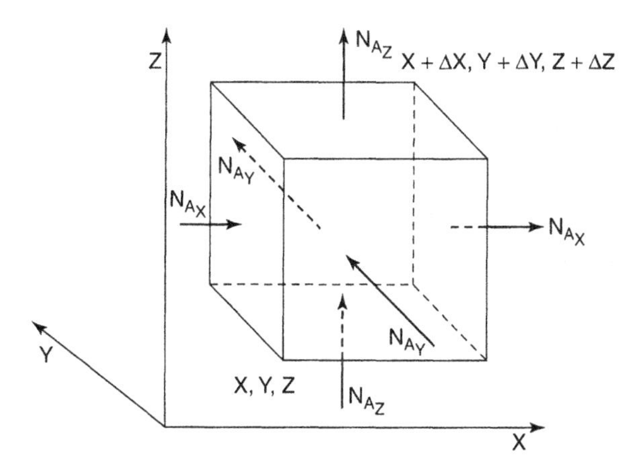

Figure 10-2. Mass balance on a volume in space. (Reproduced with permission from reference 6. Copyright 1997, American Chemical Society.)

In systems with a constant density and diffusivity, equations (10-7), (10-8), and (10-9) become equations (10-10), (10-11), and (10-12):

$$\frac{\partial c_A}{\partial t} + \left(v_x \frac{\partial c_A}{\partial x} + v_y \frac{\partial c_A}{\partial y} + \frac{\partial c_A}{\partial z} \right) = D_{AB} \left(\frac{\partial^2 c_A}{\partial x^2} + \frac{\partial^2 c_A}{\partial y^2} + \frac{\partial^2 c_A}{\partial z^2} \right) + R_A \tag{10-10}$$

$$\frac{\partial c_A}{\partial t} + \left(v_r \frac{\partial c_A}{\partial r} + v_\theta \frac{1}{r} \frac{\partial c_A}{\partial \theta} + v_z \frac{\partial cA}{\partial z} \right)$$

$$= D_{AB} \left(\frac{1}{r} \frac{\partial}{\partial r} \left(r \frac{\partial c_A}{\partial r} \right) + \frac{1}{r^2} \frac{\partial^2 c_A}{\partial \theta^2} + \frac{\partial^2 c_A}{\partial z^2} \right) + R_A \tag{10-11}$$

$$\frac{\partial c_A}{\partial t} + \left(v_r \frac{\partial c_A}{\partial r} + v_\theta \frac{1}{} \frac{\partial c_A}{\partial \theta} + v\phi \frac{1}{r \sin\theta} \frac{\partial c_A}{\partial \phi} \right)$$

$$= D_{AB} \left[\frac{1}{r^2} \frac{\partial}{\partial r} \left(r^2 \frac{\partial c_A}{\partial r} \right) + \frac{1}{r^2 \sin\theta} \frac{\partial}{\partial \theta} \left(\sin\theta \frac{\partial c_A}{\partial \theta} \right) + \frac{1}{r^2 \sin^2\theta} \frac{\partial^2 c_A}{\partial \phi^2} \right] + R_A \tag{10-12}$$

STEADY-STATE MOLECULAR DIFFUSION IN BINARY SYSTEMS

In steady-state cases the $\partial c_A/\partial t$ term is zero. A typical form (in this case for rectangular coordinates) is

$$\left(\frac{\partial N_{Ax}}{\partial x} + \frac{\partial N_{Ay}}{\partial y} + \frac{\partial N_{Az}}{\partial z} \right) = R_A \tag{10-13}$$

Likewise, if density and diffusivity are constant, we have

$$\left(v_x \frac{\partial c_A}{\partial x} + v_y \frac{\partial c_A}{\partial y} + v_z \frac{\partial c_A}{\partial z} \right) = D_{AB} \left(\frac{\partial^2 c_A}{\partial x^2} + \frac{\partial^2 c_A}{\partial y^2} + \frac{\partial^2 c_A}{\partial z^2} \right) + RA \tag{10-14}$$

In a static system where there is only molecular diffusion and chemical reaction, we have an analog to steady-state heat conduction with heat generation [see equations (5-17) through (5-20)]. Hence the applicable form of equation (10-14) is

$$D_{AB} \left(\frac{\partial^2 c_A}{\partial x^2} + \frac{\partial^2 c_A}{\partial y^2} + \frac{\partial^2 c_A}{\partial z^2} \right) + R_A = 0 \tag{10-15}$$

Furthermore, if there is no generation term (i.e., no chemical reaction), we go to a form that is analogous to steady-state heat conduction:

$$D_{AB} \left(\frac{\partial^2 c_A}{\partial x^2} + \frac{\partial^2 c_A}{\partial y^2} + \frac{\partial^2 c_A}{\partial z^2} \right) = 0 \tag{10-16}$$

Equation of the types of (10-15) and (10-16) can also be written for cylindrical and spherical coordinates.

Table 10-2 Typical Diffusivity Values for Various Systems

System	Diffusivity (m^2/sec)
Gas–gas	10^{-5} to 10^{-6}
Liquid–liquid	10^{-9} to 10^{-11}
Liquid–solid	10^{-9} to 10^{-11}
Gas–solid	10^{-11} to 10^{-14}
Solid–solid	10^{-19} to 10^{-34}

One aspect of the preceding which merits some discussion is the diffusivity D_{AB}. First of all, this transport coefficient has values that differ by many orders of magnitude for the type of system (i.e., gas–gas, liquid–liquid, etc.). The range of values found in these cases are given in Table 10-2.

As can be seen, the diffusivity values for gas–gas systems are many orders of magnitude above the other cases. This shows that molecular diffusion in a liquid or solid system will generally be the rate-controlling mass transfer step. Also, although not shown above, concentration of the diffusing species will have an effect on the diffusivity value for all of the systems except the gas–gas case. Hence, dealing with mass transport for these systems requires a knowledge of the effect of concentration.

The gas–gas case requires some additional explanation. Obviously, since diffusivity is pressure-dependent, the value will change and could be perceived in a sense as being affected by concentration. However, in the ideal gas range and probably higher, the relative molar ratios of A and B do not affect D_{AB}. Hence, for this region, regardless of whether we have 10% A–90 percent B or 50 percent A–50 percent B, the D_{AB} value is the same.

Another important aspect of binary system gas diffusivities is the relation of D_{AB} to D_{BA}. In Chapter 1 it was shown that

$$D_{AB} = 1.8583 \times 10^{-9} \frac{T^3(1/M_A + 1/M_B)}{P\sigma_{AB}^2 \Omega_{D_{AB}}} \tag{10-17}$$

Note that the temperature, pressure σ_{AB} and ΩD_{AB} will be the same for both D_{AB} and D_{BA}. Furthermore, the remaining term $(1/M_A + 1/M_B)^{1/2}$ is the same for both D_{AB} and D_{BA}. Hence, $D_{AB} = D_{BA}$ in a binary gas system.

UNSTEADY-STATE MOLECULAR DIFFUSION

For a zero-velocity, unsteady-state system without chemical reaction, equation (10-10) becomes

$$\left(\frac{\partial C_A}{\partial t}\right) = D_{AB}\left(\frac{\partial^2 C_A}{\partial x^2} + \frac{\partial^2 C_A}{\partial y^2} + \frac{\partial^2 C_A}{\partial z^2}\right) \tag{10-18}$$

This form is known as *Fick's Second Law*.

If we write a linear one-dimensional version of (10-11), we obtain

$$\frac{\partial C_A}{\partial t} = D_{AB}\frac{\partial^2 C_A}{a y^2} \tag{10-19}$$

in terms of concentration or

$$\frac{\partial y_A}{\partial t} = D_{AB}\frac{\partial^2 y_A}{\partial y^2} \tag{10-20}$$

in terms of mole fraction (y_A).

Note that heat transfer equivalent (Chapter 5) will be

$$\frac{\partial T}{\partial t} = \left(\frac{k}{pC_\rho}\right)\left(\frac{\partial^2 T}{\partial y^2}\right) \tag{10-21}$$

The physical and mathematical similarity of equations (10-19) through (10-21) mean that the extensive published solutions of Carslaw and Jaeger (2) for heat transfer and Crank (3) for mass transfer can be used interchangeably to deal with either unsteady state heat or mass transfer.

Also, the Gurney–Lurie charts of Chapter 5 (Figures 5-3 through 5-5) can be used for mass transfer by substituting analogous quantities. These substitutions would be C_{A1}, C_A and C_{A0} for T_1, T, and T_0; D_{AB} for (k/rC_p) and $D_{AB}/k_c x_1$ for k/hx_1 (where k_c is a mass transfer coefficient).

MULTICOMPONENT SYSTEM MASS TRANSFER

In binary gaseous systems the diffusivity is a *property* of the system determined by the two gases present. However, in multicomponent systems the fluxes of the various components will affect the diffusivity of any given component in the mixture.

As is apparent, multicomponent diffusion systems are extremely complicated. There are, however, some approaches that are used with certain restrictions.

In a gas system composed of a mix of ideal gases, it is possible to derive the Stefan–Maxwell equation (4,5), which solves for the mole fraction of a component in terms of the diffusivities, concentration, and diffusion velocities [velocity of a given species; U in equation (10-3)] a one-dimensional form of the equation is

$$\frac{dy_A}{dy} = \sum_{j=A}^{n} \frac{C_A C_j}{C_{\text{total}} D_{Aj}}(\overline{U}_j - \overline{U}_A) \tag{10-22}$$

This can be put into the form

$$\frac{1}{CD_{i\text{mix}}} = \frac{\displaystyle\sum_{j=1}^{n}\left(\frac{1}{CD_{ij}}\right)(X_j N_i - X_i N_j)}{N_i - X_1 \displaystyle\sum_{j=1}^{n} N_j} \tag{10-23}$$

If all of the other components are stationary or if all move with the same diffusion velocity, then

$$D_{A\text{mix}} = \frac{1 - y_A}{\sum \frac{y_j}{D_{Aj}}} \tag{10-24}$$

Some additional cases where simplified versions of (10-22) can be used are when i trace components are present in a nearly pure A

$$D_{i\text{mix}} = D_{iA} \tag{10-25}$$

or where all of diffusivities are nearly the same

$$D_{i\text{mix}} = D_{ij} \tag{10-26}$$

A useful semiempirical approach (4) for cases with large changes of $D_{i\text{mix}}$ is to use a linear relationship with either composition or position.

Liquid systems are even more complicated than the gaseous ones. Some complicated forms are available in reference 5.

WORKED EXAMPLES

Example 10-1 Find the mole fraction profiles (for A and B) and the overall flux of a component A diffusing through a stagnant layer of B (nondiffusing).

In this case we use a slab model (see Figure 10-3) of thickness y_0 or $(y_2 - y_1)$. Next we use the equation of continuity of species:

$$\underbrace{\frac{\partial C_A}{\partial t}}_{\text{Steady state}} + \underbrace{\frac{\partial N_{Ax}}{\partial x}}_{\text{No } x \text{ flux}} + \frac{\partial N_{Ay}}{\partial y} + \underbrace{\frac{\partial N_{Az}}{\partial z}}_{\text{No } z \text{ flux}} = \underbrace{R_A}_{\text{No chemical reaction}}$$

to obtain

$$\frac{\partial N_{Ay}}{\partial y} = 0$$

Then, using the definition of N_A [equation (10-6)], we obtain

$$N_{Ay} - C_A(N_{Ay}\overline{V}_A + \underbrace{N_B\overline{V}_B}_{\text{No } B \text{ flux}}) = -D_{AB}\frac{dC_A}{dy}$$

and

$$N_{Ay}(1 - C_A V_A) = N_{Ay}(1 - y_A) = -D_{AB}\frac{dC_A}{dy}$$

Next note that

$$\frac{C_A}{C_{\text{Total}}} = \frac{n_A/V}{n_{\text{Total}}/V} = y_A$$

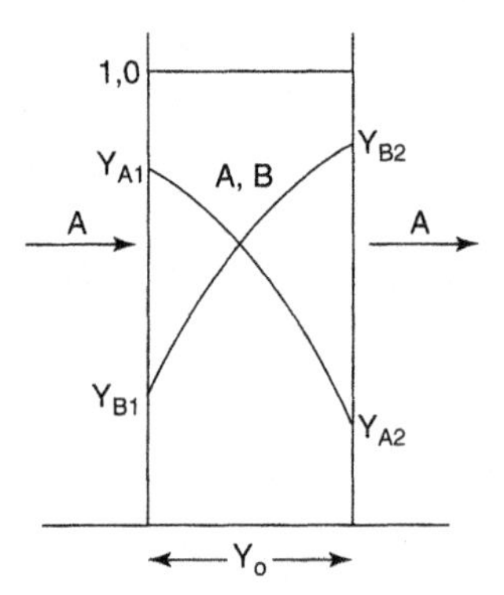

Figure 10-3. Slab model for diffusion. (Reproduced with permission from reference 6. Copyright 1997, American Chemical Society.)

but for an ideal gas

$$\frac{n_{\text{Total}}}{V} = \frac{RT}{P}$$

$$C_A = \frac{y_A P}{RT}$$

Thus,

$$N_{Ay} = -\frac{D_{AB}P}{RT(1-y_A)}\frac{dy_A}{dy}$$

The original expression from the equation of continuity of species can be written as an ordinary differential:

$$\frac{dN_{Ay}}{dy} = 0$$

Now substituting for N_{Ay} and noting that P, R, and T are constant and that D_{AB} is not a function of y, we obtain

$$\frac{d}{dy}\left(\frac{dy_A/dy}{1-y_A}\right) = 0$$

Using the boundary conditions

$$y = y, \qquad y_A = y_A$$
$$y = y_2, \qquad y_A = y_{A_2}$$

yields

$$\frac{1-y_A}{1-y_{A_1}} = \left(\frac{1-y_{A_2}}{1-y_{A_1}}\right)^{\frac{y-y_1}{y_2-y_1}}$$

$$\frac{y_B}{y_{B_1}} = \left(\frac{y_{B_2}}{y_{B_1}}\right)^{\frac{y-y_1}{y_2-y_1}}$$

The above equation give the mole fraction profiles (see Figure 10-4) within the slab.

Likewise, for flux N_{Ay} we obtain

$$NA_y(1 - y_A) = \frac{-D_{AB}P}{RT}\frac{dy_A}{dy}$$

For the slab of thickness, y_0, integration gives

$$N_{Ay} = \frac{D_{AB}P}{RTY_0}\ln\frac{Y_{B_2}}{Y_{B_1}}$$

The solutions for y_A, y_B and N_{Ay} represent a general case that can be used for various application. We will consider such a case in the next example.

Example 10-2 A graduated cylinder (cross-sectional diameter of 0.01128 m) containing chloropicrin (CCl_3NO_2) is placed in a hood. The hood has a blower system that continually circulates dry air at a constant temperature (25°C) and pressure (atmospheric).

Originally, the liquid surface is 0.0388 m from the top of the cylinder. After one day the liquid level is 0.0412 m below the top.

If the vapor pressure and density, respectively, of chloropicrin are 3178.3 N/m^2 and 1650 kg/m^3, estimate the substances diffusivity in air. Molecular weight of chloropicrin is 164.39.

This case corresponds to the stagnant film system considered in Example 10-1. Hence, we use the flux equation:

$$N_{Ay} = \frac{D_{AB}P}{RTY_0}\ln\frac{Y_{B_2}}{Y_{B_1}}$$

For this equation we designate chloropicrin as "A" and air as "B." Furthermore, we assume that any chloropicrin diffusing to the top of the graduated cylinder is swept away (i.e., $y_{A_2} = 0$, $y_{B_2} = 1.0$). Also we average the y_0 values (0.0388 and 0.0412 m) to get a y_0 of 0.04 m.

Note that R, P, and T are either constants or constant quantities. Furthermore, we use Raoult's Law to give us a value for the mole fraction of A at the surface:

$$y_{A_1} = \frac{3178.3 \text{ N/m}^2}{1.014 \times 10^5 \text{ N/m}^2} = 0.031$$

and

$$y_{B_1} = (1 - 0.031) = 0.969$$

Also, the flux N_{Ay} is the mass of liquid vaporized per unit area, per unit time:

$$N_{Ay} = \frac{(1650\ \text{kg/m}^3)(0.0412 - 0.0388)\text{m}\ (\pi/4)(0.0128\ \text{m})^2}{(\pi/4)(0.0128\ \text{m})^2\ 24\ \text{hr}\ (3600\ \text{sec/hr})}$$

$$N_{Ay} = 4.6 \times 10^{-5}\ \text{kg/m}^2\ \text{sec}$$

$$N_{Ay} = 4.6 \times 10^{-5}\ \text{kg/m}^2\ \text{sec}\left(\frac{1000\ \text{g}}{\text{kg}}\right)\left(\frac{\text{g mole}}{164.39\ \text{g}}\right) = 2.82 \times 10^{-4}\frac{\text{g mole}}{\text{m}^2\ \text{sec}}$$

Then,

$$D_{AB} = \frac{N_{Ay} RT y_0}{\text{P}\ln(y_{B_2}/y_{B_1})}$$

$$D_{AB} = \frac{\left(2.82 \times 10 - 4\frac{\text{g mole}}{\text{m}^2\ \text{sec}}\right)\left(8.314\frac{\text{Nm}}{\text{g mole}\ ^\circ\text{K}}\right)(298.16^\circ\text{K})(0.04\ \text{m})}{(1.014 \times 10^5\ \text{N/m}^2)\ln(1.0/0.969)}$$

$$D_{AB} = 8.75 \times 10^{-6}\ \text{m}^2/\text{sec}$$

Example 10-3 Estimate the consumption rate of a carbon particle (diameter of 3×10^{-3} m) at 1000°K and atmospheric pressure in oxygen.

The diffusivity coefficient is 1.032×10^{-4} m^2/sec.

If we assume that the spherical carbon particle is covered by a thick layer of CO_2, then we can write ($A = CO_2$)

$$\frac{W_A}{4\pi r^2} = -D_{AB}\frac{dC_A}{dr}$$

where W_A is in kg mole/sec (i.e., $W_A/4\pi r^2$ is a flux).

Then,

$$dC_A = \frac{W_A}{4\pi D_{AB}}\frac{dr}{R^2}$$

If the gas is ideal, then

$$dP_A = \frac{W_A RT}{4\pi D_{AB}}\frac{dr}{r^2}$$

Then if

$$r = r_0, \qquad P_A = P_{A0}$$
$$r = \infty, \qquad P_A = 0 \quad (\text{i.e., only } O_2 \text{ present})$$

we obtain

$$P_{A0} - 0 = \frac{W_A RT}{4\pi D_{AB}}\left(\frac{1}{r_0} - \frac{1}{\infty}\right)$$

and

$$W_A = \frac{4\pi D_{AB} P_{A0} r_0}{RT}$$

$$W_A = \frac{(4\pi)(1.032 \times 10^{-4}\ \mathrm{m^2/sec})(1.014 \times 10^5\ \mathrm{N/m^2})(1.5 \times 10^{-3}\ \mathrm{m})}{\left(8.314 \dfrac{\mathrm{N\ m}}{\mathrm{gram\ mole\ ^\circ K}}\right)(1000^\circ\mathrm{K})}$$

$$W_A = 2.37 \times 10^{-5}\ \mathrm{g\ mole/sec}$$

Also since

$$C + O_2 \longrightarrow CO_2$$

the consumption of carbon is equal to the appearance of CO_2:

$$\text{kg of carbon consumed per second} = \left(2.37 \times 10^{-5} \frac{\mathrm{g\ mole}}{\mathrm{sec}}\right)\left(\frac{\mathrm{kg}}{1000\ \mathrm{g}}\right)\left(\frac{12\ \mathrm{g}}{\mathrm{gram\ mole}}\right)$$

$$\text{kg of carbon consumed per second} = 2.84 \times 10^{-4}\ \mathrm{kg/sec}$$

Example 10-4 A given gas A can diffuse through the solid wall of a tube in which it flows. Find the rate of gas diffusion in terms of the tube dimensions (inside wall radius R_i, outside wall radius R_0), the gas–solid diffusivity, and the gas concentrations at the walls.

Equation (10-8) is the starting point for this case. Using this equation together with the system's conditions gives

$$\frac{\partial c_A}{\partial t} + \frac{1}{r}\frac{\partial}{\partial r}(rN_{Ar}) + \frac{1}{r}\frac{\partial N_{A\theta}}{\partial \theta} + \frac{\partial N_{Az}}{\partial z} = R_A$$

Steady state — Only r direction flux — No chemical reaction

$$\frac{1}{r}\frac{d}{dr}(rN_{Ar}) = 0$$

Then,

$$N_{Ar} = -D_{AB}\frac{dC_A}{dr} + x_A(N_{Ar} + N_{Br})$$

but N_{Br} is zero (B is the solid) and x_A is a small quantity. Thus,

$$N_{Ar} = -D_{AB}\frac{dC_A}{dr}$$

and

$$\frac{d}{dr}\left(r\frac{dC_A}{dr}\right) = 0$$

Boundary conditions for the system are

$$C_A = C_{Ai}, \qquad r = R_i$$
$$C_A = C_{A0}, \qquad r = R_0$$

The C_A expression is then

$$C_A = \frac{(C_{Ai} - C_{A0}) \ln(r/R_i)}{\ln(R_i/R_0)}$$

Since

$$N_{Ar} = -D_{AB}\frac{dC_A}{dr}$$

we obtain

$$N_{Ar} = \frac{D_{AB}(C_{Ai} - C_{A0})}{r \ln(R_0/R_i)}$$

The mass transfer rate W_A is

$$W_A = 2\pi r L(N_{Ar})$$
$$W_A = \frac{2\pi L D_{AB}(C_{Ai} - C_{A0})}{\ln(R_0/R_i)}$$

Example 10-5 A long, gel-like slab of material (1.5 cm thick) contains a solvent (concentration of 2×10^{-4} g mole/cm^3). The solid is placed in a fast-flowing water stream so that the solvent can diffuse in a direction perpendicular to the slab axis. The solvent diffusivity in the solid is 5.0×10^{-10} m^2/sec.

Find the centerline concentration of solvent after a day (24 hours). Also, what would the effect of a 25% increase in the slab thickness on centerline concentration be?

We can use the Gurney–Lurie chart (Figure 5-3) of Chapter 5 to solve this problem. In order to do this, we evaluate the chart parameters:

$$n = 0/0.75 \text{ cm} = 0$$
$$m = \frac{D_{AB}}{K k_c x_1}$$

where K is an equilibrium distribution coefficient.

However, the k_c (mass transfer coefficient) will be very large (i.e., fast-flowing water). Hence m is taken to be zero.

The chart's abscissa X is

$$X = \frac{D_{AB}t}{x_1^2} = \frac{(5 \times 10^{-10} \text{ m}^2/\text{sec})(24)(3600 \text{ sec})}{(7.5 \times 10^{-3} \text{ m})^2}$$
$$X = 0.768$$

The corresponding y value from Figure 5-3 (m, n both zero) is

$$y = 0.2$$

and

$$0.2 = \frac{C_1 - C}{C_1 - C_0} = \frac{0 - C}{0 - (2 \times 10^{-4} \text{ g mole/m}^3)}$$

$$C = 4 \times 10^{-5} \text{ g moles/cm}^3$$

In the second case x_1 is changed. This affect only the abscissa X (i.e., m and n are still both zero).

$$X = \frac{D_{AB}t}{x_1^2} = \frac{(5 \times 10^{-10} \text{ m}^2/\text{sec})(24)(3600 \text{ sec})}{(9.375 \times 10^{-3} \text{ m})^2}$$

$$X = 0.491$$

This gives a y value of 0.4 from Figure 5-3.

Then,

$$C = C_0(y) = 8 \times 10^{-5} \text{ g mole/cm}^3$$

Example 10-6 A gaseous reaction (A_1, A_2, A_3, A_4) system contains four components, only three of which react:

$$A_1 + 3A_2 \rightarrow A_3$$

In the system (A_1, A_2, A_3, A_4 are represented by the subscripts 1, 2, 3, 4) the binary CD_{ij} values (all in kg mole/m sec) are

$$CD_{12} = CD_{21} = 24.2 \times 10^{-11}$$

$$CD_{13} = CD_{31} = 1.95 \times 10^{-11}$$

$$CD_{14} = CD_{41} = 6.21 \times 10^{-11}$$

$$CD_{23} = CD_{32} = 20.7 \times 10^{-11}$$

$$CD_{24} = CD_{42} = 41.3 \times 10^{-11}$$

$$CD_{34} = CD_{43} = 5.45 \times 10^{-11}$$

Hence the N summation term is

$$\sum_{j=1}^{4} N_j = 3N_1$$

If the mole fractions of A_1, A_2, A_3, and A_4 are, respectively, 0.10, 0.8, 0.05, and 0.05, find values of the component diffusivities in the mixture.

In order to do this, we use equation (10-23), which we rewrite in the form

$$\frac{1}{CD_{i\text{mix}}} = \frac{\sum_{j=1}^{n}\left(\frac{1}{CD_{ij}}\right)(X_j N_i - X_i N_j)}{N_i - X_1 \sum_{j=1}^{n} N_j}$$

All of the fluxes (N values) can be determined or related to N_1.
From the reaction we obtain

$$N_2 = -3N_1$$
$$N_3 = -N_1$$

Also, because N_4 doesn't participate, we have $N_4 = 0$.

This means that all of the N terms can be expressed as either zero or related to N_1. The N_1 values in the numerator and denominator can be cancelled. Using the case of $CD_{i\text{mix}}$, we see that

$$\frac{1}{CD_{i\text{mix}}} = \frac{\frac{X_1 - X_1}{CD_{11}} + \frac{X_2 - 3X_1}{CD_{12}} + \frac{X_3 + X_1}{CD_{13}} + \frac{X_4 - 0}{CD_{14}} N_1}{N_1 - X_1(3N_1)}$$

$$\frac{1}{CD_{i\text{mix}}} = \frac{1}{1 - 3X_1} \frac{X_2 - 3X_1}{CD_{12}} \frac{X_3 + X_1}{CD_{13}} \frac{X_4}{CD_{14}}$$

$$\frac{1}{CD_{i\text{mix}}} = \frac{1}{1 - 0.3} \frac{0.8 - 0.3}{24.2} + \frac{0.05 + 0.10}{1.95} + \frac{0.05}{6.21} 10^{11} \frac{\text{kg mole}}{\text{m sec}}$$

$$CD_{i\text{mix}} = 6.63 \times 10^{-11} \text{ kg mole/m sec}$$

Likewise for $CD_{2\text{mix}}$, we have

$$\frac{1}{CD_{2\text{mix}}} = \frac{1}{3 - 3X_2} = \frac{3X_1 - X_2}{CD_{12}} + \frac{3X_2 - 3X_2}{CD_{22}} + \frac{3X_3 + X_2}{CD_{23}} + \frac{3X_4}{CD_{44}}$$

Substituting the appropriate values gives

$$CD_{2\text{mix}} = 20.8 \times 10^{-11} \text{ kg mole/m sec}$$

Values of $CD_{i\text{mix}}$ for components 3 and 4 (obtained in the same manner) are

$$CD_{3\text{mix}} = 8.71 \times 10^{-11} \text{ kg mole/m sec}$$
$$CD_{4\text{mix}} = 60 \times 10^{-11} \text{ kg mole/m sec}$$

As can be seen, the only values of $CD_{i\text{mix}}$ close to its binary values is that of component 2; this is explained by the fact that it has a much greater mole fraction value than the other components.

PROBLEMS

10-1. Oxygen diffuses through carbon monoxide (steady state). The carbon monoxide is stagnant. Temperature and pressure are 0°C and 1×10^5 N/m^2. Oxygen partial pressures are 13,000 and 6500 N/m^2 at two planes 3 mm apart. Mixture diffusivity is 1.87×10^{-5} m^2/sec. What is the oxygen rate of diffusion per plane square meter?

10-2. A droplet of material C (radius r_1) is suspended in a gas stream of D which forms a stagnant film around the droplet (radius r_2). If the C concentrations are Xc_1 and Xc_2 at r_1 and r_2, find the flux of C. Let D_{CD} be the diffusion coefficient.

10-3. Hydrogen gas is stored at high pressure in a rectangular container (10-mm-thick walls). Hydrogen concentration at the inside wall is 1 kmole/m^3 and essentially negligible on the outside wall. The D_{AB} for hydrogen in steel is 2.6×10^{-13} m^2/sec. What is the molar flux of the hydrogen through the steel?

10-4. An industrial pipeline containing ammonia gas is vented to the atmosphere (a 3-mm-tube is inserted into the pipe and extends for 20 m into the air). If the system is at 25°C, find the mass rate of ammonia lost from vent.

10-5. Oxygen at 250°C is at pressures of 1 and 2 bars, respectively, on each side of an elastomeric membrane (0.5 mm thick). Find the molar diffusion flux of oxygen and the concentrations on each side of the membrane.

10-6. Ammonia gas diffuses at a constant rate through 1 mm of stagnant air. Ammonia is 50 percent (by volume) at one boundary. The gas diffusing to the other boundary is rapidly absorbed. Concentration of ammonia at the second boundary is negligible. Ammonia diffusivity is 0.18 cm^2/sec at the system conditions (295°K, 1 atm). Determine the rate of diffusion of the ammonia.

10-7. What is the effect of increasing total pressure from 100 to 200 kN/m^2 for absorption of ammonia from 10 percent (by volume) in air with water as the solvent. Assume that the gas phase constitutes the principal resistance to mass transfer. What would the result be if the water solution exerted an ammonia partial pressure of 5 kN/m^2?

10-8. Two milliliters of acetone (mol. wt. of 58) and 2 ml of dibutyl phthalate are combined and placed in a 6-mm-diameter vertical glass tube placed in a 315°K bath. Air at 315°K and atmospheric pressure flows across the top of the tube. The liquid mixture is initially at 1.15 cm below the tube top. What is the amount of time needed for the liquid level to fall to 5 cm below the tube top? Vapor pressure and diffusivity of acetone are 60.5 kN/m^2 and 0.123 cm^2/sec. Acetone and dibutyl phthalate liquid densities are 764 and 1048 kg/m^3.

10-9. Benzene at 22°C is open to the atmosphere in a circular tank (6.10 m in diameter). Vapor pressure and specific gravity for benzene are 0.132 atm and 0.88. An air film of 5-mm thickness is above the benzene. What is the cost of evaporated benzene per day (assume value of benzene is $2 per gallon)?

10-10. An ethanol–water solution forms a 2.0-mm-thick layer that contacts an organic solvent (ethanol- but not water-soluble). The concentrations of ethanol (wt. %) are 16.8 and 6.8. Corresponding solution densities are 972.8 and 988.1 kg/m^3. If the diffusivity of ethanol (293°K) is 0.74×10^{-9} m/sec, find the steady-state ethanol flux.

10-11. A substance A diffuses to a catalyst surface where it is instantaneously polymerized (i.e., $nA \rightarrow An$). Find the expression for N (z the direction through an imaginary gas film).

10-12. Two agitated solutions of urea in water are connected by a 0.08-m-long tube of an agar gel (1.05 wt. % agar in water; 278°K). Urea concentrations in each solution are 0.2 and 0 g mole/liter (solution). If the urea diffusivity is 0.727×10^{-9} m^2/sec, what is the molar flux of the urea?

10-13. Vapor in humidified rooms frequently diffuse through a plaster wall and condenses in the surrounding insulation. Estimate the mass diffusion rate of water through 3 by 5 m wall 10 mm thick. Room and insulation water vapor pressures are 0.03 and 0 bar. Diffusivity of water in the dry wall is 10^{-9} m^2/sec. Water solubility in the solid is 5×10^{-3} kmole/mbar.

10-14. Carbon dioxide and nitrogen counterdiffuse in a circular tube (1 m long, diameter 50 mm) at 25°C and 1 atm. The tube ends are connected to large chambers where the species concentrations are kept at fixed values. Partial pressures of carbon dioxide are 0.132 and 0.066 atm at each tube end. What is the carbon dioxide mass transfer rate through the tube.

10-15. Ammonia is absorbed from air into water at atmospheric pressure and 20°C. Gas resistance film is estimated to be 1 mm thick. If ammonia diffusivity in air is 0.20 cm^2/sec and the partial pressure is 0.066 atm, what is the transfer rate. If the gas pressure is increased to 2 atm, what would the effect be on the mass transfer rate?

10-16. Hydrogen at 17°C and 0.01 atm partial pressure diffuses through neoprene rubber 0.5 mm thick. Calculate the flux (only resistance is neoprene). Diffusivity and solubility of hydrogen in neoprene are 1.03×10^{-10} m^2/sec and 0.0151 m^3 (gas 0°C, 1 atm)/m (solid) atm.

10-17. A liquid film of C flows down a vertical solid wall where for the vertical dimension z there is no dissolution for $z < 0$, but dissolution of a slightly soluble species for $0 < z < L$. The fluid velocity depends only on the y dimension (horizontal). Derive the appropriate equations and boundary conditions that will yield the concentration profiles and average mass transfer rate. In the foregoing the use of the groupings $a = \rho g\delta/\mu$ and $y(a/9Dz)$ are helpful.

10-18. Find the appropriate differential equation and boundary conditions for the injection of a small amount (W_A) of a chemical species A into a flowing stream of velocity V. Using cylindrical coordinates, A is taken downstream in the z direction but diffuses both radially (r direction) and axially (z direction.)

10-19. A gas A contacts a static liquid B in a tall vertical container. When the gas A diffuses, it also reacts (irreversible first-order reaction). Find the concentration profile of A in the liquid as well as its molar flux.

10-20. An alcohol and water vapor mixture is being separated by contact with an alcohol–water liquid solution. Alcohol is transferred from gas to liquid, and water is transferred from liquid to gas. Temperature and pressure are 25°C and 1 atm. The components diffuse through a gas film 0.1 mm thick at equal flow rates. Mole percents of alcohol are 80 and 10 on either side of the film. Find the rate of diffusion of both components through a film area of 10 m^2.

10-21. A 10 percent ammonia–air mixture is scrubbed with water to a 0.1 percent concentration in a packed column operating at 295°K and atmospheric pressure. Assume that a thin gas film constitutes the principal resistance to mass transfer. At a point where the ammonia concentration is reduced to 5 percent and where its partial pressure in equilibrium with the aqueous solution is 660 N/m^2 with a transfer rate of 10^{-3} kmole/m^2sec, what is the hypothetical gas film (ammonia diffusivity in air is 0.24 cm^2/sec)?

10-22. A 20-mm-thick rubber plug (surface area of 300 mm) contains carbon dioxide at 298°K and 5 bar in a 10-liter container. What is the mass loss rate of carbon dioxide? Also find the pressure reduction in 24 hr.

10-23. An open pan (0.2-m diameter, 80 mm high) contains water (27°C) and is exposed to air (27°C, 25 percent relative humidity). Find the evaporation rate based on diffusion alone.

10-24. The pores of a 2-mm-thick sintered silica (porosity 0.3) are filled with water at 25°C. At one side of the solid a potassium chloride solution (0.10 g mole/liter) is placed. Pure water flows past the other side. Experimental observation shows the potassium chloride flux to be 7.01×10^{-9} kg mole/sec. Based on this result, derive an appropriate equation to describe this system. (*Hint*: Alter the diffusion equation to reflect the porous medium.)

10-25. A sphere of naphthalene is suspended in dry still air at 318°K and atmospheric pressure. The vapor pressure and diffusivity of naphthalene are 7.3×10^{-4} atm and 6.92×10^{-6} m^2/sec. What is the rate of the naphthalene sublimation?

10-26. A liquid A evaporates into a vapor B in a tube of infinite length. The system is at constant temperature and pressure. The vapor is an ideal gas mixture. Furthermore, B is not soluble in A. Set up necessary equations and boundary conditions to find the rate of evaporation of A. Assume liquid level is maintained at axial dimension of $z = 0$ for any time.

10-27. Gas A diffuses through the cylindrical wall of a plastic tube. As it diffuses, it reacts at a rate R. Find the appropriate differential equation for this system.

10-28. Carbon dioxide is a necessary part of nature's photosynthesis process. If a body of water is contacted by a carbon dioxide source, diffusion will occur. However, the photosynthesis reaction of aquatic plants will retard the gas concentration in water. Find a differential equation to describe this system (assume the reaction rate constant to be k).

10-29. A large slab of salt is placed at the bottom of a tank containing water. The salt solid density is 2165 kg/m^3. The density of the solution at the surface is 380 kg/m^3. Diffusivity of salt water is 1.2×10^{-9} m^2/sec. Find the density distribution in the water and the rate of salt surface dissolution as a function of time.

10-30. A flask consists of a bulb with a cylindrical tube (15 cm long). The liquid in the bulb portion (well-mixed) is a saturated solution of salt (NaCl) in water. The entire apparatus is immersed in water. Find the diffusion rate of the salt. Density and diffusivity (in water) of the salt are 2163 kg/m^3 and 1.35×10^{-5} cm^2/sec (at 20°C).

10-31. A porous solid slab (50 percent void space) is soaked in ethanol. The slab is placed in a well-mixed container of water at 25°C. The effective diffusivity of ethanol in water can be taken as 1×10^{-6} cm^2/sec. Assume that the water and ethanol densities are essentially the same. If the concentration of ethanol in the water is initially zero, find how long it will take the ethanol mass fraction at the slab center to decrease to 0.009.

10-32. A mixture of benzene and toluene is fed as a vapor to the bottom of a distillation column. At a given point in the unit the vapor contains 80 mole % benzene while the corresponding liquid is 70 mole % benzene. Vapor pressure and diffusivity for benzene are 1.3 atm and 5.92×10^{-4} m^2/sec. The molal latent heats of vaporization are essentially the same. Find the rate of interchange of benzene and toluene, assuming a stagnant vapor layer of 0.254 cm.

10-33. Hydrogen gas (27°C; 10 bars) is stored in a 100-mm-diameter spherical tank (2-mm-thick wall). Molar concentrations of hydrogen at the inner and outer wall are 1.5 kg mole/m^3 and 0. Diffusivity of hydrogen in steel is 0.3×10^{-12} m^2/sec. Find the initial rate of hydrogen loss through the wall as well as the initial rate of pressure drop in the tank.

10-34. Helium gas (25°C, 4 bars) is contained in a glass tube (100-mm inside diameter; 5-mm wall thickness). Find the rate of mass lost per unit length.

10-35. A solid sphere (0.01 m in diameter) immersed in stagnant water at 25°C has its surface continually supplied with benzoic acid (i.e., sphere of benzoic acid with a constant diameter diffusing into water of infinite volume). If the aqueous solution at the surface is kept at 0.0278 g mole/liter, how long will it take for the benzoic acid flux to attain 99 percent of its steady-state value.

10-36. A condenser operates with a feed vapor of ammonia, water, and hydrogen (3.36 atm). At a given point in the unit the respective mole fractions are 0.3, 0.4, and 0.3. The liquid on the condenser is at 37.8°C (0.10 ammonia, 0.90 water). Estimate the rate of condensation of water relative to ammonia.

10-37. Oxygen gas diffuses through a 2-to-1 volume ratio of methane and hydrogen. The oxygen partial pressures (2.0 mm at points apart) are 13,000 and 6500 N/m^2. Diffusivity of oxygen, respectively, in hydrogen and methane are 6.99×10^{-5} and 1.86×10^{-5} m^2/sec. Calculate the mass transfer flux of the oxygen.

10-38. What is the diffusion rate of acetic acid across a 12-mm-thick layer of water at 17°C. Acetic acid concentrations are 9 and 3 wt. % acid. The diffusivity of acetic acid is 0.95×10^{-9} m^2/sec.

10-39. Oxygen is transferred from the inside of the lung through the lung tissue to blood vessels. Assume the lung tissue to be a plane wall of thickness L and that inhalation maintains a constant oxygen molar concentration at the inner wall as well as another constant oxygen molar concentration at the outer wall. Additionally, oxygen is consumed in the lung tissue by a metabolic reaction (zeroth order). Determine the distribution of oxygen in the tissue and the rate of assimilation of oxygen by the blood.

10-40. Carburization of steel is a high-temperature process. At a temperature of 1273°K, how much time would be required to raise the steel carbon content at a depth of 1 mm from 0.1 to 1.0 percent (carbon mole fraction on steel surface is 0.02)?

10-41. A droplet of a liquid (C) is suspended in a gas (D). The droplet (radius r_1) is surrounded by a spherical stagnant gas film (radius r_2). Obtain the net surface flux of C (concentrations at r_1 and r_2 are x_{C_1} and x_{C_2}).

Also, if this flux is equal to k_P $(P_{A_1} - P_{A_2})$, find the k_P value if $r_2 \rightarrow \infty$.

10-42. Methane (298°K, 1 atm) diffuses through argon and helium. At the base position ($z = 0$) the methane, argon, and helium partial pressures are 0.4, 0.4, and 0.2. At a position of $z = 0.005$ m they are 0.1, 0.6, and 0.3. Binary diffusivities and helium are 2.02×10^{-5} m^2/sec and 6.75×10^{-5} m^2/sec. Diffusivity of argon in helium is 7.29×10^{-5} m^2/sec. Determine the flux of methane.

REFERENCES

1. L. Onsager, *Phys. Rev.* **37**, 405, **38**, 2265 (1931).
2. H. S. Carslaw and J. C. Jaeger, *Heat Conduction in Solids*, Oxford University, Oxford, England (1959).
3. J. Crank, *The Mathematics of Diffusion*, Oxford University, Oxford, England (1956).
4. C. F. Curtiss and J. O. Hirschfelder, *J. Chem. Phys.* **17**, 550 (1949).
5. J. O. Hirschfelder, C. F. Curtiss, and R. B. Bird, *Molecular Theory of Gases and Liquids*, Wiley, New York (1954).
6. R. G. Griskey, *Chemical Engineering for Chemists*, American Chemical Society, Washington, D.C. (1997).

11

CONVECTIVE MASS TRANSFER COEFFICIENTS

INTRODUCTION

Earlier we saw that situations involving convective heat transfer, complicated geometries, and flows required the use of the heat transfer coefficient. The same situation applies for mass transfer. In the preceding chapter we essentially treated situations involving molecular diffusion (the counterpart to conduction heat transfer). Extension of mass transfer to complicated cases produces situations where the solutions of the equations of change (developed in the preceding chapter) are extremely difficult to attain.

Although there is analogy between the heat and mass transfer coefficient (which we will explore in more detail), a basic difference does occur. This difference is that there are a number of possible mass transfer coefficients depending on the driving forces chosen.

Some examples are shown in equation (11-1):

$$N_A = k_c(C_{A1} - C_{A2}) = k_y(y_{A1} - y_{A2}) = k^* \left(\frac{y_{A1} - y_{A2}}{y_{BM}} \right) = k_g(P_{A1} - P_{A2}) \tag{11-1}$$

The k_c, k_y, k^*, and k_g are all mass transfer coefficients that, respectively, have the units cm/sec, g mole/sec cm^2, g mole/sec cm^2 and g mole/sec cm^2 atm. The P_A's are partial pressure for component A, and Y_{BM} is the logarithmic mole fraction of the nondiffusing component (analogous to the logarithmic mean temperature difference):

$$y_{BM} = \frac{y_{A1} - y_{A2}}{\ln\left(\dfrac{y_{B2}}{y_{B1}}\right)} \tag{11-2}$$

It might at first seem strange that we can have a number of different mass transfer coefficients in contrast to essentially one heat transfer coefficient. However, consider the situation for chemical equilibrium constants where we also have a number of different constants (based on activities, fugacities, partial pressures, concentrations, etc). In dealing with the mass transfer coefficient it becomes very important to clearly know the characteristic driving force used for the system. While the units of the coefficients are helpful, they are not foolproof (*example*: both k_y and k^* have same units).

BASIC RELATIONSHIPS FOR THE MASS TRANSFER COEFFICIENT

In Figure 11-1, we illustrate a situation where mass is transferred from a soluble wall into a flowing liquid. The soluble material has a mole fraction x_{A0} at the wall.

If the bulk concentration of A is taken to be x_{Ab}, then

$$N_A - x_{Ao}(N_A + N_B) = k_x(x_{A0} - x_{Ab}) \tag{11-3}$$

or

$$W_A - x_{Ao}(W_A + W_B) = k_x \pi DL(x_{Ao} - x_{Ab}) \tag{11-4}$$

As in the case of heat transfer, we can write the expression

$$W_A - x_{Ao}(W_A + W_B) = \int_0^L \int_0^{2\pi} \left(CD_{AB} \frac{\partial x_A}{\partial r} \right)_{r=R} R \, d\theta dz \tag{11-5}$$

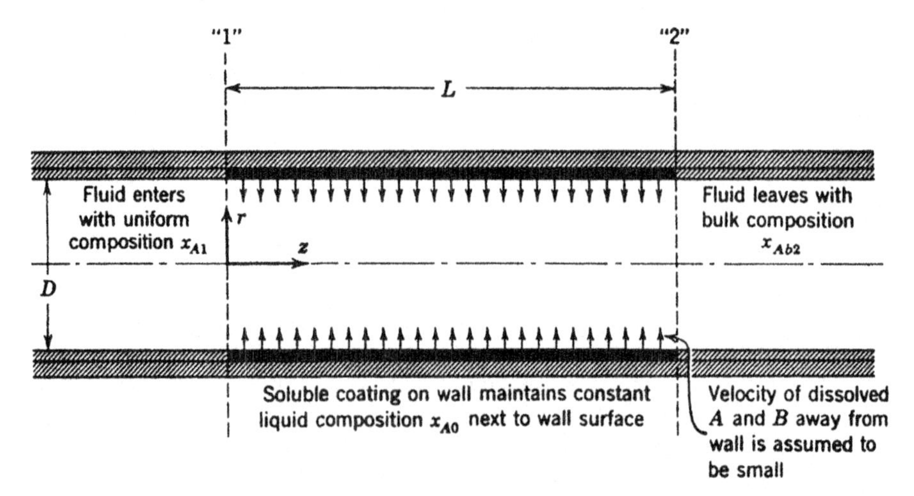

Figure 11-1. Mass transfer model; solid dissolving in a flowing liquid. (Adapted with permission from reference 21. Copyright 1960, John Wiley and Sons.)

Substituting for k_x gives

$$k_x = \frac{1}{(\pi DL)(X_{A0} - X_{Ab})} \int_0^L \int_0^{2\pi} \left(CD_{AB} \frac{\partial x_A}{\partial r} \right) \bigg|_{r=R} R \, d\theta dz \qquad (11\text{-}6)$$

Next we make the above equation dimensionless by using the quantities

$$r^* = r/D, \qquad z^* = z/D, \qquad X_A^* = \left(\frac{X_A - X_{A0}}{X_{Ab} - X_{A0}} \right)$$

to give

$$\frac{k_x D}{CD_{AB}} = \frac{1}{2\pi L/D} \int_0^{L/D} \int_0^{2\pi} \left(\frac{-\partial X_A^*}{\partial r^*} \right) \bigg|_{r=R} d\theta dz^* \qquad (11\text{-}7)$$

The $k_x D/CD_{AB}$ term is a mass transfer Nusselt number.

Additional analysis reveals that the dimensionless mole fraction is a function of various groups (r^*, θ, z^* the Reynolds number) and an additional term, the Schmidt number:

$$\text{Sc} = \left(\frac{\mu}{\rho D_{AB}} \right) \qquad (11\text{-}8)$$

The net result is that

$$\text{Nu}_{x,\text{mass}} = \phi(\text{Re, Sc}, L/D) \qquad (11\text{-}9)$$

Alternatively, the relation in equation (11-9) can be found empirically. Here we start by using the ratio technique illustrated earlier:

$$\text{Schmidt no.} = \text{Sc} = \left(\frac{\text{Momentum diffusivity}}{\text{Mass diffusivity}} \right) \qquad (11\text{-}10)$$

and

$$\text{Sc} = \frac{\mu/\rho}{D_{AB}} = \frac{\mu}{\rho D_{AB}} \qquad (11\text{-}11)$$

If we multiply the Schmidt number by the Reynolds number and numerator and denominator by a concentration driving force, we obtain

$$\text{(Schmidt number)(Reynolds number)} = \text{(Sc)(Re)} = \frac{V(C_0 - C_1)/D}{D_{AB}(C_0 - C_1)/D^2}$$

This combination of ScRe is analogous to that of PrRe and as such is a mass transfer Peclet number.

Empirically, we also note that in laminar flow (i.e., layers of fluid) or turbulent entrance flow the L/D is important. Hence, we obtain from various mass transfer Nusselt numbers (i.e., $k_c Y_{Bm} D/D_{AB}$, etc.) the overall relation of equation (11-13):

$$\frac{k_c Y_{BM} D}{D_{AB}} = \frac{k_y RT y_{BM} D}{P D_{AB}} = \frac{k^* RT D}{P D_{AB}} = \frac{k_g y_{BM} RT D}{D_{AB}} = f(\text{Re}, \text{Sc}, L/D) \tag{11-13}$$

Note that unlike the heat transfer case no viscosity ratio μ_b/μ_w is used. This is because the effect of mass transfer does not usually alter the properties used in the dimensionless groups. One additional point that should be noted is that the mass transfer coefficient can be altered when a large amount of mass is transferred. This will be dealt with later.

The mass transfer cases discussed above are for forced convection. Free convection also exists for mass transfer when the equation of motion involves a term $-\rho\psi g(X_A - X_{A0})$ instead of $-\nabla P + \rho g$. The ψ term is

$$\psi = -\frac{1}{\rho}\left(\frac{\partial \rho}{\partial X_A}\right)_{P,T} \tag{11-14}$$

The mass transfer Nusselt number is then

$$\text{Nu}_{\text{mass}} = k_x D/C D_{AB} = \phi(\text{Gr}_{\text{mass}}, \text{Sc}) \tag{11-15}$$

where

$$\text{Gr}_{\text{mass}} = \frac{D^3 \rho^2 g \psi \Delta X_A}{2} \tag{11-16}$$

Furthermore, since

$$\text{Gr}_{\text{mass}} = \left(\frac{\text{Mass transfer buoyancy forces}}{\text{Viscous forces}}\right)\left(\frac{\text{Inertial forces}}{\text{Viscous forces}}\right) \tag{11-17}$$

$$\frac{\text{Gr}_{\text{mass}}}{\text{Re}^2} = \frac{\text{Mass transfer buoyancy forces}}{\text{Inertial forces}} \tag{11-18}$$

THE CONCEPT OF ANALOGY

Before undertaking the discussion of the mass transfer coefficients for various cases, it is important to first consider the concept of analogy. In nature many processes are analogous in that they can be described by similar physical models, common mathematics, and other aspects. Earlier, for example, we used the mechanical analogs of a dashpot (viscous fluids) and a spring (elastic solid) to describe the viscoelastic material. Later, we used an electrical analog to describe heat conduction in a system involving many different materials. We have already used the analogous behavior of heat and mass transfer by employing the Gurney–Lurie charts for unsteady-state mass transfer situations.

Analogy between heat and mass transfer has long been recognized in the engineering and scientific community. This analogy has been demonstrated experimentally, empirically, and analytically (i.e., the elegant Onsager approach of coupling fluxes of the same order).

As we will see without analogy, our knowledge of mass transfer would be much more limited. The use of the much more widely studied field of heat transfer (by employing analogy) in essence gives us a decisive technical edge.

Among the earliest of the analogies (still used today) were those developed by Chilton and Colburn (1). The j_D and j_H factors were defined as

$$j_D = \frac{k_c}{V}\left(\frac{\mu}{\rho D_{AB}}\right)^{2/3} \tag{11-19}$$

$$j_H = \frac{h}{C_p \rho V}\left(\frac{C_p \mu}{k}\right)^{2/3} \tag{11-20}$$

Note that in essence

$$j_D = j_H = \phi(\text{Re, geometry boundary conditions}) \tag{11-21}$$

Basically then for similar cases of flow, geometry, and boundary conditions we can use the analogy between heat and mass transfer. It is also possible in certain flow situations to extend the analogy to momentum transfer as well. However, this additional analogy is not a general one.

Furthermore, there are a number of cases where geometries that appear quite different to the unpracticed eye are actually similar. This, of course, makes analogy possible for such situations.

MASS TRANSFER COEFFICIENTS

As has been mentioned, solutions to the equations of change for mass transfer are limited by geometry, flow, and other complexities. Furthermore, the relation for mass transfer coefficients are determined from experiment by analogy with the much more studied field of heat transfer.

In this section, representative correlations will be presented for various physical situations. Obviously, not all existing mass transfer correlations can be covered.

Let us begin with the system of mass transfer for flow in a conduit: Figure 11-2 illustrates the correlation for heat and mass transfer. Note that the additional parameter as L/D is used in the regimes of laminar and transition flow. Also note that the heat and mass transfer can be described by one correlation at high Reynolds numbers (i.e., 5000 or more). This means that the turbulent mass transfer Nusselt number (by analogy) to equation (6-21) is

$$\frac{k_c D P_{BM}}{D_{AB}} = 0.023\text{Re}^{0.8}\text{Pr}^{1/3} \tag{11-22}$$

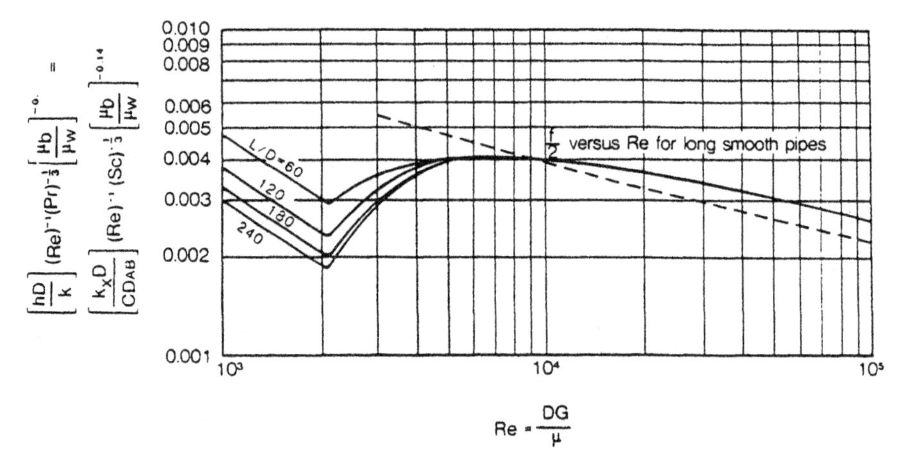

Figure 11-2. Analogy between heat and mass transfer for flow in tubes. (Adapted with permission from references 2 and 22. Copyrights 1936 and 1997, American Chemical Society.)

Also see that for Reynolds numbers above 10,000 the $f/2$ for smooth tubes is also analogous to heat and mass transfer. This occurs because the system's velocity profile becomes blunter (i.e., more like plug flow). As such, the usually more complex momentum transfer, while not truly one-dimensional, does closely resemble such a more simplified situation. This makes it possible to include momentum transfer in the overall analogy.

Flow over a flat plate is another case where the analogy also carries over to momentum transfer. Here for gases (3)

$$J_D = (k_c)_{\text{average}} \frac{P_{BM}}{VP} (\text{Sc})^{2/3} \tag{11-23}$$

and

$$J_D = J_H = \frac{f}{2} = 0.037(\text{Re})^{-0.2} \tag{11-24}$$

where $8000 < \text{Re} > 300{,}000$ and $Re = X_t V\rho/\mu$. The X_t is the plate length.

Mass transfer in a wetted wall column (liquid film flowing down tube walls that absorbs a gas from a counter- or co-current gas flow stream) is given by (4)

$$\frac{k_c D P_{BM}}{D_{AB} P} = 0.023(\text{Re})^{0.83}(\text{Sc})^{0.44} \tag{11-25}$$

for $\text{Re} > 2000$.

The reason that the Reynolds and Schmidt numbers have different exponents than equations (11-22) is that the velocity used for equation (11-25) is the gas velocity relative to the tube wall and not to the liquid film's velocity.

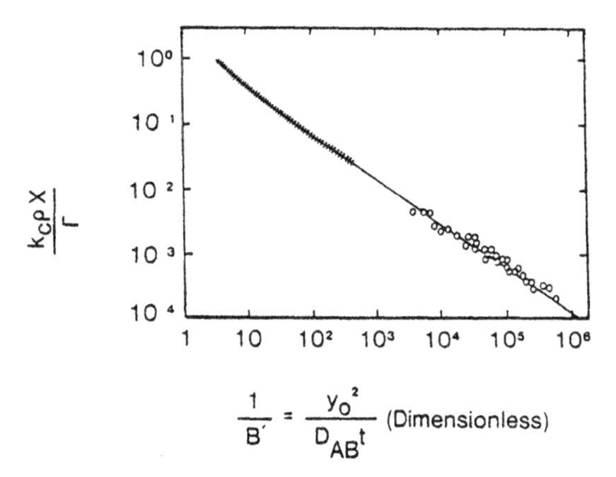

Figure 11-3. Mass and heat transfer analogy for solids dissolving into a flowing liquid (5–7).

In some cases the analogy between different mass transfer cases is more subtle. This is so for the data of Figure 11-3 (Γ is the mass flow rate per unit perimeter or width, X is the vertical distance, ρ is the density, Y_O is the film thickness, t is the residence time, and D_{AB} is the diffusivity), where mass is transferred from a solid surface to a falling film. Three separate cases are included in the correlation. The data in the upper left-hand corner represent the dissolution of the metal surface of a vertical tube by a falling acid film (5). Dissolution of benzoic acid from a flat plate at an angle of 45° is represented by the circles in the lower right-hand corner of the plot (6). The line connecting all of the data was obtained by analogy with heat transfer (7).

The mass transfer situation involving flow over a solid sphere was obtained by analogy with heat transfer (8, 9):

$$\frac{k_c D\rho}{D_{AB}} = 2 + 0.60\left(\frac{D_p V_\infty \rho}{\mu_f}\right)^{1/2}\left(\frac{\mu}{\rho D_{AB}}\right)^{1/3} \tag{11-26}$$

where

$$D_p = \text{particle diameter}$$

$$V_\infty = \text{approach velocity}$$

If flows are very low, then (10)

$$\frac{k_c D_p}{D_{AB}} = \left[4.0 + 1.21(\text{ReSc})^{2/3}\right]^{1/2} \tag{11-27}$$

Note that equations (11-26) and (11-27) both give the same intercept if Re is zero (i.e., 2.0).

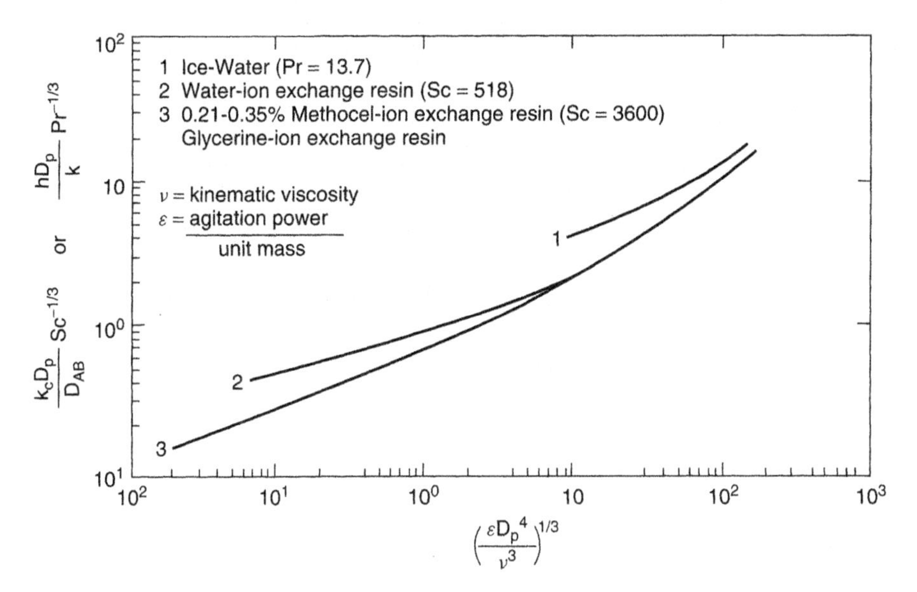

Figure 11-4. Heat and mass transfer analogy for solid particles suspended in an agitated fluid. (Adapted with permission from references 9 and 11. Copyright 1969 and 1962, American Institute of Chemical Engineers.)

A much more complicated system (solid particles suspended in agitated fluids) also lends itself to analogy (Figure 11-4). Here experimental heat (10) and mass transfer (9) data all correlated with the cube root of ε, the agitation power per unit mass times the particle diameter to the fourth power divided by the cube of the kinematic viscosity. In addition, the Prandtl or Schmidt number must be used for each case.

Both drops and bubbles are widely used in processing operations. The former are liquids and hence essentially incompressible; on the other hand, the bubble is a gas and as such subject to distortion, eccentric flow, and mass gradients within the bubble. This means that a correlation used for drops (11)

$$\frac{k_c D_p}{D_{AB}} = 1.13\mathrm{Re}^{1/2}\mathrm{Sc}^{1/2} \tag{11-28}$$

is altered to

$$\frac{k_c D_p}{D_{AB}} = 1.13\mathrm{Re}^{1/2}\mathrm{Sc}^{1/2}\left(\frac{D_p}{0.45 + 0.02D_p}\right) \tag{11-29}$$

where D_p is in centimeters

There is excellent analogy between heat and mass transfer for flows perpendicular (normal to cylinder). The plot shown in Figure 11-5 includes both heat transfer data (curves B and C and reference 12 and 13) and mass transfer data

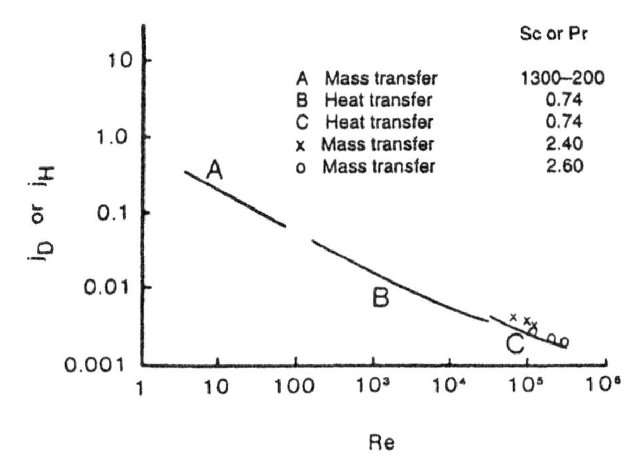

Figure 11-5. Mass and heat transfer for normal flow to a cylinder. (Curve A, with permission from reference 14. Curves B and C, data with permission from references 12, 13, and 15. Copyright 1940, 1969, 1970, American Society of Mechanical Engineers.)

(curve C and the points in the lower right-hand corner; references are 14, 15, and 16).

Actually, correlation of heat and mass transfer for *any* orientation of flow to a cylinder is excellent. This situation is covered in reference 17.

The rotating disk is a system used extensively in electrochemical systems.

For the case of rotating disks and laminar flow (18) we have

$$\frac{k_c D}{D_{AB}} = 0.879 \text{Re}^{1/2} \text{Sc}^{1/2} \tag{11-30}$$

In turbulent flow (19) we have

$$\frac{k_c D}{D_{AB}} = 5.6 \text{Re}^{1.1} \text{Sc}^{1/3} \tag{11-31}$$

where $6 \times 10^5 < \text{Re} < 2 \times 10^6$ and $120 < \text{Sc} < 1200$.

A widely used device in industry is the packed bed. For the case of single phase flow (i.e., with a gas or liquid) through a packed bed we can use the form (20)

$$j_D = 1.17 \left(\frac{D_p U_{AV} \rho}{\mu} \right)^{-0.415} \tag{11-32}$$

for $10 < \text{Re} > 2500$.

In the above, the D_p is the average particle diameter, U_{AV} is the superficial velocity (velocity if there were no bed in the column), and μ and ρ are the viscosity and density of the fluid. Note that we must use an indirect approach (i.e., U_{AV}) because there is no way up actually determining the fluid velocities in the bed.

EFFECT OF HIGH MASS TRANSFER RATES ON MASS TRANSFER

High mass transfer rates will influence not only the mass transfer coefficient but also the heat transfer coefficients and friction factor. Analysis of film theory penetration theory and boundary layer theory (21) show that the relation of the various coefficients at high (k_x^*) and low mass transfer (k_x) can be given by θ's:

$$\theta_V = f^*/f \tag{11-33}$$

$$\theta_T = h^*/h \tag{11-34}$$

$$\theta_{AB} = k_x^*/k_x \tag{11-35}$$

The θ's are related to diffusivity parameters Λ and rate factors ϕ.

$$\Lambda_V = 1 \tag{11-36}$$

$$\Lambda_T = \text{Pr} \tag{11-37}$$

$$\Lambda_{AB} = \text{Sc} \tag{11-38}$$

and

$$\phi_V = \frac{N_{A0}M_A + N_{B0}M_B}{0.5\rho V_\infty f} \tag{11-39}$$

$$\phi_T = \frac{N_{A0}C_{Pa} + N_{B0}C_{Pb}}{h} \tag{11-40}$$

$$\phi_{AB} = \frac{N_{A0} + N_{B0}}{k_x} \tag{11-41}$$

The zero subscripts indicate molar mass flux at the surface. V_∞ is the main stream or approach velocity.

Figure 11-6 interrelates Λ, θ, and ϕ for the various theories.

WORKED EXAMPLES

Example 11-1 Air passes through a naphthalene tube that has an inside diameter of 0.0254 m and a length of 1.83 m. The velocity is 15.24 m/sec, and the air is at 10°C and atmospheric pressure.

Air Properties	Naphthalene Properties
$\rho = 1.249$ kg/m^3	Vapor pressure = 2.79 N/m^2
$\mu = 0.000018$ kg/m-sec	$D_{AB} = 5.2 \times 10^{-6}$ m^2/sec
	mol. wt. = 128.2

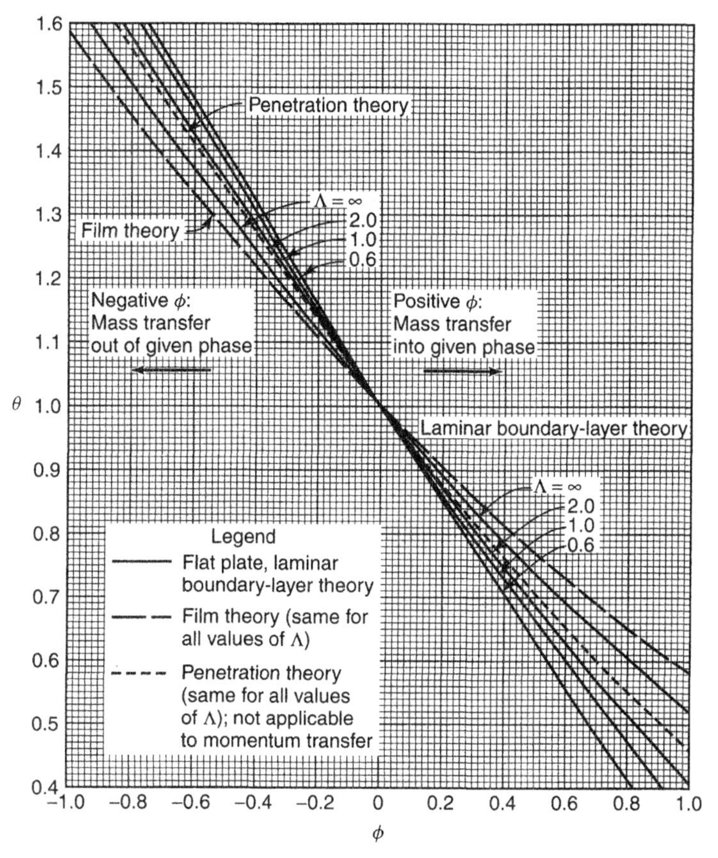

Figure 11-6. Correction of mass transfer coefficients, heat transfer coefficients, and friction factors due to high mass transfer rates. (Reproduced with permission from reference 21. Copyright 1960, John Wiley and Sons.)

Determine percent saturation of air and rate of naphthalene sublimation at 15.24 m/sec:

$$\text{Re} = \frac{(0.0254)(15.24\ \text{m/sec})(1.249\ \text{kg/m}^3)}{(0.0018\ \text{kg/m-sec})}$$

$$\text{Re} = 26{,}860$$

$$\text{Sc} = \frac{(0.000018\ \text{kg/m-sec})}{(1.249\ \text{kg/m}^3)(5.2 \times 10^{-6}\ \text{m}^2/\text{sec})}$$

$$\text{Sc} = 2.77$$

In working this problem we see that saturations are expressed as densities. Furthermore, the use of a k_c would introduce problems because of the need for P_{BM} (air).

We therefore use a k_ρ:

$$k_\rho = (k_c)\left(\frac{P_{BM}}{P}\right)$$

and

$$\frac{k_\rho D}{D_{AB}} = \frac{k_c D P_{BM}}{D_{AB} P} = 0.023\text{Re}^{0.80}\text{Sc}^{0.33}$$

$$k_\rho = 0.0229 \text{ m/sec}$$

By mass balance we equate the mass taken up by the air to that transferred from the tube wall:

$$(\pi/4)(0.0254 \text{ m})^2(15.24 \text{ m/sec})(d_{\rho_A})_{\text{BULK}} = k_\rho(\pi(0.0254 \text{ m})dx(\rho_{A \text{ SAT}} - \rho_{\text{BULK}})$$

$$\int_0^{\rho_{A \text{ BULK}}} \frac{d\rho_{A \text{ BULK}}}{\rho_{A \text{ SAT}} - \rho_{A \text{ BULK}}} = 2.58\, k_\rho \int_0^{1.83} dx$$

Then

$$-\ln \frac{\rho_{A \text{ SAT}} - \rho_{A \text{ BULK}}}{\rho_{A \text{ SAT}}} = 0.43$$

$$\rho_{A \text{ BULK}} = 0.35\rho_{A \text{ SAT}}$$

Hence, percent saturation is 35 percent.
At 10°C, saturation of naphthalene in air is 1.52×10^{-4} kg/m^3 since

$$\rho_{A \text{ SAT}} = \frac{(2.79 \text{ N/m}^2)(128)(1.249 \text{ kg/m}^3)}{1.01 \times 10^5 \text{ N/m}^2(2 \text{ g})}$$

$$\text{Total evaporation rate} = (\pi/4)(0.0254 \text{ m})^2(15.24 \text{ m/sec})(0.35) \times (1.52 \times 10^{-4} \text{ kg/m}^3)$$

$$\text{Total evaporation rate} = 4.2 \times 10^{-7} \text{ kg/sec}$$

Example 11-2 A spherical drop of water (0.05 cm in diameter) is falling at a velocity of 215 cm/sec through dry, still air at 1 atm. Estimate the instantaneous rate of evaporation from the drop if the drop's surface is at 21.1°C and the air is at 60°C.

In order to solve the problem, assume ideal gas behavior, insolubility of air in water, equilibrium at the interface, and pseudo-steady-state conditions. The last is reasonable if the drop slowly evaporates.

For a small evaporation rate we have

$$W_A - X_{A0}(W_A + W_B) = k^* A(X_{A0} - X_{A\infty})$$

where W_A is the molar rate of exchange of A.

$$W_A = k^* A \frac{X_{A0} - X_{A\infty}}{1 - X_{A0}}$$

$$W_A = k^* \pi D^2 \frac{X_{A0} - X_{A\infty}}{1 - X_{A0}}$$

$$T_f = \frac{21.1 + 60}{2} = 40.6^{\circ}\text{C}$$

$$X_{A0} = \frac{2.495 \times 10^3 \text{ N/m}^2}{1.01 \times 10^5 \text{ N/m}^2} = 0.0247$$

$$X_{A\infty} = 0$$

Also

$$C_f = 3.88 \times 10^{-5} \text{ g mole/cm}^3$$

$$\rho f = C_f M = 1.12 \times 10^{-3} \text{ g/cm}^3$$

$$\mu f = 1.91 \times 10^{-4} \text{ g/cm sec}$$

$$D_{AB} = 0.292 \text{ cm}^2/\text{sec}$$

$$\text{Sc} = \frac{\mu}{\rho D_{AB}} = \frac{1.91 \times 10^{-4} \text{ g/cm sec}}{(1.12 \times 10^{-3} \text{ g/cm}^3)(0.292 \text{ cm}^2/\text{sec})} = 0.58$$

$$\text{Re} = \frac{D V \rho_f}{\mu_f} = \frac{(0.05 \text{ cm})(215 \text{ cm/sec})(1.12 \times 10^{-3} \text{ g/cm}^3)}{1.91 \times 10^{-4} \text{ g/cm sec}}$$

$$\text{Re} = 63$$

$$k^* = \frac{C_f D_{AB}}{D}\left[2 + 0.60\text{Re}^{1/2}\text{Sc}^{1/3}\right]$$

$$= \frac{(3.88 \times 10^{-5} \text{ g mole/cm}^3)(0.9292 \text{ cm}^2/\text{sec})}{0.05 \text{ cm}} \times \left[2 + 0.60(63)^{1/2}(0.58)^{1/3}\right]$$

$$k^* = 1.35 \times 10^{-3} \text{ g mole/sec cm}^2$$

Then

$$W_A = k^* \pi D^2 \frac{X_{A0} - X_{A\infty}}{1 - X_{A0}}$$

$$W_A = (1.35 \times 10^{-3} \text{ g mole/sec cm}^2)(\pi)(0.05 \text{ cm})^2 \frac{0.0247 - 0}{1 - 0.0247}$$

$$W_A = 2.70 \times 10^{-7} \text{ g mole/sec}$$

which amounts to a decrease of 1.23×10^{-3} cm/sec in drop diameter. Hence, a drop would fall a considerable distance before evaporating, and the evaporation estimate is reasonable.

Example 11-3 Consider two systems involving mass transfer with a bed void fraction of 0.40. In the first system, water at room temperature with a superficial velocity of 0.45 m/sec flows through a bed of 0.005 m spheres of benzoic acid. The diffusivity of benzoic acid in water is 7.7×10^{-10} m^2/sec.

The second system involves dry air (atmospheric) flowing through a bed of porous spheres saturated with water. Superficial velocity of the air is also 0.35 m/sec. Diffusivity of the air–water vapor system is 2.33×10^{-5} m^2/sec. Sphere diameter is also 0.005 m.

Calculate column heights for 90 percent saturation of flowing streams.

For both systems the mass taken up by the fluid flowing equals the mass transferred.

Hence,

$$U\,dc = k_c a (C_{\text{int}} - C)\,dh$$

where U is the superficial velocity, C_{int} is the interface concentration, h is the height, a is the surface per unit volume, and k_c is the mass transfer coefficient.

Solving the above balance gives

$$h = \frac{U}{k_c a} \ln \frac{C_{\text{int}} - C_1}{C_{\text{int}} - C_2}$$

But for both cases $C_2 = 0.9C_{\text{int}}$ and C_1 is zero so

$$h = \frac{U}{k_c a} \ln\ 10$$

Also, for both cases since the bed is made up of spheres we have

$$a = \frac{6(1 - 0.40)}{D_\rho}$$

$$a = \frac{6(1 - 0.40)}{0.005\ \text{m}} = 720\ \text{m}^2/\text{m}$$

Now considering the first case

$$\text{Re} = \frac{D_\rho U \rho}{\mu}$$

Here we treat the dilute stream of benzoic acid in water as if it were water:

$$\text{Re} = \frac{(0.005\ \text{m})(0.35\ \text{m/sec})(1000\ \text{kg/m}^3)}{0.001\ \text{kg/m sec}} = 1750$$

$$\text{Sc} = \frac{(0.001\ \text{kg/m sec})}{(1000\ \text{kg/m}^3)(7.7 \times 10^{-10}\ \text{m}^2/\text{sec})} = 1299$$

Then using the fit of the j_D - Re data

$$j_D = 1.17(\text{Re})^{-0.415}$$
$$j_D = 0.053$$

From the definition of j_D we have

$$k_c = j_D U \text{Sc}^{-2/3}$$
$$k_c = (0.053)(0.35 \text{ m/sec})(1299)^{-2/3} = 1.56 \times 10^{-4} \text{m/sec}$$

Now returning to the solution for h we obtain

$$h = \frac{0.35 \text{ m/sec}}{(1.56 \times 10^{-4} \text{ m/sec})(720 \text{ m}^2/\text{m}^3)} \ln 10$$
$$h = 7.18 \text{ m}$$

For the air–water vapor case (with $\rho = 1.21 \text{ kg/m}^3$; $\mu = 1.72 \times 10^{-5}$ kg/m sec) we have

$$\text{Re} = \frac{D_p U \rho}{\mu} = \frac{(0.005 \text{ m})(0.35 \text{ m/sec})(1.21 \text{ kg/m}^3)}{1.72 \times 10^{-5} \text{ kg/m sec}} = 123$$

$$\text{Sc} = \frac{\mu}{\rho D_{AB}} = \frac{1.72 \times 10^{-5} \text{ kg/m sec}}{(1.21 \text{ kg/m}^3)(2.33 \times 10^{-5} \text{ m}^2/\text{sec})} = 0.61$$

Then j_D is given again by

$$j_D = 1.17(\text{Re})^{-0.415} = 0.159$$
$$k_c = jdU\text{Sc}^{-2/3} = \frac{(0.159)(0.35 \text{ m/sec})}{(0.61)^{2/3}} = 0.0773 \text{ m/sec}$$

Finally, the h value is

$$h = \frac{0.35 \text{ m/sec}}{(0.0773)(720 \text{ m}^2/\text{m}^3)} \ln 10$$
$$h = 0.0144 \text{ m}$$

Example 11-4 Two engineers are engaged in a heated discussion about a required calculation for a given system.

The system involves a large amount of water in parallel flow to a benzoic acid plate. Since benzoic acid is soluble in water (solubility of 0.02595 kg mole/m^3), the mass flux must be determined.

Engineer A argues that equation (11-24) can be used by simply letting P_{BM}/P be unity because the system involves a liquid. The other engineer (B) counters that what should be used is the relation

$$j_D = 0.99\text{Re}^{-1/2}$$

developed for flowing liquids in the range (600 < Re < 50,000).

Water velocity over the 0.3-m-long plate is 0.05 m/sec. Diffusivity of benzoic acid in water is 1.24×10^{-9} m/sec.

The properties needed for the calculation of the dimensionless groups will be taken as those of water since the flowing liquid will have a dilute concentration of benzoic acid. Hence, viscosity and density are, respectively, 8.71×10^{-4} kg/m sec and 996 kg/m^3.

Then

$$\text{Re} = \frac{X_t V \rho}{\mu} = \frac{(0.3\text{ m})(0.05\text{ m/sec})(996\text{ kg/m}^3)}{(8.71 \times 10^{-4}\text{ kg/m sec})}$$

$$\text{Re} = 17{,}153$$

Next

$$\text{Sc} = \frac{\mu}{\rho D_{AB}} = \frac{(8.71 \times 10^{-4}\text{ kg/msec})}{(996\text{ kg/m}^3)(1.24 \times 10^{-9}\text{ m}^2/\text{sec})} = 705$$

Taking equation (11-24), we obtain

$$jd = \frac{kcPBM}{VP}(\text{Sc})^{2/3} = 0.037\text{ Re}^{-0.2}$$

Then, if $P_{BM}/p = 1.0$

$$k_c = \frac{0.037\text{ V}}{(\text{Sc})^{2/3}\text{Re}^{0.2}} = \frac{(0.037)(0.3\text{ m/sec})}{(705)^{2/3}(17{,}153)^{0.2}}$$

$$k_c = 3.32 \times 10^{-6}\text{ m/sec}$$

The liquid correlation gives

$$jd = \frac{k_c}{V}(\text{Sc})^{2/3} = \frac{0.99}{\text{Re}^{0.5}}$$

$$k_c = 4.77 \times 10^{-6}\text{ m/sec}$$

Next, the flux N_A is given by

$$N_A = k_c \Delta\, C$$

The ΔC driving force is taken to be (0.0295 kg mole/m^3-0)—that is, the wall concentration minus zero (since the solution is dilute).

For Engineer A the calculated flux would be

$$N_A = (3.32 \times 10^{-6}\ \text{m/sec})(0.0295\ \text{kg mole/m}^3)$$

$$N_A = 9.79 \times 10^{-8}\ \text{kg mole/m}^2\ \text{sec}$$

while for Engineer B the calculated flux would be

$$N_A = (4.77 \times 10^{-6}\ \text{m/sec})(0.0295\ \text{kg mole/m}^3)$$

$$N_A = 14.1 \times 10^{-8}\ \text{kg mole/m}^2\ \text{sec}$$

The percent difference would be 30.1 percent, a sizeable amount.

Example 11-5 An oddly shaped object is to be used to transfer ammonia into air. The flow velocity of the air (temperature of 40°C and atmospheric pressure) is 17 m/sec.

Your assignment is to find the mass transfer coefficient for the system. Unfortunately, the only available data are for heat transfer between flowing air and the object. The empirical result of this work is the relation

$$h = 22G^{0.55} = 22(V\rho)^{0.55}$$

when G has a value of 20 kg/m^2 sec.

We start by using the concept of analogy with the implication that

$$j_H = j_D$$

We will also use the concept that the object will behave in a similar manner to the flat plate. Hence,

$$j_H = \frac{h}{Cp\rho V}\,\text{Pr}^{2/3} = \frac{h}{CpG}\,\text{Pr}^{2/3} = C_1\text{Re}^n$$

where C_1 and n are constants.

Then,

$$h = \frac{CpGC_1\text{Re}^n}{\text{Pr}^{2/3}} = 22G^{0.55}$$

Writing a Reynolds number with some characteristic length of the object Lc, we obtain

$$h = \frac{CpGC_1\dfrac{(LcV\rho)^n}{\mu}}{\text{Pr}^{2/3}} = \frac{CpGC_1\dfrac{(LcG)^n}{\mu}}{\text{Pr}^{2/3}}$$

$$h = \frac{CpC_1\dfrac{(Lc)^n}{\mu}G^{1+n}}{\text{Pr}^{2/3}} = 22G^{0.55}$$

From this we deduce that

$$1 + n = 0.55$$
$$n = -0.45$$

and

$$\frac{C_1 C_\rho}{\mathrm{Pr}^{2/3}} \frac{(Lc)^{-0.45}}{\mu} = 22$$

so that

$$C_1 = 22 \frac{\mathrm{Pr}^{2/3}}{C_\rho} \frac{(\mathrm{Lc})^{0.45}}{\mu}$$

Then, for air at 40°C and 1 atm the Cp value is 1002 J/kg°K Pr is 0.68 and viscosity is 1.85×10^{-5} kg/m sec.

This gives

$$C_1 = 1.33 Lc^{0.45}\ (\text{meters})^{-0.45}$$

By substituting, we obtain

$$jh = \frac{1.33 Lc^{0.45}}{\mathrm{Re}^{0.45}}\ (\text{meter})^{-0.45}$$

Then

$$jd = \frac{kc P_{BM} \mathrm{Sc}^{2/3}}{VP} = \frac{1.33 Lc^{0.45}}{\mathrm{Re}^{0.45}}\ (\text{meters})^{-0.45}$$

Next, if we take P_{BM}/P to be approximately unity, we obtain

$$k_c = \frac{1.33 Lc^{0.45} V}{\mathrm{Sc}^{2/3} (LcV\rho/\mu)^{0.45}}\ (\text{meters})^{-0.45}$$

Then with a density of 1.13 kg/m^3 and a diffusivity of 2.27×10^{-5} m^2/sec, we have

$$k_c = \frac{(1.33)(17\ \text{m/sec})(Lc/Lc)^{0.45}\ (\text{meter})^{-0.45}}{\left[\frac{1.85 \times 10^{-5}\ \text{kg/m sec}}{(1.13\ \text{kg/m}^3)(2.27 \times 10^{-5}\ \text{m}^2/\text{sec})}\right]^{2/3} \left[\frac{(17\ \text{m/sec})(1.13\ \text{kg/m}^3)}{1.85 \times 10^{-5}\ \text{kg/m sec}}\right]^{0.45}}$$

$$k_c = 0.05542\ \text{m/sec}$$

Example 11-6 Carbon dioxide is absorbed from air in a wetted wall column (liquid is water). The gas stream moves at a velocity of 1 m/sec in the 0.05-m-diameter column. Temperature and pressure for the system are, respectively, 25°C and 1.013×10^6 N/m^2.

At a certain location in the column the carbon dioxide mole fraction in the gas is 0.1. The carbon dioxide has a mole fraction of 0.005 in the water.

The Henry's Law constant for the system carbon dioxide–water is

$$1.66 \times 10^8 \frac{\text{N/m}^2}{\text{mole fraction CO}_2} \quad (\text{i.e., } P\text{CO}_2 = \text{H}X_{\text{CO}_2})$$

Diffusivity for the gas system is 1.64×10^{-5} m^2/sec at 25°C and 1.013×10^5 N/m^2 pressure.

We commence by computing the Reynolds and Schmidt numbers (properties are assumed to be those of air):

$$\text{Re} = \frac{DV_\rho}{\mu} = \frac{(0.05\ \text{m})(1\ \text{m/sec})(11.9\ \text{kg/m}^3)}{1.8 \times 10^{-5}\ \text{kg/m sec}}$$

$$\text{Re } 33{,}056$$

The diffusivity, D_{AB}, for carbon dioxide–air has to be corrected for pressure:

$$(D_{AB})_1 P_1 = (D_{AB})_2 (P_2)$$

$$(D_{AB})_2 = (1.64 \times 10^{-5}\ \text{m}^2/\text{sec}) \left(\frac{1.013 \times 10^5\ \text{N/m}^2}{10.13 \times 10^5\ \text{N/m}^2} \right)$$

$$(D_{AB})_2 = 1.64 \times 10^{-6}\ \text{m}^2/\text{sec}$$

$$\text{Sc} = \frac{\mu}{\rho D_{AB}} = \frac{1.8 \times 10^{-5}\ \text{kg/m sec}}{(11.9\ \text{kg/m}^3)(1.64 \times 10^{-6}\ \text{m}^2/\text{sec})}$$

$$\text{Sc} = 0.92$$

We use equation (11-25) for the wetted wall column:

$$\frac{k_c P_{BM} D}{D_{AB} P} = 0.023 \text{Re}^{0.83} \text{Sc}^{0.44}$$

Note that P_{BM} is based on the air's partial pressure values.

Calculating the value of P_B's used in finding P_{BM}, we obtain

$$(P_B)_1 = [P - (P_A)_1]$$

The value of $(P_A)_1$ is obtained from the Henry's Law relation:

$$(P_A)_1 = (1.66 \times 10^8\ \text{N/m}^2)(0.005) = 8.3 \times 10^5\ \text{N/m}^2$$

and

$$(P_B)_1 = (10.13 - 8.3) \times 10^5\ \text{N/m}^2 = 1.83 \times 10^5\ \text{N/m}^2$$

Likewise, the $(P_A)_2$ value is 1.013×10^5 N/m$_2$, and $(P_B)_2$ is

$$(P_B)_2 = -(10.13 - 1.013) \times 10^5 \text{ N/m}_2 = 9.12 \times 10^5 \text{ N/m}^2$$

The P_{BM} value is then

$$P_{BM} = \frac{(1.83 - 9.12)10^5 \text{ N/m}^2}{\ln(1.83/9.12)}$$

$$P_{BM} = 4.54 \times 10^5 \text{ N/m}^2$$

Now returning to the equation for k_c, we obtain

$$k_c = \left(\frac{P}{P_{BM}}\right)\left(\frac{D_{AB}}{D}\right) 0.023\text{Re}^{0.83}\text{Sc}^{0.44}$$

$$k_c = \left(\frac{10.13}{4.54}\right)\left(\frac{1.64 \times 10^{-6} \text{ m}^2/\text{sec}}{0.05 \text{ m}}\right) 0.023(33{,}056)^{0.83}(0.92)^{0.44}$$

$$k_c = 0.398 \text{ m/sec}$$

PROBLEMS

11-1. Find the mass transfer coefficient for water evaporating into air in a wetted wall column. The air at 298°K and 2 atm flows at 0.20 kg/sec. Diffusivity value is 1.3×10^{-5} m^2/sec.

11-2. A 1-mm-diameter droplet of water falls through dry still air (1 atmosphere, 37.8°C). Find the drops velocity and surface temperature. Also, find the rate of change of the drop diameter. A film temperature (dry air) of 26.7°C should be used for this case.

11-3. A wet bulb thermometer is a device in which the instrument is encased in a wetted cloth cover with dry atmospheric air flowing across it. If such a thermometer reads 65°F, what is the dry air temperature?

11-4. Atmospheric air at 298°K flows at a velocity of 1.5 m/sec across a 30 × 30-cm square piece of ice. If the air is dry and the ice is insulated (except for exposed surface), find the moisture evaporated per second.

11-5. In order to extend the temperature measuring range of a thermometer, it is used in the wet bulb mode before it is placed in a flowing air stream. If the thermometer reads 32°C, what is the correct air temperature (assume atmospheric pressure and dry air)?

11-6. Water at 27°C flows over a flat plate (0.244 by 0.244 m) of solid benzoic acid at a velocity of 0.061 m/sec. Find the molar flux of the benzoic acid. Diffusivity and solubility of benzoic acid in water are 1.24×10^{-9} m^2/sec and 0.0295 kg mole/m^3.

11-7. Find the molar flux for a sphere of naphthalene (2.54-cm diameter) to flowing air (45°C; 1 atm) at a velocity of 0.305 m/sec. Diffusivity and vapor pressure of naphthalene at 45°C are 6.92×10^{-6} m^2/sec and 7.3×10^{-4} atm.

11-8. Find the mass transfer coefficients for water into an air stream (37.8°C, atmospheric pressure) flowing at 3.048 m/sec for two cases. Case 1 is for flow over a horizontal flat plate 0.3048 m long. Case 2 is for a wetted wall column (0.0254 m in diameter).

11-9. A streamlined shape with airfoil cross section is to be used to absorb ammonia from a flowing air stream (velocity of 4.5 m/sec) on its wetted surface. No mass transfer data exist for this given shape. However, heat transfer experiments with the same shape and air velocity show the heat transfer coefficient to be 52.2 W/m^2 °K.

11-10. A wetted wall column (at 0.68 atm) evaporates water into an air stream (2×10^{-3} kg/sec). Water partial pressure and vapor pressure are 0.1 and 0.182 atm. The observed rate of water evaporation is 2.18×10^{-3} kg/sec. The same unit (at a pressure of 1.08 atmospheres) uses air (1.67×10^{-3} kg/sec) to evaporate *n*-butyl alcohol. What is the rate of vaporization? Alcohol partial and vapor pressure are 0.04 and 0.07 atm.

11-11. Air (superficial velocity of 1.524 m/sec; 37.8°C; 2 atm) flows through a shallow bed of naphthalene spheres (1.27-cm diameter) with a porosity of 0.40. If the naphthalene vapor pressure is 0.154 atm, how many kg/sec will evaporate from 0.0929 m^2 of bed.

11-12. Small spheres (100-μm diameter) of benzoic acid are dissolved in water in an agitated tank. How much time would be required for complete dissolution at 25°C? The solubility and diffusivity of the benzoic acid in water are 0.43 kg/100 kg water and 1.21×10^{-9} m^2/sec.

11-13. Air (3 m/sec) flows perpendicularly to a uranium hexafluoride cylinder (6-mm diameter) whose surface is at 43°C. Vapor pressure and partial pressure of the solid are 0.53 and 0.26 atm. Diffusivity is 0.09 cm^2/sec. The air is at atmospheric pressure and 60°C. Find the rate of sublimation of the solid.

11-14. What is the rate of carbon dioxide absorption into a water film flowing down a vertical wall (1 m long) at 0.05 kg/sec per meter of width. The gas (pure carbon dioxide) is at 25°C and atmospheric pressure. Solubility and diffusivity are 0.0336 kilomoles/m^3 solution and 1.96×10^{-5} cm^2/sec.

11-15. Water evaporates from a strangely shaped surface into a flowing stream of hydrogen (15 m/sec, 38°C, 1 atm). Heat transfer studies for air flowing past a similarly shaped object at a superficial mass velocity of 21.3 kg/m^2 sec show that $h = 2.3G^{0.6}$, where h is the heat transfer coefficient and G is the mass velocity. Find the water evaporation rate into the hydrogen if hydrogen–water vapor diffusivity is 0.775 cm^2/sec.

11-16. A solid sphere falls at its terminal velocity in a liquid (density of 1000 kg/m^3, viscosity of 1.0 cP). The solid's diameter and density are 100 μm and 2000 kg/m^3. If the diffusivity of the solid in the liquid is 10^{-9} m^2/sec, find the mass transfer coefficient.

11-17. Repeat Problem 11-8 for flow past a single sphere (0.0254-m diameter) and normal to a 0.0254-m-diameter cylinder.

11-18. The Sherwood number (kD_p/D_{AB}) (k is mass transfer coefficient, D_p is the particle diameter, D_{AB} is the diffusivity) for equimolar counter diffusion from a sphere to a surrounding infinite medium approaches a constant value. What is this value?

11-19. A shallow bed of water-saturated solid flakes is dried by blowing dry air at 4.57 m/sec and 836 mm Hg pressure. The bed solids surfaces are kept at 15.6°C. Find the air temperature.

11-20. Determine the rate of drying (i.e., water removal) of the system of problem 11-19 if the particle surface area/unit bed volume is 54.9 m^{-1}.

11-21. A radiation correction is necessary for the case of a wet bulb and dry bulb thermometer placed in a long duct with constant inside surface temperature and a small gas velocity. Obtain relations for the gas temperature and the vaporation rate. (*Hint*: Use T, h, emissivity, and absorptivity of the dry bulb thermometer and the surface temperature.)

11-22. Nickel and carbon monoxide react to form nickel carbonyl:

$$Ni + 4CO \longrightarrow Ni(CO)_4$$

The reaction is carried out by flowing CO down through a bed of nickel spheres (12.5-mm diameter, 0.1-m bed cross section, 30 percent voids) CO conditions are 50°C, atmospheric pressure, and a rate of 2 $\times 10^{-3}$ kg mole/sec. Find the bed depth needed to reduce CO gas content to 0.5 percent if the reaction is very rapid (CO partial pressure at nickel surface is zero), pressure and temperature are constant; and gas viscosity and Schmidt number are 2.4 $\times$ 10^{-5} kg/m sec and 2.0. Assume that nickel spheres remain constant.

11-23. A wet bulb thermometer's cloth cover is saturated with benzene and contacted with flowing dry air. The thermometer reads 26°C. If the vapor pressure and enthalpy of vaporization of benzene are 11.3 kN/m^2 and 377 kJ/kg find the air temperature.

11-24. An open container (0.15-m diameter; 0.075 m deep) is filled with 25°C water. If atmospheric air is at 25°C and 50 percent relative humidity, calculate the water evaporation rate.

11-25. A test tube (0.0125-m diameter, 0.15 m deep) contains benzene at 26°C. What is the benzene evaporation rate to 26°C dry air? Use the benzene properties of Problem 11-23.

11-26. The liquid A evaporates from a wetted porous slab submerged in a tangentially flowing stream of pure gas B (noncondensable). At a specific point on the surface, $X = 0.80$ and the local k_X is 1.17×10^{-6} kg mole/sec m^2. Calculate the local rate of evaporation.

11-27. Apply the mass transfer correction factor to the results of Example 11-26.

11-28. If the Schmidt number in Problem 11-26 is 0.6, calculate the corrected mass flux using boundary layer theory.

11-29. A fixed-bed catalyst reactor is regenerated by using air (nitrogen diluted) to burn off carbon deposits by the reaction

$$12O_2 + C_{10}H_8 \longrightarrow 10CO_2 + 4H_2O$$

All of the above are gases except the solid with the empirical formula $C_{10}H_8$. The gas stream (oxygen and nitrogen mole fractions of 0.01 and 0.099) is at 537.8°C with a C_p of 7.54 calories/g mole °K. Heat of combustion is -100 kilocalories per mole of oxygen consumed. If Sc is 0.8 and Pr 0.7, find the maximum possible particle temperature.

11-30. A 0.05-m-diameter pipe's inside wall is coated with a thin film of water at 25°C. Dry air (atmospheric pressure, 25°C) flows through the pipe at 3 m/sec. What is the water vapor concentration in the air if the pipe is 3 m long?

11-31. A human forearm is approximated by a cylinder 0.102 m in diameter and 0.3048 m long. A wind (16.1 km/hr; dry air at 46.1°C) blows across the perspiring arm. In addition there is a radiant heat flux of 1103.9 W/m^2 (cylinder view area is length times diameter; water film emissivity is one). Estimate the arm's temperature.

11-32. Repeat Problem 11-31 with a heat generation of 1860 W/m^3 for the human body.

11-33. A 30 by 30-cm plate with a thin film of water is placed in a wind tunnel (walls at 10°C). Dry air (43°C, 1 atm) flows over the plate at 12 m/sec. Emissivity of the water film is 1. Find the plate's equilibrium temperature.

11-34. Repeat Problem 11-33 and find the evaporation rate of water.

11-35. Dry air (atmospheric pressure, 65°C) flows over a 20 by 20-cm plate at 6 m/sec. The plate is coated with a smooth porous material to which water is supplied at 25°C. How much water must be supplied to keep the plate at 38°C. Also, surroundings radiation temperature is 65°C and the porous surface emissivity is 1.

11-36. A droplet of A is suspended in a gas B. The drop (radius r_1) is surrounded by a stagnant film (radius r_2). Concentrations of A at r_1 and r_2 are x_{A1}

and x_{A2}, respectively. Find the radial molar flux of A; and if the flux is defined as

$$N_{Ar} = k_P(P_{A1} - P_{A2}), \text{ obtain the } k_P \text{ if } r_2 \longrightarrow \infty.$$

11-37. A drop of water (0.3-mm diameter) falls through the air (20°C, 50 percent relative humidity at 5 m/sec).
Estimate the drops steady state temperature.

11-38. Ammonia from an ammonia–air mixture is absorbed into a vertical wetted wall column down which dilute sulfuric acid flows. The column dimensions are diameter of 0.015 m and length of 0.826 m. Air flow rate is 0.0197 g moles/sec. Ammonia partial pressures at inlet and outlet are 56.6 and 14.5 mm Hg. Compute the mass transfer coefficient.

11-39. A Lister bag is a porous canvas device used to store drinking water. Water diffuses through the canvas and evaporates to cool the bag's surface. If the bag can be simulated by a 0.762-m sphere and it hangs in 0.805 kilometer/hr wind at 32.2°C, what are the heat and mass transfer coefficients?

11-40. An equation used to give mass transfer coefficients for fine bubbles or particles (22) is

$$k = 2D_{AB}/D_P + 0.31\text{Sc}^{-2/3}\left(\frac{\Delta\rho\mu_c g}{\rho_c^2}\right)^{1/3}$$

where μ_c is the viscosity of the continuous phase and ρ_c is its density. Also ρ_p is the particle density. Calculate the rate of oxygen absorption in an agitated fermenter from fine air bubbles (100 μm) at 37°C into deaerated water. Solubility and diffusivity of oxygen are, respectively, 2.26×10^{-4} kg mole/m^3 and 3.25×10^{-9} m^2/sec.

REFERENCES

1. T. H. Chilton and A. P. Colburn, *Ind. Eng. Chem.* **26**, 1183 (1934).
2. E. N. Sieder and G. E. Tate, *Ind. Eng. Chem.* **28**, 1429-1435 (1936).
3. T. K. Sherwood and R. L. Pigford, *Absorption and Extraction.*, 2nd edition; McGraw-Hill, New York (1954).
4. E. R. Gilliland and T. K. Sherwood, *Ind. Eng. Chem.* **26**, 516 (1934).
5. H. Hikita, K. Nakanishi, and S. Asai, *Kagaku Kogaku* **23**, 28 (1959).
6. H. Kramers and P. J. Kreyger, *Chem. Eng. Sci.* **6**, 42 (1956).
7. W. Z. Nusselt, *VDI-Z* **67**, 206 (1923).
8. N. Frossling, *Gerlands Beitr. Geophys.* **52**, 170 (1938).
9. P. L. T. Brian and H. B. Hales, *AIChE J.* **15**, 419 (1969).

10. T. K. Sherwood, R. L. Pigford, and C. R. Wilke, *Mass Transfer*; McGraw-Hill, New York (1975).
11. P. Harriott, *AIChE J.* **8**, 93 (1962).
12. W. H. McAdams, T. B. Drew, and G. S. Bays, Jr., *Trans. ASME* **62**, 627 (1940).
13. T. R. Johnson and P. N. Joubert, *J. Heat Transfer* **February**, 91 (1969).
14. P. H. Vogtlander and C. A. P. Bakker, *Chem. Eng. Sci.* **18**, 583 (1963).
15. J. Kestin and R. T. Wood, *Am. Soc. Mech. Eng. [Pap.]* 70-WA/HT-3 (1970).
16. Sogin, H. H. Subramanian, V. S. *Am. Soc. Mech. Eng. [Pap.]* 60-WA-193 (1960).
17. R. G. Griskey and R. E. Willins, *Can. J. Chem. Eng.* **53**, 500 (1975).
18. V. G. Levich, *Physicochemical Hydrodynamics*, Prentice-Hall, Englewood Cliffs, NJ (1962).
19. I. Cornet and U. Kaloo, *Tr. Mezhdunar, Kongr. Korroz. Met.*, 3rd **3**, 83 (1966).
20. M. Eisenberg, C. W. Tobias, and C. R. Wilke, *Chem. Eng. Prog. Symp. Ser.* **51**(16), 1 (1955).
21. R. B. Bird, W. E. Stewart, and E. N. Lightfoot, *Transport Phenomena*, John Wiley and Sons, New York (1960).
22. C. J. Geankoplis, *Transport Processes and unit Operations*, R309, Allyn and Bacon, Boston, MA (1978).

12

EQUILIBRIUM STAGED OPERATIONS

INTRODUCTION

The diffusional mass transfer approach is useful for many situations. However, when there is a need to design large-scale mass transfer or separation equipment, we find that such an approach is quite difficult and indeed in some cases not possible. In these instances, we use an approach based on the concept of an ideal or equilibrium stage. This device is one in which perfect mixing occurs, and as a result the streams leaving the unit are in equilibrium (phase) with each other. The principle is illustrated in Figure 12-1.

A number of questions can be raised about such a device. Once concerns the nature of the driving force that brings about mass transfer. The driving forces are the differences in concentration between the entering streams and their equillibrium concentrations. Another important question relates to the concept of perfect mixing. As we have seen earlier, mixing is not only nebulous but also imperfect. However, note that even though the equilibrium stage truly does not exist, a real stage will at least perform at some efficiency and bring about a separation. The parallels to this are the efficiencies used for mechanical devices such as pumps, compressors, and so on. Also, note that the net result of putting together a series of even real stages will be to effect separation or purification of process streams.

Basically, the design of equilibrium stages requires the following:

1. A material or mass balance
2. An enthalpy or energy balance
3. Appropriate equilibrium data

The type of equilibrium data needed is dependent on the process. A listing of some typical cases are shown in Table 12-1.

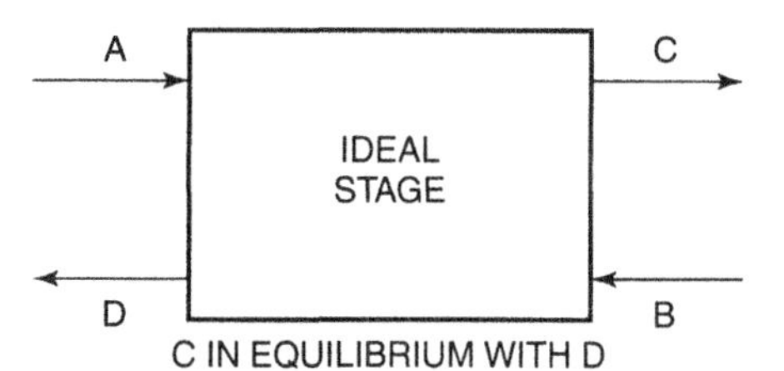

Figure 12-1. Schematic of ideal or equilibrium stage. (Reproduced with permission from reference 18. Copyright 1997, American Chemical Society.)

Table 12-1 Types of Equilibrium Data Needed

Process	Data
Distillation	Vapor–liquid
Extraction	Liquid–liquid
Absorption	Vapor–liquid
Leaching	Solid–liquid
Adsorption	Solid–liquid or solid–gas
Crystallization	Solid–liquid

The forms of the data vary with the processes. For example, note that both distillation and absorption require vapor–liquid equilibrium data. There is, however, a difference between the two processes. In absorption, where a gas diffuses into a liquid, the situation is often decribed by Henry's Law:

$$P_A = HX_A \qquad (12\text{-}1)$$

or

$$Y_A = HX_A \qquad (12\text{-}2)$$

Equations (12-1) and (12-2) deal with a gas being absorbed by a large amount of liquid at a given temperature. Typical absorption data (for the ammonia–water system) are shown in Figure 12-2.

On the other hand, distillation involves the interaction of a system where a number of temperatures are involved. Data for a binary system (benzene–toluene) is shown in Figure 12-3 in the form of a boiling point diagram. This form can be converted to a plot of mole fraction of the more volatile in the vapor versus mole fraction of the more volatile in the liquid (Figure 12-4).

All of the cases found in equilibrium behavior can be presented in the form of Figure 12-4. An example of an Y–X diagram for an azeotrope is shown in Figure 12-5 (the azeotrope occurring where the curve crosses the $Y = X$ line). Also, the low x region of Figure 12-4, if expanded, can be represented by a straight line ($Y_A = bX_A$, that is, Henry's Law).

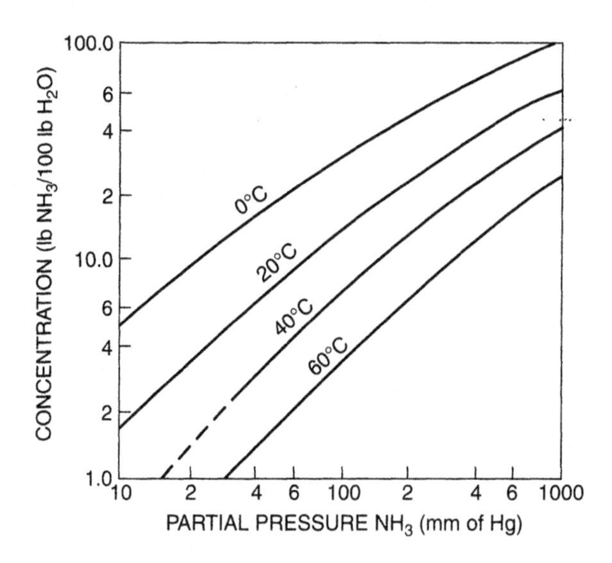

Figure 12-2. Ammonia solubility in water. (Reproduced with permission from reference 8. Copyright 1997, American Chemical Society.)

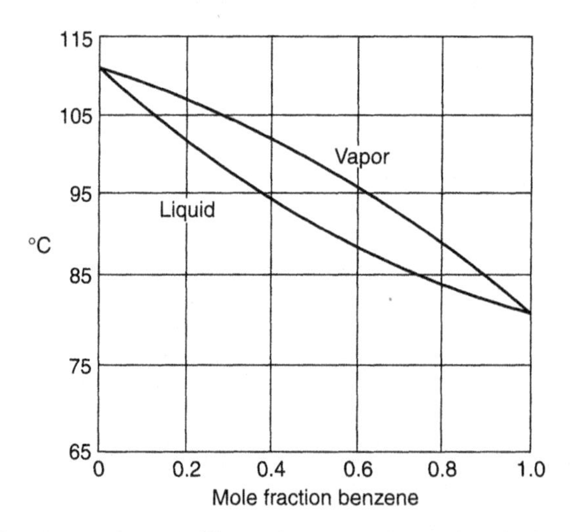

Figure 12-3. Benzene–toluene boiling point curve. (Reproduced with permission from reference 8. Copyright 1997, American Chemical Society.)

All of the other cases for the various processes will have some graphical representation. One additional plot is shown in Figure 12-6. The plot is a ternary diagram for the system acetone–water–methylisobutylketone (MIK). This situation represents the data used for a liquid–liquid extraction. The equilibrium region is under the dome. The straight lines joining the sides of the dome are called tie lines. Their endpoints represent equilibrium concentrations.

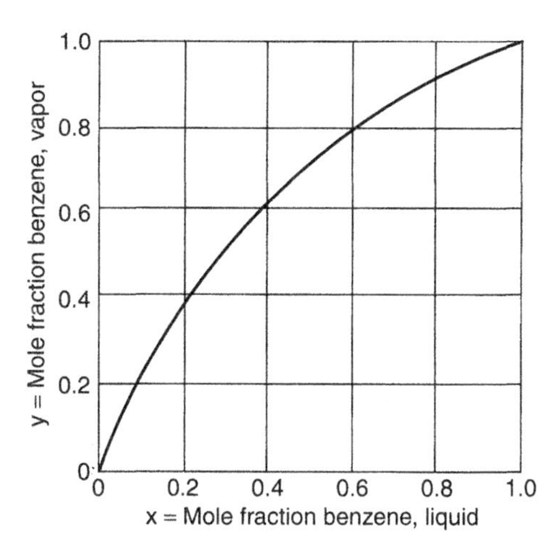

Figure 12-4. Benzene–toluene system equilibrium curve. (Reproduced with permission from reference 8. Copyright 1997, American Chemical Society.)

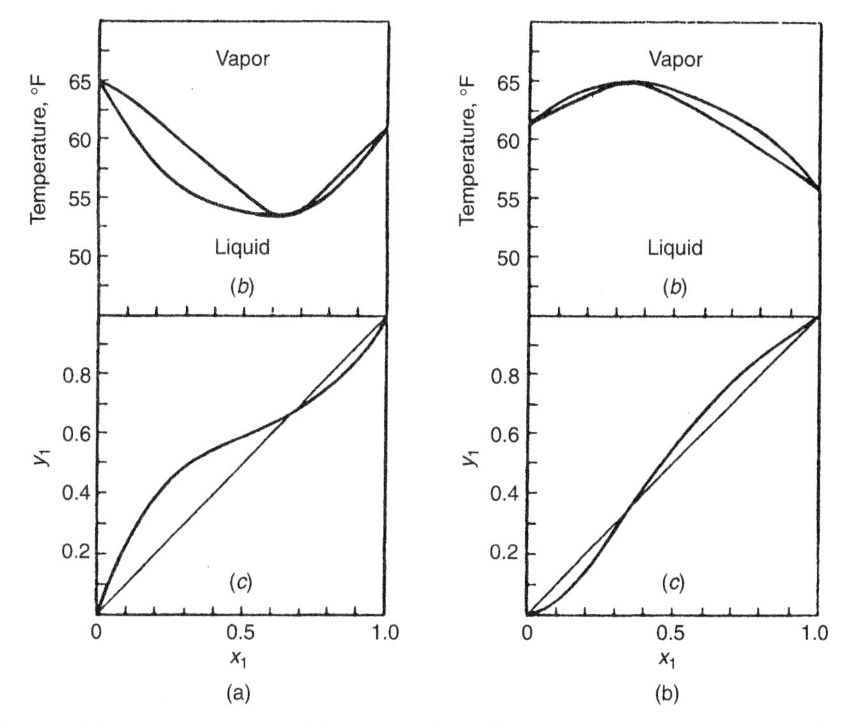

Figure 12-5. Minimum (a) and Maximum (b) boiling azeotropes. System a (chloroform 1, and methanol 2 system b (acetone 1, and chloroform, 2) (19).

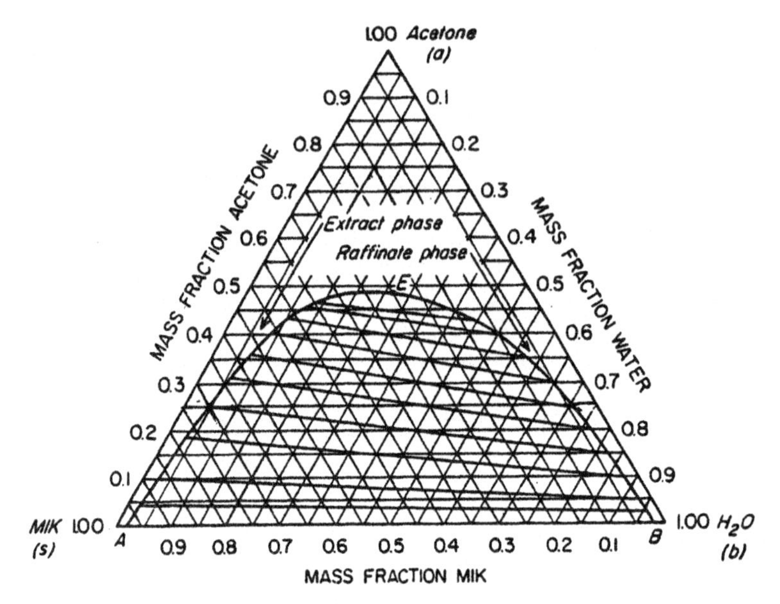

Figure 12-6. Ternary diagram for the system acetone–water–methylisobutylketone. Point E is the plait point. (Reprinted with permission from reference 17. Copyright 1941, American Chemical Society.)

MASS BALANCES IN THE DESIGN OF AN EQUILIBRIUM STAGE DISTILLATION COLUMN

As has been mentioned, the combination of a mass balance, enthalpy balance, and equilibrium data are used in the design of equilibrium stages. We will now consider such a design for a continuous binary distillation column (shown schematically in Figure 12-7). At first it might seem that this might be a more difficult undertaking than a batch distillation column. However, note that at steady state the concentration of any component in the vapor or liquid at a given stage (i.e., $m, m+1, n$, etc.) in the column will be constant. That is, concentration changes from stage to stage but is constant for any given stage. In contrast, the concentration in batch column will be changing from the start to the finish of the distillation.

Before undertaking the column design, it is worthwhile to consider the column itself. The feed point in the column is used as a dividing line. The portion of the column above it is called the enriching or rectifying section, while the part below is termed the stripping section. Vapor leaving the top of the column is condensed into a liquid without a change in concentration. A portion of the liquid effluent from the condenser is recycled to the column (i.e., the reflux). The liquid leaving the bottom of the column goes to a reboiler.

Before we start the design process, we recall the general rule with regard to mass balances — namely, that only the streams crossing our chosen boundary are

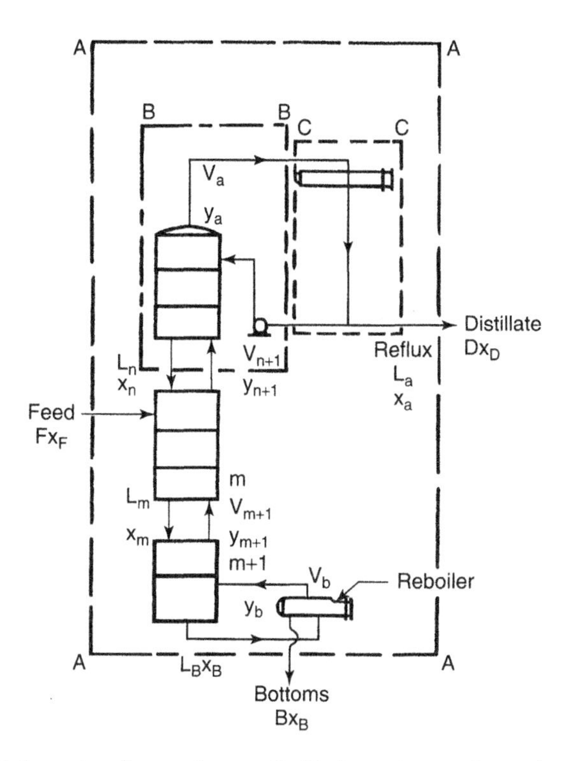

Figure 12-7. Schematic of a continuous distillation column. (Reproduced with permission from reference 18. Copyright 1997, American Chemical Society.)

considered. Hence, for the boundary A we can write two mass balances:

$$F = D + B \tag{12-3}$$

$$FX_F = DX_D + DX_B \tag{12-4}$$

The first [equation (12-3)] is an overall balance while the second [equation (12-4)] is for the more volatile component (A). In the above,

F, D, B = moles/hour of feed, overhead product, and bottoms product, respectively

X_F, X_D, X_B = fractions of component A

If equations (12-3) and (12-4) are solved simultaneously, the results are

$$\frac{D}{F} = \frac{X_F - X_B}{X_D - X_B} \tag{12-5}$$

$$\frac{B}{F} = \frac{X_D - X_F}{X_D - X_B} \tag{12-6}$$

These represent the ratio of top and bottom product to feed. Note that we can easily determine these ratios because the various mole fractions are either given (i.e., X_F) or set (X_D, X_B).

Next, we consider the rectifying or enriching portion of the column. First, we use boundary B to get the balance

$$(V_{n+1})(Y_{n+1}) + (L_a X_a) = (V_a Y_a) + (L_n X_n) \tag{12-7}$$

Likewise, using boundary C (about the condenser), we find that

$$DX_D + (L_a X_a) = (V_a Y_a) \tag{12-8}$$

Combining equations (12-7) and (12-8) yields

$$Y_{n+1} = \frac{L_n X_n}{V_{n+1}} + \frac{DX_D}{V_{n+1}} \tag{12-9}$$

which represent the mass balance for the enriching section of the column.

At this point, we make one alteration of equation (12-9) which involves a boundary that cuts the L_n, V_{n+1}, and D stream (not shown). This gives

$$V_{n+1} = L_n + D \tag{12-10}$$

which transforms equation (12-9) into

$$Y_{n+1} = \frac{L_n X_n}{L_n + D} + \frac{DX_D}{L_n + D} \tag{12-11}$$

Similar equations can be found for the stripping section.

$$Y_{m+1} = \frac{L_m}{V_{m+1}} X_m - \frac{B}{V_{m+1}} X_B \tag{12-12}$$

$$Y_{m+1} = \frac{L_m}{L_m - B} X_m - \frac{B}{L_m - B} X_B \tag{12-13}$$

A profitable exercise for the reader is to use the procedure described above to check equations (12-12) and (12-13).

ENTHALPY AND ENERGY BALANCES IN THE DESIGN OF A CONTINUOUS DISTILLATION COLUMN

The enthalpy or energy balance is essentially an application of the First Law of Thermodynamics. If we choose a stage n in the column, we can carry out our solution using liquid at T_n (the stage's temperature) as our base point (i.e., enthalpy of zero).

Table 12-2 Items that Make Up the Enthalpy Balance

Item	Symbol
Latent heat for vapor V_{n+1}	a
Sensible heat for vapor V_{n+1}	b
Sensible heat for liquid L_{n-1}	c
Latent heat for vapor V_n	d
Heat of mixing	e
Convection and radiation loss	g

The items that make up the enthalpy balance are shown in Table 12-2 together with symbols.

In the above, a and d represent the enthalpy needed to vaporize V_{n+1} and V_n at T_n. The sensible heats for V_{n+1} and L_{n-1} are given by

$$\int_{T_n}^{T_{n+1}} C_p \, dT \tag{12-14}$$

$$\int_{T_n}^{T_{n-1}} C_p \, dT \tag{12-15}$$

Note that the sensible heat for L_n is zero since it is given by

$$\int_{T_n}^{T_n} C_p \, dT \tag{12-16}$$

The heat of mixing refers to any energy involved with interaction in stage n, while the last terms covers heat losses.

Using the appropriate signs, the enthalpy balance is

$$a + b - c - d + e - g = 0 \tag{12-17}$$

Now if we consider the relative magnitude of the terms in equation (12-17), we see that a and d (the latent heats) will be much larger than the other terms. Furthermore, the remaining terms are all of about the same magnitude. Hence by neglecting $(b + e - c - f)$, we obtain

$$a = d \tag{12-18}$$

Rewriting using enthalpies of vaporization λ_n and λ_{n+1}, we obtain

$$V_n \lambda_n = (V_{n+1})(\lambda_{n+1}) \tag{12-19}$$

However, because V_n and V_{n+1} are close in composition, the λ's are about the same and

$$V_n = V_{n+1} \tag{12-20}$$

Likewise,

$$L_{n-1} = L_n \tag{12-21}$$

while for the stripping section we have

$$L_{m-1} = L_m \tag{12-22}$$

The overall implication of these results is a condition in the column known as constant molal overflow. In essence, all of the V's in the rectifying section (V_n, V_{n+1}, etc.) are the same. The liquid values (L_n, L_{n-1}, etc., including L_a the reflux) are also the same (though not equal obviously to the V's). Similarly, in the stripping section all of the V's are the same as are the L's (L_m, L_{m-1}, etc.)

Note that while composition continues to change throughout the column, the overall rates are constant. That is, a fixed number of moles per unit time pass through the rectifying or stripping sections.

This behavior of the vapor and liquid streams has an important effect on equations (12-9), (12-11), (12-12), and (12-13). The effect is that they are all straight lines when constant molal overflow occurs.

This overall approach involving the enthalpy balance with resultant constant molal overflow is known as the McCabe–Thiele method.

THE MCCABE–THIELE METHOD

The McCabe–Thiele method is a technique that combines the mass balance, the enthalpy balance, and equilibrium data, either analytically or graphically. The latter method is especially useful because it enables the user to visualize the design and operation of the column in a meaningful manner.

If we consider the vapor–liquid equilibrium of a given binary system (as, for example, Figure 12-4), we realize that equations (12-9), (12-11), (12-12), and (12-13) will appear as straight lines. The problem, of course, is that we don't know exactly where these lines will be placed. We can, however, as we will show find points though which the straight lines will pass. For example, in the case of the enriching mass balance line, either equation (12-9) or (12-11), we realize that $X_D = Y_D$ (i.e., in Figure 12-7 at the column top $Y_A = X_D = X_A$). Hence, equations (12-9) and (12-11) must go through the point (X_D, Y_D).

The location of one point, however, still does not solve our problem of placing the line. In order to do this, we need an additional point. If we consider equation (12-11), we see that one possibility is 0 where $x = 0$ (the y axis). Hence for the y-axis intercept (y_{int}) we find that

$$y_{\text{int}} = \frac{DX_D}{L_n + D} \tag{12-23}$$

$$Y_{\text{int}} = \frac{X_D}{\dfrac{L_n}{D} + 1} \tag{12-24}$$

However, for constant molal overflow we have $L_n = L_a$, where the L_a represents the reflux. Furthermore, the ratio of L_a/D is the reflux ratio (the ratio of reflux to top product):

$$y_{\text{int}} = \frac{X_D}{\dfrac{L_a}{D} + 1} = \frac{X_D}{R_D + 1} \tag{12-25}$$

where $R_D = L_a/D$ = reflux ratio. This ratio is either given or can be estimated by a method that will be discussed later.

The determination of Y_{int} allows us to definitively place equation (12-11) on the diagram (see Figure 12-8). Furthermore, we also know from Figure 12-7 that $X_B = Y_B$. This gives us one of the points on the lines given by equations (12-12) and (12-13). Unfortunately, we have nothing equivalent to the Y intercept for the enriching mass balance line. We do, however, know that the stripping mass balance line must pass through (X_B, Y_B) and join the enriching mass balance line somewhere between (X_B, Y_B) and (X_D, Y_D). Exactly where is the problem?

If we again consider Figure 12-7, we see that an item that both the enriching and stripping sections have in common is the feed. As such, the feed should be able to be used as a junction for the two column sections.

This indeed is what the McCabe–Thiele method used as shown in Figure 12-9. In this figure, we see a series of lines that pass through the point (X_F, Y_F). These lines are at various slopes labeled liquid below the bubble point, saturated liquid, liquid and vapor, and so on. Note that the intersection of the given feed line with the enriching mass balance is joined to (X_B, Y_B) to give the stripping mass balance line. Junctions are shown for the cases of liquid below the boiling point

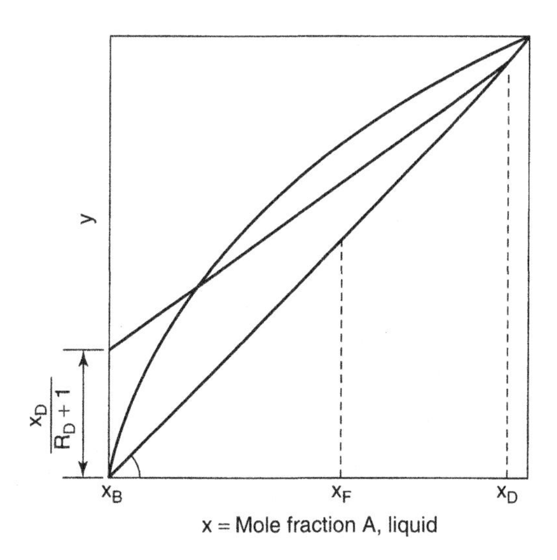

Figure 12-8. Placement of enriching or rectifying mass balance line on equilibrium diagram.

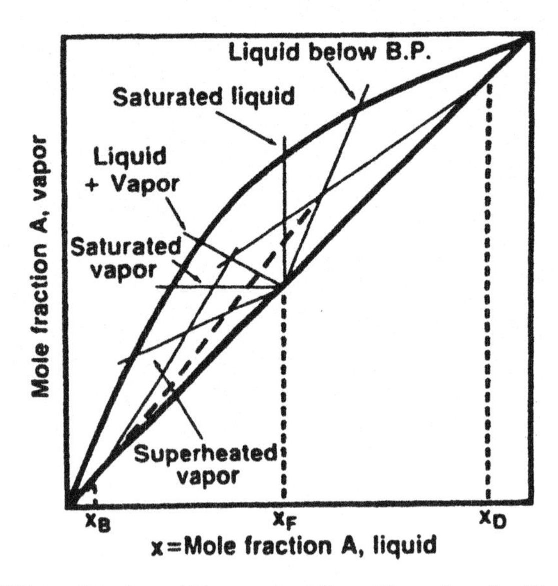

Figure 12-9. Effect of feed condition on feed line. (Reproduced with permission from reference 18. Copyright 1997, American Chemical Society.)

and a liquid–vapor mix in Figure 12-9. The reader should satisfy himself or herself as to where the functions would occur for the other cases.

The equation of the feed line that passes through (X_F, Y_F) is given by

$$Y = -\left(\frac{1-f}{f}\right)X + \frac{X_F}{f} \tag{12-26}$$

where xF is the feed mole fraction of the more volatile component and f is the feed conditions. Note that the feed line is a straight line passing through (xF, y_F) with a slope of

$$-\left(\frac{1-f}{f}\right)$$

Feed conditions are given in Table 12-3.

In the Table 12-3, C_{liquid} and C_{vapor} are the C_p values for the liquid and vapor feed, l is the heat of vaporization, and T_b and T_d respectively are the bubble point and dew point temperatures. The former is the temperature at which the first vapor leaves the mixture, and the latter is the temperature at which the first vapor condenses.

We now have all that is needed to determine the number of ideal stages for a given binary distillation separation. In order to do this, we follow the procedure below (see Figure 12-10).

1. Plot the equilibrium Y–X relation.
2. Draw the $Y = X$ (45° straight line) and locate X_B, X_F, and X_D.

Table 12-3 Feed Conditions

Feed Condition	f Value
Cold feed	$f < 0$ and $f = -\dfrac{C_{\text{liquid}}(T_b - T_f)}{\lambda}$
Saturated liquid feed (bubble point)	$f = 0$
Mixture of liquid and vapor	f = fraction vapor
Feed saturated vapor	$f = 1.0$
Feed superheated vapor	$f = 1 + C_{\text{vapor}}\left(\dfrac{T_f - T_d}{\lambda}\right)$

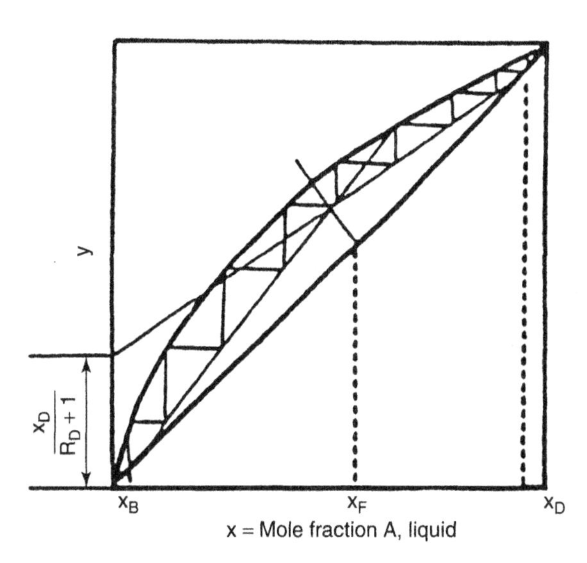

Figure 12-10. McCabe–Thiele solution for a number of stages. (Reproduced with permission from reference 18. Copyright 1997, American Chemical Society.)

3. Determine the Y-axis intercept by calculating $X_D/R_D + 1$.
4. Connect the Y-axis intercept and the point (X_D, Y_D). This gives the enriching or rectifying section mass balance line.
5. Next calculate the feed-line slope $-(1 - f/f)$ and plot the straight line through (X_F, Y_F).
6. Connect the function of the feed line and the enriching mass balance line to the point (X_B, Y_B). This gives the stripping mass balance line.
7. Now determine the number of ideal stages by moving horizontally from (X_D, Y_D) to the equilibrium line then by dropping a vertical to the enriching mass balance line.

8. Continue to repeat the stepping procedure until the feed line is crossed. At this point use the stripping mass balance line.
9. Repeat the alternate lines between equilibrium and the stripping mass balance line until (X_B, Y_B) is reached or passed (i.e., sometimes a partial step is needed to reach the $X_B - Y_B$ point).

As can be seen in Figure 12-10, a total of eight ideal stages would be required. Furthermore, the feed would be fed to the fifth stage from the top. Also, note that only seven stages would be used because the reboiler itself acts as a stage.

Note that the McCabe–Thiele technique gives us the ability to quickly evaluate a staged column. Also, we not only readily obtain the number of ideal or equilibrium stages and the point at which the feed enters but also can find the concentration on each stage (the intercept with the equilibrium curve) and the concentration of either liquid or vapor leaving one stage and going on to the next. In this last case the vapor leaving a stage is that corresponding to the equilibrium intercept (i.e., horizontal line to mass balance line) while the liquid is the X value corresponding to the equilibrium intercept (i.e., vertical line to the mass balance line).

LIMITING CASES, EASE OF SEPARATIONS, EFFICIENCIES

The operation of a binary distillation has two important limiting cases: minimum and maximum stages for a given separation. Minimum stages occur when the column is operated with total reflux (i.e., no top product is withdrawn). This situation, when applied to the McCabe–Thiele method, uses the $y = x$ line as the mass balance for both rectifying and stripping. The situation is depicted in Figure 12-11.

Maximum stages do not occur as might be expected at zero reflux but rather at some minimum reflux value. In this regard, consider Figure 12-12. If we proceed to carry out a McCabe–Thiele solution for this case, we first plot the feed line through (X_F, Y_F). Recalling that the the rectifying mass balance line always passes through (X_D, Y_D), we change the slope from total reflux ($Y = X$ line) until we reach the condition shown in Figure 12-12. Here the equilibrium curve, rectifying mass balance, and feed line meet at point N. When this occurs the number of ideal stages will become infinite in number. Hence, the y intercept gives us $R_{\min}$, the minimum reflux possible in the column (i.e., any less reflux will still give infinite stages).

It is possible to attain minimum reflux if the shape of the equilibrium curve causes the occurrence of infinite stages. This case is shown in Figure 12-13.

Both limiting conditions are important quantites. This is particularly so for minimum reflux, because 1.2 to 1.5 times this value is usually a reasonable value for the operation of a column. Hence, a determination of $R_{\min}$ will give us a value of reflux ratio that can be used for overall column design.

From all the foregoing it becomes apparent that the equilibrium behavior of a system is a necessary ingredient for distillation column design. Data of this type

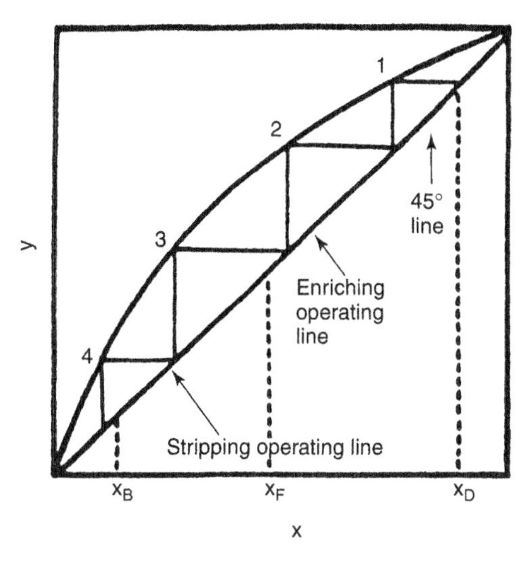

Figure 12-11. Total reflux and minimum stages with McCabe–Thiele method. (Reproduced with permission from reference 18. Copyright 1997, American Chemical Society.)

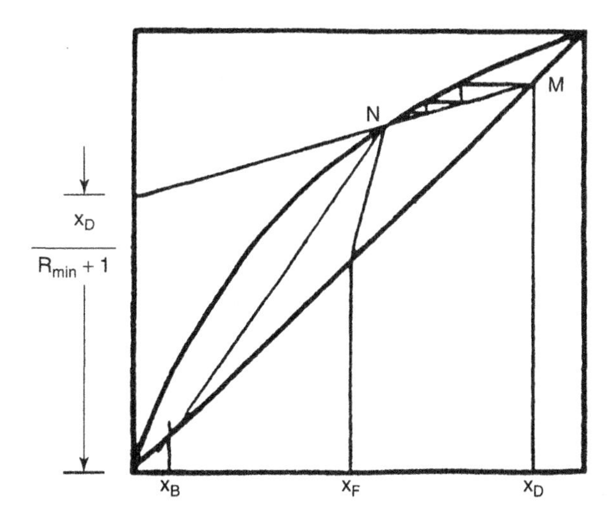

Figure 12-12. Infinite stages and minimum reflux with intersection of equilibrium curve feed line and rectifying mass balance. (Reproduced with permission from reference 18. Copyright 1997, American Chemical Society.)

are available for certain systems. In other cases it can be estimated (3) using thermodynamics. A useful and empirical relationship in this regard is the relative volatility α_{AB} given below:

$$\alpha_{AB} = \frac{Y_A/X_A}{Y_B/X_B} \tag{12-27}$$

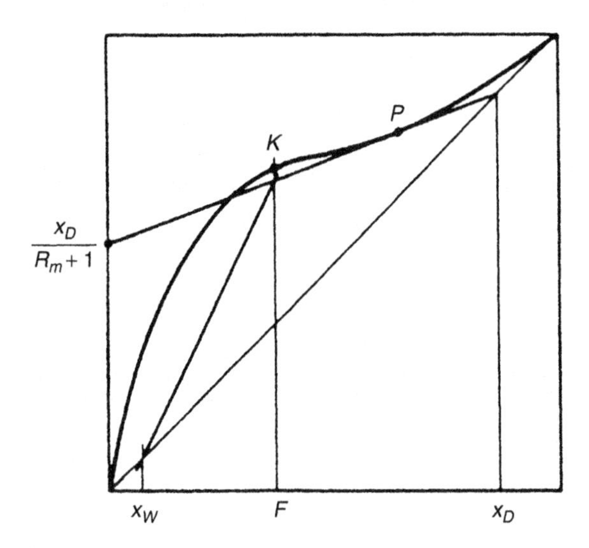

Figure 12-13. Tangency of rectifying mass balance lines with equilibrium curve giving minimum reflux. (Adapted from reference 13.)

The relative volatility is a ratio that remains reasonably constant at a given pressure, hence the ratio will hold for a given column even though temperature can vary. Also note that the ratio can also be written as

$$\alpha_{AB} = \frac{Y_A(1 - X_A)}{X_A(1 - Y_A)} \tag{12-28}$$

This means that with a single relative volatility value we can estimate an entire equilibrium curve (i.e., assume an X_A and calculate the corresponding Y_A).

The relative volatility has another use; namely, it indicates the ease of separation for given system. If the value of α_{AB} were unity, no separation would be possible. On the other hand, large or small values of α_{AB} would indicate a relatively easy separation.

We have continually stated and restated the fact that the stages determined either analytically or graphically by the McCabe–Thiele method are *ideal* or *equilibrium stages*. A better determination (i.e., calculation of stages) can be made if we have available combined equilibrium and enthalpy data for a given system (see Figure 12-14). For such a situation, we can calculate analytically the stages for a proposed separation. The stages so determined will be much closer to real stages because they do not assume constant molal overflow (since the enthalpy balance is included). This technique, while superior to the McCabe–Thiele, has a severe drawback, namely, the lack of good and reliable data.

What is basically done, therefore, is to use the concept of efficiencies to convert ideal stages to real stages. Such a method, while empirical, can yield reasonable results. The drawback to this approach is that although detailed studies

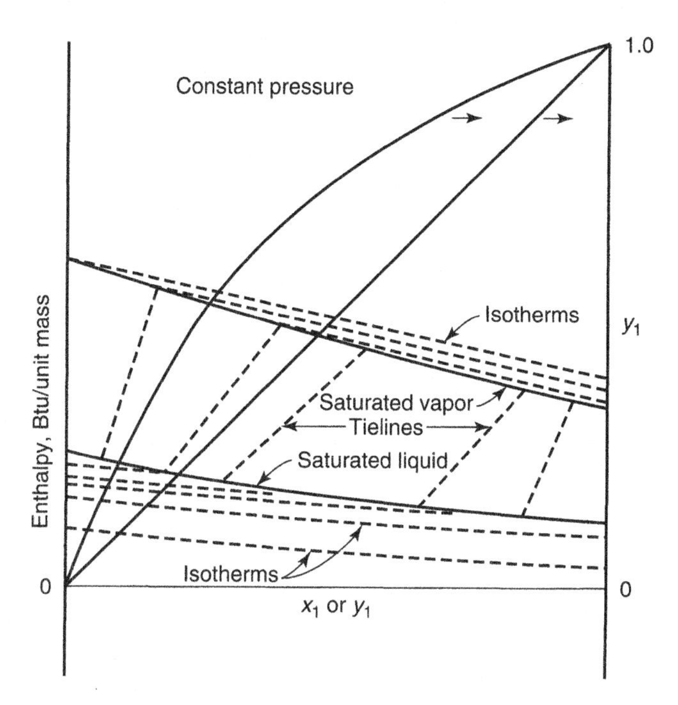

Figure 12-14. Construction of an enthalpy-composition plot (constant pressure) (19).

(5–7) have been made of stage efficiencies, totally reliable generalized correlations are not available. If a system has been widely studied, it would obviously be possible to derive stage efficiencies. However, in the absence of such data, stage efficiencies are estimated to be between 0.6 and 0.8 with the lower value favored.

MULTICOMPONENT, SPECIAL, AND BATCH DISTILLATIONS

We have considered only binary distillation to this point because multicomponent systems increase complexity by orders of magnitude. Furthermore, we continue to fight the battle of too little data. In order to properly design a multicomponent column, we would need equilibrium data for the multicomponent system and enthalpy data. The latter is usually not even available for binaries. Even with all of the required data available, the column design would be an extraordinarily complicated calculation requiring a stage-by-stage determination.

There is fortunately a shorthand method available that can give reasonable estimates of ideal stages for a multicomponent system. The first step is to compare the relative volatilities of all component to the least volatile as shown below for a five-component system.

Component	Volatility Relative to Least Volatile
A	5.1
B	3.6
C	1.7
D	1.3
E	1.0

In the above, we can see that in any effective separation the break would occur between B and C (most of B in overhead, little in bottoms; most of C in bottoms, little of C in overhead). This in essence treats the system as if it were a pseudobinary. Components B and C are called the light and heavy keys.

For this case the *minimum steps* or *stages* are given by (8)

$$S_{\text{minimum}} = S_{\min} = \frac{\log\left[\left(\frac{X_{\text{lk}}}{X_{\text{hk}}}\right)_D \left(\frac{X_{\text{hk}}}{X_{\text{lk}}}\right)_W\right]}{\log \alpha'} \tag{12-29}$$

where the X's are the mole fractions of the heavy or light keys in top (D) and bottom (W) products. The α' is the relative volatility of the light key to the heavy key (in the case cited above, $\alpha' = 3.6/1.7$). This case for $S_{\min}$ involves (as for binaries) total reflux.

Next for minimum reflux we have

$$R_{\min} + 1 = \frac{\alpha_A x_{D_A}}{\alpha_A - \theta} + \frac{\alpha_B x_{D_B}}{\alpha_B - \theta} + \cdots + \frac{\alpha_Z x_{D_Z}}{\alpha_Z - \theta} \tag{12-30}$$

where the α's are the relative volatilities to the least volatile, the x_D's are the mole fraction in the overhead product, and θ is an empirical constant. The latter is obtained by trial and error from

$$\frac{\alpha_A x_{F_A}}{\alpha_A - \theta} + \frac{\alpha_B x_{F_B}}{\alpha_B - \theta} + \cdots + \frac{\alpha_Z x_{F_Z}}{\alpha_Z - \theta} = f \tag{12-31}$$

where f is the feed condition and the X_F's are the mole fractions of the components in the feed. Again minimum reflux (just as with binaries) implies infinite stages.

If $S_{\min}$ and $R_{\min}$ are known, it is then possible (for any given reflux ratio R) to determine the stages needed by using the graphical correlation shown in Figure 12-15.

The treatment of batch distillation is usually a difficult and complex problem. There is, however, one aspect of such distillation that can be treated more simply. This situation is that of *differential distillation*. In such a distillation the liquid is vaporized and each segment of vapor is removed from liquid contact as it is formed. This means that although the vapor can be in equilibrium with the liquid

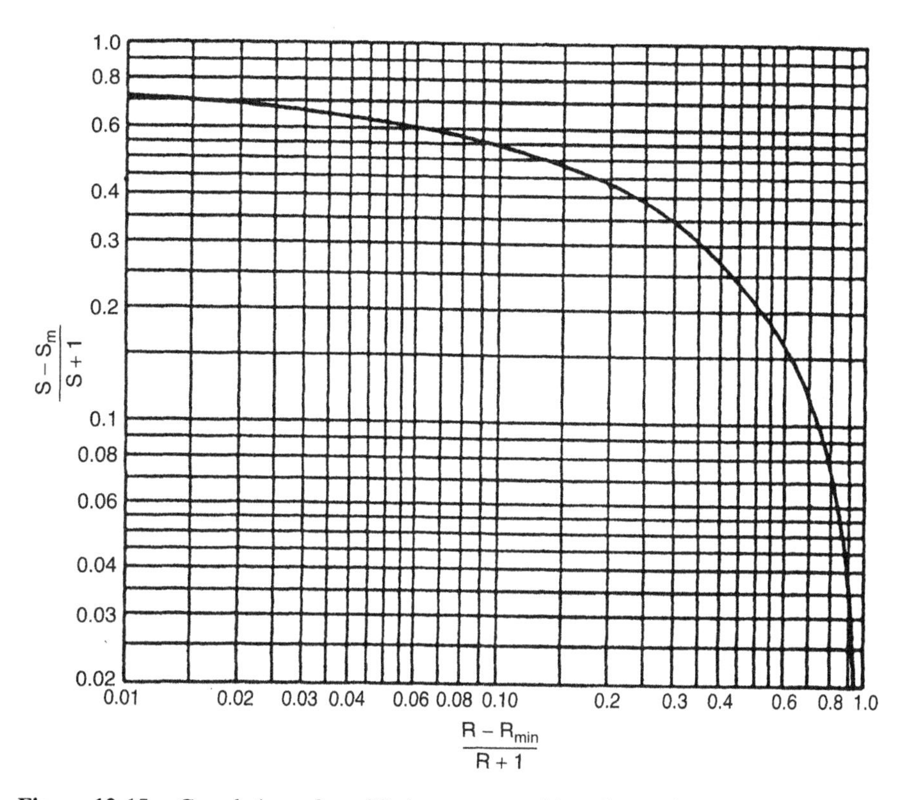

Figure 12-15. Correlation of equilibrium stages with reflux ratio. (Reproduced with permission from reference 10. Copyright 1940, American Chemical Society.)

as it is formed, the *average vapor formed* will not be in equilibrium with the liquid residue.

The original treatment is due to Lord Rayleigh (11). For an amount $-dW$ to be vaporized we have

$$-ydW = -(Wdx) \tag{12-32}$$

and then

$$\frac{Wdx}{dW} = y - x \tag{12-33}$$

so that

$$\int_{W_0}^{W} \frac{dW}{W} = \int_{x_0}^{x} \frac{dx}{y - x} \tag{12-34}$$

Finally,

$$\ln\left(\frac{W}{W_0}\right) = \int_{X_0}^{X} \frac{dx}{y - x} \tag{12-35}$$

Similarly, for any two components in a differential distillation we have

$$\frac{A}{A_0} = \left(\frac{B}{B_0}\right)^{\alpha} \tag{12-36}$$

where A, and B are the moles in the still at time t, A_0, and B_0 are the original moles, and the α is the relative volatility of A to B.

In our earlier discussion of relative volatility, we indicated that such values near unity would make separations almost impossible under normal circumstances. For such systems, we resort to the addition of a third component that enables us to carry out an *extractive* or an *azeotropic distillation*.

The former case involves adding an *extractive agent* that imbalances the relative volatility of the two components to be separated. In essence, one of the components is then separated from the mix of the other component and extractive agent. A schematic of an extractive distillation system is shown in Figure 12-16. As can be seen, the separated extractive agent is recycled to the column.

Some typical extractive agents are given in Table 12-4.

In azeotropic distillation, the third component forms an azeotrope with the system that becomes either the top or bottom product. The azeotrope is then separated into the agent and component. Sometimes such separation must be done using another process such as liquid extraction. Some typical systems (the azeotroping agent in parentheses) are acetic acid–water (butyl acetate), and ethanol–water (benzene).

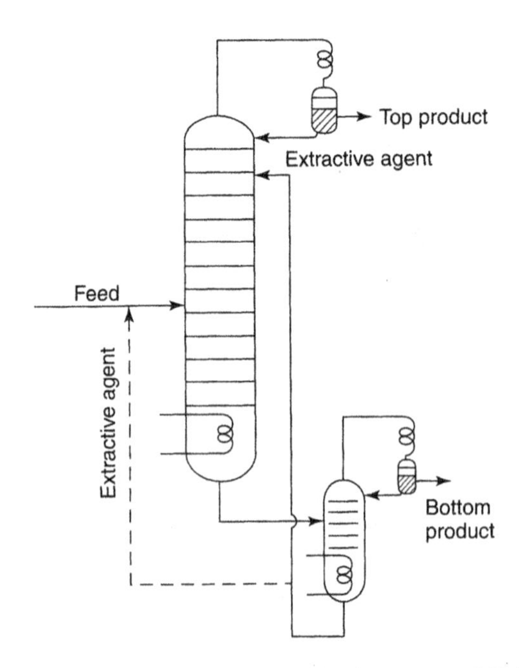

Figure 12-16. Extractive distillation system (11).

Table 12-4 Some Typical Extractive Agents

Original System	Extractive Agent
$HCl-H_2O$	H_2SO_4
HNO_3-H_2O	H_2SO
Ethanol–water	Glycerin
Butane–butene	Acetone or furfural
Butadiene–butene	Acetone or furfural
Isoprene–pentene	Acetone
Toluene–paraffinic hydrocarbons	Phenol
Acetone–methanol	Water

DISTILLATION COLUMN STAGES

While we have considered column design in relation to the number of ideal stages required for a given separation, we have not discussed the stages themselves. Such stages must, of course, give the best possible mixing (i.e., approach equilibrium). There are many systems in use.

One such system is the bubble cap tray (see Figure 12-17). As can be seen, the vapor from the stage below flows up into the slotted cap where it is bubbled into the liquid coming from the stage above. The liquid after contacting flows

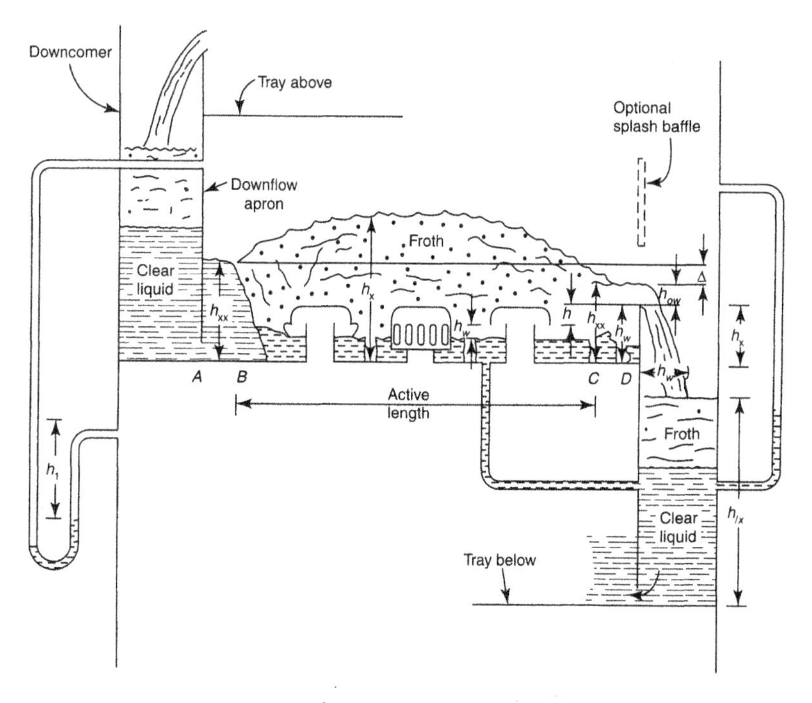

Figure 12-17. Bubble cap tray fluid dynamics (19).

over a weir to a downcomer which tapes it to the stage below. Bubble cap trays were once the most widely used contacting device in distillation columns. While no longer the dominant tray, they are still found in many operations.

Sieve trays (see Figure 12-18) are simpler and less expensive than bubble cap trays. In these units the gas flowing upward through the holes mixes with the liquid to form a frothy mass.

Valve trays (see Figure 12-19) are units in which holes are covered with movable caps whose rise varies with gas flow rate. The valve tray is a very widely used device because it represents a useful compromise between the bubble cap and sieve trays.

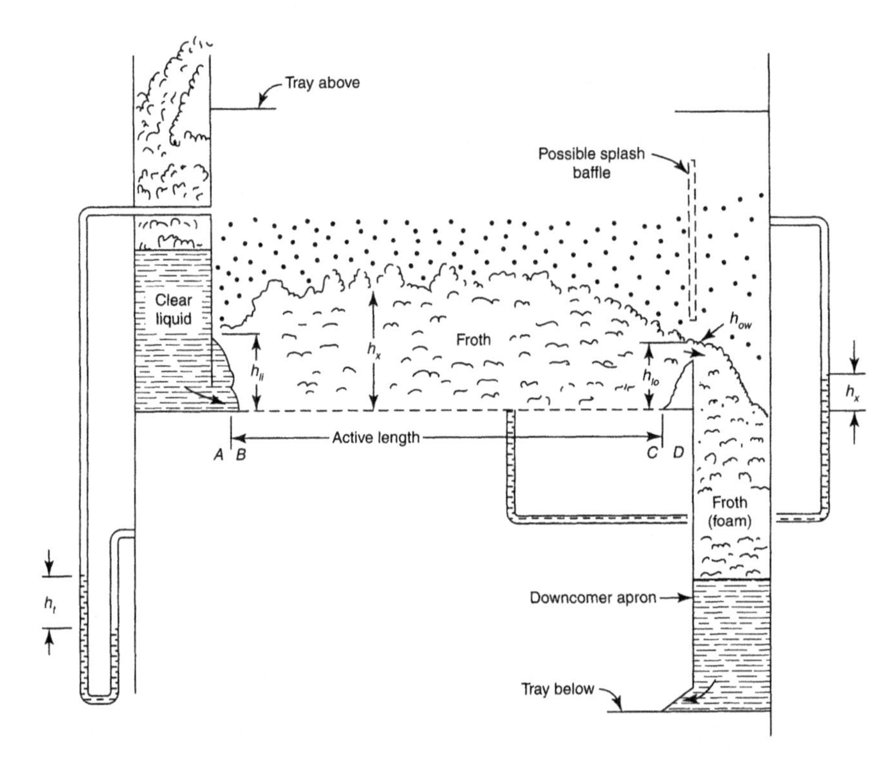

Figure 12-18. Sieve tray fluid dynamics (19).

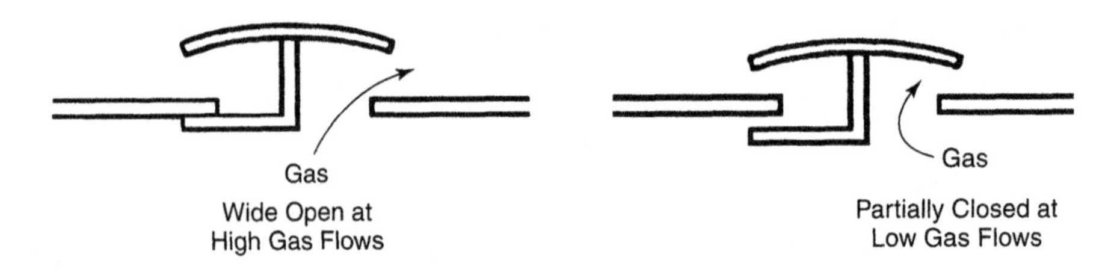

Figure 12-19. Valve tray cap. (Adapted from reference 12.)

Figure 12-20. Column packings. (Reproduced with permission from reference 18. Copyright 1997, American Chemical Society.)

The physical design of any of the tray columns represents a complex hydrodynamic problem. For example, in bubble cap trays pressure drops through the caps, tray pressure drop, flow over the weir, hydraulic gradients and tray spacing must all be considered. In sieve trays such items as the perforations, pressure drops, liquid gradients, and entrainment are important. An additional factor that must be dealt with for sieve trays is the process known as *weeping* (liquid leaking through sieve holes). General discussions of the physical designs for column trays can be found elsewhere (1, 2, 4).

Distillations can also be carried out in packed columns. These devices use commerical packings (see Figure 12-20) to break up the flows and thus bring about intimate contact between gas and liquid. Design of such units involves combining equilibrium stage and diffusional approaches. The height of the column is found by multiplying an HETP (height equivalent to a theoretical plate) or HETS (height equivalent to a theoretical stage) times the number that will be discussed in the next chapter because the packed column is very widely used with the absorption process.

DISTILLATION COLUMN EFFICIENCIES

In actual distillation column operation, an efficiency must be applied to the ideal stages determined. Two reasonably simple correlations can be used to estimate column efficiencies.

The first, that of Drickamer and Bradford (15), correlates the efficiency with the molar average viscosity (at average tower temperature) in centipoise (see Figure 12-21).

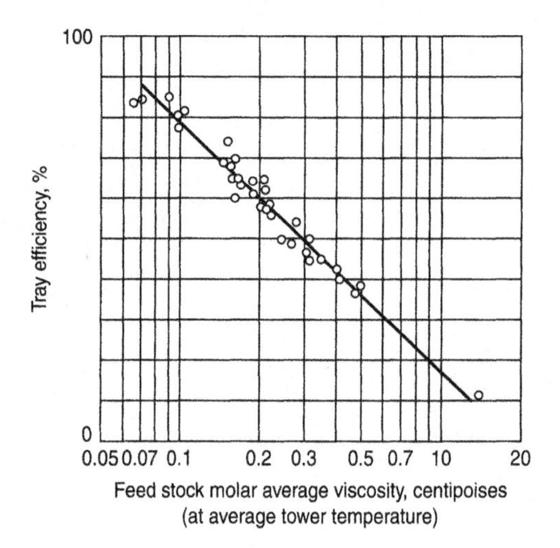

Figure 12-21. Drickamer–Bradford correlation for overall column efficiency. (Reproduced with permission from reference 15. Copyright 1943, American Institute of Chemical Engineers.)

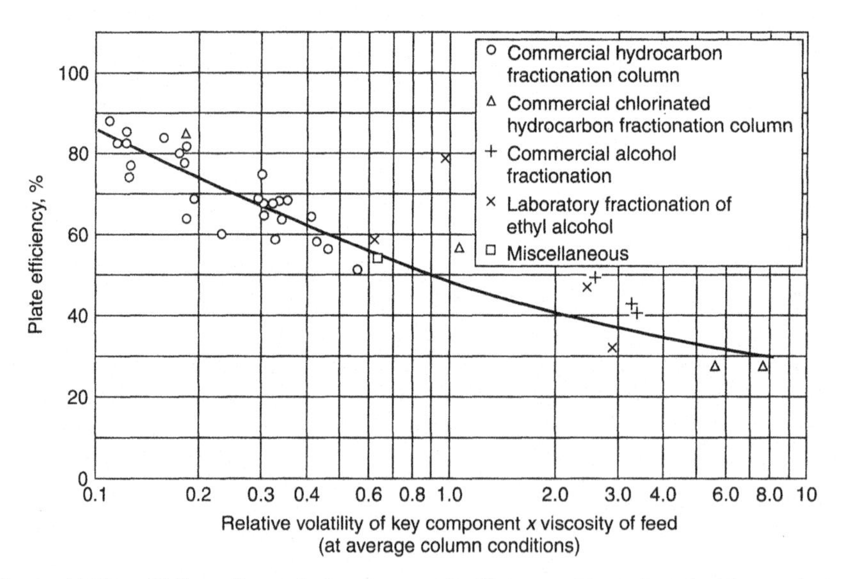

Figure 12-22. O'Connell correlation for overall efficiency. (Reproduced with permission from reference 16. Copyright 1946, American Institute of Chemical Engineer.)

The second derived by O'Connell (Figure 12-22) correlates efficiency with the product of relative volatility times the feed viscosity (at average column temperature).

In addition, the impact of column capacity on efficiency has also been presented. This correlation is presented in Chapter 13 of this text.

WORKED EXAMPLES

Example 12-1 Design a continuous fractionating distillation column to separate 3.78 kg/sec of 40 percent benzene and 60 percent toluene into an overhead product containing 97 mass percent benzene and a bottom product containing 98 mass percent toluene. Use a reflux ratio of 3.5 mole to 1 mole of product. Latent molal heat of both benzene and toluene is about 357.1 kJ/kg.

Find the amounts of top and bottom product; also, consider the following cases for the design:

1. Liquid feed at its boiling point.
2. Liquid feed at 20°C ($C_{\text{liquid}} = 1.75$ kJ/kg °C).
3. Feed is two-thirds vapor.

First determine the necessary material balances, mole fractions, and so on:

$$X_F = \frac{\frac{40}{78}}{\frac{40}{78} + \frac{60}{92}} = 0.44$$

$$X_D = \frac{\frac{97}{78}}{\frac{97}{78} + \frac{3}{92}} = 0.974$$

$$X_B = \frac{\frac{2}{78}}{\frac{2}{78} + \frac{98}{92}} = 0.0235$$

The feed rate F is

$$F = \frac{\frac{3.78 \text{ kg/sec}}{100}}{\frac{40}{78} + \frac{60}{92}} = 0.0441 \text{ kg moles/sec}$$

$$D = 0.0441 \left(\frac{X_F - X_B}{X_D - X_B} \right) = 0.0193 \text{ kg mole/sec}$$

$$B = F - D = (0.0441 - 0.0193) \text{ kg mole/sec} = 0.0248 \text{ kg mole/sec}$$

Now the graphical solution can be approached. First plot, the equilibrium line and ($x = y$) line together with the vertical X_B, X_F, and X_D (see Figure 12-23). Next locate the rectifying operating line (passes through $X_D = Y_D$ and y-axis

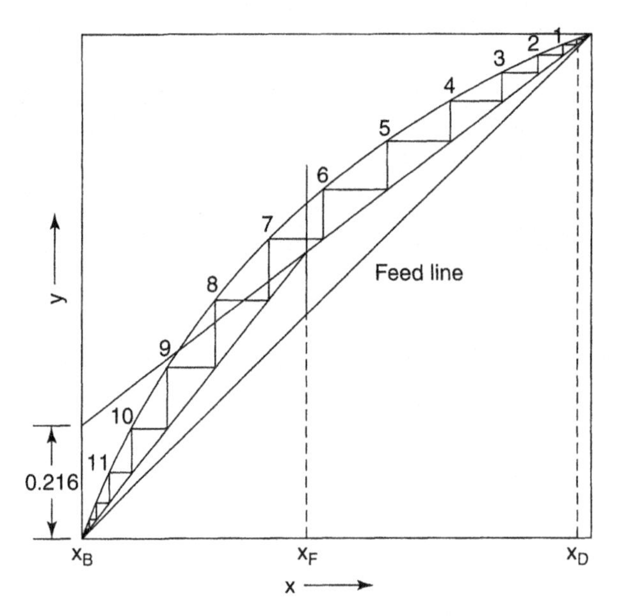

Figure 12-23. Solution for saturated liquid feed. (Reproduced with permission from reference 18. Copyright 1997, American Chemical Society.)

intercept of $X_D/R_D + 1)$ or

$$\frac{0.974}{3.5 + 1} = 0.216.$$

The f values for the first case (liquid feed at the boiling point) is zero, giving a vertical feed line. The intercept of the feed line and the rectifying operating line gives one intercept for stripping operating line (the other intercept is $X_B = Y_B$).

The solution Figure 12-23 gives 11 ideal stages plus a reboiler. Feed enters in the seventh stage from the top.

For case 2, f is given by (the bubble point temperature T_b is obtained from Figure 12-3)

$$f = \frac{-C \text{ liquid } (T_b - T_F)}{\lambda}$$

$$f = \frac{-1.75(95 - 20)}{357.1}$$

$$f = -0.370$$

and the slope for the feed line is

$$\text{slope} = -\left(\frac{1 - f}{f}\right) = -\left(\frac{1 - (-0.370)}{-0.370}\right) = 3.70$$

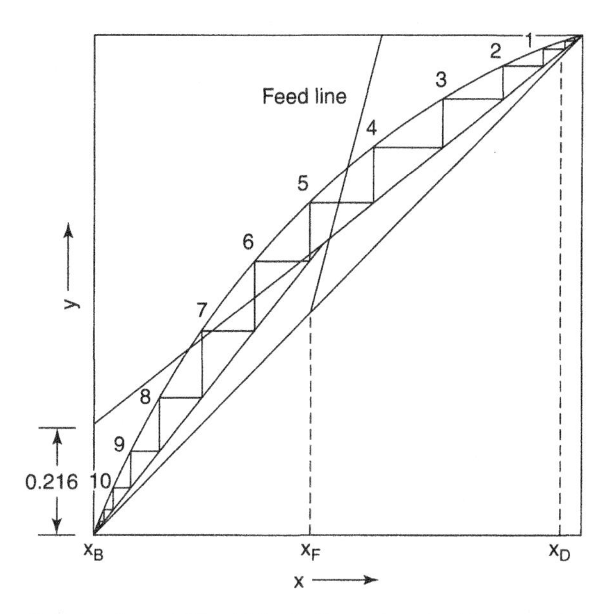

Figure 12-24. Solution for cold liquid feed. (Reproduced with permission from reference 18. Copyright 1997, American Chemical Society.)

A line with this slope is drawn through X_F, Y_F (Figure 12-24). Note that the mass balance line for the rectifying section remains the same (because reflux ratio and X_D are unchanged). The function of the feed and rectifying mass balances is then joined to X_B. The ideal stages are then determined (ten plus a reboiler with the feed entering on the fifth plate from the top).

Finally, in case 3, the f value is 2/3 (since 2/3 of the feed is vapor), thus giving a slope of $-1/2$ for the feed line. Here again the feed line intersects the rectifying mass balance at a different point. This function is connected to X_B, and the resultant solution is shown in Figure 12-25. There are 12 ideal stages, plus a reboiler with feed entering on the seventh plate from the top.

Example 12-2 A distillation column (having three ideal stages) is used to separate ammonia and water.

The feed (saturated vapor) consists of 0.5 mole percent ammonia (A) and 99.5 mole percent water (B) and enters between the second and third stage. Reflux to the column top is 1.3 moles per mole of feed. Likewise, the vapor from the reboiler is 0.6 moles per mole of feed.

Ammonia–water equilibrium is given by $Y_A = 12.6X_A$

Find X_B, X_D, and the mole fraction of ammonia in the overflow from the feed.

Use 100 moles of feed as the basis for the calculation.

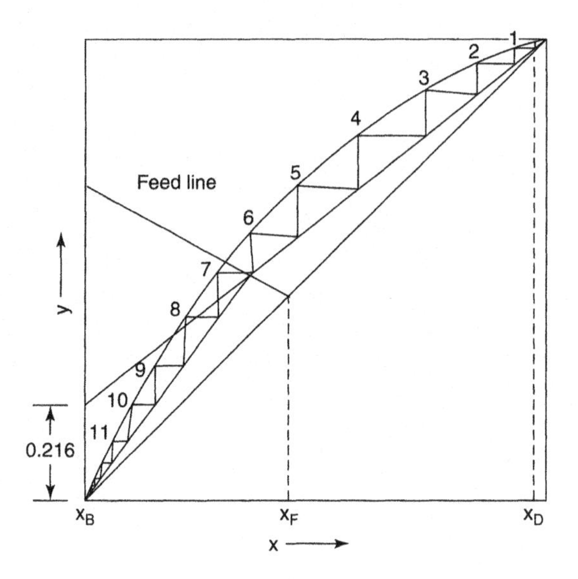

Figure 12-25. Solution for feed that is two-thirds vapor. (Reproduced with permission from reference 18. Copyright 1997, American Chemical Society.)

The liquid reflux and liquid overflow above the feed plate are both 1.3 (100) or 130 moles.

Likewise, the vapor in the rectifying section is 100 + 0.6(100) — the feed plus the vapor from the reboiler.

Then,

$$D = 160 - 130 = 30$$
$$B = 130 - 60 = 70$$

Using equation (12-10) with ammonia, we obtain

$$Y_{n+1}V = X_n L + X_D D$$

because L and V are constant.

$$Y_{n+1} = \frac{L}{V}X_n + \frac{D}{V}X_D$$
$$Y_{n+1} = \frac{130}{160}X_n + \frac{30}{130}X_D$$
$$Y_{n+1} = 0.813X_n + 0.188X_D$$

Then, if we do a stage-by-stage calculation, at the first plate (topmost) we have

$$X_1 = \frac{Y_1}{12.6} = \frac{X_D}{12.6}$$

since $X_D = Y_1$

Then for plate 2 we have

$$Y_2 = 0.813X_1 + 0.188X_D$$
$$Y_2 = \left(\frac{0.813}{12.6} + 0.188\right) X_D$$
$$Y_2 = 0.252X_D$$
$$X_2 = 0.252X_D/12.6 = 0.020X_D$$

Next for plate 3 we have

$$Y_3 = 0.813X_2 + 0.188X_D$$
$$Y_3 = (0.813)(0.020) + 0.188)X_D = 0.204X_D$$
$$X_3 = 0.204X_D/12.6 = 0.0162X_D$$

Using a mass balance around the reboiler (liquid with X_3 in, vapor with Y_0 out, liquid product X_B out), we obtain

$$130X_s = 60Y_0 + 70X_B$$

Solving simultaneously gives

$$X_D = 393X_B$$

An overall ammonia balance is

$$(0.005)(100) = 30XD + 70X_B$$
$$0.5 = [(30)\ (393)\ + 70]X_B$$
$$X_B = 0.000042$$
$$X_D = (393)\ X_B = 0.166$$
$$X_3 = 0.01617X_D = 0.000268$$

Example 12-3 A mixture of 35 mole percent A and 65 mole percent B is to be separated in a distillation column. The mole fraction of A in the distillate is 0.93 and 96 percent of A is in the distillate. The feed is half vapor and half liquid. Reflux ratio is 4 and the relative volatility of A to B (αAB) is 2. How many equilibrium stages are needed for this separation?

We first use the concepts of the relative volatility to obtain equilibrium data.

$$\alpha_{AB} = \frac{Y_A/X_A}{Y_B/X_B} = \frac{Y_A X_B}{Y_B X_A} = \frac{Y_A(1 - X_A)}{X_A(1 - Y_A)}$$

$$\alpha_{AB} = \alpha_{AB} Y_A = Y_A \frac{1 - X_A}{X_A} \qquad \text{if } \alpha_{AB} = 2.0$$

$$2.0 - 2.0Y_A = Y_A \left(\frac{1 - X_A}{X_A}\right)$$

$$Y_A \frac{1 - X_A}{X_A} + 2 = 2$$

$$\left(\frac{1 + X_A}{X_A}\right) Y_A = 2$$

Assuming values of X_A yields Y_A's.

X_A	0	0.2	0.4	0.50	0.60	0.8	1.0
Y_A	0	0.33	0.57	0.67	0.75	0.89	1.0

The above constitutes the equilibrium data for the system.

We are given $X_D(0.93)$ and $X_F(0.35)$. In order to obtain X_B, we must use a mass balance.

Assuming 100 moles of feed, for the top product using an "A" balance we have

$$(0.35)(100)(0.96) = D(0.93)$$

$$D = 36.13$$

Then, for the bottom product we obtain

$$B = 100 - 36.13 = 63.87$$

Amount of less volatile in top product is (0.07)(36.13), or 2.53. Less volatile in bottom is (65 - 2.53), or 62.47. More volatile in bottoms is (0.04)(35) , or 1.40.

$$X_B = \frac{1.40}{62.47 + 1.40} = 0.0219$$

The feed-line slope is

$$-\left(\frac{1 - f}{f}\right) = -\left(\frac{1 - 0.5}{0.5}\right) = -1$$

Furthermore, the y-axis intercept is

$$\frac{X_D}{R + 1} = \frac{0.93}{4 + 1} = 0.186$$

The graphical solution is shown in Figure 12-26. There would be 14 ideal stages (plus a reboiler). Feed is introduced on the eighth stage from the top.

Example 12-4 Distillation columns can be cooled by using a so-called pumparound stream in the columns enriching section (see Figure 12-27). Liquid and vapor rates above the pumparound stage are larger than those below because some of the vapor in the pumparound stream is condensed.

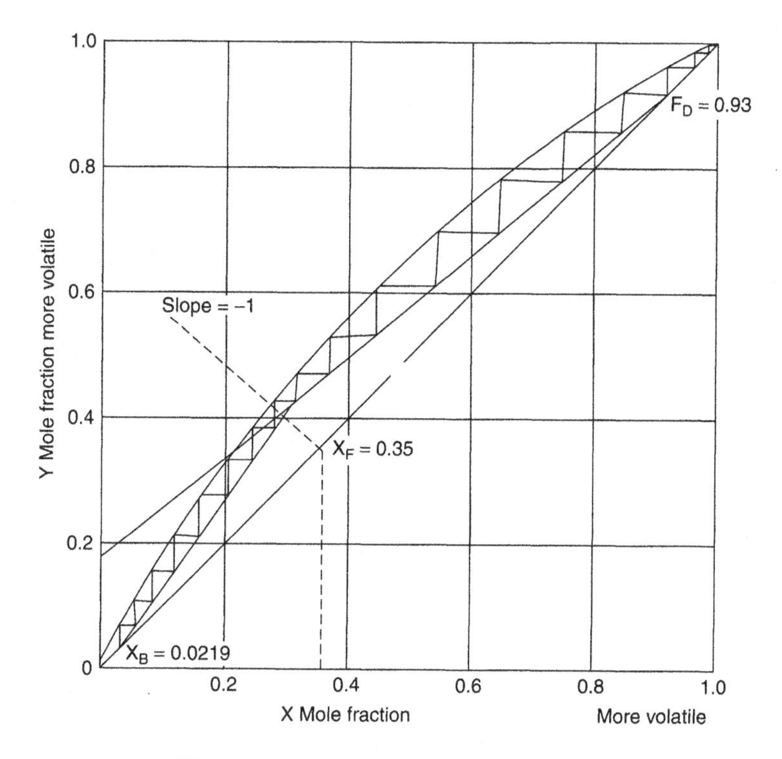

Figure 12-26. Solution for Example 12-3.

Find the total number of ideal stages, the pumparound stage, and the feed stage for the following case:

The values of X_B, X_D and X_F are 0.05, 0.95, and 0.50, respectively. Relative volatility (binary system) is 2.0. Reflux at pumparound stage is 0.7, and heat removed at the pumparound is 1.7 times that removed from the condenser.

Feed (at a rate of 100 kg mole/minute) condition is 0.5.

Let us take one minute of operation; the top and bottom products are then given by the simultaneous equations

$$100 = D + B$$

$$(0.50)(100) = 0.95D + 0.05B$$

This gives values of D and B to be 50 each.

Note that the equilibrium data of the preceding example can be used for this case because relative volatility is 2.0.

Next, the operating mass balance line for the top of the column will intercept the y axis at

$$Y_{\text{intercept}} = \frac{X_D}{R+1} = \frac{0.95}{1.5+1} = 0.38$$

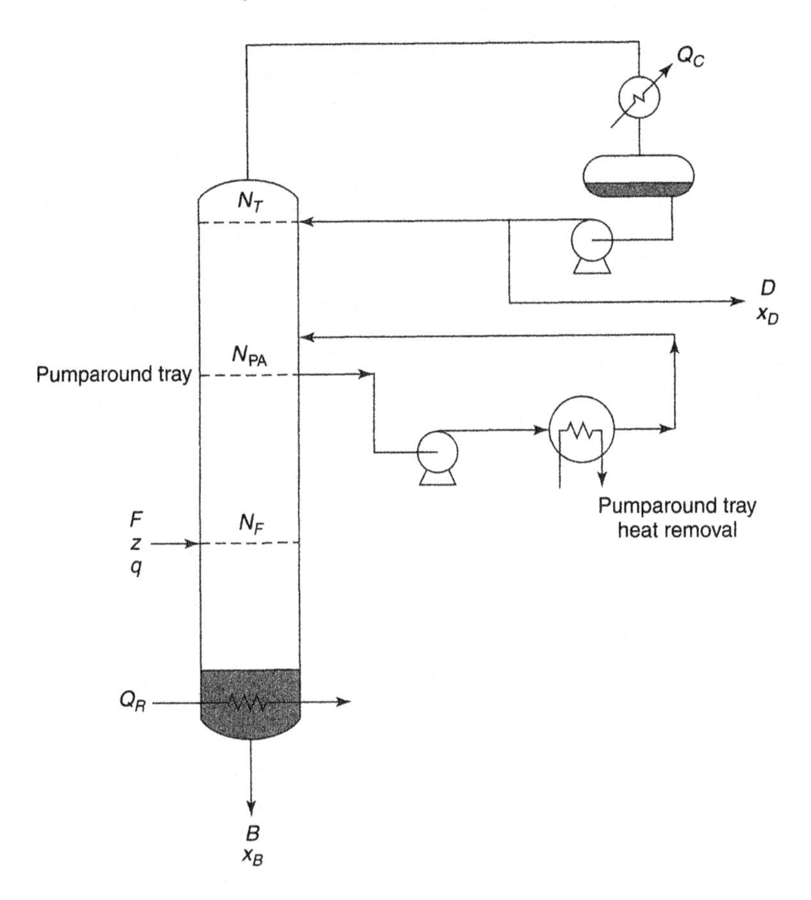

Figure 12-27. Schematic of column for Example 12-4. (Adapted from reference 12.)

This line also passes through X_D, Y_D.

The operating line (AA) in Figure 12-28 is used above the pumparound.

Next, we find the operating line below the pumparound to the feed. This can be obtained by using the energy removed in the condenser.

$$\text{Total vapor product} = D + L$$

$$\text{Total vapor product} = 50 + 75 = 125$$

$$\text{Energy removed in condenser} = 125(\Delta H)$$

$$\text{Energy removed in pumparound} = (1.7)(125)\ (\Delta H)$$

Thus, liquid at pumparound stage is (1.7) (125) , or 212.5.

Liquid overflow is then 212.5 + 75 (reflux), or 287.5. Vapor is 125 + 212.5 or 337.5.

The slope of the this operating line is then L/V or 287.5/337.5 or 0.852.

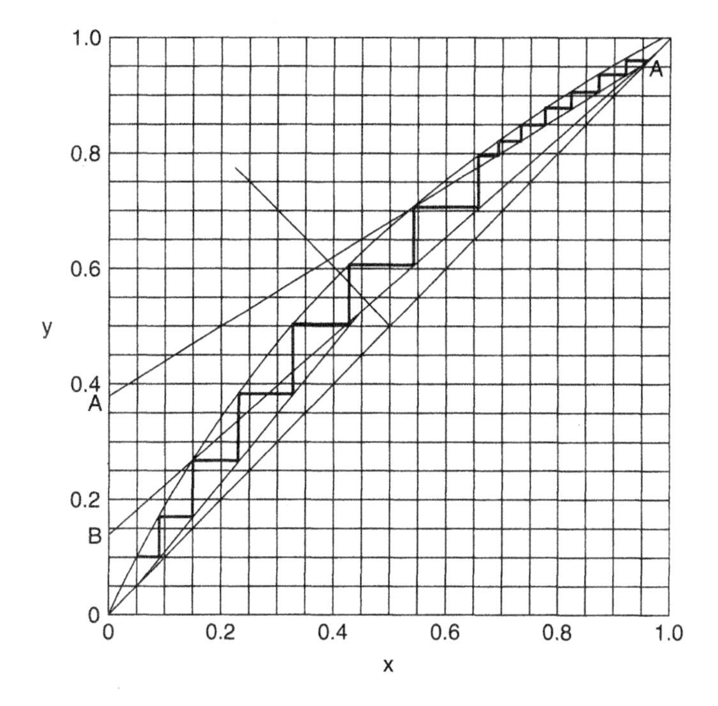

Figure 12-28. Solution for Example 12-4.

Results line is AB of Figure 12-28.

Now we can find the stages. Note that the stripping mass balance line is given by the function of the AB line and the feed line [slope of $-(1-0.5/0.5)$ or -1] and X_B.

If we step off the stages (in this case starting at the bottom of the column), we move first along the stripping mass balance to the feed stage then along the operating mass balance line below the pumparound. We switch to the top operating line after the intercept of AA with the equilibrium curve (as shown).

Hence, there are a total of 13 stages (plus a reboiler). The feed is on the ninth stage from the top. The pumparound stage is the sixth stage from the top.

Example 12-5 What are the minimum stages and reflux values for the columns of Examples 12-1 and 12-3.

The minimum stages for these cases uses the $y = x$ line as the mass balance line. For Examples 12-1 and 12-3, we can use Figures 12-23 and 12-26 to find these values.

Case	Minimum Stages
Example 12-1	8
Example 12-3	8

The minimum reflux situation can be handled by again using the equilibrium plots and determining the y-axis intercept for infinite plates.

Using the appropriate plots, we obtain the following values for the $y_{\text{intercept}}$ $(X_D/R_m + 1)$ and R_m.

Case	$y_{\text{intercept}}$	R_m
Example 12-1	0.215	3.32
Example 12-3	0.300	2.17

Example 12-6 A system to be distilled has the equilibrium diagram shown in Figure 12-29. Find minimum reflux ratios for X'_Ds of 0.8 and 0.7.

The feed (50 percent vapor) has an X_f of 0.40.

We first obtain the feed-line slope:

$$\text{Slope} = -\left(\frac{1 - 0.5}{0.5}\right) = -1$$

Next we move a straight line up the y axis using the X_D as a pivot point.

In the case of $X_D = 0.7$ the enriching mass balance line intersects the feed line and equilibrium curve (see line AA). Its intercept is 0.489.

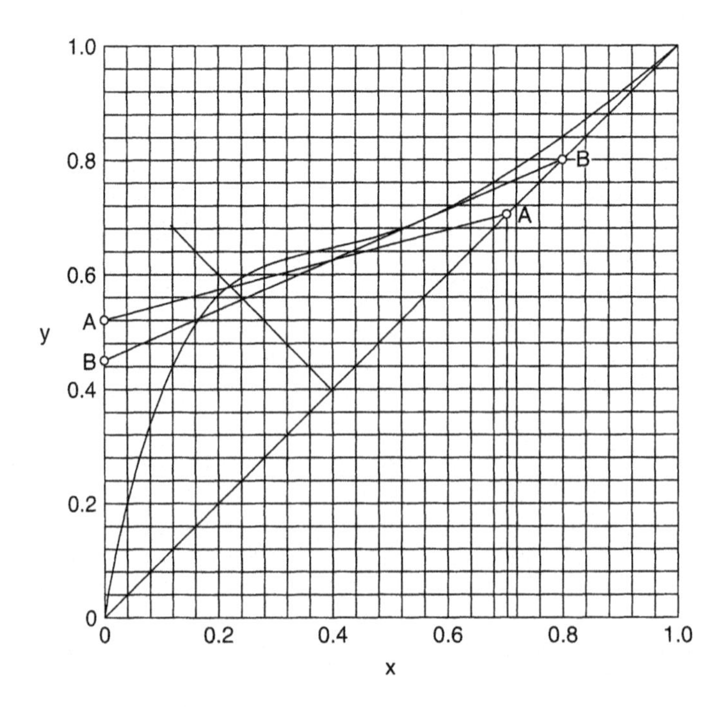

Figure 12-29. Equilibrium diagram for Example 12-6. (Adapted from reference 12.)

Such an occurrence is not possible for $X_D = 0.8$ because the mass balance line is tangent to the equilibrium curve before intersecting the feed line (BB). This gives the y intercept (0.433).

Minimum reflux values are then 0.432 (for $X_D = 0.70$) and 0.848 (for $X_D = 0.80$).

Example 12-7 It is desired to design a distillation column to separate benzene and toluene. Two separate feeds are to be used for the column (one with X_F of 0.4; the other with X_F of 0.7). These feeds are available in equal amounts as a saturated liquid ($f = 0$).

The purities of top and bottom product are $X_D = 0.98$ and $X_B = 0.03$. Reflux ratio is to be 1.5 $R_{\min}$.

The benzene–toluene system equilibrium data are given in Figures 12-4, 12-23, and 12-25.

In order to obtain the reflux ratio, let us find $R_{\min}$ for a combined feed.

If X_{F^1} is the combined feed's mole fraction

$$X_{F^1} = \left(\frac{0.4 + 0.7}{2}\right) = 0.55$$

then the feed line is vertical and the line intersecting the feed line at the equilibrium line intersects the y axis at 0.456. This gives an $R_{\min}$ of 1.15. The reflux ratio is then 1.5(1.15), or 1.73.

We can now use this value to locate the mass balance line for the top of column to the feed point for the 0.7 feed.

$$y_{\text{intercept}} = \frac{0.98}{1.73 + 1} = 0.359$$

This line goes to X of 0.70 where there is a vertical feed line (see Figure 12-30). Then, for the section between the two feeds (assuming 50 moles of each).

$$L = 50 + R = 50 + 1.73D$$

$$V = R + D = 2.73D$$

The D value can be found as follows:

$$\frac{D}{F} = \frac{D}{100} = \frac{X_F - X_B}{X_D - X_B}$$

$$D = 100\left(\frac{0.55 - 0.03}{0.98 - 0.03}\right) = 54.7$$

This gives L and V values of 144.6 and 149.3. The L/V ratio is then 0.968, which is the slope of the operating line between the 0.7 and 0.4 feed points.

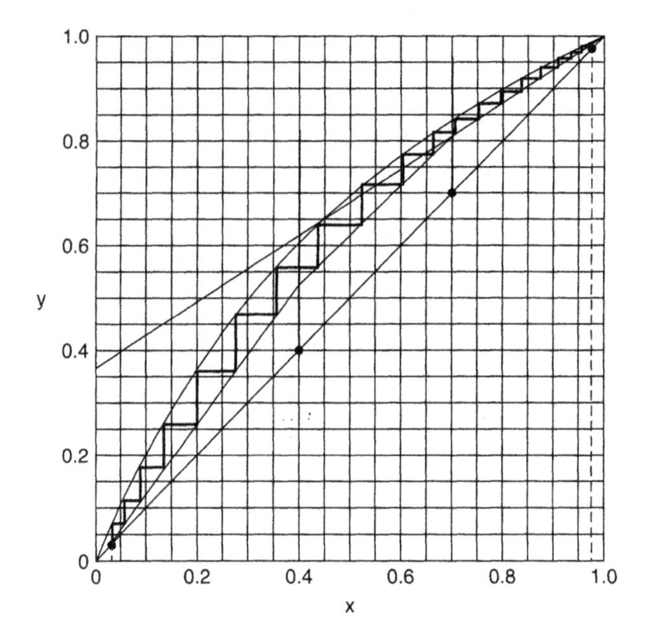

Figure 12-30. Solution for Example 12-7.

The intercept of the 0.968 slope line with the 0.4 feed line is then joined to the X_B point (0.03).

A total of 13 ideal stages plus a reboiler are needed for the column with two feeds. The feed stages are the fifth from the top ($X_F = 0.70$) and ninth from the top ($X_F = 0.40$).

Example 12-8 Use the methods of Chapter 12 together with Figure 12-15 to find the number of stages needed to separate the following multicomponent systems.

1. Benzene–toluene–xylene: relative volatilities of 5.56/2.22/1.00; feed mole fractions 0.6/0.3/0.1; top product mole fractions of 0.995/0.005/0; bottom mole fractions products of 0.005/0.744/0.251. Feed is saturated liquid.
2. Propane–normal butane–normal pentane–normal hexane; relative volatilities 10/4/2/1; feed mole fractions all 0.25; top 0.50/0.48/0.02/0 product mole fractions; bottom product mole fractions 0/0.02/0.48/0.50. Feed is saturated liquid.

Consider the second case. From equation (12-29)

$$S_{\min} = \frac{\log\left[\left(\dfrac{X_{lk}}{X_{hk}}\right)_D \left(\dfrac{X_{hk}}{X_{lk}}\right)_W\right]}{\log \alpha^1}$$

The light and heavy keys are normal butane and normal pentane. This gives an α^1 of 4/2, or 2.

$$S_{min} = \frac{\log\left[\frac{0.48}{0.02}\frac{0.48}{0.02}\right]}{\log 2}$$

$$S_{min} = 9.18$$

Next we obtain R_{min} from equations (12-30) and (12-31).
Then from equation (12-30) we obtain

$$\frac{(10)(0.25)}{10-\theta} + \frac{(4)(0.25)}{4-\theta} + \frac{(2)(0.25)}{2-\theta} + \frac{(1)(0.25)}{1-\theta} = 0$$

By trial and error, θ is found to be 2.57.
Next, using equation (12-31) we obtain

$$\frac{(10)(0.50)}{10-2.57} + \frac{(4)(0.48)}{4-2.57} + \frac{(2)(0.02)}{2-2.57} = R_{min} + 1$$

$$R_{min} = 0.945$$

Now using Figure 12-15 and taking R to be 1.5 R_{min} we obtain

$$\frac{R-R_{min}}{R+1} = \frac{(1.5)(0.945)-(0.945)}{(1.5)(0.945)+1}$$

$$\frac{R-R_{min}}{R+1} = 0.1956$$

From Figure 12-15 we have

$$\frac{S-S_{min}}{S+1} = \frac{S-9.18}{S+1} = 0.45$$

$$S = 17.5$$

This compares to an actual calculated value (stage by stage) of 15 plus a reboiler (i.e., 16).

For the first case repeating the preceding

$$S_{min} = \frac{\log \frac{0.995}{0.005}\frac{0.744}{0.005}}{\log(5.56/2.22)}$$

$$S_{min} = 11.21$$

Then for R_{min} we have

$$\frac{(5.56)(0.6)}{(5.56-\theta)} + \frac{(2.22)(0.3)}{(2.22-\theta)} + \frac{(1)\ (0.01)}{(1-\theta)} = 0$$

By trial and error we obtain $\theta = 2.81$.
Then we compute R_{min}:

$$R_{min} + 1 \frac{(5.56)(0.995)}{5.56-2.81} + \frac{(2.22)(0.005)}{2.22-2.81} + \frac{(1)\ (0)}{1-2.81}$$

$$R_{min} = 0.99$$

The factor needed for Figure 12-5 is

$$\frac{R - R_{min}}{R+1} = \frac{(1.5)(0.99) - (0.99)}{(1.5)(0.99)+1} = 0.199$$

Then from Figure 12-15 we have

$$\frac{S - S_{min}}{S+1} = 0.44$$

$$S = 20.8$$

Ideal stages by a stage-to-stage method gave 16 plus a reboiler (17 total).

PROBLEMS

12-1. A 50–50 mixture of benzene and toluene is to be separated to produce 95 percent pure products. Relative volatility for the system is 2.45. Feed condition is saturated liquid at the boiling point. Find minimum reflux, minimum stages, and ideal stages at 1.5 times minimum reflux.

12-2. An acetic acid (40 mole percent) water system is to be distilled to products of 95 percent purity. The mixture enters the column 25 percent vaporized. Reflux ratio is to be 1.3 times the minimum value. Find the number of ideal stages needed. Vapor–liquid equilibrium data (mole percent acetic acid) are

Vapor	0.0	3.7	7.0	13.6	20.5	28.4
Liquid	0.0	5.0	10.0	20.0	30.0	40.0

Vapor	37.4	47.0	57.5	69.8	83.3	90.8	100
Liquid	50.0	60.0	70.0	80.0	90.0	95.0	100

12-3. A 784 kg mole/h process stream of saturated liquid (71.6 mole percent isobutane and 28.4 mole percent propane) is to be separated. If the relative volatility is 2.4, consider the following cases and find both ideal and minimum stages:

(a) A reflux ratio of 2.774; X_D of 0.96; X_B of 0.058.

(b) All parameters the same as (a), but feed mole fraction is 0.55 propane.

12-4. A 13 ideal tray column has its feed (100 kg mole/minute of saturated vapor) introduced on tray five. Relative volatility, X_D and X_B are 2, 0.94, and 0.06. Find the column's reflux ratio. What are the various tray compositions?

12-5. A saturated liquid feed (0.5 each benzene and toluene) at the rate of 0.0126 k mole/sec. is separated at atmospheric pressure. The column contains 24 stages (50 percent efficiencies). The feed can be introduced on stage 11 or stage 17. Reboiler's maximum vaporization rate is: How many moles per hour of overhead product can be obtained?

12-6. Find the minimum number of stages and minimum reflux for the three cases of Example 12-1.

12-7. A cold feed (80 and 20 mole percent methanol and water) is separated to give X_D and X_B of 0.9999 and 0.005 (methanol).

Reflux ratio is 1.35, and 0.2 mole of vapor per mole of feed is condensed at the feed plate.

Find minimum stages, minimum reflux, and ideal stages. The vapor liquid data are given below (mole fractions):

X	0.1	0.2	0.3	0.4	0.5	0.6	0.7	0.8	0.9	1.0
Y	0.417	0.519	0.669	0.729	0.780	0.825	0.871	0.915	0.959	1.0

12-8. A column has six ideal stages and is to be used to obtain oxygen from air (5.42 atm 25 mass percent vapor). The column's reflux ratio is 2.5. Bottom product has 45 mass percent oxygen.

Enthalpy data at the feed pressure are given below:

	Liquid		Vapor	
T (°C)	N_2 (wt %)	H (kJ/g mole)	N_2 (wt %)	H (kJ/g mole)
−163	0	1.76	0	7.70
−165	7.5	1.75	19.3	7.34
−167	17.0	1.74	35.9	7.05
−169	27.5	1.72	50.0	6.80
−171	39.0	1.67	63.0	6.57
−173	52.5	1.58	75.0	6.34
−175	68.5	1.46	86.0	6.07
−177	88.0	1.26	95.5	5.96
−178	100	1.10	100	5.88

12-9. A 50-50 molal mixture of O and P (saturated liquid) is fed to a column. The relative volatitility is 2.1. There are to be three products: A distillate

($X_0 = 0.98$), bottoms ($X_B = 0.03$), and a side stream whose mole fraction of O is 0.80 (40 percent of A fed is in side stream). Find the ideal stages, amounts of product per 100 moles of feed, and minimum reflux rate.

12-10. A distillation column is used to separate methanol and water. The feed (0.0602 kmole/sec) enters the unit at 58.3°C with a 0.36 mole fraction of methanol. The X_D and X_B values are 0.915 and 0.00565. A reflux ratio of 0.908 is used. Bubble point for the feed is 76°C. Heats of vaporization for the methanol and water are 1046.7 kJ/kg and 2284 J/kg. Likewise, the specific heats are 2721 and 487 J/kg °K. Find the number of ideal stages.

12-11. Find the minimum reflux and minimum stages for the column of Problem 12-10.

12-12. Find minimum stages for X_D and X_B of 0.95 and 0.05 (relative volatility of 2). Repeat for X_D of 0.99 and X_B of 0.01. Also consider X_D and X_B of 0.95 and 0.05, for relative volatility of 1.2.

12-13. Two feeds are used in a column to yield X_D of 0.97 and X_B of 0.04. One feed (a saturated liquid) has an X_F of 0.75. The other (50 percent vaporized) has an X_F of 0.35. Reflux ratio is 1.6. Find the number of ideal stages and the locations of both feeds.

12-14. A 20 ideal stage column is to operate at total reflux. If the systems relative volatility is 2, find the values of X_D and X_B if $X_B = (1 - X_D)$.

12-15. A 13 ideal stage distillation column (relative volatility of 2) yields 162.2 kmole/hr of distillate ($X_D = 0.95$) and 254.8 kmole/hr of bottoms ($X_B = 0.05$). If the reflux ratio is 5.25, find X_F feed rate and feed condition.

12-16. Two feeds (545.5 kmole/hr with X_F of 0.70; 363.6 kmole/hr with X_F of 0.40). The larger feed is a saturated vapor and the other is a saturated liquid. Relative volatility is 3.90. Using a reflux ratio of 0.85 the unit is to give an X_D and X_B of 0.96 and 0.04. Find the total number of ideal stages and the feed plate locations.

12-17. A column that has eight ideal stages (including reboiler) is to be employed to give an X_D of 0.95 (component 1) with an X_F of 0.40 (component 1). Feed rate is 1000 mole/hr. Maximum vapor capacity at column top is 2000 mole/hr. The feed is a saturated liquid. Column reflux is 4.5. Vapor–liquid data are

Y	0.1	0.2	0.3	0.4	0.5	0.6	0.7	0.8	0.9	1.0
X	0.03	0.08	0.15	0.20	0.25	0.33	0.43	0.6	0.73	1.0

Will the column operate satisfactorily? If not, specify appropriate changes.

12-18. A binary system is to be separated in a plate column. The saturated liquid feed (X_F of 0.209) is to yield an X_D and X_B of 0.98 and 0.001. Vapor–liquid equilibrium data are

Y	0.262	0.474	0.742	0.891	0.943	0.977	0.987	0.995	1.0
X	0.0529	0.1053	0.2094	0.312	0.414	0.514	0.614	0.809	1.0

Find minimum reflux, minimum stages, and ideal stages (using 1.5 times minimum reflux).

12-19. A fifty percent vaporized feed ($X_F = 0.40$) is to yield an X_D and X_B of 0.95 and 0.05. The columns internal reflux at the top (L/V) is 0.818. A side stream is taken from the second stage from the top at rate equal to that of the overhead product. Find the total number of stages and the appropriate mole fractions associated with the side stream. The system's vapor–liquid equilibrium data are

Y	0	0.1	0.2	0.3	0.4	0.5	0.6	0.7	0.8	0.9	1.0
X	0	0.004	0.008	0.13	0.185	0.25	0.33	0.43	0.55	0.746	1.0

12-20. The two limiting reflux conditions in a distillation column are minimum and total reflux. Suppose we have an infinite staged column with a constant binary flow of constant composition.

What would happen if such a column was operated at total reflux? If a product was withdrawn from the top of this column, explain the effect of withdrawing more and more product.

12-21. A binary system (A and B) is to be separated into products of $X_D = 0.99$ and $X_B = 0.01$. For the following vapor pressure data

T (°K)		
A	B	P (atm)
256.9	291.7	0.526
272.7	309.3	1
291.96	331.2	2
323.16	365.6	5
352.66	397.9	10
389.16	437.5	20

Find the average relative volatilities for column pressures of 1, 3, and 10 atm.

What are the minimum number of ideal stages for each case?

Is there an advantage to operate above atmospheric pressure?

12-22. A plant has two process streams containing benzene and toluene (one X_F of 0.40, the other of 0.70). Equal amounts of both streams are available as saturated liquids. Contrast column design (number of ideal stages) for (a) a two-feed column and (b) a single-feed column (i.e., combining the process streams to make a single feed).

12-23. Toluene saturated with water (680 ppm at 303.16°K) is to be dried to 0.5 ppm by distillation. Feed is introduced to the top plate of the column. Overhead vapor is condensed and cooled to 30°C and then separated into two layers. Relative volatility of water is toluene is 120. If 0.2 moles of vapor are used per mole of liquid feed, how many ideal stages would be required?

12-24. A feed (half vaporized) with an X_F of 0.35 is distilled to yield an X_D of 0.95 and an X_B of 0.03. Feed rate is 3.64 k mole/min. Energy input to the reboiler is 1267.2 kJ/sec. Molecular weight of reboiler vapor is 73, and the heat of vaporization is 197.3 kJ/kg. Relative volatility is 2. Find the ideal stages needed as well as the location of the feed plate.

12-25. A system has the following vapor–liquid equilibrium data:

Y	0	0.2	0.4	0.5	0.55	0.60	0.65	0.70	0.80	0.90	1.0
X	0	0.053	0.11	0.2	0.22	0.28	0.50	0.57	0.78	0.88	1.0

The feed (50 percent vapor) has an X_F of 0.40. Find the minumum reflux ratios for two cases (X_D of 0.7, X_B of 0.04) and (X_D of 0.80, X_B of 0.04).

12-26. Air is fractionally distilled as a 75 percent liquid at atmospheric pressure into an X_D of 0.98 and an X_B of 0.03 (withdrawn from the bottom plate before the reboiler). The distillate product is taken from the condenser as a vapor. Vapor liquid data (mole fraction N_2) are:

Y	0	0.1397	0.2610	0.3660	0.4600	0.5420	0.6160
X	0	0.0385	0.0802	0.1240	0.1705	0.222	0.2773

Y	0.6795	0.7374	0.7895	0.8435	0.8895	0.9350	0.9750
X	0.338	0.4047	0.4783	0.5662	0.6665	0.7840	0.9190

For this system find the percent of oxygen fed recovered in the bottom product. Also, find minimum stages and reflux. If nine ideal stages are needed, what is the reflux ratio?

12-27. A feed (28.6 percent liquid, $X_F = 0.60$) of methanol and water is to be distilled to produce two products (saturated vapor from condenser) and a saturated liquid sidestream ($X = 0.80$). The 600 k mole/hr of feed yields 98 percent of the methanol fed to the top and sidestream products. Find minimum reflux and the effect of the sidestream, and determine the compositions and flow rates of the product streams (top, sidestream, bottom).

How many ideal stages are needed for a 2.5 reflux ratio? Specify feed and sidestream trays. Consult Problem 12-7 for pertinent vapor–liquid data.

12-28. Methanol and water are to be separated in a coupled two-column system (one at high and the other at low pressure). Relative volatilities are 2 (high pressure) and 3 (low pressure). A stream of 100 k mole/hr (saturated liquid, $X_F = 0.70$) is fed into the high-pressure column. The overhead vapor from this column is condensed in the low-pressure column's reboiler. Low-pressure column stripping section vapor rate equals the high-pressure column rectifying vapor rate. Bottoms product from the high-pressure column is the feed for the low-pressure column. The X_D values for both columns are 0.95, while X_B values are 0.54 (high pressure) and 0.05 (low pressure). Reflux ratio for the high-pressure column is 1.38 (obtained from low-pressure column reboiler). The flashing of the high-pressure bottom product into the second column gives a 20 percent vaporized stream. How many ideal stages are required for each column? Where are the feed-plate locations?

12-29. A stripping column is one where the feed enters at the top. Consider an aqueous solution of volatile (A) with the equilibrium data given below:

Y	0	0.0100	0.0200	0.0300	0.0400
X	0	0.0035	0.0077	0.0125	0.0177

Y	0.0600	0.0800	0.1000	0.1200
X	0.0292	0.0429	0.0590	0.0784

A feed ($X_F = 0.0794$) is fed as a saturated liquid to yield an X_D of 0.1125 without any reflux. The column is equipped with a still of its bottom. Find the number of ideal stages if 0.562 moles of vapor are generated per mole fed.

12-30. A saturated liquid system ($X_F = 0.70$) is fed to the top of a stripping unit at the rate of 400 kmole/hr. The column pressure is 1.013×10^5 N/m^2 and a bottom product of 60 kmole/hr with $X_B = 0.10$ is desired. How many theoretical steps are needed? What are the overhead vapor rate and composition?

12-31. A system of n-butane and a nonvolatile oil ($X_F = 0.05$) is to be stripped at 149°C and 1 atm. The feed enters at the top of the column, and steam is fed to the bottom. If 95 percent of the n-butane is removed, find the number of theoretical stages and the steam used.

12-32. Carbon dioxide is separated from methyl alcohol in a stripping column. The saturated liquid feed enters the top tray at 100 kmole/min. Equilibrium is governed by $Y = 2X$. If three ideal trays plus reboiler are used find the following items (assuming column liquid and vapor rates are constant). Feed and overhead vapor compositions when $X_B = 0.01$ and the vapor boil up rate is 59.8 kmole/min. Also, if vapor boil up goes to 100 kmole/min with the X_F previously found, what is X_D?

12-33. A mixture of isobutane and n-butane is to be stripped (i.e., concentrate the isobutane). Find the number of ideal stages needed with the following given: The relative volatility is two. A rate of 2200 kmole/hr of saturated liquid feed ($X_F = 0.45$) enters the column top. Also, vapor and liquid rates are constant in the column. Desired X_D and X_B are, respectively, 0.60 and 0.05.

12-34. A liquid feed of 100 kmole/hr made up of propane ($X_F = 0.10$) and n-hexane is to be stripped to give an X of 0.01. The reboiler generates 37.5 kmole/hr of vapor. The column uses another device (pseudo-reboiler) to vaporize an additional 37.5 kmole/hr. This unit is located on a tray where X is 0.05. If the equilibrium is given by $Y = 2X$, find theoretical stages, the tray for the pseudo-reboiler, and X_D.

12-35. A feed of propylene oxide and acetone ($X_F = 0.117$) is stripped to give an X_D of 0.153 proplylene oxide and an X_B of 0.02. The equilibrium is given by $Y = 1.5X$. The column has a cooler located at an intermediate point. This cooler removes energy and condenses a portion of the vapor. Ratio of vapor rates below and above the cooler is three. Also, the cooler is located at a plate where $X = 0.08$. Find the number of ideal stages, the cooler stage, product flow rates per 100 moles of feed, and vapor and liquid rates (per 100 moles of feed) below and above the cooler.

12-36. A feed ($X_F = 0.10$) is stripped in a column. For X_B of 0.01, find the highest possible recovery of the less volatile component (moles per mole of less volatile fed) when equilibrium is $Y = 2X$ and if $Y = 4X$.

12-37. A stripping column used to separate ethanol and water has three ideal trays. The feed (13 mole percent ethanol) enters the top as a saturated liquid. In lieu of a reboiler live steam enters below tray 1 at 1000 kmole/hour.

If 1500 kmole/hr of feed are used and $X_B = 0.02$, what is X_D? Also, if $Y = KX$, what is the K value?

12-38. A process stream from an absorber containing n pentane in an hydrocarbon oil (molecular weight 160; specific gravity of 0.84) is stripped to reduce the mole fraction of pentane in the oil from 0.1724 to 0.005. The stripper feed enters at the top and steam is fed into the units bottom. Stripped oil leaves at the bottom. Find the number of theoretical stages. [*Hint*: Use mole ratio (i.e., moles pentane per mole pentane free gas and moles pentane per mole pentane free oil).]

12-39. A system of n-pentane and n-hexane (each 50 mole percent) is to be separated by two different techniques. In the first the mixture is flash distilled at 30°C with 50 percent vaporization. For this case find the pressure as well as vapor and liquid compositions. The other case is to use a differential distillation until 50 percent is vaporized at 0.5 atm. Find initial and final temperatures as well as residue composition. Data are as follows:

	Vapor Pressure (atm)	
Temp (°C)	*n*-Pentane	*n*-Hexane
17.5	0.500	0.138
20	0.552	0.154
25	0.665	0.195
30	0.802	0.242
36	1.000	0.316
40	1.13	0.362
43	1.25	0.408
46	1.355	0.460
48.5	1.49	0.500
55	1.79	0.631
60	2.08	0.750
65	2.41	0.875
69	2.70	1.00

12-40. A system of A (most volatile) and B (least volatile) is to be separated by two techniques (at atmospheric pressure). The mixture (both components at 50 mole percent) is to be flash distilled and differentially distilled. The first case (flash) is to use the feed at 30°C and vaporize 60 mole percent of the total feed. For this case find the vapor and liquid compositions as well as the final temperature. In the second case, again distill 60 percent of the feed. Find the residue composition. System data are as follows:

Liquid T (°C)	122	115	114	112	108	106	104	101	98.5
Y	0.20	0.497	0.567	0.608	0.689	0.76	0.84	0.89	0.96
X	0.12	0.32	0.38	0.42	0.50	0.6	0.7	0.8	0.9

12-41. A system (60 mole percent benzene, 30 mole percent toluene, and 10 mole percent xylene) is to be separated such that X_D for toluene is 0.5 mole percent and X_B for benzene is 0.5 mole percent. The reflux ratio is 2. The volatilities relative to xylene are for benzene toluene and xylene (6.18, 2.50, 1.00). Feed condition is such that change in moles of overflow across the feed stage is just the moles of feed. Find the number of ideal stages for separation.

12-42. Find the minimum stages, minimum reflux, and estimated ideal stages (Gilliland correlation) for the system of Problem 12-41.

12-43. Tar acid is an industrial product that contains a variety of aromatic compounds. Consider the distillation (still pressure 250 mm Hg) of such a system with the following mole percents (phenol, 35; *o*-cresol, 30; *m*-cresol, 30; xylenols, 15; heavy components or residue, 5). Volatilities relative to the residue (130°C to 170°C) for phenol, *o*-cresol, *m*-cresol, and xylenols, and residue are 14.37, 11.49, 8.16, 4.83, and 1.00. The fractionation is to yield an overhead of 95 mole per cent phenol (90 percent phenol recovery). Reflux ratio is 10. Determine theoretical stages for the separation.

12-44. Find minimum reflux, and stages for the system of Problem 12-43. Estimate total stages needed.

12-45. A process stream has the following mole fractions for propane, *n*-butane, *n*-pentane, and *n*-hexane: 0.021, 0.285, 0.483, and 0.211. The volatilities of each relative to *n*-hexane are 31.68, 9.60, 3.00, and 1.00. The separation requires 95 percent of the *n*-butane to be in the overhead and let 95 percent of the *n*-pentane in the bottom. Feed is a saturated liquid and is used at the rate of 2000 barrels per day (60°F). Find the theoretical stages required if reflux ratio is 1.35.

12-46. For the system of Problem 12-45 find minimum stages and reflux. Estimate the number of theoretical stages at 1.4 times minimum reflux.

12-47. A four-component system (1, 2, 3, 4) is fed as a saturated liquid to a distillation column. The X_D for 2 is 0.98, while X_B for 3 is 0.975. Values of X_F for 1 through 4 are 0.05, 0.42, 0.46, and 0.07. Also, the volatilities relative to 4 are 3.23, 2.62, 1.54, and 1.0.

Find minimum plates, minimum reflux, and ideal stages at 1.2 times minimum reflux.

12-48. A five-component system (see data below) is to be distilled with 99 percent recovery of 2 in the overhead and 4 in the bottoms. What is the product composition at total reflux? If reflux ratio is reduced, how would the product compositions, change?

Component	Feed Mole Fraction	Relative Volatility
1	0.06	4.33
2	0.40	3.17
3	0.05	2.50
4	0.42	1.67
5	0.07	1.0

12-49. A saturated liquid mixture of xylenes is separated in a column. For the data given below, find minimum stages and reflux. Also estimate the reflux ratio necessary to require 100 theoretical stages of 99 percent of *m*-xylene and 3.8 percent of *o*-xylene are recovered in the overhead.

Component	X	Relative Volatility
Ethyl benzene	0.054	1.76
p-Xylene	0.221	1.64
m-Xylene	0.488	1.61
o-Xylene	0.212	1.43
n-Propyl benzene	0.025	1.0

12-50. A mixture containing water, isopropyl alcohol, and ethanol (mole fractions of 0.20, 0.40, and 0.76) is to be separated by extractive distillation into an ethanol (bottom) product containing not over 0.002 mole fraction of isopropyl alcohol on a water-free basis. The ethanol recovery is to be 98 percent. Water is the extractive agent and will be added to the reflux to make the liquid recycled to the tower 85 mole percent water. Also, the feed is diluted to make the feed water mole fraction 0.85. Furthermore, it will be heated so that the rectifying and stripping vapor rates are equal. The relative volatilities of the three components—water, ethanol, and isopropyl alcohol—are 0.202, 1.0, and 1.57. Find minimum stages and reflux. For a reflux of 1.5 times minimum, find the theoretical stages. Repeat treating the ethanol–isopropyl alcohol system as a binary.

12-51. A two-column system is used to produce anhydrous (absolute alcohol). The aqueous alcohol feed (89 percent alcohol 11 percent water) is separated in column 1 to the anhydrous product and an overhead which is condensed into a benzene and water layers. The benzene layer is used as a reflux to column 1. The water layer is the feed for column 2 (bottom product is water). Its top product is recycled to column 1's condenser. The bottom mole fractions for column 1 are 0.999 for alcohol, 0.0009 for water, and 0.0001 for benzene. Reflux to column 1 has the following mole fractions (0.217 alcohol; 0.256 water; 0.527 benzene). The relative volatilities for benzene, alcohol, and water are 3.60, 0.89, and 1.0 at the column bottom and 0.62, 0.47, and 1.00 at the reflux. Estimate minimum stages required.

REFERENCES

1. W. L. McCabe and J. C. Smith, *Unit Operations of Chemical Engineering* second edition, McGraw-Hill, New York (1967), Chapter 17.
2. W. L. McCabe, J. C. Smith, and P. Harriott, *Unit Operations of Chemical Engineering*, fourth edition, McGraw-Hill, New York (1985), p. 497.
3. J. M. Smith and H. Van Ness, *Introduction to Chemical Engineering Thermodynamics*, McGraw-Hill, New York (1987), Chapters 10-14.
4. J. M. Coulson and J. F. Richardson, *Chemical Engineering*, Volume 2, Pergamon Press, London (1968), p. 316.
5. *Bubble Tray Design Manual*, American Institute of Chemical Engineers, New York (1958).
6. S. I. Cheng and A. J. Teller, *AIChE J.* **7**, 282 (1961).
7. J. B. Jones and C. Pyle, *Chem. Eng. Prog.* **51**, 424 (1955).
8. M. Fenske, *Ind Eng. Chem.* **24**, 482 (1932).
9. J. Underwood, *J. Inst. Pet.* **32**, 614 (1946).
10. E. R. Gilliland, *Ind. Eng. Chem.* **32**, 110 (1940).
11. C. S. Robinson and E. R. Gilliland, *Elements of Fractional Distillation*, McGraw-Hill, New York (1950), p. 291.

12. W. L. Luyben and L. A. Wenzel, *Chemical Process Analysis: Mass and Energy Balances*, Prentice-Hall, Englewood Cliffs, NJ (1988).
13. R. E. Treybal, *Mass Transfer Operations*, third edition, McGraw-Hill, New York (1979).
14. T. K. Sherwood, R. L. Pigford, and C. R. Wilke, *Mass Transfer*, McGraw-Hill, New York (1975).
15. H. G. Drickamer and J. R. Bradford, *Trans. AIChE* **39**, 319 (1943).
16. H. E. O'Connell, *Trans. AIChE* **42**, 741 (1946).
17. D. F. Othmer, R. E. White, and E. Trueger, *Ind. Eng. Chem.* **33** 1240 (1941).
18. R. G. Griskey, *Chemical Engineering for Chemists*, American Chemical Society, Washington, D.C. (1997).
19. B. D. Smith, *Design of Equilibrium Stage Processes*, McGraw-Hill, New York, N.Y. (1963).

13

ADDITIONAL STAGED OPERATIONS

INTRODUCTION

The equilibrium stage approach is applicable to other separation processes (absorption, extraction, leaching, etc.) as pointed out in chapter 12. As with distillation, the combination of the mass or material balance, the energy or enthalpy balance, and appropriate equilibrium data will yield the desired design.

In this chapter, attention will be directed to several of the other separation processes and their design using the equilibrium stage technique. Most of these will involve systems where the McCabe–Thiele approach can be used since temperature effects in many of the processes are minimal.

As with distillation, real systems must be derived from the ideal stages by appropriate efficiencies.

ABSORPTION IN STAGED COLUMNS

In Chapter 12 we saw that the basic process of absorption (*ab* from the Latin meaning "into") involved vapor–liquid systems. A way of looking at the process is to consider absorbing a constituent present in a low gas concentration (dilute gas) into a large amount as solvent. This is the situation that would correspond to the lower end of the equilibrium curve of Figure 12-3 or the data themselves of Figure 12-1.

The absorption process, while similar in some respects to distillation, differs in that gas and liquid streams enter and leave at the top and bottom of the absorption column (see Figure 13-1). Because of this, there is only one mass balance equation for the entire column (i.e., one line for a McCabe–Thiele solution).

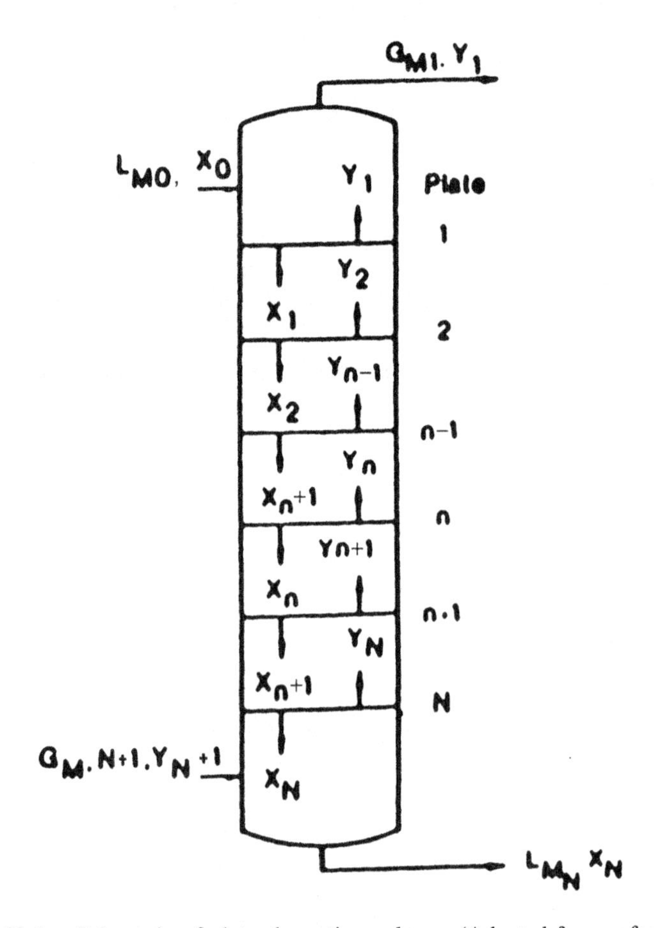

Figure 13-1. Schematic of plate absorption column, (Adapted from reference 3.)

This mass balance is sometimes based on solute free mole fractions (X', Y') as shown below:

$$Y' = \frac{Y}{1 - Y} \tag{13-1}$$

$$X' = \frac{X}{1 - X} \tag{13-2}$$

Note that $Y = Y'$ and $X = X'$ for a dilute system (i.e., $1 - Y$ and $1 - X$ are essentially 1.0).

If the mass balance approach of Chapter 12 is used, we obtain

$$Y'n + 1 = \frac{Lm}{Gm} X'n - \frac{Lm}{Gm} X'_N + Y'_{N+1} \tag{13-3}$$

which, if the gas is dilute, becomes

$$Y_{n+1} = \frac{Lm}{Gm}Xn - \frac{Lm}{Gm}X_N + Y_{N+1} \tag{13-4}$$

where *Lm* and *Gm* are the liquid and gas model mass velocities (solute free basis).

If heat effects are not large, the McCabe–Thiele method can be used. Furthermore, when the system is relatively dilute, Henry's Law applies and

$$Y_A = HX_A \tag{13-5}$$

The resultant solution for ideal stages is shown in Figure 13-2. The top line is the mass balance line (slope of $L'm/G'm$ or *Lm/Gm*). The equilibrium line is given by a Henry's Law solution ($Y = f(X)$); the point at the lower left of the mass balance line represents the top of the absorption column, while the X_N, Y_{N+1} point (at the upper right) represents the bottom of the column. Because of the nature of the absorption process, stages can be determined by starting at either the top or bottom of the column. In Figure 13-2 a total of four stages would be needed.

Infinite stages is also a limiting condition for absorption column operation. This condition coincides with minimum (L_m) liquid rate as shown in Figure 13-3.

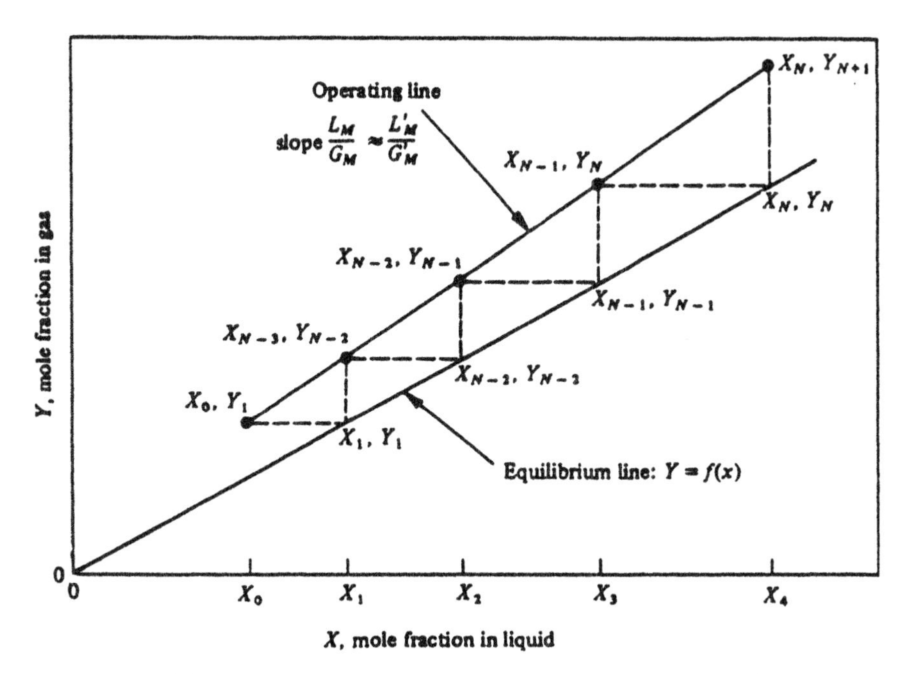

Figure 13-2. McCabe–Thiele solution for equilibrium stages in a plate absorption column. (Adapted from reference 3.)

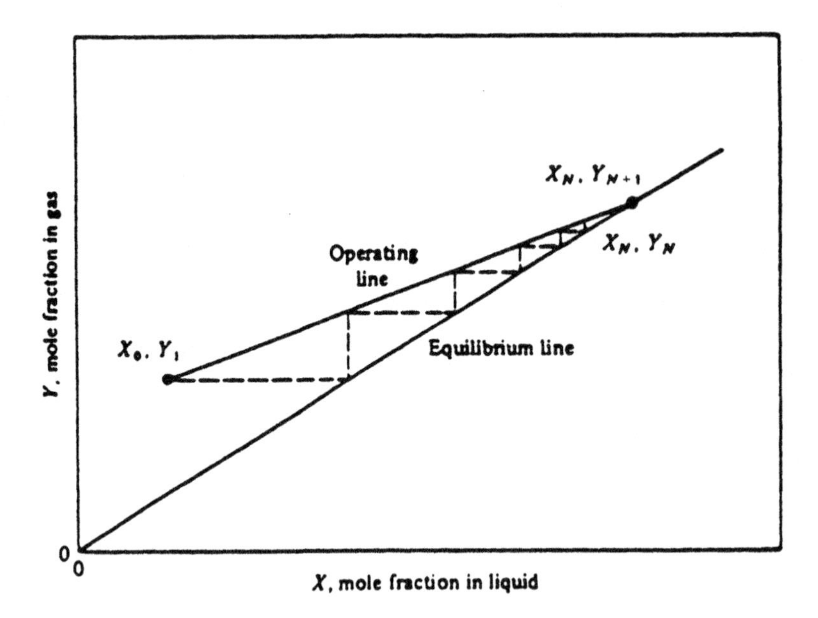

Figure 13-3. Determination of minimum liquid rate. (Adapted from reference 3.)

Note that the tangency of the mass balance line with the equilibrium line causes a "pinch point," which gives infinite stages. The significance of the minimum liquid rate is similar to that of minimum reflux in distillation; that is, 1.2 to 1.5 times this value is a reasonable operating level for the absorption column.

An algebraic solution developed by Kremser (1) gives

$$\frac{Y_{N+1} - Y_1}{Y_{N+1} - KX_0} = \frac{A^{N+1} - A}{A^{N+1} - 1} \tag{13-6}$$

where the K is a vapor–liquid equilibrium constant. Its value is m, the slope of the equilibrium line. Also, A is defined as *Lm/mGm.* A solution based on this approach is given in Figure 13-4.

If the absorption system is concentrated, Figure 13-4 can be used with the approach derived by Edmister (2):

$$\frac{Y''_{N+1} - Y''_1}{Y''_{N+1}} = \left(1 - \frac{L_0 X'_0}{A' V_{N+1} Y''_{N+1}}\right)\left(\frac{A_e^{N+1} - A_e}{A_e^{N+1} - 1}\right) \tag{13-7}$$

In the above: Y''_{N+1}, and Y'' are moles of absorbed component respectively, that enter and leave per mole of vapor that enters; X is the moles of absorbed component per mole of entering solvent; L_0 is the moles of solvent; L is the moles of solvent entering; V is the moles of entering vapor. The A_e and A are defined below:

$$A_e = \sqrt{A_N(A_1 + 1) + 0.25} - 0.5 \tag{13-8}$$

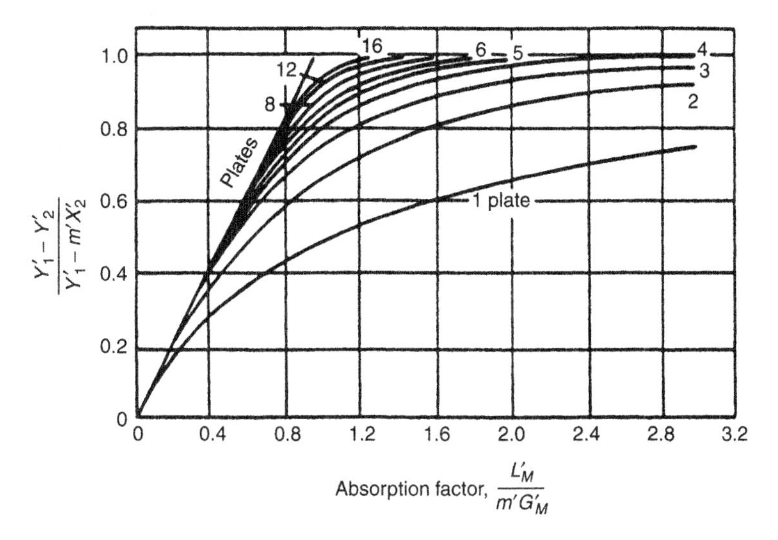

Figure 13-4. Relation between column performance and theoretical stages. (Adapted from reference 3.)

$$A' = \frac{A_N(A_1 + 1)}{A_N + 1} \tag{13-9}$$

Multicomponent systems in absorption columns can be handled by using the concept of the key component from distillation. In this case the key component (there is only one) is that material (absorbed significantly) whose equilibrium line has a slope closest to that of the mass balance line (i.e., L/G). Stages found in the manner are those ideally needed for the column. Using mass balance lines of the same slope as with the key component together with each component's equilibrium line gives the stages on which each of the other components are significantly absorbed.

The earlier-described algebraic technique can also be used for the key component with the form

$$\frac{Y_1'' - Y_2''}{Y_1'' - m'X_2'} = \frac{(L'/m'G'')^{N+1} - (L'/m'G'')}{(L'/m'G'')^{N+1} - 1} \tag{13-10}$$

The primed quantities are as defined previously. N is the number of ideal stages for the key component. The double primes are solute moles per mole rich gas and moles of gas per unit time.

ABSORPTION IN PACKED COLUMNS

One of the mainstays for absorption in industrial processing is the packed column. This unit makes use of solid packing particles that make for an excellent

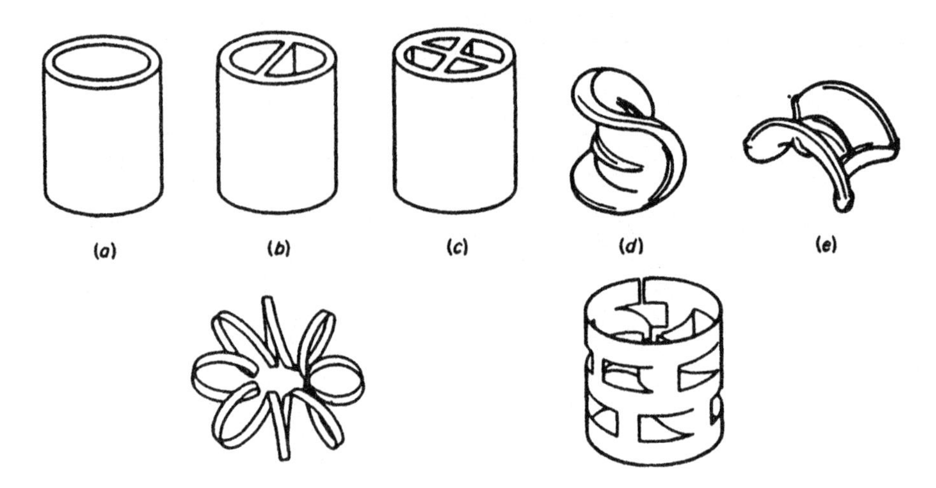

Figure 13-5. Random tower packings. (Adapted from reference 4.)

dispersion of gas and liquid in the column. The packings used can either be random or stacked.

The former (see Figure 13-5) are loaded into the tower in a completely random manner. Generally, the packings used are made for the purpose at hand. However, on occasions crushed solids (stone, gravel etc.) are used. These materials are not as preferable as those in Figure 13-5 because they give a bed which has poor fluid flow characteristics.

Stacked packings (see Figure 13-6) give lower pressure drops and allow larger fluid flow rates. However, they also have increased propensity for the channeling of the fluid flow. Furthermore, they are more expensive than random packings.

Packed column design in essence consists of treating the absorption process in a volume element $S\,dh$ (see Figure 13-7). This approach uses the mass transfer coefficient as the means of finding the mass flux N_A. The overall technique ultimately melds this mass transfer coefficient approach with the staged operation method (by using heights equivalent to theoretical stages, HETS, and numbers of stages).

Let us consider the packed column design. Again referring to Figure 13-7 we can write

$$N_A a_V = k_g a_V p(Y - Y_i) \tag{13-11}$$

$$N_A a_V = k_L a_V \overline{\rho}(X_i - X) \tag{13-12}$$

where N_A is the mass transfer flux of component A, a_V is the effective mass transfer area per unit volume, $k_g a_V$ and $k_L a_V$ are the volumetric gas and liquid phase mass transfer coefficients ((moles/m^2 sec atm, and m/sec) respectively). The X_i and Y_i are the equilibrium values at the interface. P is total gas pressure and $\overline{\rho}$ is the average molar density of the liquid.

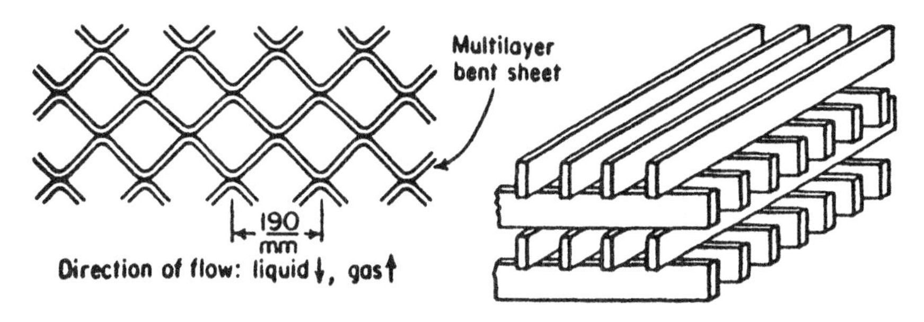

Figure 13-6. Stacked packings. (Adapted from reference 4.)

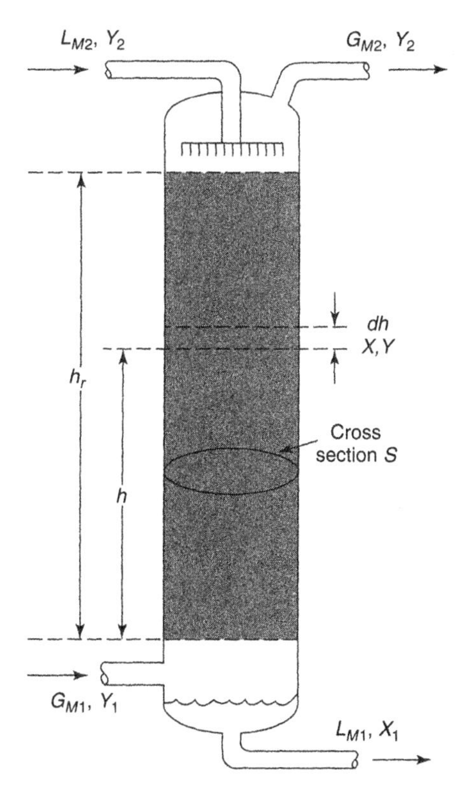

Figure 13-7. Schematic for packed column design. (Adapted from reference 3.)

If we also use a mass balance, then

$$-d(G_M Y) = N_A a_V dh = -G_M dY - Y dG_M \tag{13-13}$$

for only one component's flux

$$dG_M = -N_A a_V dh \tag{13-14}$$

so that

$$h_{\text{total}} = \int_1^2 dh = \int_{y_2}^{Y_1} \frac{G_M dY}{k_g P a_V (1-Y)(y-Y_i)} \tag{13-15}$$

or

$$h_{\text{total}} = \int_1^2 dh = \int_{X_1}^{X_2} \frac{L_M dX}{k_L a_V (1-X)(X_i - X)\overline{\rho}} \tag{13-16}$$

If we use the concept of a Y_{BM}

$$Y_{BM} = \frac{(1-Y)-(1-Y_i)}{\ln\left[\dfrac{1-Y}{1-Y_i}\right]} \tag{13-17}$$

then we can write

$$H_G = \frac{G_M}{k_g a_v P Y_{BM}} \tag{13-18}$$

and

$$N_G = \int_{Y_2}^{Y_1} \frac{Y_{BM} dY}{(1-Y)(Y-Y_i)} \tag{13-19}$$

so that

$$h_{\text{total}} = (H_G)_{\text{average}} \int_{Y_2}^{Y_1} \frac{Y_{BM} dY}{(1-Y)(Y-Y_i)} \tag{13-20}$$

The preceding method has an inherent problem—namely, the use of the interfacial compositions. This can be overcome by using overall mass transfer coefficients (analogous to the overall heat transfer coefficients, U, of Chapters 6 and 8). In order to do this we use values of X^* (liquid mole fraction in equilibrium with the vapor) and Y^* (vapor mole fraction in equilibrium with the vapor).

Then,

$$N_A = K_{OG} P (Y - Y^*) \tag{13-21}$$

$$N_A = K_{OL} \overline{\rho} (X^* - X) \tag{13-22}$$

These values of K_{OG} and K_{OL} are related to k_g and k_L as per

$$\frac{1}{K_{OG}} = \frac{1}{k_g} + m\frac{P}{\overline{\rho}}\frac{1}{k_L} \tag{13-23}$$

$$K_{OL} = \frac{1}{k_L} + \frac{\overline{\rho}}{m'P}\frac{1}{k_g} \tag{13-24}$$

where m is equilibrium line slope from X, Y^* to X_i, Y_i and m' is equilibrium line slope from X^*, Y to X_i, Y_i.

Ultimately,

$$h_{\text{total}} = \int_{Y_2}^{Y_1} \left(\frac{G_M}{K_{OG}PY^*_{BM}a_V}\right)\left(\frac{Y^*_{BM}dY}{(1-Y)(Y-Y^*)}\right) \tag{13-25}$$

where

$$Y^*_{BM} = \frac{(1-Y)-(1-Y^*)}{\ln\left[\dfrac{(1-Y)}{(1-Y^*)}\right]} \tag{13-26}$$

Similarly, for the liquid side we obtain

$$h_{\text{total}} = \int_{X_2}^{X_1} \left(\frac{L_M}{K_{OL}\overline{\rho}X^*_{BM}a_V}\right)\left(\frac{X^*_{BM}dX}{(1-X)(X^*-X)}\right) \tag{13-27}$$

$$X^*_{BM} = \frac{(1-X)-(1-X^*)}{\ln\left[\dfrac{(1-X)}{(1-X^*)}\right]} \tag{13-28}$$

and

$$h_{\text{total}} = (H_{OL})_{\text{average}} N_{OL} \tag{13-29}$$

Solutions for the foregoing involve graphical or numerical integration of the expressions for h_{total}. Such calculations are obviously complicated. There are, however, some simplified cases that can be used if we have dilute gases with straight equilibrium lines.

The first is the use of a logarithmic mean driving force. Basically,

$$L_m(X_1 - X_2) = G_m(Y_1 - Y_2) = K_{OG}a_V P h_{\text{total}}(Y - Y^*)_{\text{lm}} \tag{13-30}$$

$$(Y - Y^*)_{\text{lm}} = \frac{(Y-Y^*)_1 - (Y-Y^*)_2}{\ln\left[\dfrac{(Y-Y^*)_1}{(Y-Y^*)_2}\right]} \tag{13-31}$$

The same is true for the liquid side:

$$G_m(Y_1 - Y_2) = L_m(X_1 - X_2) = \overline{\rho}K_{OL}a_V h_{\text{total}}(X^* - X)_{\text{lm}} \tag{13-32}$$

$$(X - X^*)_{\text{lm}} = \frac{(X^*-X)_1 - (X^*-X)_2}{\ln\left[\dfrac{(X^*-X)_1}{(X^*-X)_2}\right]} \tag{13-33}$$

Another simplified method uses the concept of a dilute gas so that

$$H_{OG} = H_G + \frac{mG_m}{L_m} H_L \tag{13-34}$$

$$H_{OL} = H_L + \frac{L_m}{m, G_m} H_G \tag{13-35}$$

Likewise the N_{OG} and N_{OL} values become

$$N_{OG} = \frac{1}{1-K} \ln\left[(1-K)\left(\frac{Y_1 - mX_2}{Y_2 - mX_2}\right) + K\right] \tag{13-36}$$

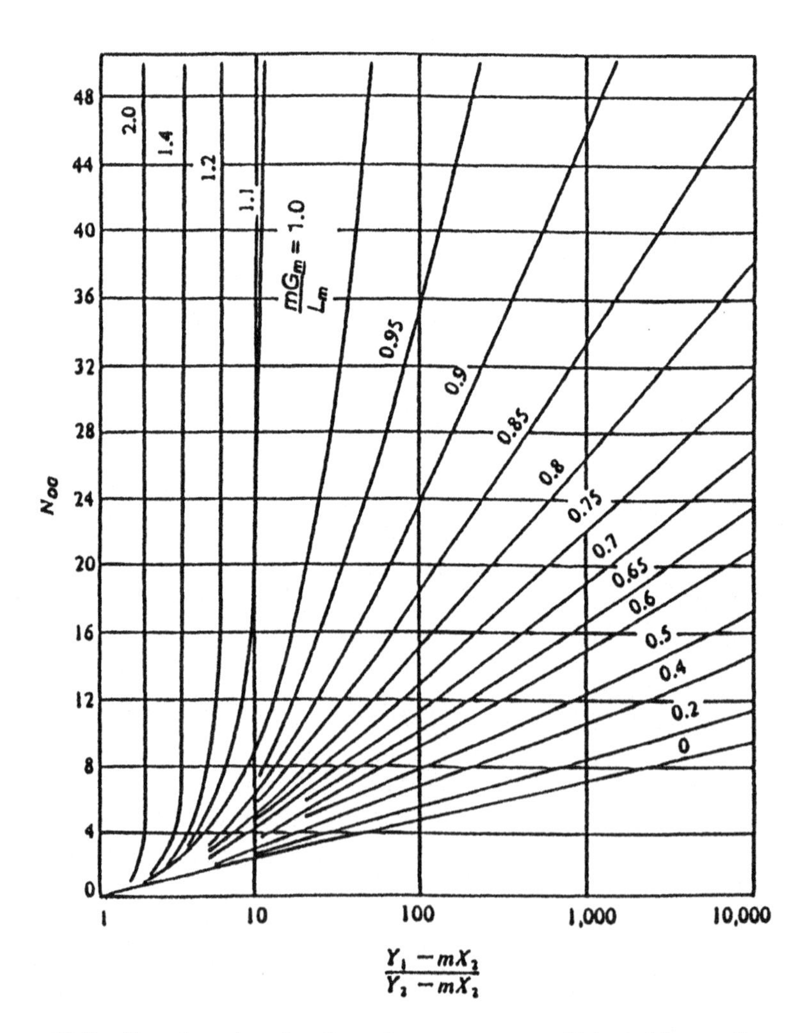

Figure 13-8. Transfer units related to column parameters. (Adapted from reference 3.)

$$N_{OL} = \frac{1}{1-K'} \ln\left[(1-K')\frac{X_1 - Y_2/m}{X_2 - Y_2/m} + K'\right] \tag{13-37}$$

where $K = mG_m/L_m$ and $K' = L_m/mG_m$. These expressions can be related in a plot as per Figure 13-8.

Complicated systems (high concentrations, heat effects, curved equilibrium lines, multicomponents, etc.) are considered elsewhere (3).

LIQUID OR SOLVENT EXTRACTIONS

Solvent or liquid extraction is another widely used industrial separation technique that lends itself to the staged operation approach. In these systems the solution to be separated is contacted with a solvent in a stage. The products of this operation are a solvent-rich phase (the extract) and a solvent-lean phase (the raffinate). Because thermal effects are not usually large, only the material or mass balance and the equilibrium relation are used to determine ideal stages.

An important parameter in extraction is the selectivity or separation factor B. For example, the effectiveness of a solvent B in separating components A and C is given by

$$\beta = \frac{(\text{wt fraction } C \text{ in } E)/(\text{wt fraction } A \text{ in } E)}{(\text{wt fraction } C \text{ in } R)/(\text{wt fraction } A \text{ in } R)} \tag{13-38}$$

where E and R are extract and raffinate.

There are, oddly enough, a number of such similarities between extraction and distillation. One aspect is the analogy between solvent in extraction and heat in distillation. Adding or removing heat in distillation is matched by the addition or removal of solvent in extraction. Furthermore, the various conditions (cold liquid, saturated vapor, etc.) in distillation can be matched by various kinds of solutions in extraction. Some of these analogies are summarized in Table 13-1.

Graphical solutions for extraction can be carried out on (a) triangular coordinates (ternary diagrams; see Figures 12-6, 13-9 through 13-12), (b) a Janecke diagram where $X_B/X_A + X_C$ is plotted versus $X_C/X_A + X_C$ and X_A, X_B, X_C are weight fractions of A, B, and C, respectively (Figure 13-10), and (c) distribution diagrams (rectangular coordinates of weight fraction of component C in component B, X_{CB} versus weight fraction of component C in component A, X_{CA}; see Figure 13-10).

Let us illustrate how we can solve for ideal stages in an extraction system. The process that we will consider (see Figure 13-13) will be for a countercurrent multiple contact system with both extract and raffinate reflux.

The reader can get a good insight into the similarity between distillation and extraction by turning the text in such a way as to make the train of Figure 13-13 vertical (the solvent separator will be on top of the train). On a one-to-one

Table 13-1 Comparison of Similarities Between Extraction and Distillation

Extraction	Distillation
Addition of solvent	Addition of heat
Solvent mixer	Reboiler
Removal of solvent	Removal of heat
Solvent separator	Condenser
Solvent-rich solution saturated with solvent	Vapor at the boiling point
Solvent-rich solution containing more solvent than that required to saturate it	Superheated vapor
Solvent-lean solution containing less solvent than that required to saturate it	Liquid below the boiling point
Solvent-lean solution saturated with solvent	Liquid at the boiling point
Two-phase liquid mixture	Mixture of liquid and vapor
Selectivity	Relative volatility
Change of temperature	Change of pressure

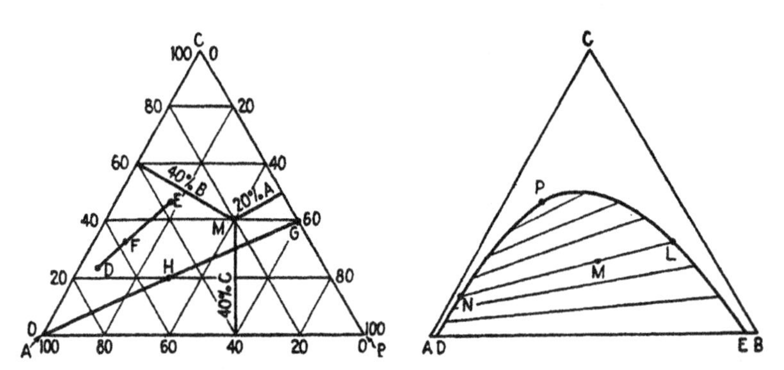

Figure 13-9. Ternary diagram. See that point *M* consists of 20 percent *A*, 40, percent *B*, and 40 percent *C*. Relative amounts of *E* and *D* in *F* are given by *DF* divided by *EF*. Line *AG* has constant ratios of *B* to *C* and varying amounts of *A*. (Adapted from reference 5.)

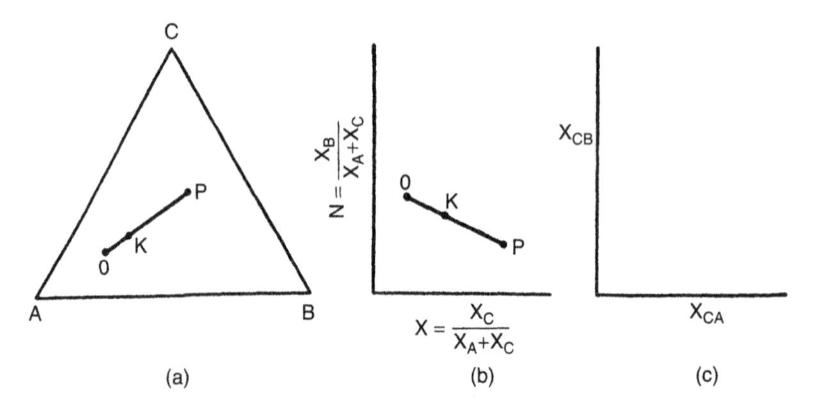

Figure 13-10. Comparison of ternary, Janecke, and distribution diagrams (5).

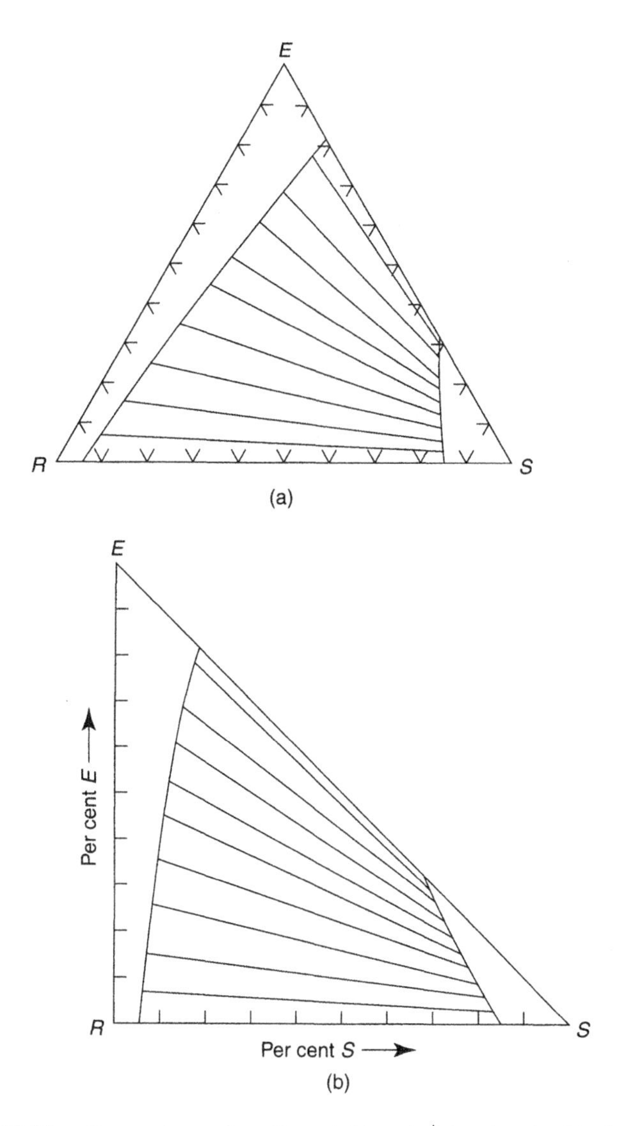

Figure 13-11. Comparison of equilateral (a) and right triangle (b) plots (55).

comparison the extract (enriching) section (above the feed) corresponds to the enriching section of the continuous distillation (above the feed). The same holds true for the extraction unit's raffinate (stripping section and the distillation column's stripping segment because both are below their respective feeds). The solvent separator (solvent removed) corresponds to the distillation column's condenser (heat removed). Likewise, the solvent mixer (solvent added) is similar to the distillation column's reboiler (heat added). Finally, the extract streams are similar to vapor, and the raffinates are similar to liquid streams.

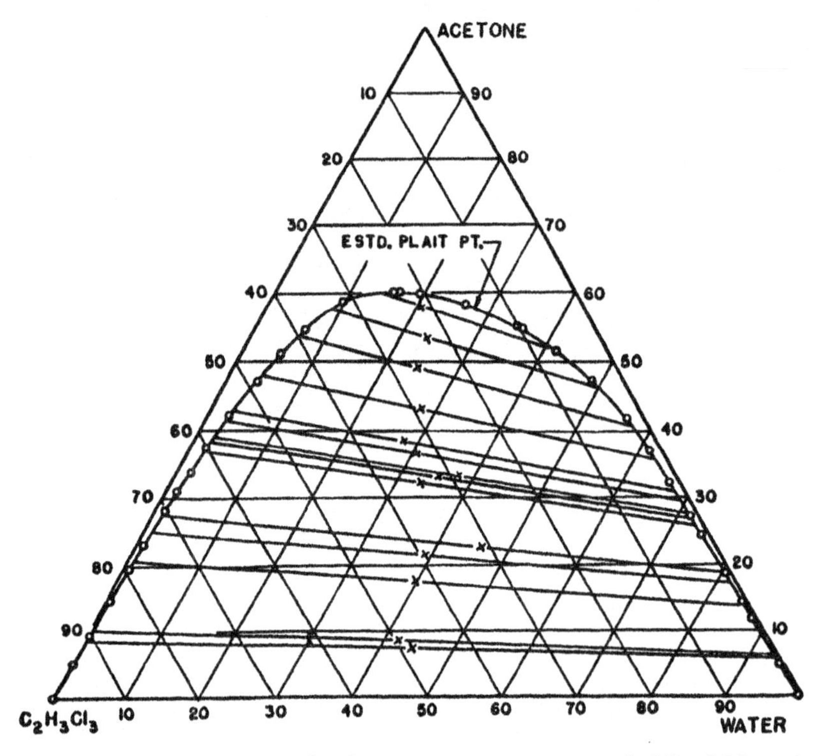

Figure 13-12. Ternary diagram for the system acetone–water –1, 1,2, trichloroethane. (Reproduced with permission from reference 9. Copyright 1946, American Chemical Society.)

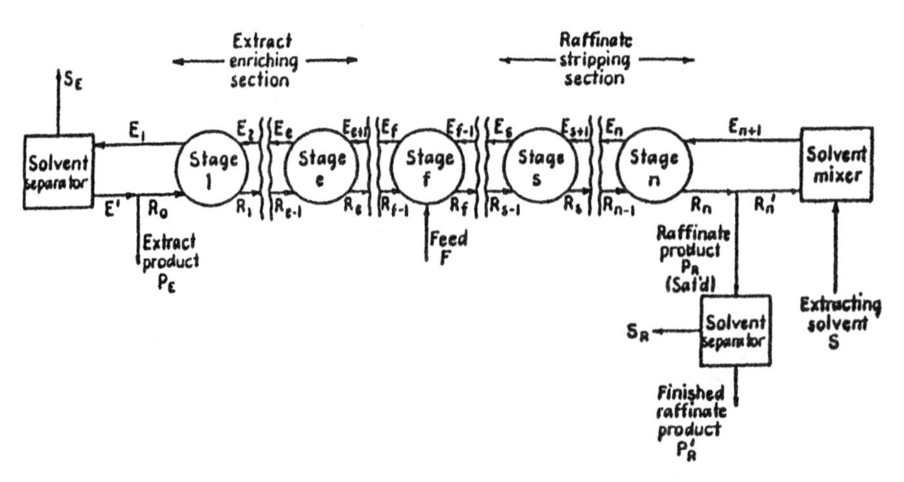

Figure 13-13. Countercurrent multiple contact system with both extract and raffinate reflux. (Adapted from reference 5.)

Let us now repeat our approach as per our earlier efforts in distillation. If we consider the extract (enriching) section in Figure 13-10, we see that by material balance

$$E_1 = S_E + E' \tag{13-39}$$

and

$$E_1 = S_E + P_E + R_0 \tag{13-40}$$

then, since

$$Q = S_E + P_E \tag{13-41}$$

$$E_1 = Q + R_0 \tag{13-42}$$

a material balance about the entire extract side of the train for any stage gives

$$E_{e+1} = S_E + P_E + R_e \tag{13-43}$$

and

$$S_E + P_E = Q = E_{e+1} - R_e \tag{13-44}$$

Equation (13-42) will be a straight line on a ternary diagram that connects the points for S and P. Q will lie on that line (i.e., lever rule). Figure 13-14 shows the location of Q.

The method used to find the stages needed for the extract section is as follows: Starting at E, use the appropriate tie line (dotted) to find R_1 (i.e., equilibrium); next, use equation (13-41) ($Q + R_1$) to find E_2; again use a tie line to find the raffinate in equilibrium which is added to Q to give a new extract. The procedure is repeated until the find point F is reached.

Likewise, for the raffinate end by material balance we have

$$S - P_R = W \tag{13-45}$$

Also, the material balance around the raffinate end gives

$$S - P_R = W = E_{S+1} - R_S \tag{13-46}$$

Figure 13-15 shows the solution for the raffinate (stripping) section. Note that point W is outside the ternary diagram. This occurs because $W + P_R = S$. The procedure starts with the R_n point which is connected to E_n by the appropriate tie line. Mass balance [equation (13-43)] is then used to locate the next raffinate point. The procedure is repeated until the point F is reached.

As with distillation, the cases of minimum reflux (infinite stages) and minimum stages (infinite reflux ratio) occur in extraction. The minimum reflux takes place in the extract section when an extended tie line meets a Q as close to S_E as possible (in Figure 13-14). Also, for the raffinate section an extended tie line

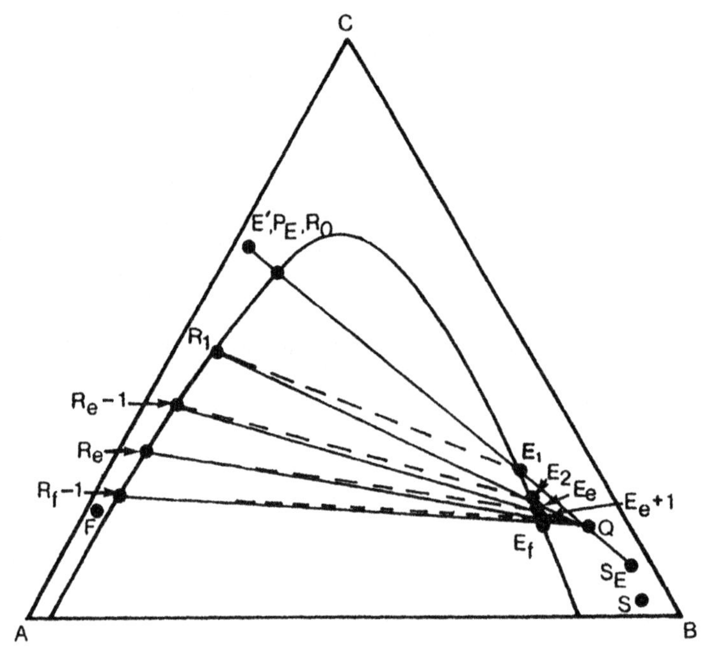

Figure 13-14. Solution for stages in extract side of system of Figure 13-13. (Adapted from reference 5.)

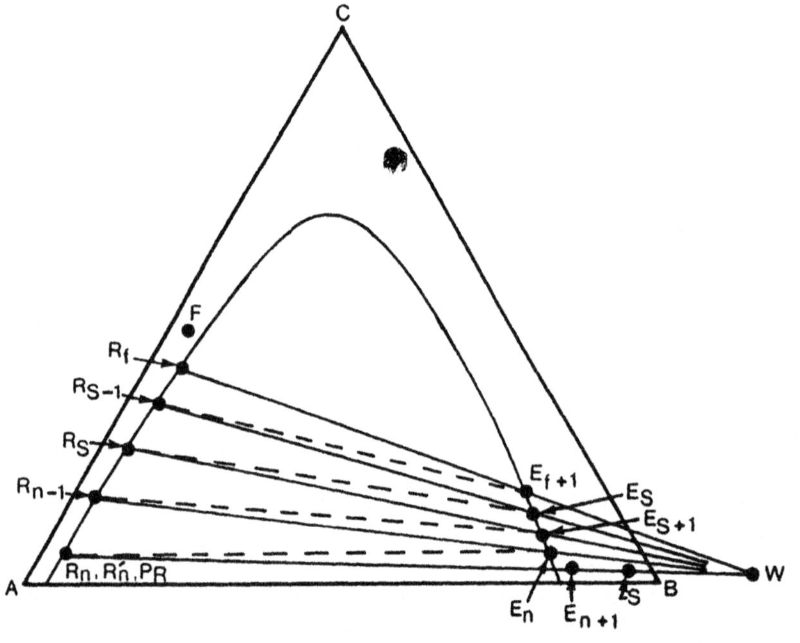

Figure 13-15. Solution for stages in raffinate side of system of Figure 13-13. (Adapted from reference 5.)

will meet the W point as close to S as possible (in Figure 13-15) For total reflux, the points S, SE, Q, and W must all coincide.

Determination of equilibrium stages, minimum reflux, and minimum stages can also be carried out on Janecke and distribution diagrams.

LEACHING; A LIQUID–SOLID SYSTEM

Leaching is the removal of a solute from a solid by means of a solvent. The solid (less the solute) can change in consistency (i.e., can become softened or slushy) but in itself is not dissolved. The diagram used for leaching calculations plot Y (the mass of solid divided by the sum of the masses of the solute and solvent) against an X (amount of solute divided by sum of the amounts of solute and solvent). A typical diagram is shown in Figure 13-16. The vertical dotted lines are the tie lines between the overflow (abscissa) and the underflow (line AB).

A solution for the number of stages for a countercurrent leaching system can be developed by using an imaginary mixer at either end of the train (see Figure 13-17). The quantity P is then the sum of V_a and L_a or V_b and L_b. Hence

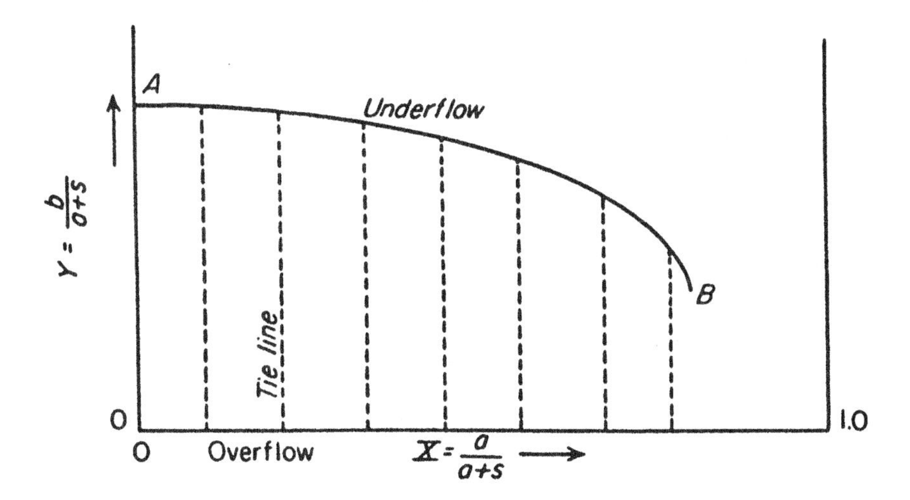

Figure 13-16. Equilibrium diagram for leaching system. (Adapted from reference 6.)

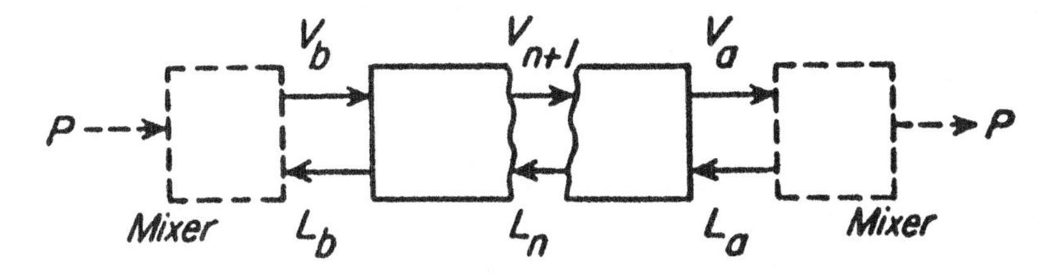

Figure 13-17. Leaching system with imaginary mixers. (Adapted from reference 6.)

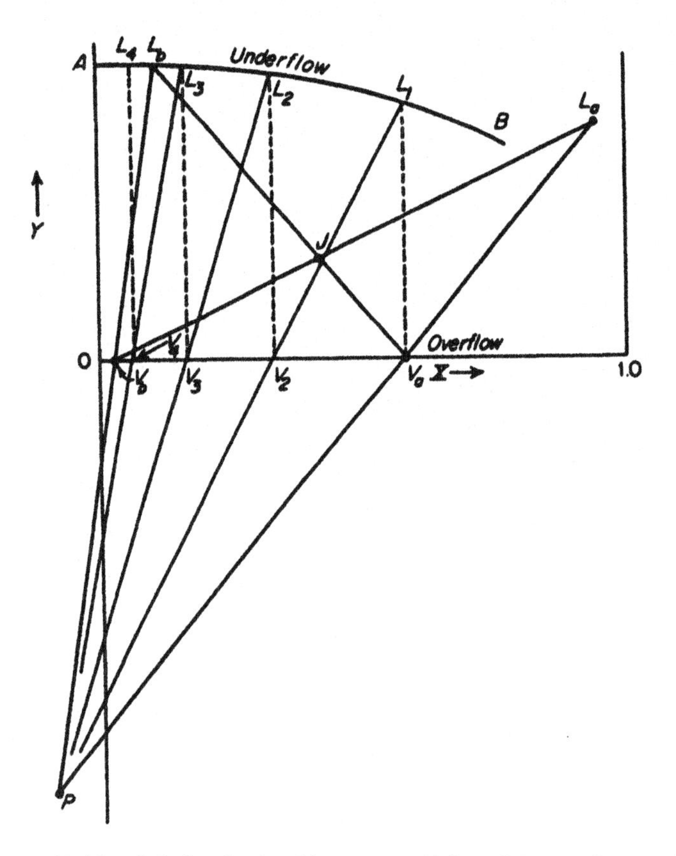

Figure 13-18. Solution for leaching stages. (Adapted from reference 6.)

if we use an *XY* leaching diagram point *P* will be located by the intersection of straight lines through L_a and V_a and L_b and V_b respectively (see Figure 13-18). The point *P* is then used as an operating point.

As Figure 13-18 shows, we use a tie line to go from V_a to L_1 and then use mass balance (straight line connecting L_1 and *P*) to obtain V_2. Continuation of the process will determine the stages needed.

Also note that the point *J* on the diagram enables us to determine the masses of the solutions in L_a, V_a, and so on.

ABSORPTION AND EXTRACTION EFFICIENCIES

Figure 13-19 gives a correlation of efficiencies for absorbers as a function of the combination of terms shown.

In addition, Figure 13-20 shows the relative effect of column capacities on efficiencies. As seen, the behavior for bubble cap and valve trays is not very sensitive to capacity, in contrast to sieve trays and packed columns.

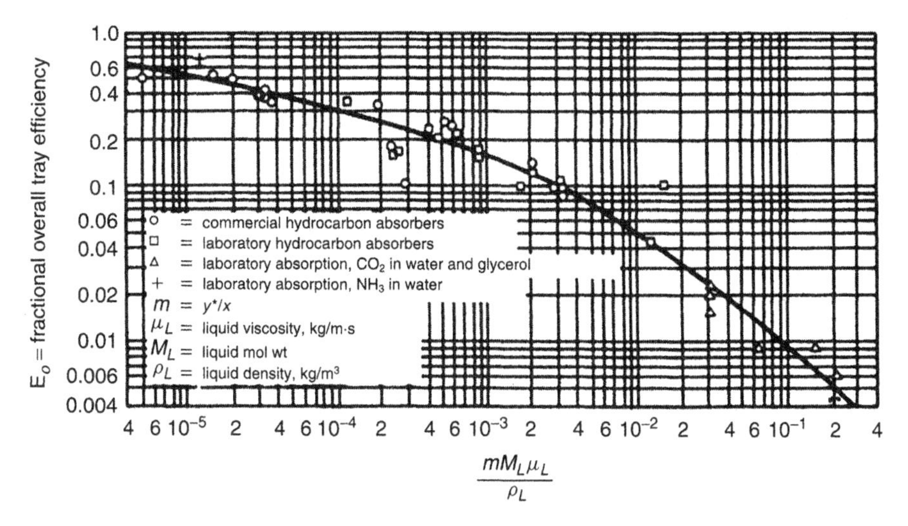

Figure 13-19. Overall efficiencies for bubble cap towers. (Reproduced with permission from reference 10. Copyright 1946, American Institute of Chemical Engineers.)

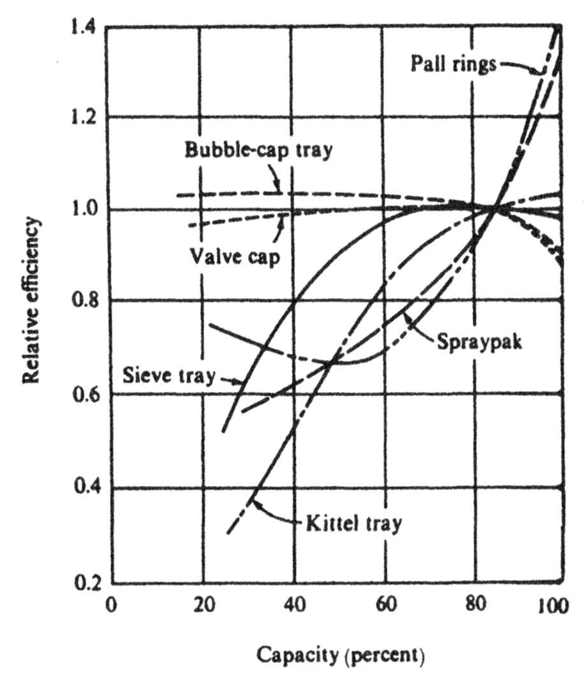

Figure 13-20. Contacting device efficiency as related to capacity. (Reproduced with permission from references 11 and 12. Copyright 1960, Institute of Chemical Engineers. Copyright 1963, Institute of Chemical Engineers.)

Table 13-2 Additional Mass Transfer Techniques

Name	Feed	Separating Agent	Products	Principle of Separation	Practical Example	References
1. Crystallization	Liquid	Cooling, or else heat causing simultaneous evaporation	Liquids and solids	Difference in freezing tendencies; preferential participation in crystal structure	sugar; *p*-xylene	18–21
2. Desublimation	Vapor	Cooling	Solid and vapor	Preferential condensation (desublimation); preferential participation in crystal structure	Purification of phthalic anhydride	18, 22, 23
3. Dialysis	Liquid	Selective membrane; solvent	Liquids	Different rates of diffusional transport through membrane (no bulk flow)	Recovery of NaOH in rayon manufacture; artificial kidneys	18, 21, 24-27
4. Dual-temperature exchange reactions	Fluid	Heating and cooling	Two fluids	Difference in reaction equilibrium constant at two different temperatures	Separation of hydrogen and deuterium	28
5. Electrodialysis	Liquid	Anionic and cationic membranes; electric field	Liquids	Tendency of anionic membranes to pass only anions, etc.	Desalination of brackish waters	18, 21, 29-31
6. Electrophoresis	Liquid containing colloids	Electric fluid	Liquids	Different ionic mobilities of colloids	Protein separation	18, 20, 32

7. Evaporation	Liquid	Heat	Liquid + vapor	Difference in volatilities (vapor pressure)	Concentration of fruit juices	40
8. Flash expansion	Liquid	Pressure reduction (energy)	Liquid + vapor	Same	Flash process for sea water desalination	29, 31
9. Flotation	Mixed powdered solids	Added surfactants; rising air bubbles	Two solids	Tendency of surfactants to adsorb preferentially on one solid species	Ore flotation; recovery of zinc sulfide from carbonate gangue	18, 33, 34
10. Freeze drying	Frozen water containing solid	Heat	Dry solid and water vapor	Sublimation of water	Food dehydration	18, 35
11. Gas permeation	Gas	Selective membrane; pressure gradient	Gases	Different solubilities and transport rates through membrane	Purification of hydrogen by means of palladium barriers	21, 36, 37
12. Gel filtration	Liquid	Solid gel (e.g., cross-linked dextran)	Gel phase and liquid	Difference in molecular size and hence in ability to penetrate swollen gel matrix	Purification of pharmaceuticals; separation of proteins	38
13. Liquid foam fractionation; Bubble fractionation	Liquid	Rising air bubbles; sometimes also complexing surfactants	Two liquids	Tendency of surfactant molecules to accumulate at gas-liquid interface and rise with air bubbles	Removal of detergents from laundry wastes; ore flotation	18, 20, 49-51

(*continued overleaf*)

Table 13-2 ***(continued)***

Name	Feed	Separating Agent	Products	Principle of Separation	Practical Example	References
14. Molecular distillation	Liquid mixtures	Heat and vacuum	Liquid and vapor	Difference in kinetic theory maximum rate of vaporization	Separation of vitamin A esters and intermediates	18, 42
15. Osmosis	Salt solution	More concentrated salt solution; membrane	Two liquids	Tendency to achieve uniform osmotic pressures removes water from more dilute-solution	Suggested for food dehydration	18, 43
16. Reverse osmosis	Liquid solution	Pressure gradient (pumping power) + membrane	Two liquid solutions	Different combined solubilities and diffusivities of species in membrane	Sea water desalination	18, 44-46
17. Thermal diffusion	Gas or liquid	Temperature gradient	Gases or liquids	Different rates of thermal diffusion	Suggested for isotope separation, etc.	18, 21, 22, 28, 47, 48
18. Ultracentrifuge	Liquid	Centrifugal force	Two liquids	Pressure diffusion	Separation of large, polymeric molecules according to molecular weight	49
19. Ultrafiltration	Liquid solution containing	Pressure gradient (pumping power)	Two liquid phases	Different permeabilities	Waste water treatment; protein	18, 25, 26, 51, 52
20. Zone melting	Solid	Heat	Solid of non-uniform composition	Same as crystallization	Ultrapurification	18, 53, 54

Various studies of stage efficiencies of liquid–liquid extraction systems found that many operations had values from 71 to 79 percent (5, 13-16). In addition, some systems ranged from 95 to 100 percent (5, 13, 14, 16).

ADDITIONAL MASS TRANSFER TECHNIQUES

Table 13-2 summarizes a number of additional mass transfer techniques together with the feed, separating agent, and principle of separation. Products, practical examples, and references are also given.

WORKED EXAMPLES

Example 13-1 We want to remove alcohol vapor (0.01 mole fraction) from a carbon dioxide gas stream. Water for the absorption contains 0.0001 mole fraction of alcohol. A total of 227 moles of gas are to be treated per hour. The equilibrium relationship for alcohol and water is given by $Y = 1.0682X$. For this case, how many theoretical plates would be required for 98 percent absorption at a liquid rate of 1.5 times minimum?

At minimum liquid rate, the operating line would intersect the equilibrium line at

$$Y = 0.01$$

and

$$X_N = 0.01/1.0682$$

Hence, point C in Figure 13-21 is (0.009362, 0.01).

Point A is determined by X_0 and Y_1. The X_0 value is given as 0.0001 in the problem statement and Y_1 by mass balance.

Location	Moles CO_2	Moles Alcohol
Bottom	224.73	2.27
Top	224.73	0.046

Then for Y_1,

$$Y_1 = 0.046/(224.73 + 0.046) = 0.000202$$

Next, the slope of AC gives the value of L_{minimum},

$$L_{\text{minimum}} = (\text{moles gas})(\text{slope } AC)$$

$$L_{\text{minimum}} = (224.78)(1.0588)$$

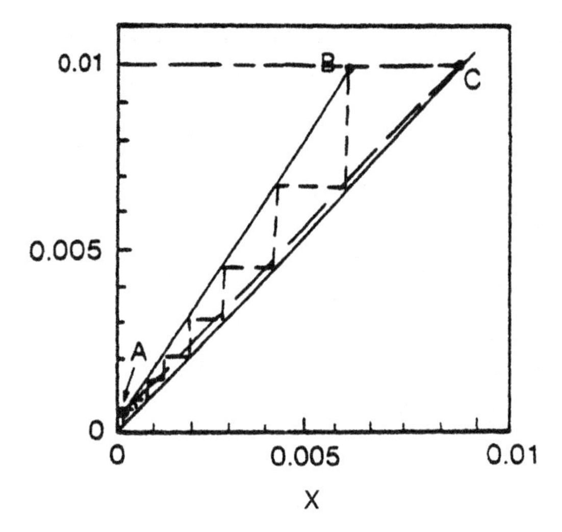

Figure 13-21. Determination of stages for Example 13-1. (Reproduced with permission from reference 17. Copyright 1997, American Chemical Society.)

and

$$L_{\text{minimum}} = 238 \text{ moles}$$

At $1.5L_{\text{minimum}}$, the L value is

$$L = (1.5)(238) = 357 \text{ moles}$$

From L we can calculate the slope of a material or mass balance line as 357/224.78 or 1.588. Point A (0.0001, 0.000202) represents one end of the mass balance as well as the composition at the top of the tower. The other end of the line is point B (composition at the bottom of the tower). For this point, Y is 0.01. Hence, line AB passes through (0.0001, 0.000202) with a slope of 1.588 and ends at $Y = 0.01$. Now the number of stages or plates can be stepped off from either A or B to yield 9 stages.

Example 13-2 Repeat Example 13-1 using Figures 13-4 and 13-8.

For Figure 13-4 the Y_1 (Y_1') value is 0.01, the Y_2 (Y_2') value is 0.000202, and the $mX_2(KX_0)$ value is (1.0682) (0.0001). Hence the abscissa is

$$0.01 - 0.000202/0.01 - (1.0682)(0.0001) = 0.990$$

Likewise, the L/mG factor is 1.588/1.0682, or 1.487.

The intersection of 0.990 and 1.487 in Figure 13-4 gives 10 stages.

Repeating for Figure 13-8, $Y_1 = 0.01$, $Y_2 = 0.000202$, and $mX_2 = (1.0682)$ (0.001) with mG/L being 1.0682/1.5888, or 0.673.

Thus,

$$0.01 - (1.0682)(0.0001)/0.000202 - (1.0682)(0.0001) = 103.9$$

Then, from Figure 13-8 N_p is approximately 10.5 stages.

Example 13-3 We want to remove alcohol vapor from carbon dioxide using a packed column containing 2.54-cm Raschig rings at 40°C. The vapor (0.01 mole fraction alcohol) needs to have 98 percent recovery of alcohol. The water used for absorption has 0.0001 mole fraction alcohol. The G_m at the tower bottom is 0.0422 kg-mol/sec m^2. The ratio of L_m to G_m is 1.5264. Equilibrium is given by $Y = 1.0682x$. Values of H_G and H_L are 0.54 and 0.30 m, respectively. First, we calculate mG_m/L_m.

$$mG_m/L_m = 1.0682/1.5264 = 0.70$$

Next, we calculate H_{OG} using equation (13-34)

$$H_{OG} = H_G + (mG_m/L_m)H_L = 0.54 \text{ m} + (0.7)(0.30 \text{ m})$$

$$H_{OG} = 0.75 \text{ m}$$

Now, we obtain X and Y values at the top and bottom of the column.

$$Y_{\text{bottom}} = 0.01(\text{given})$$

$$Y_{\text{top}} = 0.02[0.01/(1 - 0.01)] = 0.000202$$

and

$$X_{\text{top}} = 0.0001$$

The X at the bottom can be obtained by the mass balance. The alcohol in the gas stream fed to the absorber is

$$(0.000202/0.02)(0.042) \text{ kg-mole/sec-m}^2 = 0.000422 \text{ kg-mole/sec-m}^2$$

$$\text{Alcohol absorbed} = 0.000424(0.98) \text{ kg-mole/sec-m}^2$$

$$\text{Alcohol absorbed} = 0.000414 \text{ kg-mole/sec-m}^2$$

Using the alcohol absorbed, we can calculate X at the bottom as

$$X_{\text{bottom}} = [0.000414 + (0.0001)(1.5264)(0.042)] \text{ kg-mol/sec-m}^2/$$
$$[(1.5264)(0.042) + 0.000414] \text{ kg-mole/sec-m}^2$$

$$X_{\text{bottom}} = 0.006542$$

The Y^* values corresponding to the X values are

$$X_{\text{bottom}} = 0.006542$$
$$Y^*_{\text{bottom}} = (1.0682)(0.006542) = 0.006988$$
$$X_{\text{top}} = 0.0001(\text{given})$$
$$Y^*_{\text{top}} = (1.0682)(0.0001) = 0.0010682$$

With these values and equation (13-31) we obtain

$$(Y - Y^*)_{\text{lm}} = [(Y - Y^*)_{\text{bottom}} - (Y - Y^*)_{\text{top}}]/\ln(Y - Y^*)_{\text{bottom}}/(Y - Y^*)_{\text{top}}$$
$$(Y - Y^*)_{\text{lm}} = [(0.01 - 0.006988) - (0.000202 - 0.0010682)]/\ln[(0.01 - 0.006988)/(0.000202 - 0.00010682)]$$
$$(Y - Y^*)_{\text{lm}} = 0.0008443$$

Then, from equation (13-30) we have

$$(G_m Y_1 - G_m Y_2)/K_{OG}a_V P(Y - Y^*)_{\text{lm}} = h_{\text{total}}$$

However, we know [from equations (13-18) and (13-20)] that

$$K_{OG}a_V P = (G_M)_{\text{average}}/H_{OG}Y^*_{BM}$$

so that

$$K_{OG}a_V P = \frac{0.042}{(0.75)(0.9956)} = 0.0562 \text{ kg-mole/sec-m}^3$$
$$h_t = (0.0422)(0.01)/0.0562(0.008443) = 8.74 \text{ m}$$

Example 13-4 Repeat Example 13-3 using the algebraic technique and Figure 13-8.

For the algebraic method we have

$$h_t = (N_{OG})(H_{OG})$$

Then from equation (13-36) we have

$$N_{OG} = 1/(1 - 0.7)\ln(1 - 0.7)(103.94) + 0.7$$
$$N_{OG} = 11.57$$

Finally,

$$h_t = (N_{OG})(H_{OG}) = (11.57)(0.756 \text{ m})$$
$$h_t = 8.75 \text{ m}$$

Next, using Figure 13-8 we obtain

$$Y_1 - mX_2/Y_2 - mX_2 = 103.94$$

Reading to the mG/L value of 0.7 gives an N_{OG} of 11.5:

$$h_t = (N_{OG})(H_{OG}) = (11.5)(0.756 \text{ m})$$

$$h_t = 8.69 \text{ m}$$

Now comparing results, we obtain the following:

Method	ht (m)
Using equation (13-15) (not shown in text)	8.62
Example 13-4	8.74
Algebraic, this example	8.75
Figure 13-8	8.69

Note that there is excellent agreement between the various methods. This agreement occurs because of the dilute nature of the system. None the less, the methods used in this example afford an excellent starting point for packed column design.

Example 13-5 A solvent extraction system (countercurrent multistage contact process) is to reduce a 50 percent pyridine and 50 percent water (mass percents) to a 1 percent pyridine solution (i.e., saturated raffinate) by using pure benzene as a solvent.

For this case find the number of stages needed for 1.5 times the minimum solvent rate. Also find all concentrations and masses of extracts and raffinates if 0.758 kg/min of incoming solution are to be processed.

Equilibrium data are given below:

Benzene Layer		Water Layer	
Percent Pyridine	Percent Benzene	Percent Pyridine	Percent Benzene
3.28	94.54	1.17	0
9.75	87.46	3.55	0
18.35	79.49	7.39	—
26.99	71.31	13.46	0.15
31.42	66.46	22.78	0.25
34.32	64.48	32.15	0.44
36.85	59.35	42.47	2.38
39.45	56.43	48.87	3.99
39.27	55.72	49.82	4.28
48.39	40.05	56.05	19.56

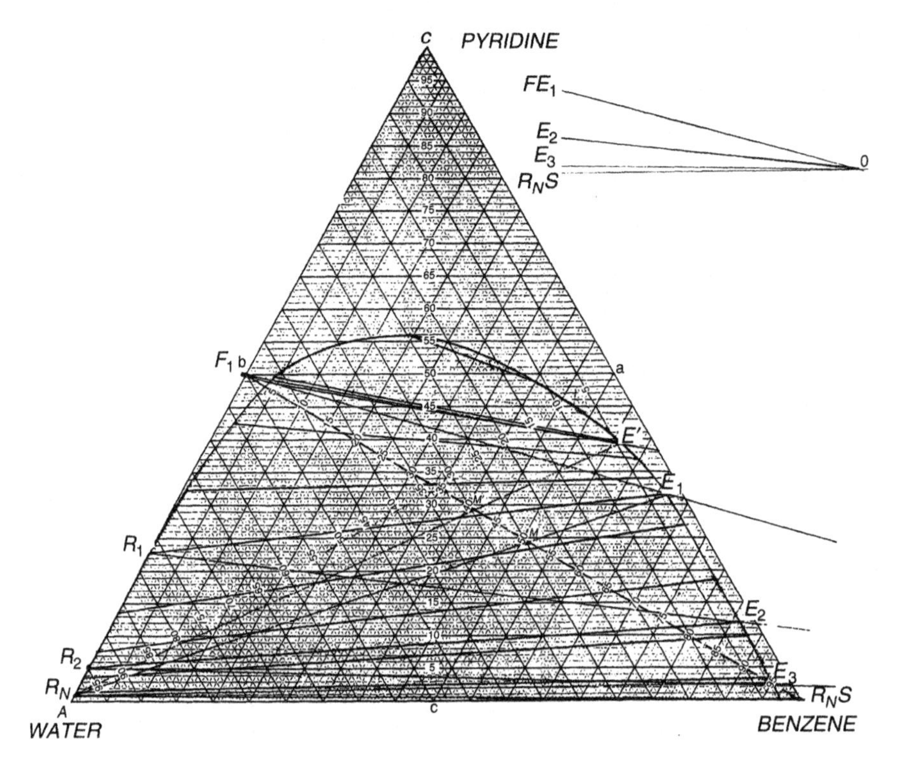

Figure 13-22. Equilibrium diagram and solution for Example 13-5.

These data can be used to plot the equilibrium diagram shown on the ternary plot (Figure 13-22).

Minimum solvent rate can be determined by extending a tie line through F (the feed point). The intercept on the extract side of the equilibrium region is then E_1^1. A line connecting E_1^1 with the final raffinate value (1 percent pyridine) R_N is then drawn. The intersection of this line ($E_1^1 R_N$) with the line FS (where S is pure benzene) gives M^1.

Then the minimum solvent rate is

$$F(X_{BM'} - X_{BF}/X_{BS} - X_{BM'})$$

$$\text{minimum solvent rate} = 0.758\ \text{kg/min}(0.411 - 0/1.0 - 0.411)$$

$$\text{minimum solvent rate} = 0.529\ \text{kg/min}$$

Actual solvent rate is 1.5 times this value or 0.794 kg/min.

Now we can solve for the number of stages. We first locate E_1 by using an M point (1.5 times M^1) that is on the FS line. Next we locate operating point O (off the ternary diagram to the right) by the intersection of lines extended through FE_1 and R_NS.

The stages are then stepped off as shown. Three stages would be required. As can be seen, a tie line through E_1 locates R_1. Next, a line from R_1 to O fixes E_2. A tie line yields R_2, which in turn (R_2, O) gives E_3.

Compositions of all of the extracts and raffinates can be determined from Figure 13-22. These together with a material balance gives the amounts of each component in the extracts and raffinates:

	E_1		E_2		E_3	
	%	kg/min	%	kg/min	%	kg/min
Pyridine	31.5	0.375	11.9	0.108	2.5	0.02
Benzene	66.4	0.789	86.1	0.781	95.2	0.77
Water	2.1	0.025	2.0	0.015	2.3	0.019
	100.0	1.189	100.0	0.904	100.0	0.809

	R_1		R_2		R_3	
	%	kg/min	%	kg/min	%	kg/min
Pyridine	22.9	0.108	5.1	0.002	1.0	0.004
Benzene	0.0	0.000	0.0	0.000	0.0	0.000
Water	77.1	0.374	94.9	0.373	99.0	0.360
	100.0	0.482	100.0	0.375	100.0	0.364

Example 13-6 Benzene is used to leach oil from meal in a continuous countercurrent unit. The system is to process 909 kg of solids per hour (based on completely exhausted solids). Meal fed to the train has 22.7 kg of benzene and 363.6 kg of oil. Solvent mixture fed to the system is made up of 595.5 kg of benzene and 9.09 kg of oil. Solids leaving the unit will have 54.5 kg of oil.

Equilibrium data for the system are as follows:

kg oil/kg solution	kg solution retained/kg solid
0.0	0.500
0.1	0.505
0.2	0.515
0.3	0.530
0.4	0.550
0.5	0.571
0.6	0.595
0.7	0.620

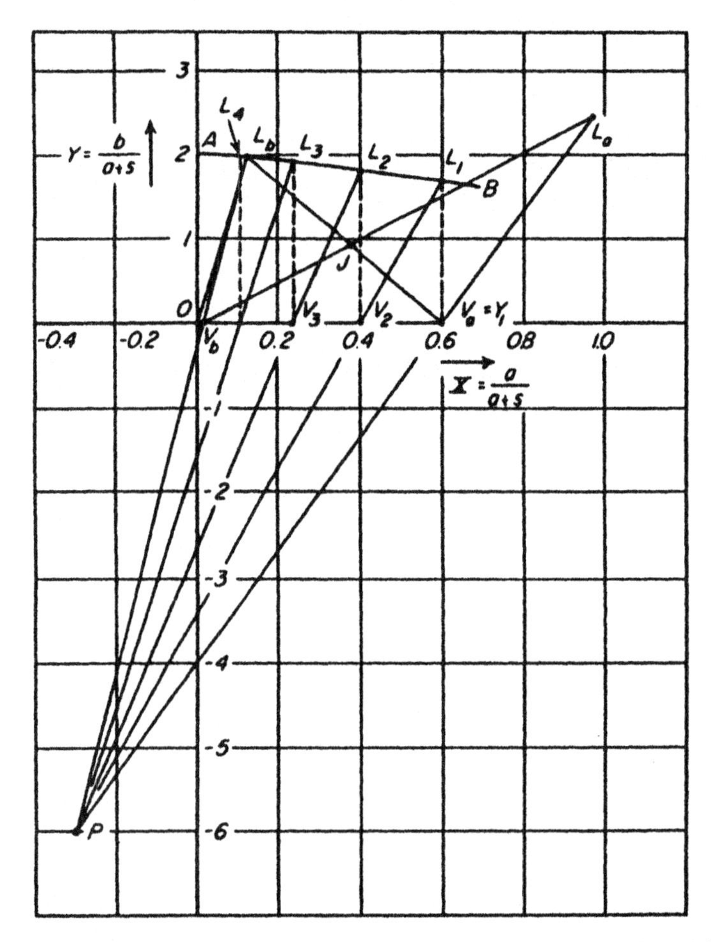

Figure 13-23. Equilibrium diagram and solution for Example 13-6. (Adapted from reference 6.)

The Y values for the plot of the data are the reciprocals of the second column, while the corresponding X values are given in the first column. Plotted data are shown in Figure 13-23.

In this problem find the stages required, strong solution concentration, mass of solution leaving with the processed meal, extract mass, and the solution concentration leaving with the processed meal.

The starting point for the stage determination is to find the operating point P. This requires finding L_a, V_a, L_b, and V_b.

The X and Y values for L_a are

$$X \text{ for } L_a = 363.6/363.6 + 22.7 = 0.941$$

$$Y \text{ for } L_a = 909/363.6 + 22.7 = 2.35$$

Likewise the X and Y values for V_b are

$$Y \text{ for } V_b = 0$$

$$X \text{ for } V_b = 9.09/9.09 + 595.5 = 0.015$$

Next we connect V_b and L_a on Figure 13-23 and note that J lies on this line. Furthermore, by the lever rule, J is $(363.6 + 22.7/363.6 + 22.7 + 9.09 + 595.5)$ of the distance between V_b and L_a (or 0.39 of line $V_b L_a$).

Furthermore, L_b will lie on the underflow curve and must be on a line of slope 909/54.4, or 16.7. Finally a straight line through L_b, and J will cut the X axis at V_a (0.592, 0). Straight lines through L_b, V_b and L_a, V_a respectively give point P.

Stages required are marked off in Figure 13-23. As can be seen, the number is four.

The strong solution concentration is the X value for V_a, or 0.592. Solution concentration leaving with the processed meal is the X value of L_b (0.12).

Solution mass leaving with the meal is the mass of L_b; total solution input is

$$L_b \text{ mass} = 990.99 \text{ kg}(0.592 - 0.372/0.592 - 0.120)$$

where 0.592 is the X value of V_a, 0.372 is the X value of J, and 0.120 is the L_b X value. The result is 461.9 kg.

The mass of V_a represents the extract mass and is (990.99 − 461.9) kg or 529.09 kg.

Example 13-7 A scrubber removes benzene from air using oil (molecular weight of 200) as a solvent. The column operates at 1 atmosphere pressure and 26.67°C. At 26.67°C the benzene vapor pressure is 0.1373 bar. The gas entering contains 5 mole percent benzene at a total flow of 0.0756 kg moles/sec.

Determine the theoretical stages needed at 1.5 times minimum liquid rate. Assume that 90 percent of entering benzene is removed.

First, we determine the equilibrium data by using Raoult's Law.

Hence,

$$\frac{Y_A}{X_A} = \frac{P_A}{P_{\text{total}}}$$

and at 26.67°CP_A is 0.1373 bar. Thus,

$$Y_A = \frac{0.1373 \text{ bar}}{1.014 \text{ bar}} X_A$$

$$Y_A = 0.136 X_A$$

$$Y_A = \frac{Y_A^1}{1 + Y_A^1}, \qquad X_A = \frac{X_A^1}{1 + X_A^1}$$

$$\frac{Y_A^1}{1 + Y_A^1} = 0.136 \frac{X_A^1}{1 + X_A^1}$$

We can then calculate Y_A^1 and X_A^1 for equilibrium relation:

X_A^1	Y_A^1	X_A^1	Y_A^1
0	0	0.272	0.03
0.038	0.005	0.396	0.04
0.0785	0.01	0.539	0.05
0.168	0.02		

Hence, we plot Y_A^1 versus X_A^1 and get the equilibrium line (AB). If we have 90 percent recovery, the benzene leaving will be

$$(0.10)(0.05)(0.0756)\ \frac{\text{kg mole}}{\text{sec}} = 0.000378\ \frac{\text{kg mole}}{\text{sec}}$$

$$\text{and } Y_1^1 = \frac{0.000378}{0.0718} = 0.00526\ \frac{\text{kg mole benzene}}{\text{kg mole air}}$$

Also $X_0^1 = 0$. This gives point $C(X_0^1, Y_1^1)$ on the diagram:

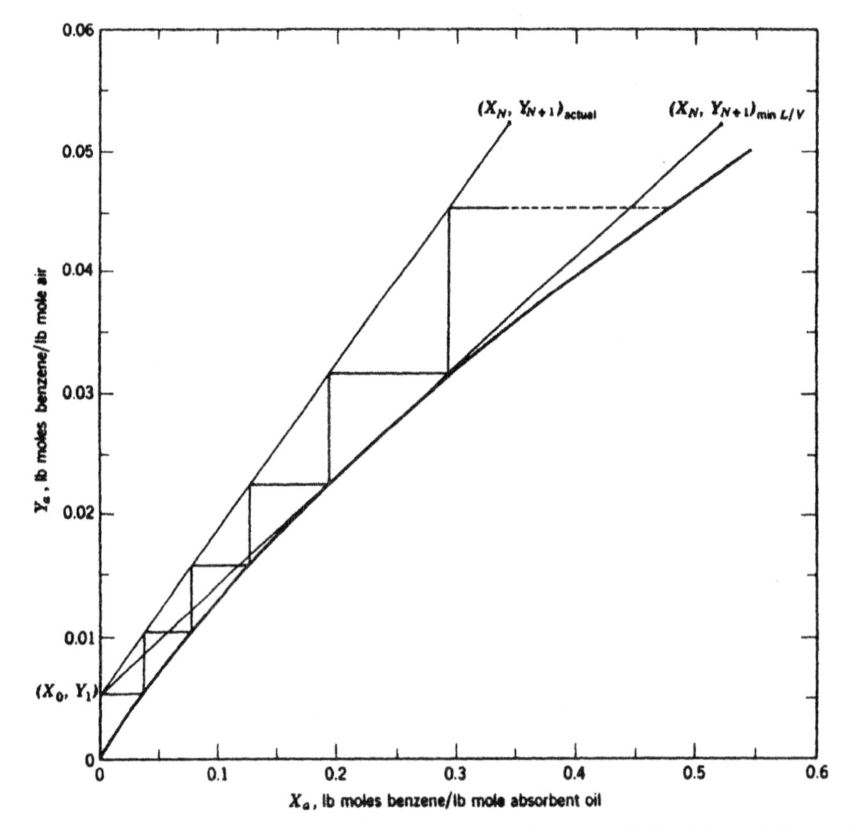

Figure 13-24. Equilibrium diagram and solution for Example 13-7. (Adapted from reference 7. Copyright 1960, John Wiley and Sons.)

Minimum liquid rate is given when a line through $C(X_0^1, Y_1^1)$ is tangential to the equilibrium line (i.e., curved line). That line CD has a slope of $(L/G_{\min}) = 0.091$ and oil has a molecular weight of 200.

$$L_{\min} = (0.091)(0.0718)\ \frac{\text{kg mole}}{\text{sec}}(200)$$

$$L_{\min} = 1.307\ \frac{\text{kg}}{\text{sec}}$$

Process L is $1.5L_{\min}$ and is equal to 1.96 kg/sec.
The slope of the mass balance line in Figure 13-24 (CE) is

$$\frac{1.96/200}{0.0718} = 0.137$$

The stages as determined in Figure 13-24 are five and a fraction. Therefore, we use six stages.

PROBLEMS

13-1. An ammonia–air mixture (0.02 mole fraction ammonia) is to be scrubbed with water at 20°C. Final mole fraction is 0.00041 ammonia. Both the gas and liquid flow rates are 1.36×10^{-3} kg/m^2 sec. Equilibrium data for partial pressures of ammonia over aqueous ammonia solutions are:

P (atm)	0.016	0.024	0.042	0.066	0.218
kg ammonia/100 kg water	2	3	5	7.5	20

Find the number of equilibrium stages needed.

13-2. An acetone–water solution (5 percent by weight of acetone) is stripped by open steam in a column. The L_M/G_M ratio is 19. The Henry's Law relation is $Y = 36.9X$. Also, the relation between X's in equilibrium with Y are

X	0	0.005	0.01	0.02
Y	0	0.152	0.262	0.405
m	36.9	30.5	26.2	20.2

The bottom mole fraction is to be 0.00025 for acetone. Find the number of theoretical plates needed.

13-3. How many theoretical stages are required for the absorption of a soluble gas in water. The equilibrium is given by $Y = 0.06X$. The mole fraction values of the absorbed gas in the water are 0 (top of column) and 0.08

(bottom). Likewise the gas mole fractions are 0.009 (bottom) and 0.001 (top). What is the number of ideal stages required?

13-4. An absorber uses an oil to recover benzene from a gas stream (0.10 benzene, 0.90 air). The gas feed rate is 0.028 kmole/sec. The oil fed (0.022 kmole/sec) to the tower top doesn't contain benzene. Equilibrium is given by $Y = 0.5X$. The desired exit gas composition is to be a mole fraction of 0.005. Find the composition of the liquid leaving. Also determine the required number of ideal stages.

13-5. Ethane is absorbed by a heavy oil in a five-plate column. The feed oil (which doesn't contain ethane) enters at the rate of 50 moles for every 100 moles of gas fed. The gas stream leaving the absorber has a mole fraction of 0.03. Liquid and gas rates in the column are constant. The equilibrium relation is 0.5 mole percent ethane in the gas per mole percent ethane in the liquid. Find ethane concentration entering as well as that leaving the first tray in the tower bottom.

13-6. Ammonia synthesis hydrogen is obtained by passing liquid nitrogen countercurrent to a carbon monoxide–hydrogen stream. A stream (0.90 mole fraction hydrogen) is used with a 100 percent liquid nitrogen stream. If both vapor and liquid streams have constant molar flows and the equilibrium relation is $Y = 1.25X$, find (exit gas carbon monoxide mole fraction of 0.01)

(a) Minimum liquid rate per mole of feed gas.

(b) Theoretical stages at 1.3 times minimum liquid flow.

13-7. Sulfur dioxide is to be absorbed into water in a plate column. The feed gas (20 mole percent sulfur dioxide) is to be scrubbed to 2 mole percent sulfur dioxide. Water flow rate is 6000 kg/hr m^2. The inert air flow rate is 150 kg air/hr m^2. Tower temperature is 293 K. Find the number of theoretical plates. Equilibrium data are

Y	0.212	0.121	0.0775	0.0513	0.0342	0.01855	0.01120
X	0.00698	0.00420	0.00279	0.001965	0.001403	0.000942	0.000564

13-8. A gas (0.01 mole fraction acetone in air) is to have 90 percent of the acetone absorbed in water. The tower (isothermal), operates at 300 K and atmospheric pressure. Inlet gas flow rate and water flow rate are 30 kmole/hr and 90 kmole/hr, respectively. If the equilibrium relation is $y = 2.53X$, find the number of ideal stages needed.

13-9. A sieve plate absorber scrubs a butane–air mixture (mole fraction 0.05 butane). The tower (15°C, atmospheric pressure) uses a heavy nonvolatile oil (molecular weight of 250; specific gravity of 0.8) as an absorbent. Butane recovery is to be 95 percent. The butane vapor pressure at 15°C is 1.92 atm. Liquid butane density is 580 kg/m^3 (at 15°C). Determine m^3 of fresh oil per m^3 of butane recovered. Repeat for a tower pressure of 3 atms.

13-10. A gas (0.02 mole fraction A, 0.01 mole fraction B) is to be scrubbed with a solvent such that A is five times as soluble as B. It is proposed to use two columns in series (separate liquid regeneration in each tower) in order to maximize the purities of A and B products. Use a $Y-X$ diagram to show the simultaneous absorption of A and B. Estimate the ratio of A and B in the liquid from the first absorber.

13-11. Zinc sulfide is oxidized (100 percent excess air) by a reaction

$$ZnS + O_2 \longrightarrow Zn + SO_2$$

that goes to completion.

In a second reaction sequence

$$SO_2 + \tfrac{1}{2}O_2 \longrightarrow SO_3$$

such that 98 percent of SO_2 reacts. The gas mixture formed (SO_3, SO_2, N_2, O_2) is then scrubbed in a stream of sulfuric acid whence 099.9 percent of the SO_3 is absorbed. A K of 0.3 is defined as the mole fraction of SO_3 (gas phase) divided by the mole fraction of SO_3 (acid phase). What ideal stages would be required at 1.5 times minimum liquid rate? Also specify the composition of the unabsorbed gas.

13-12. A device is used to purify a gaseous stream of ammonia and water (0.1 mole per 100 moles of ammonia) by the use of a cooling coil at the tower top. The result of the operation is to produce a purer ammonia as the top product and an ammonia–water liquid from the bottom. The K value for the system is 0.05. Find the minimum flow of bottom product (waste ammonia) in moles/mole feed if the top product is 0.01 moles of water per 100 moles of ammonia. What is the waste concentration if liquid flow is two times minimum?

13-13. A separation system involves the use of both an absorber and stripping column. The gas feed to the absorber is 100 kmole/hr and contains a 0.15 mole fraction of tetrahydrofuran (THF). The liquid feed to the top is THF (0.01 mole fraction remainder water). Vapor product from the top has a 0.01 mole fraction THF. Bottom liquid from the absorber is 0.40 mole fraction THF, which is then fed to the top of a stripper.

13-14. Stripper top product is 0.70 mole fraction THF. The bottom (liquid) product, which has a 0.01 mole fraction THF, is recycled back to the absorber to serve as the liquid feed (a 0.01 mole fraction THF liquid is added as makeup). Find the ideal trays for the absorber and stripper as well as all vapor and liquid flow rates.

13-15. A purification process uses both an absorber and stripper. The gas feed to the absorber (26°C, 1.06 atm, 0.25 m^3/sec) has a mole fraction of 0.02 benzene. The absorber liquid feed (0.005 mole fraction of benzene) has a molecular weight of 260. A liquid rate 1.5 times minimum is to be used. The liquid from the absorber is fed to the stripper where steam is used as the stripping agent. Liquid from the stripper (0.005 mole fraction benzene) is returned to the absorber. Steam rate is 1.5 times the minimum. Equilibrium relations are $Y'/1 + Y' = 0.125X'/1 + X'$ (absorber), and $Y'/1 + Y' = 3.16X'/1 + X'$ (stripper). The primed quantities are on a solute free basis. Find the theoretical plates for the absorber and stripper.

13-16. A three-stage absorber at 6.8 atm has a gas feed made up of methane, ethane, propane, and butane with mole fractions of 0.70, 0.15, 0.10, and 0.05. Both the gas feed and absorber oil enter at 26.7°C. The L/G value for the tower is 2.0. Equilibrium data for each of the gases as $Y = KX$ are, respectively, 23.9 (methane), 4.6 (ethane), 1.4 (propane), and 0.4 (butane). Analyze the assumptions of constant temperature and L/G rate for the column.

13-17. Air is mixed with an n-pentane gas process stream that results in a gas at 37.8°C and 1 atm with a flow rate of 0.0093 kmole/sec. The pentane mole fraction is 0.5277. Oil (molecular weight 160, specific gravity of 0.84) is used as the absorbing fluid. The liquid has a pentane mole fraction of 0.005. Liquid flow rate is 1.7 times minimum needed for 99 percent recovery of pentane (K value of 1.0). Find the number of ideal stages.

13-18. A gas mixture (0.75 mole fraction methane, 0.25 n-pentane) at 27°C and at atmospheric pressure is fed to an absorber at the rate of 1 kmole/basis time, and a nonvolatile oil (molecular weight 200) at a flow of 2 kmole/basis time enters pentane-free at 35°C. How many ideal trays are needed for recovery of 98 percent of the pentane (adiabatic operation)? The equilibrium K values ($Y = KX$) are 0.575 (20°C), 0.69 (25°C), 0.81 (30°C), 0.95 (35°C), 1.10 (40°C), and 1.25 (43°C). Enthalpies of liquid and vapor (H_L, H_G) are

$$H_L = T_L(376.8 - 199.3X) \text{ kJ/kmole solution}$$

$$H_G = T_G(35.59 + 84.16Y) + 27{,}820Y \text{ kJ/kmole gas vapor mix}$$

13-19. Sulfur dioxide is to be scrubbed from air by water. The mole fractions of sulfur dioxide in the gases entering and leaving are 0.20 and 0.005. Air flow rate is 0.272 kmole/m^3 sec. Water flow is to be twice minimum liquid rate. If 3.8 theoretical stages are needed, what is the equilibrium relation for sulfur dioxide?

13-20. Acetone is absorbed from air by a nonvolatile oil. Mole fraction of acetone in the entering gas is 0.30. The oil removes 97 percent of the acetone fed to the tower. The liquid leaving has a 0.10 mole fraction of acetone. If the acetone equilibrium relation is $Y = 1.9X$, find the number of theoretical stages needed.

13-21. Repeat the preceding for a case where the acetone entering and leaving gas mole fractions are 0.20 and 0.02; 95 percent of the acetone is absorbed; and acetone mole fraction in the leaving liquid is 0.12.

13-22. Repeat Problem 13-1 using a packed column (0.0254-m-diameter Raschig rings). The overall mass transfer coefficient (based on gas) is 0.0174 kmole/m^3 sec atm. Determine the packing height.

13-23. Find the number of liquid based transfer units if the system of Problem 13-2 is processed in a packed column.

13-24. A sulfur dioxide (mole fraction 0.06) and air mixture is scrubbed in a packed tower (0.0254-cm Raschig rings) with pure water. The exit gas has a mole fraction of 0.001. Water flow is twice minimum rate. The liquid phase Schmidt number is 570. Find the height of packing needed for a gas flow rate of 0.126 kg/sec. Equilibrium data are as follows:

Y	0.104	0.0684	0.0474	0.0259	0.0107	0.0062	0.0022
X	0.0028	0.00197	0.00141	0.00084	0.00042	0.00028	0.00014

13-25. A gas stream air and a hydrocarbon (molecular weight of 44; mole fraction 0.20) is scrubbed with an oil (molecular weight of 300) to remove 95 percent of the hydrocarbon. The operation carried out in a packed column has gas and liquid mass flow rates entering of 6.796 and 13.591 kg/m^2 sec. Find the number of transfer units based on the overall gas approach. Equilibrium data are as follows:

Y	0.15	0.082	0.052	0.038	0.021	0.004
X	0.4	0.35	0.30	0.25	0.20	0.10

13-26. A gas is absorbed in water in a packed tower. Gas mole fractions entering and leaving are 0.009 and 0.001. Mole fractions of absorbed gas in liquid are 0 (top) and 0.001 (bottom). If the equilibrium is given by $Y = 0.06X$ and H_G and H_L are 0.36 and 0.24 m, what is the height of packing.

13-27. An ammonia–air mix is scrubbed at 30°C to remove 99 percent of the ammonia fed. Mole fraction of ammonia in the entering gas is 0.3. Equilibrium data are as follows:

Y	0.0389	0.0528	0.0671	0.1049
X	0.0308	0.0406	0.0503	0.0735

13-28. A packed column strips trichloroethylene (TCE) from water containing 6 ppm TCE using air. Emission standards require less than 4.5 ppb of TCE in the product. The equilibrium m value is 417, and H_{OL} for the column is 0.914 m. Find the minimum air rate.

13-29. What is the effect on packing height for Problem 13-27 if the air rate changes successively to 1.5, 2, 3, and 5 times minimum air rate?

13-30. A gas stream (mole fraction of 0.03 benzene) is scrubbed with an oil (molecular weight 250, specific gravity of 0.875, 0.015 mole fraction benzene) in a packed column. The gas flow rate is 0.614 m^3/sec at 25°C. If the equilibrium relation is $Y = 0.125X$ and the liquid rate is 6.19×10 kmole/sec (on a benzene-free basis), find the number of transfer units needed. Also what effect would a molecular weight of 200 for the oil have?

13-31. A cyclohexane–air mixture (0.01 mole fraction cyclohexane) is scrubbed with an oil (mole fraction cyclohexane of 0.003) at 30°C in a packed tower. The gas entering flow rate is 0.1713 m^3/sec. Liquid feed is 2.52×10^{-3} kmole/sec. Ninety percent of the cyclohexane is to be removed. Additional data: k_La and k_Ga are 1.427×10^{-3} and 0.0633 kmole/m^3 sec; cross-sectional area of tower is 0.186 m^2. Find the height of packing required.

13-32. Find the height of packing needed for absorption of benzene from air (0.05 mole fraction benzene) into oil (feed rate benzene-free is 7.89 kmole/hr; benzene in at 0.047 kmole/hr).

The tower is to remove 90 percent of the entering benzene using 0.0254 m size Intalox saddles at 26.7°C. Entering gas mass velocity is 1.087 kg/m^2 sec and oil molecular weight is 230. The value of m is 0.139.

13-33. A waste stream (0.5 weight percent ammonia, remainder water) is to be stripped by air to remove 98 percent of the ammonia at 20°C. Find the minimum air rate (as kg air/kg water) and the number of transfer units if twice the minimum rate is used.

Equilibrium data are as follows:

Y	0.0239	0.0328	0.0417	0.0658
X	0.0308	0.0406	0.0503	0.0735

13-34. A solute A which is to be removed from a gas stream (0.04 mole fraction A) obeys Henry's Law as a solute. At a 1 atm tower pressure find N_{OG} (using solute-free liquid) at 1.5 times minimum liquid rate. Repeat for the same liquid rate at 2 and 4 atm.

13-35. Slaked lime, $Ca(OH)_2$, is treated with sodium carbonate solution to produce caustic soda. Resultant slurry consists of particles of calcium carbonate suspended in a 10 percent solution of sodium hydroxide. After settling, the clear sodium hydroxide solution is removed and replaced by an equal weight of water and the system is thoroughly agitated. After two such washes, what part of the original sodium hydroxide remains in the sludge?

Pertinent data are as follows:

Weight Fraction NaOH in Clear Solution	Weight Fraction NaOH in Settled Sludge Solution	kg $CaCO_3$ per kg Solution in Settled Sludge
0.0900	0.0917	0.495
0.0700	0.0762	0.525
0.0473	0.0608	0.568
0.0330	0.0452	0.600
0.0208	0.0295	0.620
0.01187	0.0204	0.650
0.00710	0.01435	0.659
0.00450	0.01015	0.666

13-36. Flaked soybeans are leached with hexane to remove soybean oil. In the process a 0.3-m-thick layer of flakes (0.25-mm flake thickness) is fed onto a slowly moving continuous perforated belt that receives liquid from sprays. After percolating through the solid, the liquid is collected in a trough under the belt and recycled to the spray system. The leached solid drains for 6 minutes between each spray. Also the solvent moves in a countercurrent direction such that each spraying and draining is a stage. Solution retention after the 6-minute drain time is a function of solution oil content:

Wt % oil in solution	0	20	30
kg solution retained per kg insoluble solid	0.58	0.66	0.70

Assume retained solution contains the only oil in the drained flakes. Entering soybean flakes containing 20 percent oil are to be leached to 0.5 percent oil (solvent-free basis). Net forward flow of hexane is 1 kg hexane/per kg flakes. Solvent draining from the flakes only contains solid in the first stage where there is 10 percent of the feed-insoluble solid as a suspended solid. Find the stages required.

13-37. Two tons per day of waxed paper is to be dewaxed by kerosene in a countercurrent system. The paper (25 percent paraffin wax, 75 percent paper) is to retain 2 lb of kerosene per pound of kerosene and waxfree paper. After drying (to remove the kerosene) the paper must not contain over 0.2 lb of wax/100 lb of wax-free paper. Fresh kerosene contains 0.05 lb of wax per 100 lb of wax-free kerosene. Process extract is to contain 5 lb of wax per 100 lb of wax-free kerosene. Find the required stages.

13-38. Ether is used to extract oil from fish livers in a countercurrent system. Cell charges are each 100 lb (based on completely exhausted livers). Before extraction the livers have 0.043 gallons oil per pound of exhausted liver. Ninety five percent oil recovery is needed. Final extract is to have 0.65 gallon of oil per gallon of extract. Pertinent data are as follows:

Gallons of solution retained per lb exhausted livers	0.035	0.042	0.05	0.058	0.068	0.081	0.099	0.120
Solution concentration gallon oil/gallon solution	0	0.1	0.2	0.3	0.4	0.5	0.6	0.68

13-39. Sludge from the reaction

$$Na_2CO_3 + CaO + H_2O \longrightarrow CaCO_3 + 2NaOH$$

is extracted in a five-stage countercurrent battery. The calcium carbonate carries with it 1.5 times its weight of solution. Sodium hydroxide recovery is to be 98 percent. Reaction products enter the first stage with 0.5 kg H_2O/kg $CaCO_3$. Find the following: solution concentration leaving each unit ($CaCO_3$ is completely insoluble); wash water per kg Ca CO_3; units added for 99.5 percent recovery (wash water constant).

13-40. If the sludge retains solution (see data below), what is the number of stages needed for 95 percent recovery of sodium hydroxide. The product is to be a 15 percent NaOH solution.

Wt % NaOH	0	5	10	15	20
kg solution/kg $CaCO_3$	1.5	1.75	2.2	2.7	3.6

13-41. A countercurrent system is to treat (on an hourly basis) 10 tons of gangue (inert material), 1.2 tons of copper sulfate, and 0.5 tons of water with water as a fresh solvent. The solution produced is 10 percent copper sulfate (remainder water). After each stage, 1 ton of the inert retains two tons of water plus dissolved copper sulfate. How many stages are needed for a 98 percent of copper sulfate recovery?

13-42. Acetaldehyde (4.5 percent solution in toluene) is to be extracted in a five-stage cocurrent system. The solution feed is 0.0126 kg/sec. Water is used at the rate of 0.00316 kg/sec. Toluene–water mixes are insoluble up to 15% acetaldehyde. If the equilibrium relation (Y is kg acetaldehyde/kg water, X is kg acetaldehyde/kg toluene) $Y = 2.15X$, find the extent of extraction.

13-43. Kerosene is used to extract nicotine from water (1% solution). Water and kerosene are essentially insoluble.

Find percentage extraction of nicotine for

(a) 100 kg feed; 150 kg solvent; one extraction

(b) 100 kg feed; 50 kg solvent three extractions

kg nicotine/kg water[a]	0	1.01	2.46	5.02	7.51	9.98	20.4
kg nicotine/kg kerosene[a]	0	0.807	1.96	4.56	6.86	9.13	18.7

[a]All values multiplied by 100.

13-44. Find the minimum kerosene rate for countercurrent extraction of nicotine from a nicotine–water solution (1 percent nicotine) to reduce nicotine content to 0.1 percent. The feed rate of nicotine–water is 0.1263 kg/sec. Also determine stages needed if kerosene rate is 0.1452 kg/sec. Equilibrium data is given in Problem 13-43.

13-45. A 50-50 solution of ethylbenzene and styrene is separated into two products containing 90 and 10 percent styrene each using diethylene glycol as the solvent. The feed is at a rate of 0.1263 kg/sec. With the given equilibrium data (below) find minimum stages, minimum reflux and stages at 1.5 times minimum reflux.
For solvent-rich solutions we have:

$\dfrac{\text{kg styrene}}{\text{kg hydrocarbons}}$	0	0.1429	0.386	0.557	0.674	0.833	1.0
$\dfrac{\text{kg glycol}}{\text{kg hydrocarbons}}$	8.62	7.71	6.04	5.02	4.37	3.47	2.69

For hydrocarbon-rich solutions we have:

$\dfrac{\text{kg styrene}}{\text{kg hydrocarbons}}$	0	0.087	0.288	0.464	0.573	0.781	1.0
$\dfrac{\text{kg glycol}^a}{\text{kg hydrocarbons}}$	6.75	8.17	10.1	12.15	14.05	18.33	25.6

[a]All values multiplied by 1000.

13-46. A countercurrent extraction system removes acetone from a mixture with water using methyl isobutyl ketone (MIK) as a solvent. The feed is 40 percent acetone (remainder water). The solvent used is equal in mass to the feed. How many ideal stages are needed to extract 99 percent of the acetone? What is the extract composition after solvent removal? See Figure 12-6 for equilibrium data.

13-47. A 40-60 mix of acetone and water is contacted with methyl isobutyl ketone (feed and solvent are equal in mass). If a single stage is used, how much acetone would be extracted? What part of the acetone would be removed if the solvent were divided into two parts used for two successive extractions? See Figure 12-6 for data.

13-48. A feed (2.78 kg/sec) of 30 percent acetone and 70 percent methyl isobutyl ketone is mixed with pure water (1.39 kg/sec). Find the compositions and flow rates, leaving a single-stage extractor (see Figure 12-6).

13-49. A countercurrent extractor is to reduce a 45-55 mix of acetone and methyl isobutyl ketone to 2.5 percent acetone using water as a solvent. Organic feed rate is 2.78 kg/sec, while the water rate is twice minimum solvent rate. Find the required stages (see Figure 12-6).

13-50. A binary mixture (35 percent acetone, 65 percent methyl isobutyl ketone) is processed in a countercurrent system. Water (containing 2 percent ketone) is the solvent at a rate 1.7 times minimum. The feed acetone concentration is to be 2 percent (feed rate is 1.389 kg/sec). Find the required theoretical stages (see Figure 12-6).

13-51. A three-stage countercurrent system removes acetone from an acetone–methyl isobutyl ketone feed. A water-based solvent (2 percent acetone) has an exit acetone concentration of 15 percent. The extracted exiting feed stream has 5 percent acetone. Find feed composition, ratio of solvent rate to feed rate, and percent recovery of acetone (see Figure 12-6).

13-52. Consider a system involving A and B with a solvent C.

Extract			Raffinate		
A	B	C	A	B	C
0	7	93	0	92	8.0
1.8	5.5	92.7	9.0	81.7	9.3
6.2	3.3	90.5	14.9	75.0	10.1
9.2	2.4	88.4	25.3	63.0	11.7
18.3	1.8	79.9	42.0	41.0	17.0
24.5	3.0	72.5	52.0	20.0	28.0
31.2	5.6	63.2	47.1	12.9	40.0

Feed is a saturated raffinate (83 percent A, 17 percent B, both solvent-free basis). Solvent-free percents in extract and raffinate are: 83 percent A, 17 percent B, 10 percent A, 90 percent B. Find the required stages at twice minimum reflux rate.

13-53. A feed of 2.525 kg/sec of 40 percent acetone, 55 percent methylisobutyl-ketone (MIK), 5 percent water is contacted with a solvent (95 percent water, 2.5 percent acetone, and 2.5 percent MIK). Raffinate product is 89.5 percent MIK, 7.5 percent acetone, and 3 percent water. Solvent-rich product is 81 percent water, 16 percent acetone, and 3 percent ketone. Find solvent flow rate, acetone recovered, and stages needed (see Figure 12-6).

13-54. A six theoretical staged extractor processes 2.147 kg/sec of a 40 percent acetone–60 percent methylisobutylketone feed. Pure water is the solvent. If raffinate concentration is 5 percent, find solvent amount used and percent recovery of acetone.

13-55. A solution (50 percent acetone; 50 percent water) of 45.46 kg is to be taken to a 10 percent acetone level by extraction with 1,1,2-trichloroethane in concurrent multiple contact system. Each stage uses 11.36 kg of solvent. Find the number of stages and concentrations of extracts (see Figure 13-12 for equilibrium data).

13-56. A countercurrent multiple contacting system is used to reduce a 50 percent acetone and 50 percent water to a 10 percent acetone level. The solvent (1,1,2-trichloroethane) rate is 3.79×10^{-3} kg/sec. Feed rate is 0.0126 kg/sec. Find concentrations and stages needed (see Figure 13-12).

13-57. Two solutions of acetone–water are to be extracted in a countercurrent system. One is a 50-50 mix and the other a 25 percent acetone–75 percent water solution. The solvent (1,1,2-trichloroethane) is fed at 6.31×10^{-3} kg/sec, while the two feeds each enter at 0.0126 kg/sec. Raffinate product is to be 10 percent acetone. Find the number of stages and the point where the 25-75 feed should be introduced. Figure 13-12 gives equilibrium data.

13-58. A continuous countercurrent system reduces a 0.0278 kg/sec solution (40 percent acetone, 60 percent water) to 10 percent acetone. The solvent is 1,1,2-trichloroethane. Find minimum solvent rate and stages at 1.6 times minimum (solvent rate)/(feed rate). Equilibrium data is given in Figure 13-12.

13-59. A system comprises picric acid, benzene, and water. The equilibrium is as follows:

C_B	0.000932	0.00225	0.01	0.02	0.05	0.10	0.18
K	2.23	1.45	0.705	0.505	0.320	0.240	0.187

where C_A and C_B are concentrations of picric acid in water (A) and benzene (B) in g mole/liter. The K values are C_B/C_A. A 0.1 mole picric acid/liter water solution is extracted with benzene to recover 75 percent of the picric acid. What quantity of benzene per liter of water solution is required for

(a) single extraction

(b) three-stage countercurrent extraction

13-60. Equilibrium between acetic acid in water and isopropyl ether is

$$C_E/C_W = 0.178 + 0.358°\text{C} + 0.819°\text{C}$$

The C_E and C_W are lb mole/ft^3 of acetic acid in ether and water, respectively. A spray column (3-inch i.d., 10 feet high, no packing) is used to extract downward flowing acetic acid–water (1.62 ft/hr) by upward flowing drops of ether (3.90 ft/hr)

The concentrations of acetic acid in water are 0.1432 lb mole/ft^3 (in) and 0.0715 lb mole/ft (out). Likewise, the concentrations in the ether are 0.0005 lb mole/ft (in) and 0.0305 lb mole/ft^3 (out). Check the material balance, find the number of stages and the HETS.

13-61. A solution of acetic acid (45.46 kg) in water (30 percent acetic acid) is to be extracted with isopropyl ether three times. Each stage will use 18.18 kg of solvent. Equilibrium data is given in Problem 13-59. Find quantities and compositions of the various streams. How much solvent would be needed if only one stage was used.

13-62. Isopropyl ether is used to remove acetic acid (30 percent) from water. The final acid concentration is 2 percent for a feed rate of 1.01 kg/sec. Equilibrium data are in Problem 13-59. Find minimum solvent rate. How many stages would be needed if 2.525 kg/sec of solvent are used?

REFERENCES

1. A. Kremser, *Nat. Petrol News* **22** (21), 42 (May 1930).
2. W. C. Edmister, *Ind. Eng. Chem.* **35**, 837 (1943).
3. T. K. Sherwood, R. L. Pigford, and C. R. Wilke, *Mass Transfer*, McGraw-Hill, New York (1975).
4. R. E. Treybal, *Mass Transfer Operations*, McGraw-Hill, New York (1980).
5. R. E. Treybal, *Liquid Extraction*, McGraw-Hill, New York (1951).
6. W. L. McCabe and J. C. Smith, *Unit Operations of Chemical Engineering*, McGraw-Hill, New York (1967).
7. A. S. Foust, L. A. Wenzel, C. W. Clump, L. Maus, and L. Anderson, *Principles of Unit Operations*, John Wiley and Sons, New York (1960).
8. W. L. Luyben and L. A. Wenzel, *Chemical Process Analysis*, Prentice-Hall, Englewood Cliffs, NJ (1988).
9. R. E. Treybal and Daley Weber, *Ind. Eng. Chem.* **38**, 817 (1946).
10. H. E. O'Connell, *Trans. AIChEJ.* **42**, 741 (1946).
11. F. J. Zuiderweg, H. Verberg, and F. A. H. Gilissen, *International Symposium on Distillation*, Institution of Chemical Engineers, London (1960), p. 151.
12. J. A. Gerster, *Chem. Eng. Prog.* **59**(3), 35 (1963).
13. V. S. Morello and N. Poffenberger, *Ind Eng. Chem.* **42**, 1021 (1950).
14. C. E. Morrell, W. J. Paltz, W. J. Packie, W. C. Asbury, and C. L. Brown, *Trans. Am. Inst. Chem. Eng.* **42**, 473 (1946).
15. D. G. Murdoch and M. Cuckney, *Trans. Inst. Chem. Eng. (London)* **24**, 90 (1946).
16. R. E. Treybal, *AIChEJ J.* **4**, 202 (1948).
17. R. G. Griskey, *Chemical Engineering for Chemists*, American Chemical Society, Washington, D.C. (1996).
18. C. J. King, *Separation Processes*, McGraw-Hill, New York (1971).

19. R. A. Findlay and A. Weedman, Separation and Purification by Crystallization, in *Advances in Petroleum Chemistry and Refining*, Vol. 1, K. A. Kobe and J. McKetta, editors, Interscience, New York (1958).
20. H. M. Schoen, editor, *New Chemical Engineering Separation Techniques*; Interscience, New York (1962).
21. A. Weissberger, editor, *Techniques of Organic Chemistry*, second edition, Vol. **3**; Interscience, New York (1956).
22. E. Rutner and Goldfinger, *Hearth Condensation and Evaporation of Solids*, Gordon and Breach, London (1964).
23. Gillot and Goldberger, *Chem. Eng. Prog. Symp. Ser.* **65** (91), 36 (1969).
24. R. N. Rickles, *Ind. Eng. Chem.* **58**, 19 (1966).
25. R. L. Dedrick, K. B. Bischoff, and E. Leonard, *Chem. Eng. Prog. Symp. Ser.* **64**, (84) (1968).
26. Li Spriggs, *Membrane Separation Processes*, Meares, editor, Elsevier, Amsterdam (1976).
27. G. B. Tuwiner, *Diffusion and Membrane Technology*; Reinhold, New York (1962).
28. M. Benedict and T. H. Pigford, *Nuclear Chemical Engineering*, McGraw-Hill, New York (1957).
29. K. S. Spiegler, *Salt Water Purification*, John Wiley and Sons, New York (1962).
30. K. S. Wilson, editor, *Demineralization by Electrodialysis*, Butterworth, London (1960).
31. K. S. Spiegler and Laird, editors, *Principles of Desalination*, Academic, New York (1980).
32. M. Bier, editor, *Electrophoresis*, Vols. 1 and 2; Academic, New York (1959).
33. D. W. Fursteneau, editor, *Froth Flotation*, AIME, New York (1962).
34. *Flotation*, AIME, New York (1976).
35. W. D. Van Arsdel, *Food Dehydration*, AVI, Westport, CT (1963).
36. R. B. McBride and D. L. McKinley, *Chem. Eng. Prog.* **61**, 81 (1965).
37. S. A. Stern, T. F. Sinclair, P. J. Gareis, N. P. Vahldieck, and P. H. Mohr, *Ind. Eng. Chem.* **57**, 49 (1965).
38. K. N. Altgelt, *Advances in Chromatography*, Vol. 7, Marcel Dekker, New York (1968).
39. R. Lemlich, editor, *Adsorptive Bubble Separation Techniques*; Academic, New York (1972).
40. R. Lemlich, *Ind. Eng. Chem.* **60**, 16 (1968).
41. B. L. Karger and D. G. DeVivo, *Separation Sci.* **3**, 393 (1968).
42. G. Burrows, *Molecular Distillation*: Oxford, London (1960).
43. W. J. Moore, *Physical Chemistry*; Prentice-Hall, Englewood Cliffs, NJ (1963).
44. Harris and Humphreys, *Spiegler Membrane Separation Processes*, Meares, editor, Elsevier, Amsterdam (1976).
45. U. Merten, editor, *Desalination by Reverse Osmosis*; M.I.T. Press, Cambridge, MA (1966).
46. A. S. Michaels, *Chem. Eng. Prog.* **64**, 31 (1968).
47. K. E. Grew and T. L. Ibbs, *Thermal Diffusion in Gases*, Cambridge, England (1952).
48. G. Vasaru, *Thermal Diffusion Column Theory and Practice*; VEB Deutscher Verlag, Berlin (1969).

49. H. K. Schachman, *Ultracentrifugation in Biochemistry*; Academic, New York (1959).
50. A. S. Michaels, *Ultrafiltration, Advances in Separations and Purifications*, E. S. Perry, editor, (Wiley, New York) (1968).
51. P. A. Schweitzer, editor, *Handbook of Separation Techniques for Chemical Engineers*, McGraw-Hill, New York (1979).
52. N. Li, editor, *Recent Developments in Separation Science*, Vol. II, CRC Press, Cleveland (1972).
53. W. G. Pfann, *Trans. Am. Inst. Mech. Eng.* **194**, 747 (1952).
54. W. G. Pfann, *Zone Melting*, second edition, John Wiley and Sons, New York (1966).
55. B. D. Smith, *Design of Equilibrium Stage Processes*, McGraw-Hill, New York (1963).

14

MECHANICAL SEPARATIONS

INTRODUCTION

As we have discussed the various aspects of mass transfer, we have emphasized its analogies to heat transfer. Basically, we can tabulate the various aspects of heat transfer and generally relate them to corresponding mass transfer situations. Such a comparison is given in Table 14-1.

There is, of course, still another heat transfer situation, namely, radiative heat transfer. Radiation has no parallel process in terms of mass transfer.

In addition, there is an aspect of mass transfer which has no parallel system in heat transfer. This additional mass transfer area is that of mechanical separation. These represent separation processes that use mechanical, electrical, or magnetic forces to bring about a separation. Mechanical (used generically to include electrical and magnetic forces) separations can be used for liquid–solid, liquid–liquid, gas–liquid, gas–solid, and solid–solid systems. Table 14-2 summarizes the principal separation processes, the phases involved, and the means of separation.

FILTRATION

Filtration is one of the most widely used of all of the mechanical separation processes. Basically, it involves the flow of a liquid–solid or gas–solid system through a porous medium.

In the case of the liquid–solid system the filtration process takes place through a filter cake made up of a filter aid and the solids themselves. The former is made up of hard, fine particles applied either as a coating on a filter cloth

Table 14-1 Comparison of Heat and Mass Transfer Operations

Heat Transfer	Mass Transfer
Steady-state heat transfer	Steady-state molecular diffusion
Unsteady-state heat transfer	Unsteady-state molecular diffusion
Convective heat transfer (heat transfer coefficient)	Convective mass transfer (mass transfer coefficients)
Convective heat transfer (heat transfer coefficient)	Equilibrium staged operations (convective mass transfer using departure from equilibrium as a driving force)
Radiative heat transfer (not analogous with other transfer processes)	Mechanical separations (not analogous with other transfer processes)

Table 14-2 Mechanical, Electrical, and Magnetic Separations

Process	Phase	How Separated
Filtration	Liquid–solid	Pressure reduction
Centrifugation	Liquid–solid or liquid–liquid	Centrifugal force
Sedimentation	Liquid–solid	Gravity
Cyclone separator	Gas–solid or Gas–liquid	Flow
Electrostatic precipitator	Gas–solid	Electric field
Demister	Gas–solid or gas–liquid	Pressure reduction
Magnetic separator	Solid–solid Solid–liquid	Magnetic field
Screening	Solid–solid	Size of particles

(see Figure 14-1) or in the liquid–solid mix (slurry). A typical filter aid is diatomaceous earth, the fossilized remains of very small marine creatures.

A general expression for the pressure gradient through a filter cake is

$$\frac{(-\Delta P)_f g_c}{L} = k\frac{(1-\epsilon)^2 \mu v_s}{\epsilon^3 D_p^2} \tag{14-1}$$

where k' is a constant, μ is the filtrate viscosity, v_s is the superficial velocity of the filtrate, ϵ is the cake porosity, and D_p is the average particle diameter.

If the flow through the cake is considered to be laminar (usually correct), then

$$\frac{(-\Delta P)_f g_c}{L} = 180\frac{(1-\epsilon)^2}{\epsilon^3}\frac{\mu v_s}{D_p^2} \tag{14-2}$$

Also,

$$D_p = \frac{6}{\frac{A_p}{V_p}} = \frac{6}{S_0} \tag{14-3}$$

where S_0 is the specific surface area of the particle. A_p and V_p are, respectively, the surface and volume of a single particle.

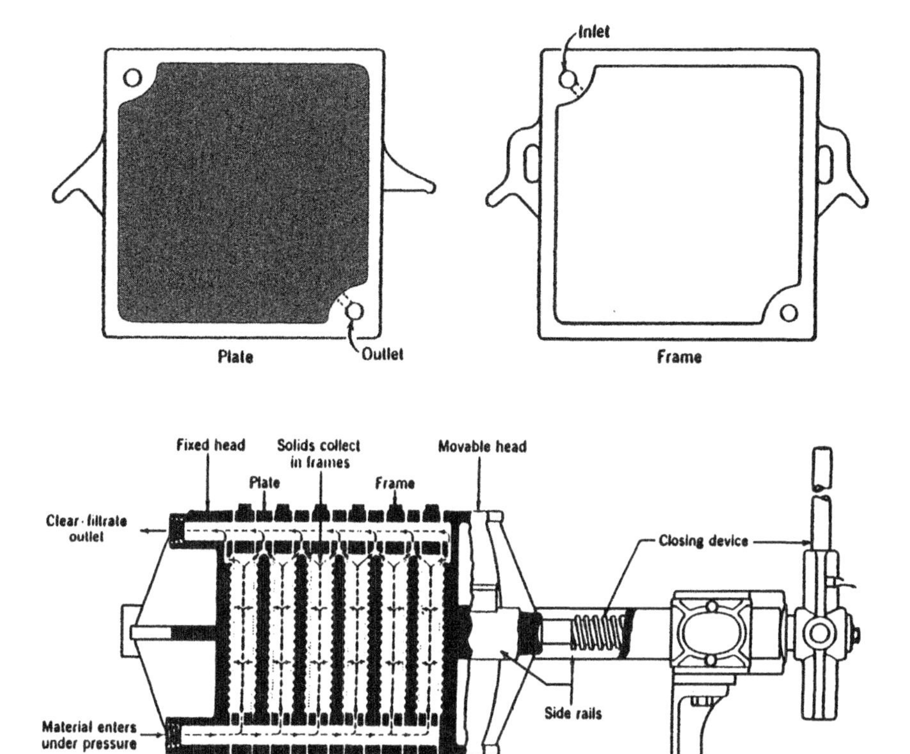

Figure 14-1. Plate and frame filter press. (Courtesy of T. Schriver and Co.)

Substituting (14-3) in (14-2) gives

$$\frac{(-\Delta P)_f g_c}{L} = \frac{5(1-\epsilon)^2 \mu v_s S_0^2}{\epsilon^3} \tag{14-4}$$

which is equation (4-1) with k' equal to 5.

If we solve for superficial velocity v_s, then

$$v_s = \frac{(-\Delta P)_f g_c \epsilon^3}{5L\mu S_0^2 (1-\epsilon)^2} = \frac{1}{A}\left(\frac{dV}{d\theta}\right) \tag{14-5}$$

where A is the filtration area and $dV/d\theta$ is the volumetric filtration rate.

By considering the relationship between the filter cake length L (thickness) and the filtrate volume, we obtain

$$LA(1-\epsilon)\rho_s = w(V+\epsilon LA) \tag{14-6}$$

where ρ_s is the solid density in the cake and w is the mass of solids in the feed slurry per volume of liquid in that slurry.

If the volume of filtrate held in the filter cake (ϵLA) is neglected, then

$$\frac{1}{A}\frac{dV}{d\theta} = \frac{(-\Delta P)_f g_c \epsilon^3}{5\dfrac{wV}{A\rho_s}\mu(1-\epsilon)S_0^2} = \frac{(-\Delta P)_f g_c}{\dfrac{\alpha\mu wV}{A}} \tag{14-7}$$

where α (the specific cake resistance) is $5(1-\epsilon)S_0^2/\rho_s\epsilon^3$.

For the actual filter unit we should also include any other flow channels that the system involves (i.e., additional segments of the equipment). If we designate these as R_M, then

$$\frac{dV}{Ad\theta} = \frac{(-\Delta P_t)g_c}{\mu\,(\alpha wV/A + R_M)} \tag{14-8}$$

where $(-\Delta P_t)$ is the *total* pressure drop.

Furthermore, if we switch to a V_e that represents the filtrate volume needed to build up a filter cake that would equal the resistance of the filter medium and any pipe conduits leading to the pressure taps, we obtain

$$\frac{dV}{Ad\theta} = \frac{(-\Delta P_t)g_c}{\dfrac{\mu\alpha w}{A}(V+V_e)} \tag{14-9}$$

Integrating equation (14-9) gives

$$\theta = \frac{\mu\alpha w}{g_c A^2(-\Delta P_t)}\left(\frac{V^2}{2} + V_e V\right) \tag{14-10}$$

Equation (14-10) allows us to find the time needed for a given volume of filtrate. In order to use equation (14-10), it is necessary to know α and V_e. The former quantity can be found when ϵ and S_0 are known. V_e, however, must be found in some other way. Generally both α and V_e are determined from pilot filtration studies. A slurry that matches the actual one is filtered under conditions that come as close to actual plant requirements as possible. Then if we plot the reciprocal of the filtration rate versus the volume of filtrate collected, we obtain a linear relation [i.e., equation (14-9) inverted].

$$\frac{d\theta}{dV} = \frac{\mu\alpha w}{g_c A^2(-\Delta P_t)}(V+V_e) \tag{14-11}$$

A plot of $d\theta/dV$ versus the volume of filtrate collected gives a straight line whose slope is $\alpha\mu w/g_c A^2(-\Delta P_t)$ with a Y-axis intercept of $[\mu\alpha w/g_c A^2(-\Delta P_t)]$ $V_e V$. Hence we find from the slope and the intercept V_e.

For the other case, that of constant rate, we use the relation

$$-\Delta P_t = \frac{\mu\alpha w}{g_c A^2}\left(\frac{dV}{d\theta}\right)(V + V_e) \tag{14-12}$$

to obtain a linear relation for $(-\Delta P_t)$ versus V. The slope then is $(\mu\alpha w/g_c A^2)/(dV/d\theta)$, and the intercept the slope times V. Both α and V_e can then be determined to find the time needed for filtration.

$$\theta = \frac{\mu\alpha w}{g_c A^2(-\Delta P_t)}\left(\frac{V^2}{2} + V_e V\right) \tag{14-13}$$

In compressible cakes the α and V_e terms become functions of pressure so that equation (14-9) becomes

$$\frac{dV}{Ad\theta} = \frac{-g_c A \rho_s}{5w\mu V}\int_0^{P_1-P_2} \frac{\epsilon^3}{(1-\epsilon)S_0^2}dP \tag{14-14}$$

where $P_1 - P_2$ is the difference between the pressure on the cake surface and that at the face of the filter medium.

The use of a specific cake resistance (α_p) transforms equation (14-14) to

$$\frac{dV}{Ad\theta} = \frac{-g_c A}{wV\mu}\int_0^{P_1-P_2} \frac{dP}{\alpha_p} \tag{14-15}$$

Experiments relating compression and permeability allow the interrelation of P and α_p, ϵ, and S_0 to be determined experimentally. Figure 14-2, 14-3, and 14-4 show typical data of this type.

Industrial filtration units can be either *batch* or *continuous* in operation. In addition to the plate and frame filter press of Figure 14-1, there are leaf filters that also function in the batch mode. Figure 14-5 shows both a vertical leaf filter and the filter leaf itself. Other leaf filters include a Niagara type (Figure 14-6), which uses a horizontal leaf filter, and the Sweetland type (Figure 14-7), which consists of two half-cylinders that allow the bottom half to be opened downward to access vertical disk-shaped leaves.

Continuous filters include rotary horizontal vacuum filters (Figure 14-8), rotary drum filters (Figure 14-9), and rotary disk units (Figure 14-10).

A different mode of filtration is used in centrifugal filtrations (Figure 14-11). These units use centrifugal force as the driving force for the fluid.

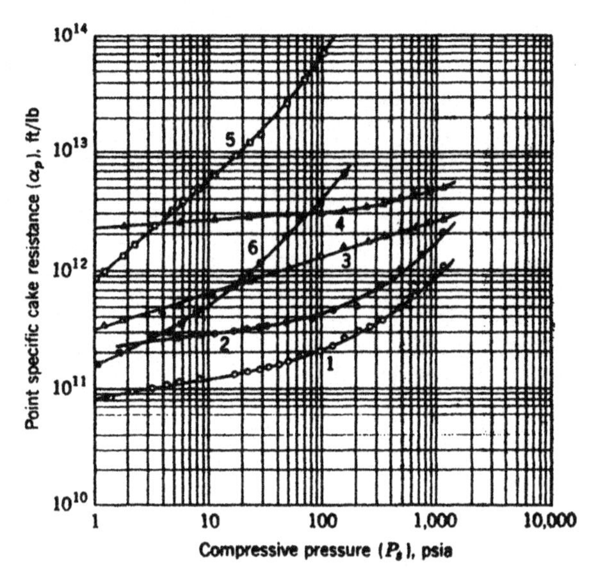

Figure 14-2. Point specific cake resistance versus compressive pressure. Codes: 1, 2 Superlite CaCO, pH values of 9.8 and 10.3; 3, 4 TiO, pH values of 7.8 and 3.5; 5,6 ZnS, pH values of 9.1. Curves 1, 3 for flocculated systems. (Reproduced with permission from reference 1. Copyright 1953, American Institute of Chemical Engineers.)

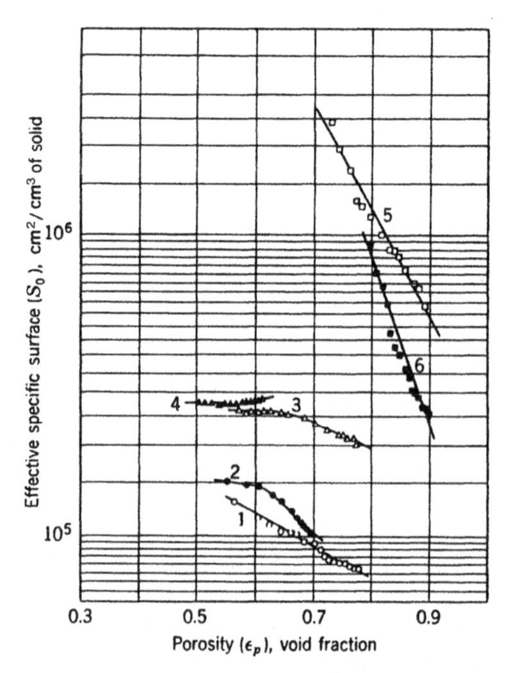

Figure 14-3. Effective surface versus porosity. Codes as in Figure 14-2. (Reproduced with permission reference 1. Copyright 1953, American Institute of Chemical Engineers.)

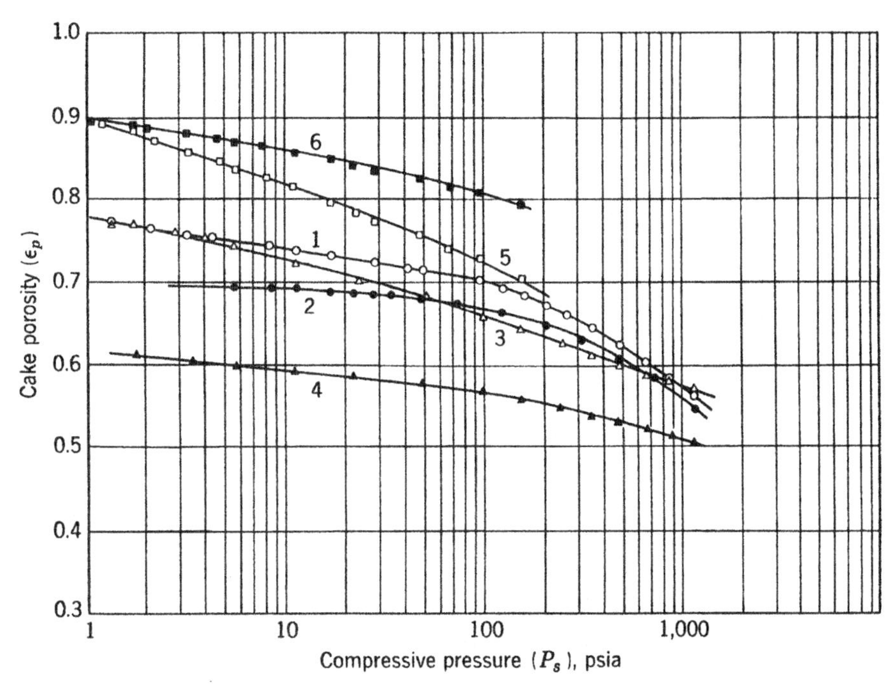

Figure 14-4. Cake porosity versus compressive pressure. Codes as in Figure 14-2. (Reproduced with permission from reference 1. Copyright 1953, American Institute of Chemical Engineers.)

Here the solution for volumetric rate gives

$$\frac{dV}{d\theta} = \frac{\pi h \rho \omega^2 (r_3^2 - r_1^2) - \dfrac{\rho (dV/d\theta)^2}{4\pi h}\left(\dfrac{1}{r_3^2} - \dfrac{1}{r_1^2}\right)}{\mu \left[\alpha \rho_s (1-\epsilon) \ln \dfrac{r_3}{r_2} + \dfrac{R'_M}{r_3}\right]} \tag{14-16}$$

where P_2, P_3 = pressures at the filter-cake surface and at the filter-cloth surface, respectively

r_2, r_3 = radii to the filter-cake surface and to the filter-cloth surface, respectively

ρ, ρ_s = density of filtrate and of solids, respectively

ϵ_p = porosity at any point in the bed

α_p = specific cake resistance at any point in the bed

h = height of cylindrical surface on which the cake is being built

$(dV/d\theta)$ = filtration rate

ω = rate of rotation, radians/sec

R'_M = resistance of the filter medium

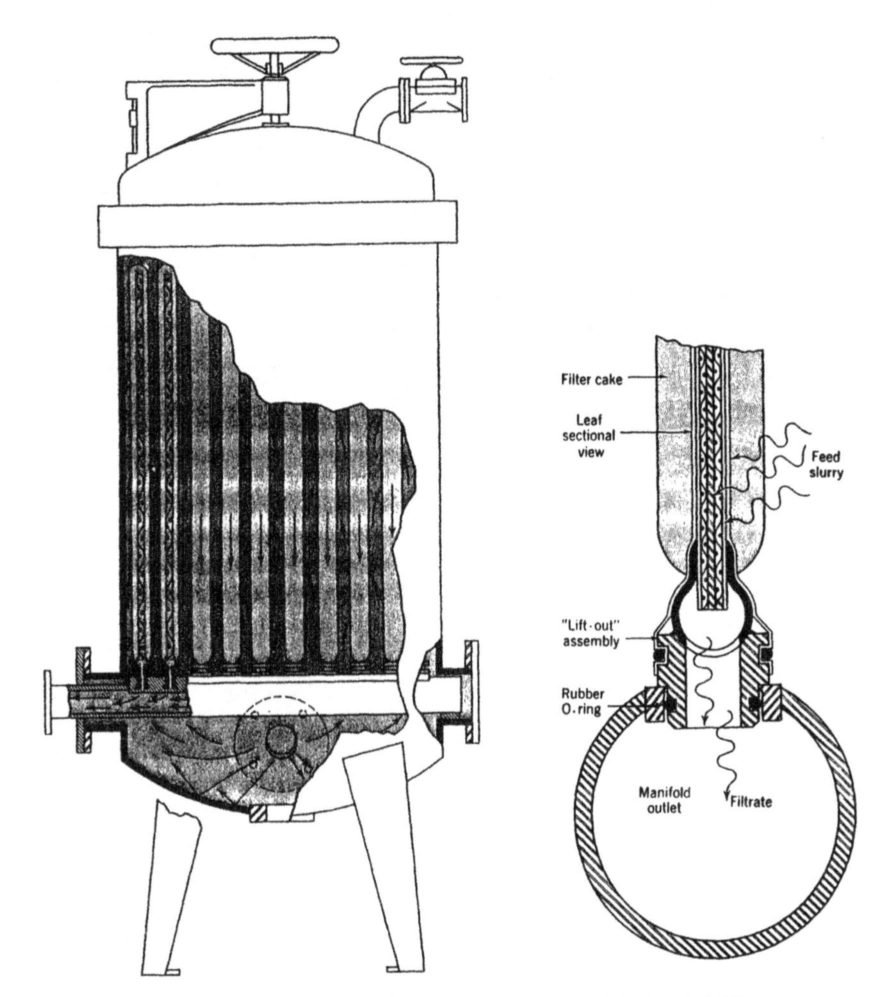

Figure 14-5. Vertical leaf filter and filter leaf. (Courtesy of Industrial Filter and Pump Mfg. Co.)

If we neglect the $(dV/d\theta)^2$ term (i.e., kinetic energy changes) in the numerator of equation (14-16) and use $\omega = \pi N/30$ (where N is the speed in revolutions per minute), we obtain

$$\frac{dV}{d\theta} = \frac{\pi^3 N^2 \rho h (r_3^2 - r_1^2)}{(30)^2 \mu \left[\alpha \rho_s (1 - \epsilon) \ln \dfrac{r_3}{r_2} + \dfrac{R'_M}{r_3} \right]} \tag{14-17}$$

The pressure in the unit is given by

$$dP_g = \frac{\rho \omega^2 r\, dr}{g_c} \tag{14-18}$$

Figure 14-6. Horizontal leaf filter. (Courtesy of Niagara Filter Division, American Machine and Metals Inc.)

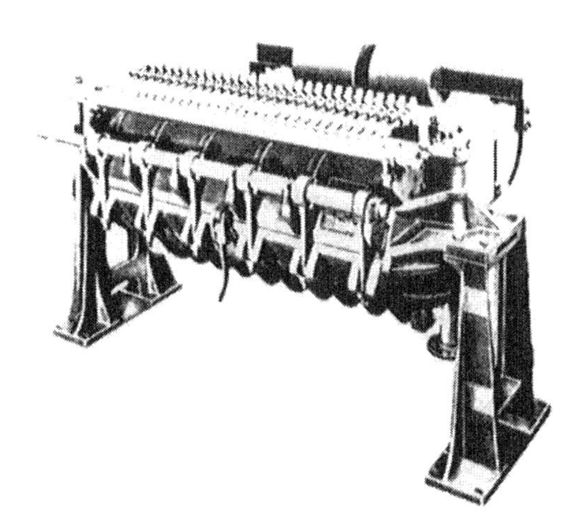

Figure 14-7. Sweetland pressure filter. (Courtesy of GL & V/Dorr-Oliver.)

which, when integrated for the filter cake and the liquid above it, yields

$$\Delta P_g = \frac{\rho \omega^2}{g_c} \frac{r_3^2 - r_1^2}{2} \tag{14-19}$$

If we combine equations (14-17) and (14-19) with a solids mass balance

$$\rho_s(1 - \epsilon) = \frac{wV}{\pi h(r_3^2 - r_2^2)} \tag{14-20}$$

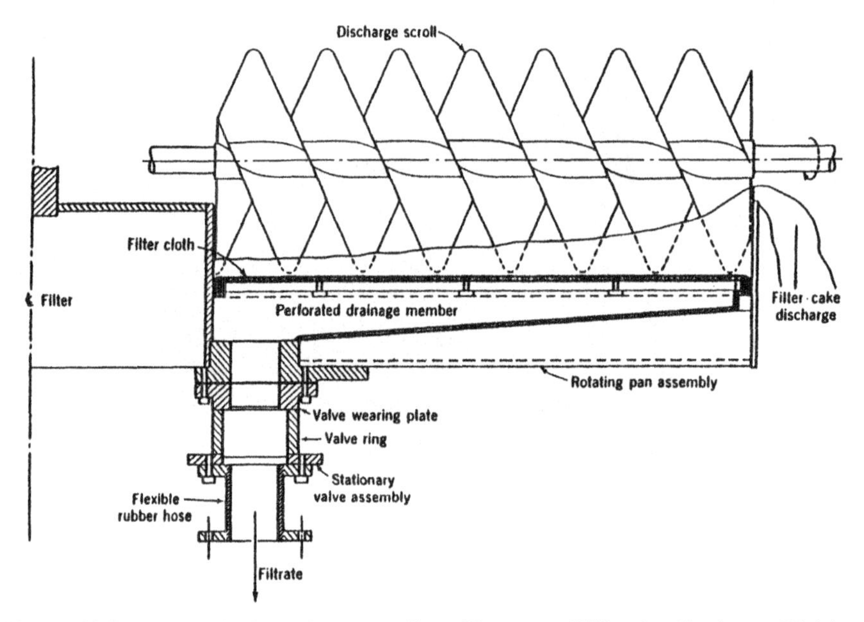

Figure 14-8. Rotary horizontal vacuum filter. (Courtesy of Filtration Engineers Division, American Machine and Metals Inc.)

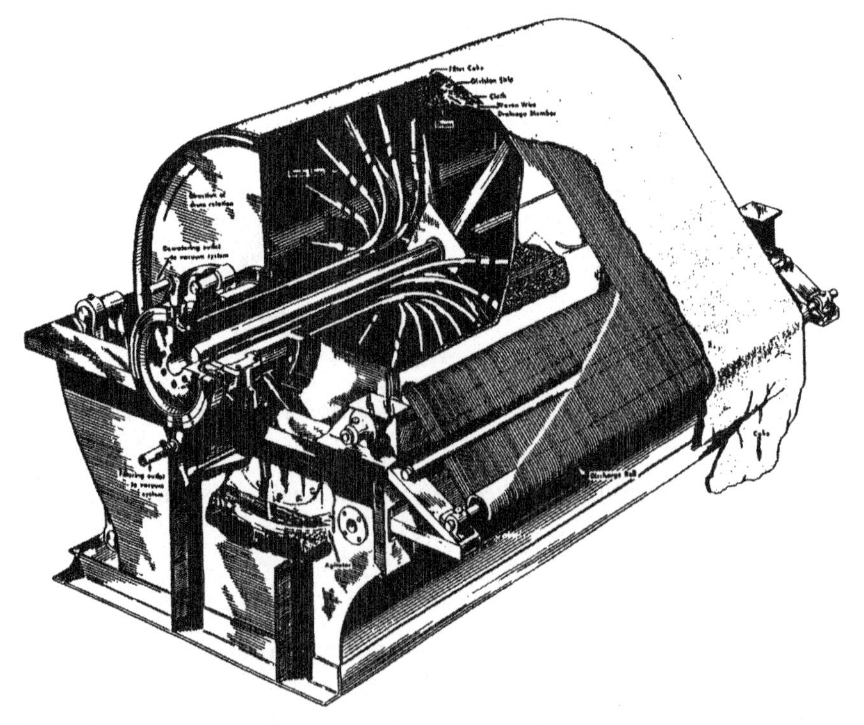

Figure 14-9. Rotary drum vacuum filter. (Courtesy of filtration Engineers Division, American Machine and Metals Inc.)

Figure 14-10. Rotary disk vacuum filter. (Courtesy GL & V/Dorr-Oliver.)

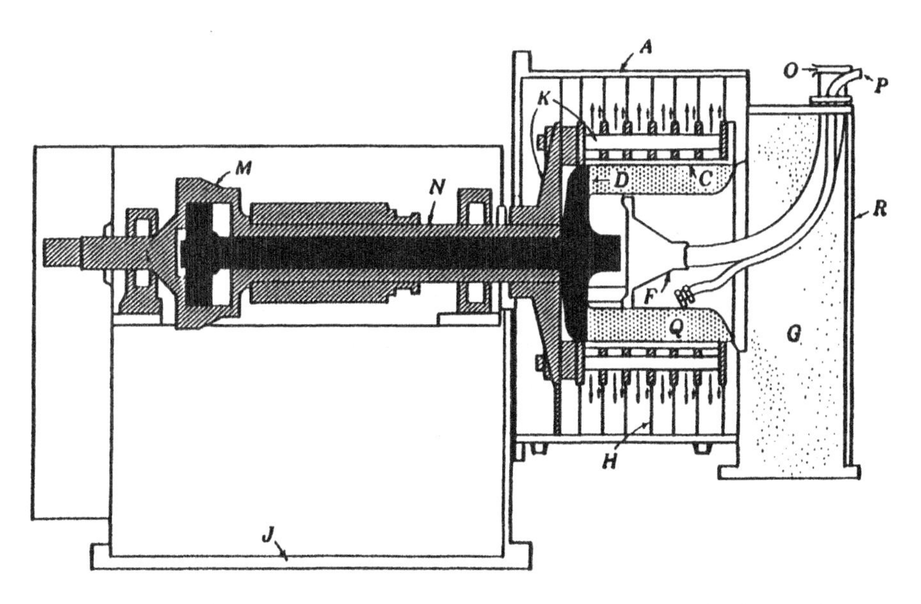

A—filtrate housing *C*—filter screen *D*—cake pusher *F*—feed funnel *G*—cake-discharge chute *H*—wet housing separators *J*—base *K*—basket *M*—hydraulic servo motor *N*—piston rod *O*—slurry feed pipe *P*—wash feed pipe *Q*—filter cake *R*—access door.

Figure 14-11. Continuous centrifugal filter. (Courtesy of Baker-Perkins Inc.)

we obtain

$$\frac{dV}{d\theta} = \frac{\Delta P_g g_c}{\mu\left(\dfrac{\alpha w V}{\left(\dfrac{2\pi h(r_3 + r_2)}{2}\right)\left(\dfrac{2\pi h(r_3 - r_2)}{\ln(r_3/r_2)}\right)} + \dfrac{R'_M}{2\pi h r_3}\right)} \tag{14-21}$$

Substituting for the various areas finally yields

$$\frac{dV}{d\theta} = \frac{\Delta P_g g_c}{\mu\left(\dfrac{\alpha w V}{A_m A_{lm}} + \dfrac{R'_M}{A_c}\right)} \tag{14-22}$$

where A_m, A_{lm}, and A_c are, respectively, the mean, logarithmic mean, and filter cloth areas.

CENTRIFUGATION

The centrifuge is a widely used piece of process equipment that can separate either liquid–solid or liquid–liquid systems. Centrifugal force is the means used to effect separation. The force balance for a particle moving in a centrifugal field is

$$\frac{dv}{d\theta} = r\omega^2\left(\frac{\rho_s - \rho}{\rho_s}\right) - \frac{C_D \rho v^2 S}{2\,m} \tag{14-23}$$

where C_D is the drag coefficient, S is the area of the solid projected normal to flow, and m is the particle mass. If we take C_D for the rate of fall of the smallest particles, we have Stokes Law flow and

$$C_D = 24/\mathrm{Re}_p \tag{14-24}$$

Also, for a spherical particle we have

$$\frac{m}{S} = \frac{\rho_s V}{S} = \frac{\dfrac{\pi}{6} D_p^3 \rho_s}{\dfrac{\pi}{4} D_p^2} = \frac{4\rho_s D_p}{6} \tag{14-25}$$

If we substitute equations (14-24) and (14-25) into equation (14-23), we obtain

$$\frac{dv}{d\theta} = r\omega^2\left(\frac{\rho_s - \rho}{\rho_s}\right) - \frac{18\mu v}{\rho_s D_p^2} \tag{14-26}$$

The centrifugal field strength changes radially. This means that particle terminal settling velocities become functions of the radial positions. At any given position, $dv/d\theta = 0$ is zero; and for any instant for the movement of a single particle, dv/dr is positive.

For a given position $dv/d\theta = 0$ we have

$$v_R = \frac{r\omega^2(\rho_s - \rho)D_p^2}{18\mu} \tag{14-27}$$

where v_R = the terminal falling velocity of spherical particles of diameter D_p at radius r in a centrifugal field rotating at rate ω

For a differential time $(d\theta)$

$$v_R d\theta = dr = \frac{r\omega^2(\rho_s - \rho)D_p^2}{18\mu} d\theta \tag{14-28}$$

Integration yields

$$\ln\frac{r_2}{r_1} = \frac{\omega^2(\rho_s - \rho)D_p^2}{18\mu}\theta = \frac{\omega^2(\rho_s - \rho)D_p^2}{18\mu} \cdot \frac{V}{Q} \tag{14-29}$$

where V = volume of material held in the centrifuge
Q = volumetric feed rate to the centrifuge
V/Q = residence time of a particle in the centrifuge

When the liquid layer in a centrifuge is very narrow compared to the radius, the centrifugal field can be taken as constant and

$$v_R\theta = x = \frac{r\omega^2(\rho_s - \rho)D_p^2}{18\mu} \cdot \frac{V}{Q} \tag{14-30}$$

The x is the radial distance traveled by a particle (D_p) during the residence time.

When x is taken as $[(r_2 - r_1)/2]$, half of the particles of a given diameter D'_p will remain in the suspension and half will separate at the wall. The D'_p is then given by

$$D'_p = \sqrt{\frac{9\mu Q}{(\rho_s - \rho)\omega^2 V} \cdot \frac{r_2 - r_1}{r}} \tag{14-31}$$

where $r_2 - r_1$ = the thickness of the liquid layer
D'_p = critical particle diameter

If the liquid layer thickness is such that the change of centrifugal field with radius becomes important, an effective value of $r_2 - r_1/r$ is used, $2\ \ln(r_2/r_1)$.

Some typical centrifuges are shown in Figures 14-12, 14-13, and 14-14. The disk bowl centrifuge (Figure 14-12) is equipped with vanes through which the solids move. Internal solid motion is shown in Figure 14-13 for a solid bowl centrifuge. The tubular bowl centrifuge (Figure 14-14) can be used to separate

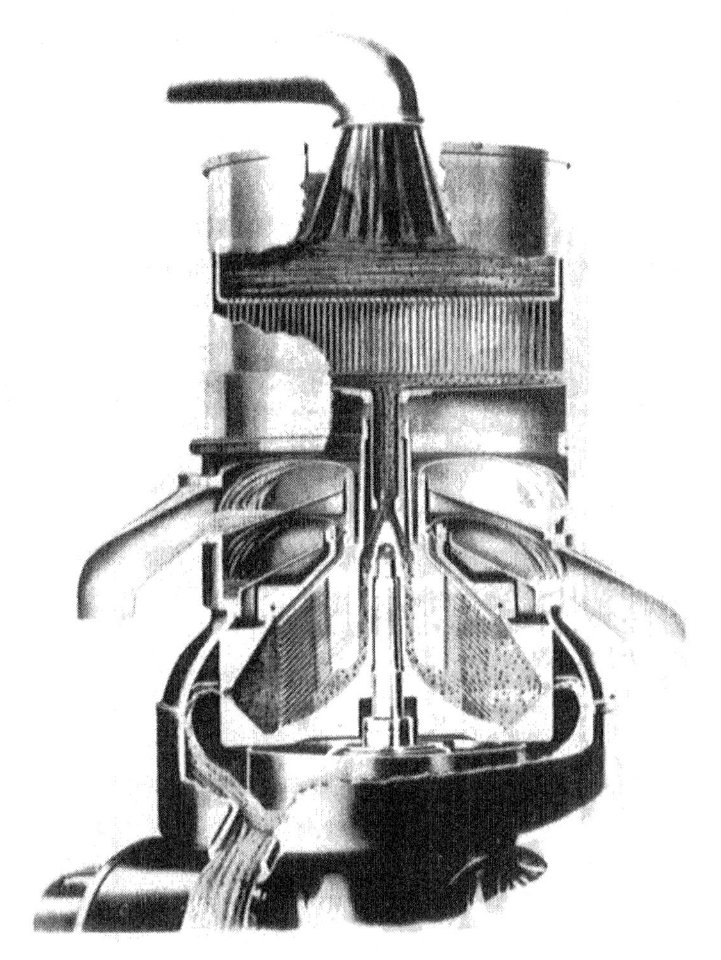

Figure 14-12. Disk bowl centrifuge. (Courtesy of DeLaval Separator Company.)

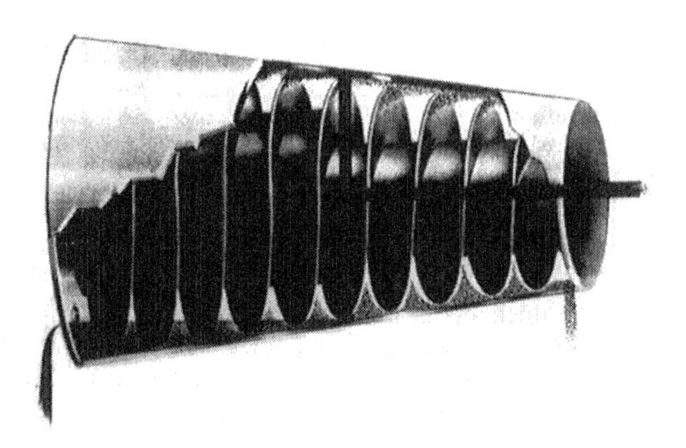

Figure 14-13. Solid bowl centrifuge. (Courtesy of Sharples Corp.)

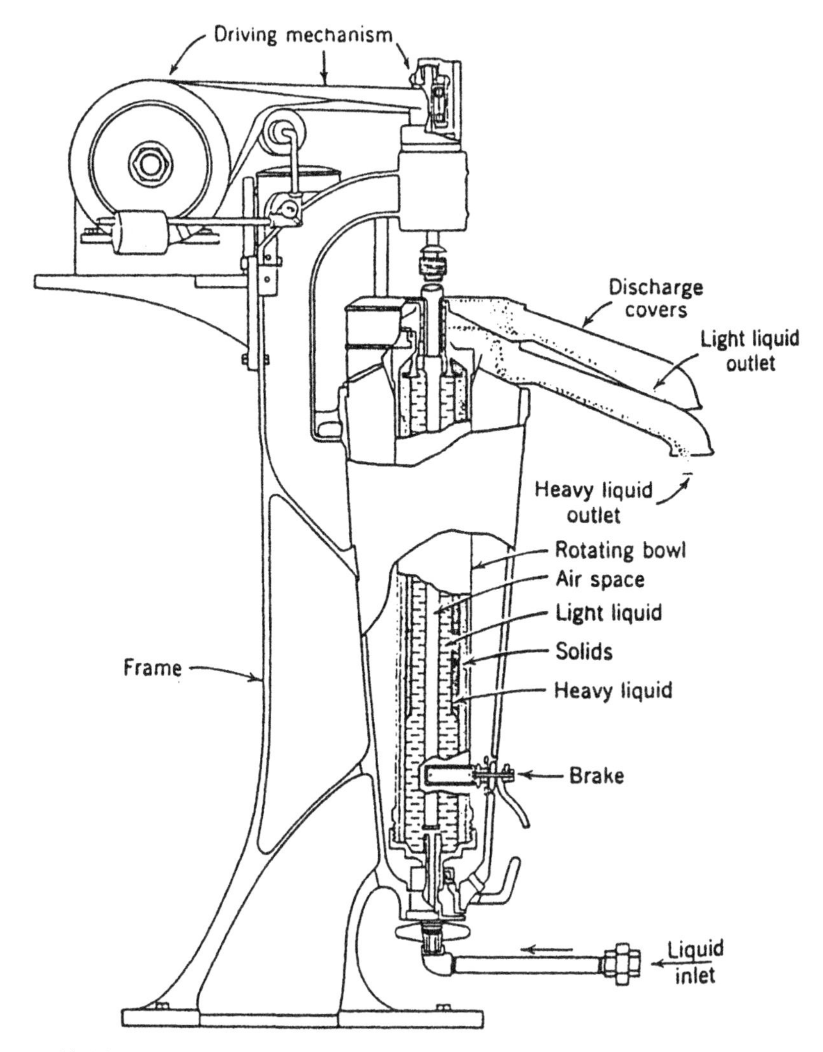

Figure 14-14. Tubular bowl centrifuge. (Adapted with permission from reference 2. Copyright 1960, John Wiley and Sons, Inc.)

either liquid–solid or liquid–liquid systems. Different internal configuration are used (Figure 14-15) for removing solids or separating liquids. In the latter case the radii of Figure 14-15 can be related to the densities of the light and heavy liquid phases;

$$\frac{r_2^2 - r_4^2}{r_2^2 - r_1^2} = \frac{\rho_l}{\rho_h} \tag{14-32}$$

where ρ_l = density of the light phase
ρ_h = density of the heavy phase

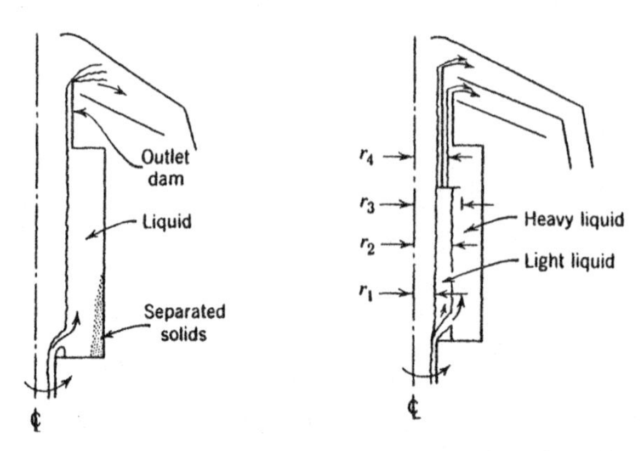

Figure 14-15. Tubular bowl centrifuge internal configurations for solid–liquid and liquid–liquid separations. (Adapted with permission from reference 2. Copyright 1960, John Wiley and Sons, Inc.)

An interesting scale-up technique exists for centrifuges. Essentially, it indirectly compares a given centrifuge to a settling tank that will perform the same function. This is done by using the cross-sectional surface area, $\sum$, of such a settling tank. The

$$\frac{Q_1}{\sum_1} = \frac{Q_2}{\sum_2} \tag{14-33}$$

$\sum$ values can be calculated (1) for tubular bowl

$$\sum = \frac{\pi \omega^2 l}{g} \frac{(r_2^2 - r_1^2)}{\ln \dfrac{r_2^2}{r_1^2}} \tag{14-34}$$

where l = the bowl length

and for disk bowl centrifuges

$$\sum = \frac{2n\pi(r_2^3 - r_1^3)\omega^2}{3\ g \tan \Omega} \tag{14-35}$$

where n = number of spaces between disks in the stack
r_2, r_1 = outer and inner radii of the disk stack
Ω = the conical half-angle

SEDIMENTATION TANKS AND THICKENERS

Sedimentation is a process that uses gravity as a separating force. The process that produces a solid and a liquid product also functions as a thickener. In

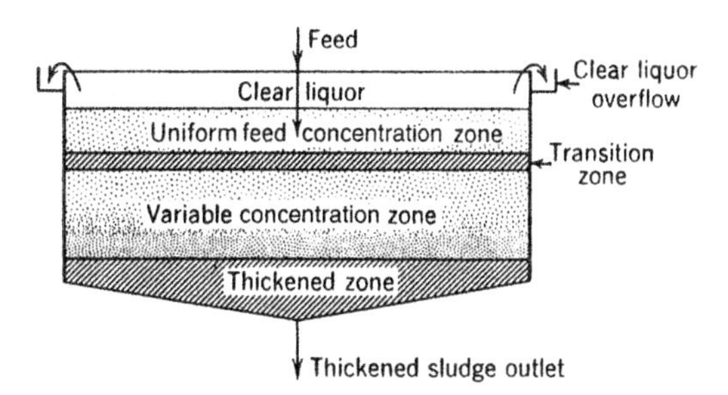

Figure 14-16. Schematic of continuous sedimentation process. (Adapted with permission from reference 2. Copyright 1960, John Wiley and Sons, Inc.)

practice, sedimentation can be carried out either as a continuous or batch process. Figure 14-16 is a schematic of the settling zones in a continuous device.

Design of a thickener is based on determining the minimum cross-sectional area giving solid passage with a limiting intermediate concentration. In the thickener, the upward velocity of the limiting layer is a constant value. This together with an experimentally determined plot of interface height Z versus time t (Figure 14-17), and the values of Z_0, Z_i and Z_L make the design possible

The settling velocity ν_L can be found from the equation

$$\frac{z_i - z_L}{\theta_L} = \nu_L \tag{14-36}$$

This equation together with the balance equation

$$c_L = \frac{c_0 z_0}{z_L + \nu_L \theta_L} \tag{14-37}$$

give

$$c_L z_i = c_0 z_0 \tag{14-38}$$

where C_L is the concentration at Z_L and θ_L while C_0 and Z_0 are initial conditions.

Thickener area is given by the relation

$$\frac{L_L c_L}{S} = \frac{\nu}{\left[\dfrac{1}{c_L} - \dfrac{1}{c_u}\right] \dfrac{\rho_{av}}{\rho_w}} \tag{14-39}$$

The depth of a thickener is determined by first estimating the volume of the compression zone of the thickened mass:

$$V = \frac{L_0 c_0}{\rho_s}(\theta - \theta_c) + \frac{L_0 c_0}{\rho} \int_{\theta_c}^{\theta} \frac{W_l}{W_s} d\theta \tag{14-40}$$

where V = compression zone volume

$L_0 c_0$ = mass of solids fed per unit time to thickener

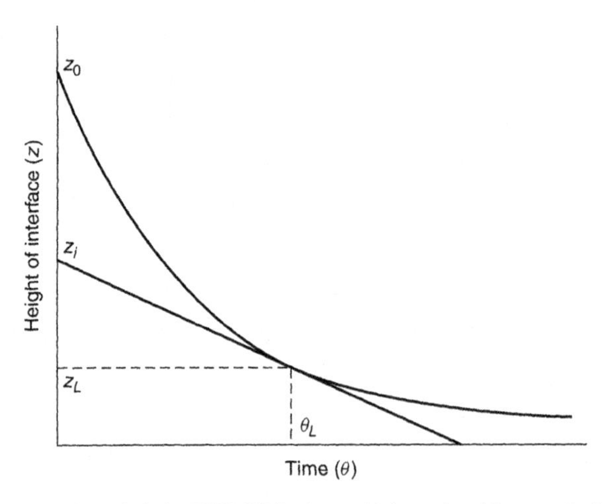

Figure 14-17. Interface height VERSUS time. (Adapted with permission from reference 2. Copyright 1960, John Wiley and Sons.)

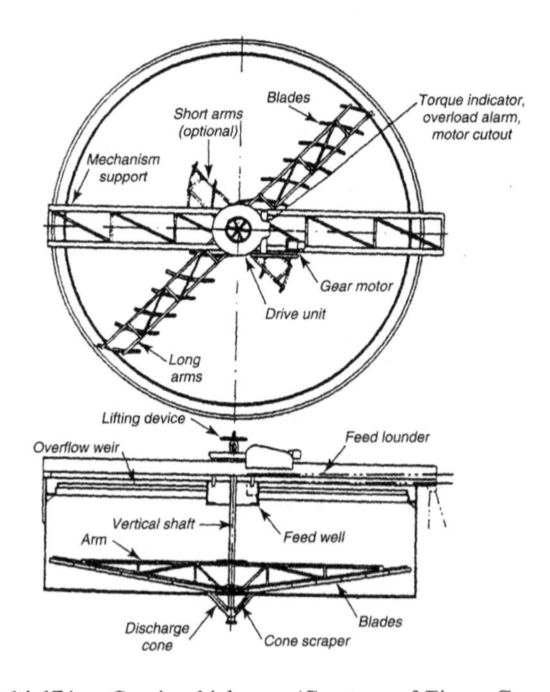

Figure 14-17A. Gravity thickener. (Courtesy of Eimco Corporation.)

W_l = mass of liquid in compression zone
W_s = mass of solids in compression zone
$(\theta - \theta_c)$ = compression zone retention time
ρ_s = density of solid phase
ρ = density of liquid phase

Table 14-3 Estimated Heights for Thickener Design

Item	Height Ranges (Meters)
Bottom pitch	0.3048 to 0.6096
Storage capacity	0.3048 to 0.6096
Feed submergence	0.3048 to 0.9144

The height needed for the compression zone is then obtained by dividing V by S. In addition, heights are added for bottom pitch, storage capacity, and feed submergence. These are given in Table 14-3.

CYCLONE SEPARATORS, ELECTROSTATIC PRECIPITATORS, OTHER DEVICES

The cyclone separator (Figure 14-18) is a device that uses centrifugal forces to accelerate the settling of solid particles. Basically, a particulate-laden gas stream is introduced tangentially in the unit. Particles move to the cyclone separator walls and then to the bottom of the device. The gas first moves downward, then up to the top.

Design of the unit involves a force balance between the centrifugal force on a particle at r and the frictional drag of the gas. This yields

$$\frac{V_t}{r} = \frac{\mu 18 V_r}{\rho_s D_p^2} \tag{14-41}$$

where V_t and V_r are, respectively, the tangential and radial velocity components of the gas; μ is the gas viscosity; ρ_s is the solid density; D_p is the particle diameter, and r is the radius in the separator.

The particle's Stokes' Law velocity is

$$V_0 = \frac{D_p^2 g \rho_s}{18\mu} \tag{14-42}$$

so that

$$V_0 = \frac{V_r}{V_t^2} r g \tag{14-43}$$

Assuming that a particle is separated if it rotates outside a core of diameter of 0.4 d_0 (the outlet diameter), then since $r = 0.2d_0$ we obtain

$$V_0 = \frac{V_r}{V_t}(0.2d_0)g \tag{14-44}$$

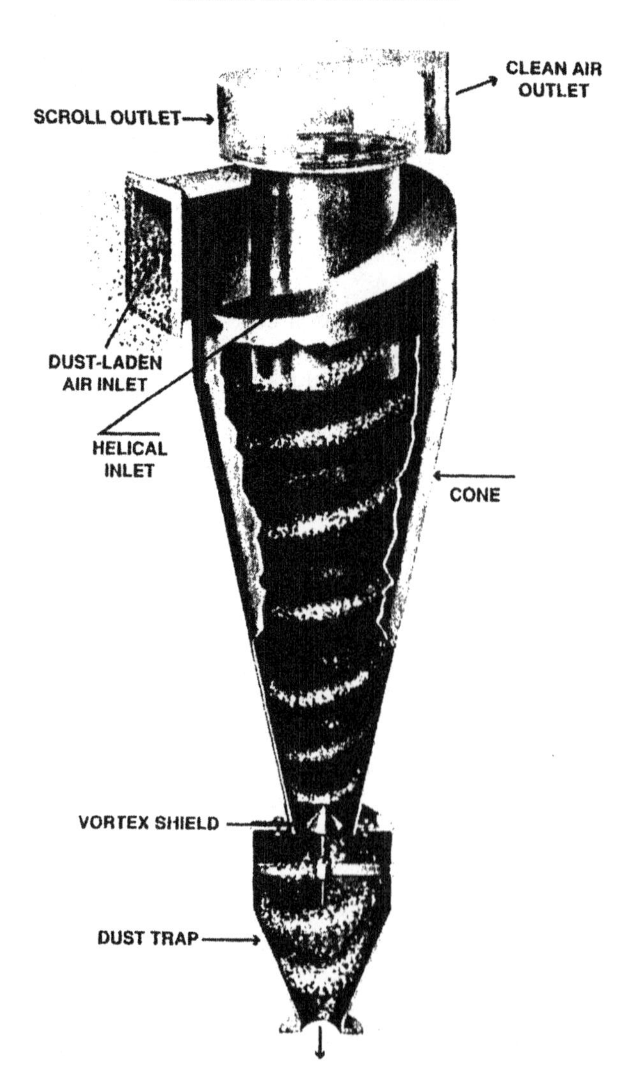

Figure 14-18. Cyclone separator. (Courtesy of Ducon Technologies, Inc.)

Next we write expressions for V_r and V_t:

$$V_r = \frac{\mathrm{W}}{2\pi r Z \rho} \tag{14-45}$$

$$V_t = V_{t0}\sqrt{\frac{d_t}{2r}} \tag{14-46}$$

where W is the gas mass flow rate through the unit, Z is the separator depth, and d_t is the diameter of the cyclone.

Combining equations (14-44), (14-45), and (14-46) gives an equation for the V_0:

$$V_0 = \frac{0.2A_i^2 d_0 \rho g}{\pi Z d_t G} \tag{14-47}$$

where A_i is the inlet cross-sectional area.

Cyclone separators operate as do all mechanical separators with a given efficiency. This is shown in Figure 14-19. A selected diameter called the *cut diameter* is the point at which 50 percent of the given material is removed. Note that only about 90 percent of particles four times the cut diameter are removed. Indeed particles up to eight times the cut diameter find their way through the system.

The electrostatic precipitator (see Figure 14-20) operates by ionizing the inlet gas stream. These ions become attached to the particles that are carried by means of electric field to the collector. Electrostatic precipitators, while highly efficient for small particles, have high capital and operating costs. As such, they are usually put in place after the cyclone separator, which removes larger particles.

Magnetic separators use a magnetic field to transfer mass. Materials attracted by such a field are termed *paramagnetic*. Those substances repelled by a magnetic field are diamagnetic. The separations can be carried out on either a dry or wet basis. Dry separations require a system that flows freely and free of dust. The wet processes are used mainly for fine particles.

Screening involves the separation of solid–solid systems by size. The systems used for larger particles (larger than 0.0254 m) are shaking screens and grizzlies. Vibrating screens or oscillating screens and used for finer particles.

Grizzlies are sets of bars arranged in parallel manner. The overall unit is sloped at an angle between 20 and 50 degrees. The motion of shaking screens

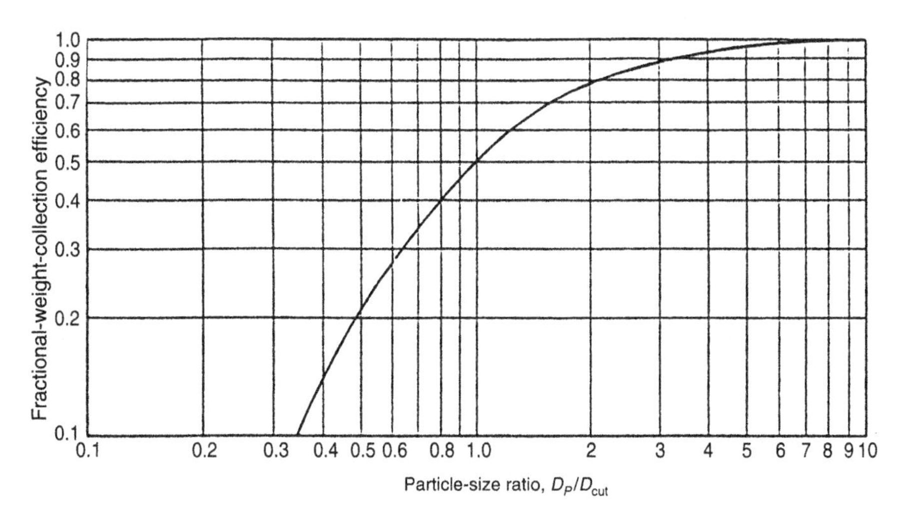

Figure 14-19. Cyclone separator efficiency curve. (Adapted from reference 4. Courtesy of VDI-Verlag, Dusseldorf, Germany.)

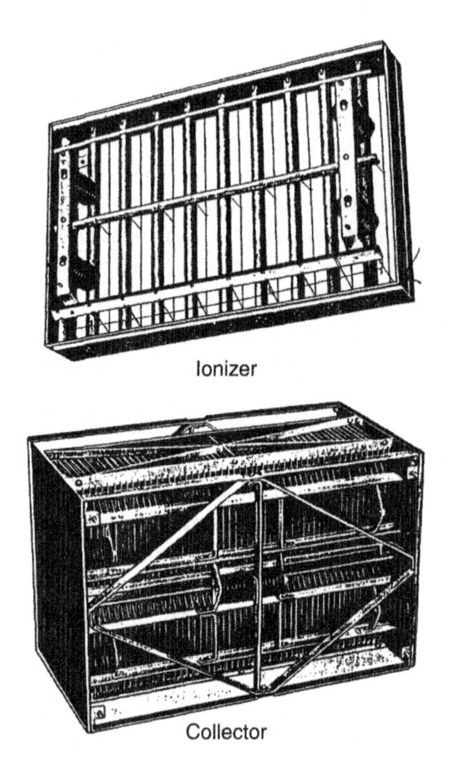

Figure 14-20. Electrostatic precipitator. (Adapted from reference 5.)

makes the material move forward. These screens are slightly sloped (18 to 20 degrees). Their speeds range from 60 strokes (0.229-m amplitude) to 800 strokes (0.019-m amplitude).

Vibrating screen systems operate at high speeds (up to 1800 rpm). They are widely used for wet systems. The oscillating screens are low-speed (300 to 400 rpm) systems.

EXAMPLES

Example 14-1 A laboratory plate and frame filter press (single frame) is used to filter a system (water with a mass fraction of 0.0723 calcium carbonate). Tests (5) at 18.89°C and $\Delta P = 2.76 \times 10^5$ pascals gave the following results:

Filtrate Volume (liters)	Time (seconds)
0.2	1.8
0.4	4.2
0.6	7.5

Filtrate Volume (liters)	Time (seconds)
0.8	11.2
1.0	15.4
1.2	20.5
1.4	26.7
1.6	33.4
1.8	41.0
2.0	48.8
2.2	57.7
2.4	67.2
2.6	77.3
2.8	88.7

The unit was 0.03 m thick and had 0.0263-m^2 filtering area. Dried cake density was 1603 kg/m^3.

Determine the filtrate volume equivalent in resistance to the filter medium and piping (V_s), the specific cake resistance (α), the cake porosity (ϵ), and the cake specific surface (S_0).

Rewriting equation (4-9) in the form of $\Delta\theta/\Delta V$, we obtain

$$\frac{\Delta\theta}{\Delta V} = \frac{\mu\alpha w}{g_c A^2(-\Delta P_t)}(V + V_e)$$

Next we can find values of $\Delta\theta/\Delta V$ that correspond to V's. For example, in the first interval, $\Delta\theta$ is 1.8 sec and $\Delta\theta/\Delta V$ is 9.0 sec/liter. Repeating for all of the other data yields the following:

V (liters)	θ_{sec},	$\Delta\theta$	$\dfrac{\Delta\theta}{\Delta V}$
0	0	1.8	9.0
0.2	1.8	2.4	12.0
0.4	4.2	3.3	16.5
0.6	7.5	3.7	18.5
0.8	11.2	4.2	21.0
1.0	15.4	5.1	25.5
1.2	20.5	6.2	31.0
1.4	26.7	6.7	33.5
1.6	33.4	7.6	38.0
1.8	41.0	7.8	39.0
2.0	48.8	8.9	44.5
2.2	57.7	9.5	47.5
2.4	67.2	10.1	50.5
2.6	77.3	11.4	57.0
2.8	88.7		

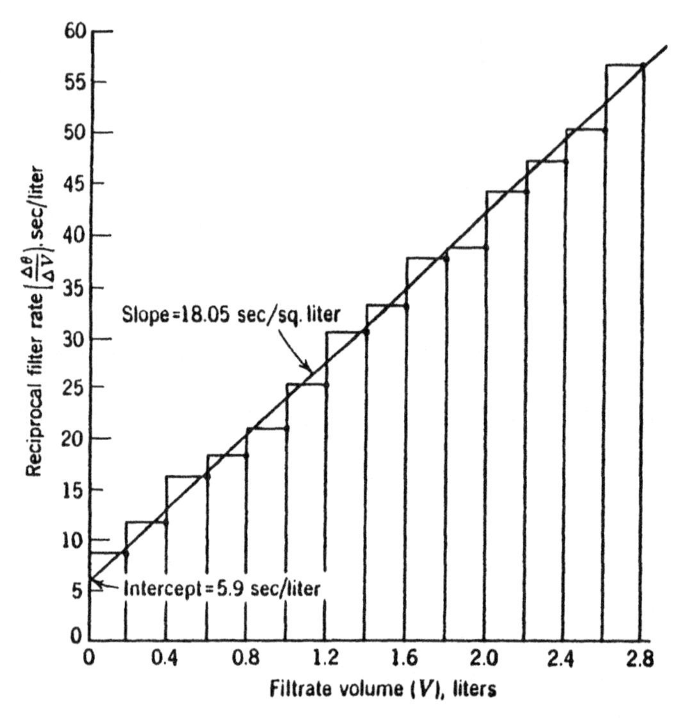

Figure 14-21. Filtration rate reciprocal versus filtration volume. (Adapted with permission from reference 2. Copyright 1960, John Wiley and Sons, Inc.)

A bar chart (Figure 14-21) can then be plotted. The slope and intercept of the straight line fitted through the data are 18.05 sec/liter^2 and 5.9 sec/liter^2, respectively.

$$V_E = \frac{\text{intercept}}{\text{slope}} = \frac{5.9\text{sec/liter}}{18.05\text{sec/liter}^2} = 0.3271$$

$$\mu\alpha w = (18.05\text{sec/liter}^2)(g_c)(A^2)(\Delta P_T)$$

$$\mu\alpha w = (18.05\text{sec/liter}^2)(\text{liter}^2/10^{-6}\ \text{m}^6)(\text{kg/Nsec})$$
$$\times (2 \times 0.0263\ \text{m}^2)^2(2.76 \times 10^5\ \text{N/m}^2)$$

$$\mu\alpha w = 1.38 \times 10^{10}\ \text{kg/m}^3\text{sec}$$

$$W = \frac{(0.0723)(1000\ \text{kg})}{0.9277\ \text{m}^3} = 77.94\ \text{kg CaCO}_3/\ \text{m}^3\ \text{H}_2\text{O}$$

$$\alpha = (1.38 \times 10^{10}\ \text{kg/m}^3\text{sec})(1/\mu\ \text{W})$$

$$\alpha = (1.38 \times 10^{10}\ \text{kg/m}^3\text{sec})(1/77.94\ \text{kg CaCO}_3/\ \text{m H}_2\text{O})$$
$$\times (1/1.1 \times 10^{13}\ \text{kg/msec}$$

$$\alpha = 1.6 \times 10^{11}\ \text{m/kg}$$

Density ρ_s of solid $CaCO_3$ is 2930 kg/m^3.
Then, porosity ϵ_i is

$$\epsilon = \frac{2930 - 1603}{2930} = 0.453$$

Also, S_0 is given by

$$\alpha = \frac{5(1-\epsilon)S_0^2}{\rho_s \epsilon^3}$$

$$S_0 = 3.99 \times 10 \frac{\text{m}^2}{\text{m}^3}$$

Example 14-2 A 20-frame (plate and frame filter) has dimensions (per frame) of 0.75 by 0.75 m and a thickness of 0.064 m. If this system is used for the slurry of Example 14-1 find the volume of slurry and time for filtration at a constant pressure of 2.76×10^5 pascals. Effective filtering area per frame is 0.87 m^2.

$$\text{Frame volume} = \left(\frac{0.87\ \text{m}^2}{2}\right)(0.064\ \text{m})(20) = 0.557\ \text{m}^3$$

$$\text{Cake solids mass} = (0.557\ \text{m}^3)(1603\ \text{kg/m}^3) = 892.9\ \text{kg}$$

$$\text{Mass of slurry fed} = \frac{892.9\ \text{kg}}{0.0723} = 12{,}350\ \text{kg}$$

The slurry density calculated for 92.77 percent water and 7.23 percent CaCO is 1050 kg/m^3:

$$\text{Slurry volume} = \frac{12{,}350\ \text{kg}}{1050\ \text{kg/m}^3} = 11.76\ \text{m}^3$$

We now use equation (14-10) with the α and V_e values of the preceding example. In order to get the filtrate volume, we use a mass balance:

$$V = \frac{(\text{Volume of frames})(1-\epsilon)\rho_s}{\dfrac{(\text{Mass})(CaCO_3)}{\text{Volume } H_2O}} - (\epsilon)(\text{Volume of frames})$$

$$V = \frac{0.557\ \text{m}^3)(1-0.453)(2933)}{77.94 \dfrac{\text{kg } CaCO_3}{\text{m}^3\ H_2O}} - (0.453)(0.557\ \text{m}^3)$$

$$V = 11.22\ \text{m}^3$$

Then

$$\theta = \frac{\mu \alpha w}{g_c A^2(-\Delta P_t)}\left(\frac{V^2}{2} + V_e V\right)$$

$$\theta = \frac{(1.38 \times 10^{10}\ \text{kg/m}^3\text{sec})}{(1\ \text{kg m/Nsec})(20 \times 0.87\ \text{m}^2)^2(2.76 \times 10^5\ \text{N/m}^2)}$$

$$\times \left[\frac{(11.42\ \text{m}^3)^2}{2} + (3.27 \times 10^{-7}\ \text{m}^3)(1)\right]$$

$$\theta = 10{,}769\text{sec or } 2.99\ \text{hr.}$$

Example 14-3 A type B zinc sulfide slurry is filtered with a pressure differential of 4.83×10 pascals. Data for compression permeability is given in Figure 14-22. What is the average specific cake resistance (α) for this system?

From equation (4-15) we have

$$\frac{1}{A}\frac{dV}{d\theta} = \frac{-g_c A}{wV\mu}\int_0^{P_1-P_2}\frac{dp}{\alpha_p} = \frac{-g_c A}{wV\mu}\frac{(\Delta P)}{\alpha}$$

and

$$\text{Average specific cake resistance} = \alpha = \frac{\Delta P}{\displaystyle\int_0^{P_1-P_2}\frac{dp}{\alpha_p}}$$

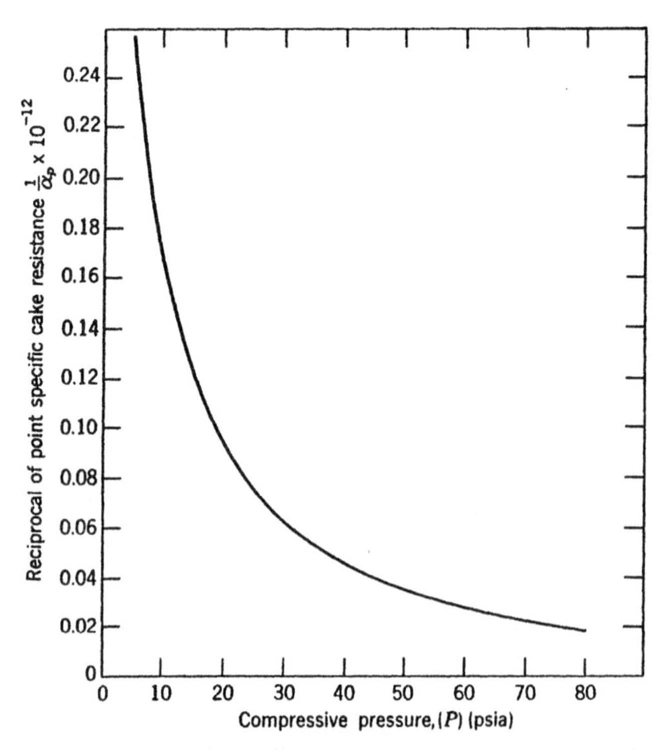

Figure 14-22. Reciprocal of point specific cake resistance versus pressure. (Adapted with permission from reference 2. Copyright 1960, John Wiley and Sons, Inc.)

Figure 14-22 gives α_p as a function of pressure. Values from this figure are

$p \times 10^{-5}$ (pascals)	$\alpha_p \times 10^{-12}$ m/kg
0.0690	0.570
0.690	4.03
1.38	7.05
2.07	10.74
3.45	19.46
4.83	30.20

These data can be used to obtain an extrapolated value of (0.168×10^{12} m/kg) at zero pressure.

In order to evaluate the integral, we need the reciprocal of α_p as a function of pressure.

$p \times 10^{-5}$ (pascals)	$\frac{1}{\alpha_p} \times 10^{12}$ kg/m
0	5.95
0.069	1.75
0.690	0.248
1.38	0.142
2.07	0.093
3.45	0.051
4.83	0.033

Then the integral can be evaluated (either numerically or graphically):

$$\int_0^{P_1-P_2} \frac{dp}{\alpha_p} = 1.59 \times 10^{-7} \frac{\text{pascals kg}}{\text{m}}$$

and

$$\text{Average specific cake resistance} = \alpha = \frac{4.83 \times 10^5 \text{ pascals}}{1.59 \times 10^{-7} \text{ pascal kg/m}}$$

$$= 3.04 \times 10^{12} \text{ m/kg}$$

Example 14-4 A centrifugal filter (0.620-m inside basket diameter and a 0.30-m height) is used to separate the calcium carbonate slurry of Examples 14-1 and 14-2. The rotational speed of the basket is 1000 rpm.

For a 0.03-m-thick cake, what is the filtration rate?

Assume that the cake is incompressible and that the liquid surface and the filter cake surface correspond.

The applicable equation is (14-17). Because the cake is incompressible, we can use the parameters found in Example 14-1 (i.e., α, ϵ):

$$\frac{dV}{d\theta} = \frac{\pi^3 N^2 \rho h (r_3^2 - r_1^2)}{(30)\ ^2\mu \left[\alpha\rho_s(1-\epsilon)\ln\frac{r_3}{r_2} + \frac{R_M^1}{r_3}\right]}$$

$$\frac{dV}{d\theta} = \frac{\dfrac{\pi^3(1000\ \text{min}^{-1})^2(1000\ \text{kg/m}^3)(0.3\ \text{m})\,[(0.3\ \text{m})^2 - (0.27\ \text{m})^2](\text{min}/60\text{sec})^2}{900(1\times10^{-3}\ \text{kg/msec})}}{\left((1.62\times10^{11}\ \text{m/kg})(2933\ \text{kg/m}^3)(1-0.453)\ln\left[\frac{0.3}{0.27}\right]\right)}$$

$$\frac{dV}{d\theta} = 1.794\times10^{-9}\ \text{m}^3/\text{sec}$$

Example 14-5 Sodium sulfate crystals are to be removed from a liquid solution (viscosity of 0.1 kg m sec, density of 810 kg/m^3). The crystal's density is 1460 kg/m^3.

Tests in a laboratory supercentrifuge operating at 23,000 rpm give a satisfactory separation for a rate of 6.3×10^{-4} kg/sec of solution.

The centrifuge dimensions are as follows:

$$\text{Bowl length} = 0.197\ \text{m}$$
$$r_2 = 0.0222\ \text{m}$$
$$r_2 - r_1 = 0.0151\ \text{m}$$

For the system find the critical particle diameter. Also, if the separation is done with a no. 2 disk centrifuge (50 disks at a 45° angle), what is the rate of throughput?

The critical diameter can be found from equation (14-31):

$$D_p' = \sqrt{\frac{9\mu Q}{(\rho_2-\rho)\omega^2 V}\cdot\frac{r_2-r_1}{r}}$$

Because the liquid layer is thick, we must have an effective value of $(r_2 - r_1)/r$ which is $2\ln\ r_2/r_1$.

The values of Q, ω, and V are

$$V = (0.197\ \text{m})2\pi\left[(0.0222\ \text{m})^2 - (0.007\ \text{m})^2\right]$$
$$V = 2.74\times10^{-4}\ \text{m}^3$$

$$\omega' = (2\pi)(23{,}000\ \overset{-1}{\text{min}})\ \text{min}/60\text{sec}$$

$$\omega = 2408.6\text{sec}^{-1}$$

$$Q = \frac{6.3 \times 10^{-4}\ \text{kg/sec}}{810\ \text{kg/m}^3} = 7.79 \times 10^{-7}\ \text{m}^3/\text{sec}$$

Thus

$$D'_p = \sqrt{\frac{(9)(0.1\ \text{kg/m sec})(7.79 \times 10^{-7}\ \text{m}^3/\text{sec})2\ln(0.0222/0.071)}{(1460 - 810)\ \text{kg/m}^3(2408.6\ \text{sec}^{-1})^2(2.74 \times 10^{-4}\ \text{m}^3)}}$$

$$D'_p = 1.243 \times 10^{-6}\ \text{m}$$

Literature values (3) for the small and large centrifuge Σ are 120 m^2 and 6745 m^2, respectively.

Then the large-scale Q is

$$Q_2 = Q_1\Sigma_1\Sigma_2$$

$$Q_2 = (7.79 \times 10^{-7}\ \text{m}^3/\text{sec})\left(\frac{6745\ \text{m}^2}{120\ \text{m}^2}\right)$$

$$Q_2 = 4.38 \times 10^{-5}\ \text{m}^3/\text{sec}$$

or

$$\text{Mass flow rate} = (4.38 \times 10^{-5}\ \text{m}^3/\text{sec})(810\ \text{kg/m}^3).$$

$$\text{Mass flow rate} = 0.0355\ \text{kg/sec}$$

Example 14-6 A vegetable oil is reacted with sodium hydroxide. The reacted oil is then separated from the overall system with a tubular bowl centrifuge. The product has a density of 920 kg/m^3 and a viscosity of 0.02 kg/m sec. The reaction mass has a density of 980 kg/m^3 and a viscosity of 0.3 kg/m sec.

The centrifuge, operating at 18,000 rpm, has a bowl 0.762 m long and a 0.051-m inside diameter. The light-phase radius is 0.0127 m, while the heavy-phase radius is 0.0130 m.

Find the locus of the liquid–liquid interface. Also, if a feed of 5.25×10^{-5} m^3/sec is used (12 volume percent heavy phase), what is D'_p?

Equation (14-32) gives the interface locus:

$$\frac{r_2^2 - r_4^2}{r_2^2 - r_1^2} = \frac{\rho_1}{\rho_4}$$

$$\frac{r_2 - (0.0130\ \text{m})^2}{r_2 - (0.0127\ \text{m})^2} = \frac{920\ \text{kg/m}^3}{980\ \text{kg/m}^3}$$

Solving for r_2, we obtain

$$r_2 = 0.0172 \text{ m}$$

Next, the volume of the heavy phase is

$$= (0.762 \text{ m})2\pi\left[(0.0254)^2 - (0.0172)^2\right] \text{ m}^2$$
$$= 8.35 \times 10^{-4} \text{ m}^3$$

the residence time is

$$t = \frac{8.35 \times 10^{-4} \text{ m}^3}{5.25 \times 10^{-5} \text{ m}^3/\text{sec}} = 159 \text{ sec}$$

Then using equation (14-31), we obtain

$$D'_p = \sqrt{\frac{9(0.3 \text{ kg/m sec})2\ln(0.0254/0.0172)}{(980 - 920 \text{ kg/m}^3)(1885 \text{ sec}^{-1})^2(15.9 \text{ sec})}}$$

$$D'_p = 2.493 \times 10^{-5} \text{ m}$$

Example 14-7 Limestone slurry (236 kg limestone per m^3 of slurry) is tested in a single-batch settling system. The data from the test are as follows:

t (sec)	Interface Height (m)
0	0.36
900	0.324
1800	0.286
3600	0.210
6300	0.147
10,800	0.123
17,100	0.116
43,200	0.098
72,000	0.088

Compare settling rate and concentration.

From the above, Z_0 is 0.36 m and then C at any time [equation (14-38)] is

$$C = \frac{(236 \text{ kg/m}^3)(0.36 \text{ m})}{Z_i}$$

$$C = \frac{84.96 \text{ kg/m}^2}{Z_i}$$

Plotting the data gives Figure 14-23, which can then be used to find Z_i values and in turn the C values. For example, at 7200 sec (2 hr) the Z_i (from the

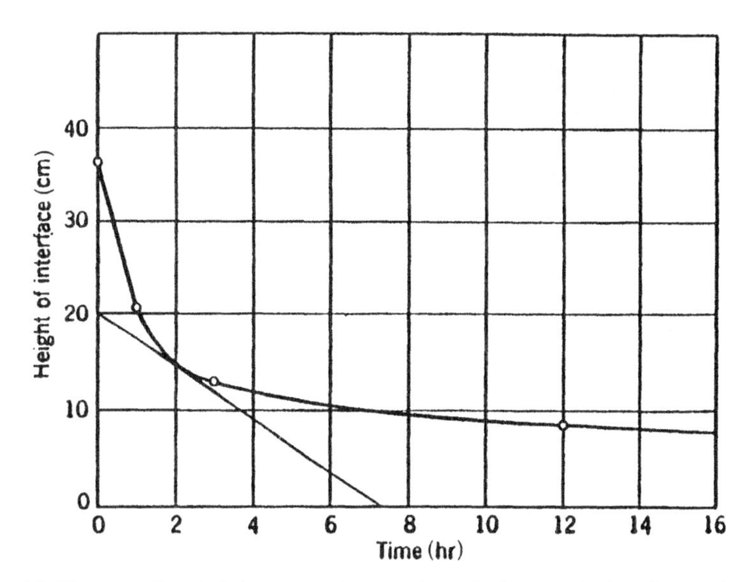

Figure 14-23. Interface height versus time. (Adapted with permission from reference 2. Copyright 1960, John Wiley and Sons, Inc.)

tangent) is found to be 0.20 m and C is 424.8 kg/m^3. The settling velocity is the slope of the tangent, 7.72×10^{-6} m/sec.

The relationship between t, settling velocity, and C is given below:

t (sec)	C (kg/m^3)	Settling Velocity (m/sec $\times 10^5$)
1800	236	4.35
3600	236	4.35
5400	358	1.39
7200	425	0.77
10,800	525	.35
14,400	600	0.18
28,800	714	0.04

These data are also plotted in Figure 14-24.

Example 14-8 A thickener produces a thickened limestone sludge whose concentration is 550 kg/m^3. The feed of slurry (limestone–water) to the thickener is 12.626 kg/sec. Initial slurry concentration is 236 kg/m^3. For these conditions find the thickener cross-sectional area. Assume that the limestone–water slurry is similar to that in Example 14-7.

We begin by using the data of Figure 14-23 and equation (14-34)

$$\frac{L_L c_L}{S} = \frac{\nu}{\left[\dfrac{1}{c_L} - \dfrac{1}{c_u}\right]\dfrac{\rho_{av}}{\rho_w}}$$

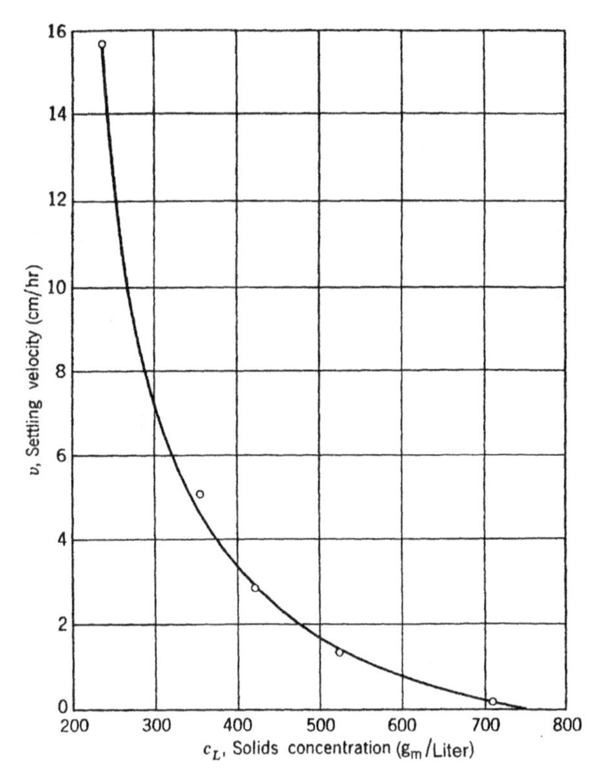

Figure 14-24. Settling velocity versus concentration. (Adapted with permission from reference 2. Copyright 1960, John Wiley and Sons, Inc.)

v (m/sec) $\times 10^5$)	C(kg/m^3)	$(1/C_L - 1/C_U)$ m^3/kg	$L_L C_L/S$(kg/m^{2}sec)
2.78	265	0.00195	0.01428
2.22	285	0.00169	0.01317
1.67	325	0.00125	0.0133
0.83	415	0.00059	0.01414
0.56	465	0.00033	0.01603
0.28	550	0	0

A plot of $L_L C_L/S$ versus v gives a minimum value of 0.01314 (m/sec)/m^3 kg at v of 1.917×10^{-5} m/sec.

The corresponding concentration C_L is 310 kg/m^3.

The cross-sectional area of the thickener is then

$$S = (L_L C_L)/(L_L C_L/S)$$

$$S = \frac{(12.625 \text{ kg/sec})}{0.01314(\text{m/sec})/\text{m}^3/\text{kg}}$$

$$S = 962.71 \text{ m}^2$$

PROBLEMS

14-1. A slurry composed of $Fe(OH)_3$, 0.01180 kg/kg water, filter aid 0.03646 kg/kg water, and a soluble salt 0.00686 kg/kg water is tested in a laboratory with the following results:

Time (min)	Filtrate Collected (m)($\times 10^4$)
0	0
0.167	4.89
0.333	7.73
0.500	9.96
0.666	11.8
0.833	13.54
1.000	15.01
1.50	19.03
2.00	22.43
2.50	25.40
3.00	28.07
3.50	30.59
4.00	32.94

Filter area is 0.043 m^2. The cake obtained was 19.2 percent solids. Determine the appropriate characteristics for a constant pressure filtration (at 1.05 atm).

14-2. How long a time would be needed if the filter in the previous problem is washed with water equal to the total amount of filtrate collected.

14-3. A solid is to be recovered from a slurry (80.13 kg dry solid per m^3 of solid free liquid). Liquid viscosity is 1 centipoise. The filter should deliver 11.33 m^3 of solid-free filtrate over a 2-hr period. System pressure differential is 1.7 atm.

Find the total area of filtering surface required.

Laboratory-scale data (plate and frame filter press with an area of 0.7432 m^2) are as follows:

Total Filtrate Volume (m^3)	Times from start hours $\Delta P = 1.36$ (atm)	$\Delta P = 2.04$ (atm)	$\Delta P = 2.72$ (atm)
0.1416	0.34	0.25	0.21
0.2266	0.85	0.64	0.52
0.2832	1.32	1.00	0.81
0.3398	1.92	1.43	1.17

14-4. A small-scale leaf filter gives the following results:

Time (minutes)	1	2	3	5	10	20
Pressure Drop (atm)	0.476	0.755	1.02	1.58	2.93	5.56

for a constant filtration rate of 1.136×10^{-3} m^3/min.

Find the capacity (volume/time) for the following cycle:

(a) Constant rate of 1.893×10^{-3} m^3/min until pressure change is 3.4 atm

(b) Continued at constant pressure until total volume collected is 0.0568 m^3

(c) Wash of 0.0114 m^3 for 10 min.

14-5. A leaf filter is used to process a suspension (0.225 kg carbonyl iron powder per liter of 0.01 *N* sodium hydroxide). The slurry forms an incompressible cake (porosity of 0.40, average cake resistance of 1.386×10^{10} m/kg). The resistance of the medium is 0.254 cm of cake.

Find the filter area needed to obtain 45.46 kg of dry cake in 1 hr at a pressure drop of 1.36 atm.

14-6. The filter in Problem 14-5 is to be operated at maximum capacity. Dumping and cleaning requires 15 min. Wash water volume in any cycle is 1/10 of filtrate volume collected. Develop an appropriate relation that describes system capacity.

14-7. A suspension of TiO_2 (curve 4 in Figures 14-2 14-3 and 14-4) is to be filtered with a filter system of 2500 ft^2. The suspension is 50 grams per liter of water. Find the time needed to obtain 20,000 lb of dry cake. The resistance of the medium is 2×10^{11} ft^{-1}.

14-8. Filtration studies at constant pressure changes are given below:

Time (in seconds) with Respect to Constant Pressure Change

Filtrate volume (liters)	0.456 atm	1.10 atm	1.92 atm	2.47 atm	3.34 atm
0.5	17.3	6.8	6.3	5.0	4.4
1.0	41.3	19.0	14.0	11.5	9.5
1.5	72.0	34.6	24.2	19.8	16.3
2.0	108.3	53.4	37.0	30.1	24.6
2.5	152.0	76.0	51.7	42.5	34.7
3.0	201.7	102.0	69.0	56.8	46.1
3.5		131.2	88.8	73.0	59.0
4.0		163	110.0	91.2	73.6
4.5			134.0	111.0	89.4
5.0			160.0	133.0	107.3
5.5				156.8	
6.0				182.5	

If the solid mass per unit volume of filtrate is 23.5 kg/m^3 and the filter area is 0.044 m^2, find the filter medium resistance and the average specific cake resistance.

14-9. A rotary drum filter (30 percent submergence) is used for a concentrated slurry (235.6 kg/m^3). If the filter cake contains 50 percent moisture, what filter area is needed to handle 6.31×10^{-4} m^3/sec for a 5-min cycle? Assume that the specific cake resistance is the same as that in Problem 14-8.

14-10. A constant rate filtration gives the following results:

Time (sec)	Pressure Change (atm)
10	0.299
20	0.340
30	0.435
40	0.510
50	0.592
60	0.694
70	0.803
80	0.918
90	1.034
100	1.197
110	1.361

The rate was 489.3 kg/m^2 sec. Viscosity of the filtrate was 0.92 cP, and slurry concentration was 17.30 kg/m^3. Find the filter medium resistance and the average specific cake resistance.

14-11. A test filter press gives a production rate of 1500 cm^3/min when 1000 cm^3 were collected. The rate after 5000 cm^3 were collected was 600 cm^3 /min. If the filter was run until the rate fell to 750 cm^3/min at the same constant pressure as the test and then at a constant rate, how much filtrate could be collected in 10 min?

14-12. A slurry consisting of a dihydrate $Ca(OCl)_2 \cdot 2H_2O$ and its aqueous mother liquor is filtered. Plant results are as follows:

	Mother liquor	Slurry	Cake	Actual Filtrate
% Water	71.8	58.36	47.50	69.20
% $Ca(OCl)_2$	10.20	27.77	42.35	12.92

The calcium hypochlorite dihydrate is the only solid phase present in the mother liquor. Find the percent dihydrate in the original slurry. Observation of the actual filtrate shows that it is cloudy. Determine percent dihydrate in the actual filtrate. What fraction of the total dihydrate is lost?

14-13. A sludge is filtered in a plate and frame press (equipped with 0.0254-m frames). In the first 10 min of operation, the feed pump operates at maximum and the pressure increases to 4.08 atm. One-fourth of the total filtrate is obtained for this segment. Another hour (at constant pressure) is required to complete the filtration. Also 15 min is required to empty and reset the press. When the cloths are precoated with filter aid (depth of 0.159 cm), cloth resistance is reduced to a quarter of its original value. How much will the press overall throughput increase? Assume that application of precoat takes 3 min.

14-14. A slurry (0.2 kg solid/kg water) is fed to a rotary drum filter (0.610-m diameter, 0.610 m length). Drum speed is 0.167 rpm, and 20 percent of the filter surface contacts the slurry. What thickness of cake (porosity of 0.5) results if filtrate is produced at a rate of 0.126 kg/sec using a vacuum of 0.67 atm. The solid's specific gravity is 3.0.

14-15. If the process of Problem 14-14 is carried out in a plate and frame press (frames of 0.0929 m^2), determine the number and thickness of such frames. The operation is to be carried out at the same overall rate with an operating pressure of 1.7 atm. Also, the press takes 2 min to dismantle and 2 min to reassemble. Cake removal from each frame requires 2 min.

14-16. Find the filter medium resistance and the specific cake resistance for a constant pressure filtration (pressure change of 3.33 atm) Slurry concentration is 23.47 kg/m^3. Filter area of the plate and frame press is 0.0439 m^2. Data from the filtration are as follows:

Time (sec)	4.4	9.5	16.3	24.6	34.7	46.1	59.0	73.6	89.4	107.3
Volume ($m^3 \times 10^4$)	5	10	15	20	25	30	35	40	45	50

14-17. The slurry of Problem 14-16 is to be filtered in a plate and frame press (20 frames, 0.873-m^2 area per frame). If filter cake and cloth properties as well as pressure are unchanged, find the time needed to recover 3.37 m^3 of filtrate.

14-18. A material has the following relation for specific cake resistance and pressure drop (in lbf/ft^2).

$$\alpha = 8.8 \times 10^{10}[1 + 3.36 \times 10^{-4}(\Delta\rho)^{0.86}]$$

How many square feet of filter surface (21.1°C) are needed to yield 200.5 ft^3 of filtrate in 1 hr? Additional data: Slurry is 3.0 lb of solid/ft^3 filtrate, pressure drop is 5.44 atm, filter medium resistance is 1.5×10^{10} ft^{-1}, and fluid viscosity is same as water.

14-19. The filter in the preceding problem is operated at a constant rate of 0.5 gallons/ft^2 min from the start until the pressure drop reaches 5.44 atm. Next operation is at a constant pressure drop until 200.5 ft^3 of filtrate is produced. How long is the total required filtration time?

14-20. If the filter in Problem 14-18 is washed at 5.44 atm with wash water volume equal to one-third of the filtrate, what is the required time?

14-21. A continuous filter is used to produce 200.5 ft^3 of filtrate from the system of Problem 14-18. Pressure drop is limited to a maximum of 3.4 atm.

Submergence is 50 percent, and cycle time is to be 3 min. Find the filter area that must be available.

14-22. A 0.1524-m filter press (area of 0.0929 m^2) is used to filter a slurry (0.139 mass fraction solids). Find the specific cake resistance, the filter medium resistance, and the cake thickness for the four experiments described below.

Pressure drop 0.34 atm, mass dry cake/mass wet cake of 1.59, density of 1018 kg/m^3:

Time (sec)	0	24	71	146	244	372	524	690	888	1,188
Filtrate (kg)	0	0.91	1.82	2.73	3.64	4.55	5.45	6.36	7.27	8.18

Pressure drop 1.02 atm, mass dry cake/mass wet cake of 1.47, density of 1170 kg/m^3:

Time (sec)	0	50	181	385	660	1009	1443	2117
Filtrate (kg)	0	2.27	4.55	6.82	9.09	11.36	13.63	15.91

Pressure drop 2.04 atm, mass dry cake/mass wet cake of 1.47, density of 1170 kg/m^3:

Time (sec)	0	26	98	211	361	555	788	1083
Filtrate (kg)	0	2.27	4.55	6/82	9.09	11.36	13.63	15.91

Pressure drop 3.4 atm, mass dry cake/mass wet cake 1.47, density of 1179 kg/m^3:

Time (sec)	0	19	68	142	241	368	524	702
Filtrate (kg)	0	2.27	4.55	6.82	9.09	11.36	13.63	15.91

14-23. The material described in Problem 14-22 is to be processed in a press (total area 10 m^2, frames 40 mm thick). Constant pressure drop is 2 atm. Determine filtration time and filtrate volume obtained in one cycle.

14-24. A continuous rotary filter (submergence of 25 percent, speed 2 rpm) with a pressure drop of 0.7 atm is used to obtain the same production as that of Problem 14-23 (slurry as described in Problem 14-22).

14-25. Mineral particles (density of 2800 kg/m^3 and equivalent diameter of 5μm) are all separated from water at a volumetric rate of 0.25 m^3/sec. What corresponding size cut will result for a suspension of coal particles (density of coal of 1300 kg/m^3) in oil (density of 850 kg/m^3, viscosity of 0.01 N sec/m) Assume Stokes' Law holds.

14-26. A centrifuge with a perforated basket (300-mm diameter, 200 mm deep) is to replace a filtration unit. The centrifuge has a speed of 3900 rpm, and its radius of inner surface of slurry is 75 mm. How much time will be required to achieve the same result as using a plate and frame filter (two frames 50 mm thick and 150 mm square) operating at a pressure change of 350 kN/m^2? The filter press frames are filled in 1 hr. Filter cake is incompressible, and cloth resistance is equivalent to 3 mm of cake.

14-27. A centrifuge is used to clarify a viscous solution (density 801 kg/m^3, viscosity of 100 cP). The solid density is 1461 kg/m^3. Dimensions of the centrifuge are: 0.00716 and 0.02225 m for the radii and 0.1970 m for height. The flow rate is 0.002832 m^3/hr, and speed is 23,000 rpm. What is the diameter of the largest particles in the exit stream?

14-28. In an extract oil refining process, oil and aqueous phases are to be separated in a centrifuge. Densities of oil and aqueous phases are, respectively, 919.5 kg/m^3 and 980.3 kg/m^3. Overflow radius for light liquid is 10.16 mm. Outlet for the heavier liquid is 10.414 mm. Find interface location.

14-29. Find the capacity of a clarifying centrifuge (bowl diameter of 0.61 m, liquid layer thickness of 0.0762 m, depth of 0.406, speed of 1000 rpm). Specific gravities of solid and liquid are 1.6 and 1.3. Particle cut size is to be 30 microns.

14-30. A batch centrifugal filter (bowl diameter of 0.762 m, height of 0.4572 m, speed of 2000 rpm) is used to process a suspension (60 g/liter concentration, dry solid density of 2000 kg/m^3, liquid is water). Cake parameters are: porosity of 0.435, cake thickness of 0.1524 m, specific cake resistance of 1.316×10^{10} m/kg, filter medium resistance of 8.53×10^{10} m^{-1}. Final cake is washed such that inner surface of liquid is 0.2032 m. Find wash water rate if it equals final filtrate rate of flow.

14-31. An aqueous suspension (solids of 2500 kg/m^3 density, size range of 1 to 10 m) is centrifuged in a unit (basket diameter of 450 mm, speed of 4800 rpm, layer of 75 mm thick). How long will it take for the smallest particle to settle?

14-32. A centrifuge (100-mm diameter, 600-mm length) has a 25-mm discharge weir. Find the maximum volumetric flow of liquid through the centrifuge. When operated at 12,000 rpm the unit retains all particles greater than 1 μm in diameter. What is the maximum volumetric flow rate through the unit? Specific gravities of solid and liquid are 2 and 1. Liquid viscosity is 1 m N sec/m^2. Retarding force is $3\pi \mu D$, where μ is particle velocity relative to liquid. Neglect particle inertia.

14-33. How long will it take spherical particles as listed below to settle under free settling conditions through 1.524 m of water at 21,1°C.

Material	Specific Gravity	Diameter (cm)
Galena	7.5	0.0254
Galena	7.5	0.00254
Quartz	2.65	0.0254
Quartz	2.65	0.00254

14-34. A slurry (5 kg water/kg solids) is to be thickened to a product of 1.5 kg water/kg solids in a continuous unit. Find the minimum thickener area to separate 1.33 kg/sec of solids. Laboratory test data are given below:

kg water/kg solid	5.0	4.2	3.7	3.1	2.5
sedimentation rate (mm/sec)	0.20	0.12	0.094	0.07	0.05

14-35. The ratio of sedimentation velocity to free-falling particle velocity is $(1 - C)^{4.8}$, where C is the fractional volume concentration. For a system of 0.1-mm-diameter glass spheres (density 2600 kg/m^3) settling in water (density of 1000 kg/m^2, viscosity of 1 mNsec/m^2) find the concentration at which the rate of deposition of particles per unit area will be a maximum.

14-36. A circular basin is to treat 0.1 m^3/sec of a solids slurry (150 kg/m^3). What is the underflow volumetric flow rate with total separation of all solids and a clear overflow when underflow concentration is 1290 kg/m^3? Batch settling test data are as follows:

Solid Concentration (g/cm)	Velocity Settling (μm/sec)
0.1	148
0.2	91
0.3	55.3
0.4	33.3
0.5	21.4
0.6	14.5
0.7	10.3
0.8	7.4
0.9	5.6
1.0	4.20
1.1	3.3

14-37. A bulk settling test gives the following results:

Time (min)	Interface (ml)
0	1000
No interface	900
0.65	800
0.80	700
1.0	600
1.15	500
1.25	400

Design a circular basin that will process 136.2 m^3/hr.

14-38. A system is compression tested and yields the following results:

Time (hr)	Interface Level (ml)
0	1000
0.25	875
0.5	767
1.0	560
1.5	382
1.75	315
2	302
3	270
4	248
5	235
6	225
7	218
8	214
9	211
10	208[a]
15	206

[a]Dry solid weight at this point is 218 g.

Find the volume of the compression zone per ton per day of solids fed. What is the unit area (ft^2/ton solids/day)?

14-39. A mixture of silica and galena with a size range of 5.21×10^{-6} to 2.50×10^{-5} m are separated at 20°C in water by free settling. Specific gravities of silica and galena are 2.65 and 7.5. What are the size ranges of the various fractions obtained?

14-40. A study was made of silica (2.54 specific gravity, concentration in water percent by volume of solid is 21 percent, average particle diameter is 0.00174 cm) settling in water. The initial height was 100 cm, and the height after 1000 sec was 60 cm. Bulk density is 1.16 g/ml. Viscosity data are as follows:

Sludge Conc. %	10	20	40	60	80	95
Suspension Viscosity (cP)	1.1	1.25	2.2	4.4	8.3	14

Find the settling curve for silica particles (0.000895-cm diameter, initial concentration 14.4 percent particles bulk density is 0.935 g/cm^3). Plot the range from height of 200 cm (initial) to 100 cm.

14-41. Dust particles (50-μm diameter, particle density of 2404 kg/m^3) are to be removed from 3.78 m^3/sec of air in a settling chamber. The air contains 1.15×10^{-3} kg/m^3. What are the dimensions of the settling chamber?

14-42. Particles (density of 1600 kg/m^3, average diameter of 20μm) in air enter a cyclone at a linear velocity of 15 m/sec. What fraction of particles would be removed?

14-43. A cyclone separator has a collection efficiency of 45 percent (particles 0 to 5 μm), 80 percent (particles 5 to 10 μm), and 96 percent for all larger particles. Determine overall efficiency for a dust that is 50 percent (0–5 μm), 30 percent (5–10 μm), and 20 percent (above 10 μm).

14-44. A cyclone separator (0.3-m diameter; 1.2 m long; circular inlet and outlet both 25-mm diameters) is used to remove particles (2700 kg/m^3 density) from a gas (air) that enters at 1.5 m/sec. What is the particle size for the theoretical cut? Air properties are as follows: density 1.3 kg/m^3, viscosity 0.018 m N sec/m^2.

14-45. A sludge is filtered at a constant rate until pressure reaches 400 kN/m^2. The remainder of the filtration is carried out at this pressure. Time for the constant rate filtration that accounts for 1/3 of total filtrate is 900 sec. Find total filtration time, and also determine the filtration cycle if the time needed for cake removal and filter reassembling is 1200 sec.

14-46. A plate and frame press filtering a slurry yielded 8 m^3 of filtrate in 1800 sec and 11 m^3 in 3600 sec when filtration ended. What is the washing time if 3 m^3 of wash water is used? Process is constant pressure, and cloth resistance can be neglected.

14-47. A plate and frame press (0.45-m square frames; working pressure of 450 kN/m^2) used for a slurry yields 2.25 m^3 per 8-hr day. Pressure is built up slowly for 300 sec. Rate of filtration during this period is constant. A laboratory test using a single leaf filter (35 kN/m^2 of pressure, area of 0.05 m^2) collected 400 cm^3 of filtrate in 300 sec and collected another 400 cm^3 in the next 600 sec. Dismantling, removing the filter cake, and reassembling requires 300 sec as well as an additional 180 sec for each cake produced. What is the minimum number of frames required (filter cloth resistance is the same for both laboratory tests and large unit)?

14-48. A sludge is filtered in a plate and frame press (50-mm frames). Pressure rises to 500 kN/m^2 in the first 3600 sec (one-fourth total filtrate obtained). The remaining filtration (at constant pressure) requires another 3600 sec. Cleaning and reassembling requires 900 sec. When a filter aid is used (cloths precoated to a depth of 1.6 mm), cloth resistance is reduced to one-fourth its former value. How will the overall throughput increase if precoat is applied in 180 sec?

14-49. A slurry of a solid (3.0 specific gravity, concentration 100 kg/m^3 water) is filtered in a plate and frame unit which takes 900 sec to dismantle, clean, and reassemble. If the cake is incompressible (porosity of 0.4, viscosity of slurry is 1×10^{-3} N sec/m^2), find the optimum cake thickness for a pressure of 1000 kN/m^2.

14-50. A rotary filter operating at 1.8 rpm filters 0.0075 m^3/sec. What speed would be needed if filtration rate is 0.0160 m^3/sec?

14-51. A slurry is tested using a vacuum leaf filter (30 kN/m^2 vacuum) with an area of 0.05 m^2. In two successive 300-sec periods, filtrate volumes of 250 cm^3 and 150 cm^3 were collected. The slurry is then filtered in a plate and frame unit (12 frames, 0.3-m square, 25 mm thick). Filtration pressure in 200 sec is slowly raised to 500 kN/m^2 at a constant rate. The filtration is then carried out at constant pressure, and the cakes all completely formed after an additional 900 sec. Cakes are then washed (at 375 kN/m^2) for 600 sec. Cloth resistance is the same in both units. What is the filtrate volume collected during each cycle? How much wash water is used?

14-52. A 20-frame (0.3-m square, 50-mm frame thickness) plate and frame filter press is operated first at constant rate for 300 sec to a pressure of 350 kN/m^2 (one-fourth total filtrate obtained). Filtration then takes place at 350 kN/m^2 for 1800 sec until the frames are full. Total filtrate volume per cycle is 0.7 m^3. Breaking down and reassembly requires 500 sec. A replacement rotary drum filter (1.5 m long, 2.2-m diameter) is to give the same overall filtration rate as the plate and frame unit. Assume that resistance of the filter cloth is the same and that the filter cakes are incompressible. What is the speed of rotation if pressure difference is 70 k N/m^2 and drum submergence is 25 percent?

14-53. A rotary drum filter 1.2 m in diameter and 1.2 m long can handle a slurry (10 percent solids) at 6 kg/sec with a speed of 0.3 rpm. If the speed is increased to 0.48 rpm, the system can handle 7.2 kg/sec of the slurry. What are the limitations to the effect of increasing speed on capacity? What is the maximum rate of slurry that can be handled?

14-54. The frames of a plate and frame press (two 50-mm-thick frames, 150-mm squares) fill with a liquid slurry (1000 kg/m^3 density) in 3.5 ksec at a pressure of 450 kN/m^2. If the slurry is then filtered in a perforated basket centrifuge (300-mm diameter, 200 mm deep) with an inner slurry surface of 75 mm, how long will it take (speed of 3900 rpm) to produce the same amount of filtrate from a cycle from the filter press? Cloth resistance is equal to 3 mm of incompressible cake.

14-55. A small laboratory filter (area of 0.023 m^2) is run at a constant rate of 12.5 cm^3/sec. Pressure differences were 14 kN/m^2 (after 300 sec) and 28 kN/m^2 (after 500 sec). When the latter pressure difference is reached, cake thickness is 38 mm. It is desired to filter the slurry with a continuous rotary filter (pressure difference of 70 kN/m^2) to produce 0.002 m^3 of filtrate. If the rotary filter's cloth resistance is half that of the test filter, suggest suitable dimensions and operating conditions for the larger unit.

14-56. A rotary drum filter (area 3 m^2, 30 percent submergence; internal pressure of 30 kN/m^2) operates at 0.5 rpm. If the filter cake is incompressible and the filter cloth has a resistance equal to 1 mm of cake, find the rate of production of filtrate and the cake thickness.

14-57. A rotary drum filter (2-m diameter, 2 m long, 40 percent submergence, pressure of 17 kN/m^2) is to filter a slurry (40 percent solids). Laboratory data (leaf filter of 200 cm^2, cloth similar to the drum filter) produces, respectively, 300 cm^3 (60 sec) and 140 cm^3 (next 60 sec) at a pressure of 17 k N/m^2. Minimum cake thickness is 5 mm. Bulk dry cake density and filtrate density are 1500 and 1000 kg/m^3. Specify drum speed for maximum throughput. Also determine that throughput in terms of weight of slurry fed per unit time.

14-58. If the thinnest cake that can be removed from the drum in Problem 14-56 is 5 mm, what is the maximum rate of filtration? What is the speed of rotation at this rate?

REFERENCES

1. H. P. Grace, *Chem Eng. Prog.* **49**, 303, 367, 427 (1953).
2. A. S. Foust, L. A. Wenzel, C. W. Clump, L. Maus, and L. Andersen, *Principles of Unit Operations*, John Wiley and Sons, New York (1960).
3. C. M. Ambler, *Chem. Eng. Prog.* **48**, 150 (1952).
4. P. Rosin, E. Rammler, and W. Intelmann, *V.D.I.* **76** 433 (1932).
5. J. M. Coulson and J. F. Richardson, *Chemical Engineering*, Pergamon, London (1968).
6. B. F. Ruth and L. L. Kempe, *Trans, Am. Inst. Chem. Eng.* **33**, 34 (1937).

APPENDIX A

A-1 UNITS

$$\text{meter (m)} = 100 \text{ cm} = 3.2808 \text{ ft}$$

$$\text{m}^2 = 10^4 \text{ cm}^2 = 10.7637 \text{ ft}^2$$

$$\text{m}^3 = 10^6 \text{ cm}^3 = 35.3147 \text{ ft}^3 = 264.17 \text{ gallons}$$

$$\text{kg} = 1000 \text{ g} = 2.2046 \text{ lbm}$$

$$\text{newton (N)} = 10^5 \text{ dynes} = 0.2248 \text{ lb force}$$

$$\text{joule (J)} = \text{N m} = 10^7 \text{ ergs} = 0.239 \text{ calories} = 0.7376 \text{ ft lbf}$$
$$= 9.4783 \times 10^{-4} \text{ Btu}$$

$$\text{watt} = \text{J/sec} = 0.239 \text{ cal/sec} = 0.7376 \text{ ft lbf/sec}$$
$$= 9.4783 \times 10^{-4} \text{ Btu/sec} = 1.341 \times 10^{-3} \text{ horsepower}$$

$$\text{pascal} = \text{N/m}^2 = 10 \text{ dyne/cm}^2 = 1.4504 \times 10^{-4} \text{ lbf/m}^2$$
$$= 10^{-5} \text{ bar} = 9.8687 \times 10^{-6} \text{ atm}$$

$$\text{pascal-second} = 10 \text{ poise} = 1000 \text{ cP} = 0.672 \text{ lbm/ft sec}$$

$$\text{watt/m}^\circ\text{K} = 0.5778 \text{ Btu/ft hr}^\circ\text{F} = 2.39 \times 10^{-3} \text{ cal/cm}^\circ\text{K}$$

$$\text{m}^2/\text{sec} = 10^4 \text{ cm}^2/\text{sec} = 10.7637 \text{ ft}^2/\text{sec}$$

$$^\circ\text{R} = 1.8(^\circ\text{K})$$

$$(\Delta^\circ\text{F}) = 1.8(\Delta^\circ\text{F})$$

$$°F = 1.8(°C) + 32°F$$

$$°K = °C + 273.16$$

$$°R = °F + 460$$

A-2 CONSTANTS

Gas constant*(R) $= 8.314$ J/a $= 8.314$ m^3 Pa/a $= 83.14$ cm^3 bar/a

$= 82.06$ cm^3 atm/a $= 1.987$ cal/a $= 1.987$ Btu/b

$= 0.7302$ ft^3 atm/b $= 10.73$ ft^2(psia)/b $= 1545$ ft lbf/b

where a = g mole° K

b = lb mole° R

Acceleration of gravity $= 9.8067$ m/sec^2 $= 32.1740$ ft/s^2

Avogadro's number $\tilde{N} = 6.02_3 \times 10^{23}$ molecules g-mole^{-1}

Boltzmann's constant $K = R/\tilde{N} = 1.380_5 \times 10^{-16}$ erg molecule$^{-1°}$ K^{-1}

Faraday's constant $F = 9.652 \times 10^4$ abs-coulombs g-equivalent^{-1}

Planck's constant $h = 6.62_4 \times 10^{-27}$ erg sec

Stefan–Boltzmann constant $\sigma = 1.355 \times 10^{-12}$ cal sec^{-1} cm$^{-2°}$ K^{-4}

$= 0.1712 \times 10^{-8}$ Btu hr^{-1} ft$^{-2°}$ R^{-4}

Electronic charge $e = 1.602 \times 10^{-19}$ abs-coulomb

Speed of light $c = 2.99793 \times 10^{10}$ cm sec^{-1}

A-3 BASIC DATA

Table A-3-1 Ratios, Forms and Dimensionless Groups (92)

Ratio	Form	Dimensionless Number
Inertia forces / Viscous forces	$\frac{DV\rho}{\mu}$	Reynolds
Inertia forces / Gravity forces	$\frac{V^2}{gD}$	Froude (Fr)
Inertia forces / Surface tension forces	$\frac{DV^2\rho}{\gamma}$	Weber (We)

(*continued overleaf*)

Table A-3-1 ***(continued)***

Ratio	Form	Dimensionless Number
$\frac{\text{Surface tension forces}}{\text{Gravity forces}}$	$\frac{\gamma}{g\rho L^2}$	Eotvos (Eo)
$\frac{\text{Buoyancy forces}}{\text{Viscous forces}} \frac{\text{Inertia forces}}{\text{Viscous forces}}$	$\beta g \frac{\Delta T}{\mu^2} \frac{L^3 \rho^2}{}$	Grashof (Gr)
$\frac{\text{Momentum diffusivity(kinematic viscosity)}}{\text{Thermal diffusivity}}$	$\frac{\mu/\rho}{k/\rho C_p} = \frac{C_p \mu}{k}$	Prandtl (Pr)
$\frac{\text{Forced convection heat transfer}}{\text{Conduction heat transfer}}$	$\frac{D V \rho C_p}{k}$	Peclet (Pe) also, (Pe = RePr)
$\frac{\text{Heat produced by viscous heating}}{\text{Conduction heat transfer}}$	$\frac{\mu V}{k \Delta T}$	Brinkman (Br)
$\frac{\text{Free convection heat transfer}}{\text{Conduction heat transfer}}$	$\frac{\beta g \Delta T L^3 \rho^2 C_p}{\mu}$	Gr Pr
$\frac{\text{Convective heat transfer}}{\text{Radiative heat transfer}}$	$\frac{\rho C_p V}{\sigma \epsilon T^3}$	Radiation group
$\frac{\text{Momentum diffusivity(kinematic viscosity)}}{\text{Molecular diffusivity}}$	$\frac{\mu/\rho}{D_{AB}} = \frac{\mu}{\rho D_{AB}}$	Schmidt (Sc)
$\frac{\text{Convective mass transfer}}{\text{Molecular diffusion}}$	$\frac{DV}{D_{AB}}$	Mass transfer Peclet (Pe_{AB}) or (Pe_{AB} = Re Sc)
$\frac{\text{Homogeneous chemical reaction}}{\text{Bulk flow}}$	$\frac{rL}{cV}$	Damkohler (homogeneous)
$\frac{\text{Homogeneous chemical reaction}}{\text{Bulk flow}}$	$\frac{k_r C^{2n-2} L^2 T}{V^2}$	Damkohler (homogeneous based on k_r)
$\frac{\text{Batch chemical reaction}}{\text{Time}}$	$k_r C^{2n-2} L^2 T$	Unnamed
$\frac{\text{Heterogeneous chemical reaction}}{\text{Bulk flow}}$	$\frac{r}{\text{Sc}}$	Damkohler (heterogeneous)

In the above, L (a characteristic) length can be used for D (diameter).

Table A-3-2 Nomenclature for Table A-3-1 (92)

c = concentration moles/unit	T = temperature
C_p = specific heat	ΔT = temperature difference
g = acceleration of gravity	V = average velocity
k = thermal conductivity	β = coefficient of thermal expansion

Table A-3-2 (*continued*)

k_r = specific reaction rate	γ = surface tension
t = time	ϵ = emissivity
D = diameter	σ = Stefan-Boltzmann constant
D_{AB} = diffusivity	ρ = density
L = characteristic length	μ = viscosity
S = space velocity	

References 1–5 discuss the concepts of similitude and dimensionless groups.

Table A-3-3 Intermolecular Force Parameters and Critical Properties (1. 9–14)

		Lennard-Jones Parameters (9, 10)		Critical Constants (11–14)				
Substance	Molecular Wt. M	σ (A)	ϵ/K (°K)	T_c (°K)	P_c (atm)	V_c (cm^3 g-mole^{-1})	μ_c (g cm^{-1} sec^{-1}) $\times 10^4$	k_c (cal s^{-1} cm^{-1} °K^{-1}) $\times 10^4$
Light Elements:								
H_2	2.016	2.915	38.0	33.3	12.80	65.0	34.7	—
He	4.003	2.576	10.2	5.26	2.26	57.8	25.4	—
Noble Gases:								
Ne	20.183	2.789	35.7	44.5	26.9	41.7	156	79.2
Ar	39.944	3.418	124	151	48.0	75.2	264	71.0
Kr	83.80	3.498	225	209.4	54.3	92.2	396	49.4
Xe	131.3	4.055	229	289.8	58.0	118.8	490	40.2
Simple polyatomic substances:								
Air	28.97	3.617	97.0	132	36.4	86.6	193	90.8
N_2	28.02	3.681	91.5	126.2	33.5	90.1	180	86.8
O_2	32.00	3.433	113	154.4	49.7	74.4	250	105.3
O_3	48.00	—	—	268	67	89.4	—	—
CO	28.01	3.590	110	133	34.5	93.1	190	86.5
CO_2	44.01	3.996	190	304.2	72.9	94.0	343	122
NO	30.01	3.470	119	180	64	57	258	118.2
N_2O	44.02	3.879	220	309.7	71.7	96.3	332	131
SO_2	64.07	4.290	252	430.7	77.8	122	411	98.6
F_2	38.00	3.653	112	—	—	—	—	—
Cl_2	70.91	4.115	357	417	76.1	124	420	97.0
Br_2	159.83	4.268	520	584	102	144	—	—
I_2	253.82	4.982	550	800	—	—	—	—
Hydrocarbons:								
CH_4	16.04	3.822	137	190.7	45.8	99.3	159	158.0
C_2H_2	26.04	4.221	185	309.5	61.6	113	237	—
C_2H_4	28.05	4.232	205	282.4	50.0	124	215	—
C_2H_6	30.07	4.418	230	305.4	48.2	148	210	203.0
C_3H_6	42.08	—	—	365.0	45.5	181	233	—

(*continued overleaf*)

Table A-3-3 (*continued*)

Substance	Molecular Wt. M	Lennard-Jones Parameters (9, 10) σ (A)	ϵ/K (°K)	Critical Constants (11–14) T_c (°K)	P_c (atm)	V_c (cm^3 g-mole^{-1})	μ_c (g cm^{-1} sec^{-1}) $\times 10^4$	k_c (cal s^{-1} cm^{-1} °K^{-1}) $\times 10^4$
C_3H_8	44.09	5.061	254	370.0	42.0	200	228	—
n-C_4H_{10}	58.12	—	—	425.2	37.5	255	239	—
i-C_4H_{10}	58.12	5.341	313	408.1	36.0	263	239	—
n-C_5H_{12}	72.15	5.769	345	469.8	33.3	311	238	—
n-C_6H_{14}	86.17	5.909	413	507.9	29.9	368	248	—
n-C_7H_{16}	100.20	—	—	540.2	27.0	426	254	—
n-C_8H_{18}	114.22	7.451	320	569.4	24.6	485	259	—
n-C_9H_{20}	128.25	—	—	595.0	22.5	543	265	—
Cyclohexane	84.16	6.093	324	553	40.0	308	284	—
C_6H_6	78.11	5.270	440	562.6	48.6	260	312	—
Other organic Compounds:								
CH_4	16.04	3.822	137	190.7	45.8	99.3	159	158.0
CH_3Cl	50.49	3.375	855	416.3	65.9	143	338	—
CH_3Cl_2	84.94	4.759	406	510	60	—	—	—
$CHCl_3$	119.39	5.430	327	536.6	54	240	410	—
CCl_4	153.84	5.881	327	556.4	45.0	276	413	—
C_2N_2	52.04	4.38	339	400	59	—	—	—
COS	60.08	4.13	335	378	61	—	—	—
CS_2	76.14	4.438	488	552	78	170	404	—

Table A-3-4 Functions for Prediction of Transport Properties of Gases (Low Density) (1, 15)

KT/e or KT/e_{AB}	$\Omega\mu = \Omega k$ (for viscosity and thermal conductivity)	ΩDAB (for mass diffusivity)	KT/e or KT/e_{AB}	$\Omega\mu = \Omega k$ (for viscosity and thermal conductivity)	ΩDAB (for mass diffusivity)
0.30	2.785	2.662	2.60	1.081	0.9878
0.35	2.628	2.476	2.70	1.069	0.9770
0.40	2.492	2.318	2.80	1.058	0.9672
0.45	2.368	2.184	2.90	1.048	0.9576
0.50	2.257	2.066	3.00	1.039	0.9490
0.55	2.156	1.966	3.10	1.030	0.9406
0.60	2.065	1.877	3.20	1.022	0.9328
0.65	1.982	1.798	3.30	1.014	0.9256
0.70	1.908	1.729	3.40	1.007	0.9186
0.75	1.841	1.667	3.50	0.9999	0.9120
0.80	1.780	1.612	3.60	0.9932	0.9058
0.85	1.725	1.562	3.70	0.9870	0.8998
0.90	1.675	1.517	3.80	0.9811	0.8942

Table A-3-4 (*continued*)

KT/e or KT/e_{AB}	$\Omega\mu = \Omega k$ (for viscosity and thermal conductivity)	ΩDAB (for mass diffusivity)	KT/e or KT/e_{AB}	$\Omega\mu = \Omega k$ (for viscosity and thermal conductivity)	ΩDAB (for mass diffusivity)
0.95	1.629	1.476	3.90	0.9755	0.8888
1.00	1.587	1.439	4.00	0.9700	0.8836
1.05	1.549	1.406	4.10	0.9649	0.8788
1.10	1.514	1.375	4.20	0.9600	0.8740
1.15	1.482	1.346	4.30	0.9553	0.8694
1.20	1.452	1.320	4.40	0.9507	0.8652
1.25	1.424	1.296	4.50	0.9464	0.8610
1.30	1.399	1.273	4.60	0.9422	0.8568
1.35	1.375	1.253	4.70	0.9382	0.8530
1.40	1.353	1.233	4.80	0.9343	0.8492
1.45	1.333	1.215	4.90	0.9305	0.8456
1.50	1.314	1.198	5.0	0.9269	0.8422
1.55	1.296	1.182	6.0	0.8963	0.8124
1.60	1.279	1.167	7.0	0.8727	0.7896
1.65	1.264	1.153	8.0	0.8538	0.7712
1.70	1.248	1.140	9.0	0.8378	0.7556
1.75	1.234	1.128	10.0	0.8242	0.7424
1.80	1.221	1.116	20.0	0.7432	0.6640
1.85	1.209	1.105	30.0	0.7005	0.6232
1.90	1.197	1.094	40.0	0.6718	0.5960
1.95	1.186	1.084	50.0	0.6504	0.5756
2.00	1.175	1.075	60.0	0.6335	0.5596
2.10	1.156	1.057	70.0	0.6194	0.5464
2.20	1.138	1.041	80.0	0.6076	0.5352
2.30	1.122	1.026	90.0	0.5973	0.5256
2.40	1.107	1.012	100.0	0.5882	0.5170
2.50	1.093	0.9969			

APPENDIX B

Table B-1 Coordinates for Use in Nomograph (Figure B-1)

Liquid	X	Y	Liquid	X	Y
Acetaldehyde	15.2	4.8	Butyl alcohol	8.6	17.2
Acetic acid, 100%	12.1	14.2	Butyric acid	12.1	15.3
Acetic acid, 70%	9.5	17.0	Carbon dioxide	11.6	0.3
Acetic anhydride	12.7	12.8	Carbon disulfide	16.1	7.5
Acetone, 100%	14.5	7.2	Carbon tetrachloride	12.7	13.1
Acetone, 35%	7.9	15.0	Chlorobenzene	12.3	12.4
Acetonitrile	14.4	7.4	Chloroform	14.4	10.2
Acrylic acid	12.3	13.9	Chlorosulfonic acid	11.2	18.1
Allyl alcohol	10.2	14.3	*meta*-Chlorotoluene	13.3	12.5
Allyl bromide	14.4	9.6	*ortho*-Chlorotoluene	13.0	13.3
Allyl iodide	14.0	11.7	*para* -Chlorotoluene	13.3	12.5
Ammonia, 100%	12.6	2.0	*meta*-Cresol	2.5	20.8
Ammonia, 26%	10.1	13.9	Cyclohexane	9.8	12.9
Amyl acetate	11.8	12.5	Cyclohexanol	2.9	24.3
Amyl alcohol	7.5	18.4	Dibromomethane	12.7	15.8
Aniline	8.1	18.7	Dichloroethane	13.2	12.2
Anisole	12.3	13.5	Dichloromethane	14.6	8.9
Arsenic trichloride	13.9	14.5	Diethyl ketone	13.5	9.2
Benzene	12.5	10.9	Diethyl oxalate	11.0	16.4
Brine, 25% $CaCL_2$	6.6	15.9	Diethylene glycol	5.0	24.7
Brine, 25% NaCl	10.2	16.6	Diphenyl	12.0	18.3
Bromine	14.2	13.2	Dipropyl ether	13.2	8.6
Bromotoluene	20.0	15.9	Dipropyl oxalate	10.3	17.7
Butyl acetate	12.3	11.0	Ethyl acetate	13.7	9.1
Butyl acrylate	11.5	12.6	Ethyl acrylate	12.7	10.4

Table B-1 (*continued*)

Liquid	X	Y	Liquid	X	Y
Ethyl alcohol, 100%	10.5	13.8	Methyl formate	14.2	7.5
Ethyl alcohol, 95%	9.8	14.3	Methyl iodide	14.3	9.3
Ethyl alcohol, 40%	6.5	16.6	Methyl isobutyrate	12.3	9.7
Ethyl bromide	14.5	8.1	Methyl propionate	13.5	9.0
Ethyl chloride	14.8	6.0	Methyl propyl ketone	14.3	9.5
Ethyl ether	14.5	5.3	Methyl sulfide	15.3	6.4
Ethyl formate	14.2	8.4	Naphthalene	7.9	18.1
Ethyl iodide	14.7	10.3	Nitric acid, 95%	12.8	13.8
Ethyl propionate	13.2	9.9	Nitric acid, 60%	10.8	17.0
Ethyl propyl ether	14.0	7.0	Nitrobenzene	10.6	16.2
Ethyl sulfide	13.8	8.9	Nitrogen dioxide	12.9	8.6
Ethylbenzene	13.2	11.5	Nitrotoluene	11.0	17.0
2-Ethylbutyl acrylate	11.2	14.0	Octane	13.7	10.0
Ethylene bromide	11.9	15.7	Octyl alcohol	6.6	21.1
Ethylene chloride	12.7	12.2	Pentachloroethane	10.9	17.3
Ethylene glycol	6.0	23.6	Pentane	14.9	5.2
2-Ethylhexyl acrylate	9.0	15.0	Phenol	6.9	20.8
Ethylidene chloride	14.1	8.7	Phosphorus tribromide	13.8	16.7
	13.7	10.4	Phosphorus trichloride	16.2	10.9
Formic acid	10.7	15.8	Propionic acid	12.8	13.8
Freon 11	14.4	9.0	Propyl acetate	13.1	10.3
Freon 12	16.8	15.6	Propyl alcohol	9.1	16.5
Freon 21	15.7	7.5	Propyl bromide	14.5	9.6
Freon 22	17.2	4.7	Propyl chloride	14.4	7.5
Freon 113	12.5	11.4	Propyl formate	13.1	9.7
Glycerol, 100%	2.0	30.0	Propyl iodide	14.1	11.6
Glycerol, 50%	6.9	19.6	Sodium	16.4	13.9
Heptane	14.1	8.4	Sodium hydroxide, 50%	3.2	25.8
Hexane	14.7	7.0	Stannic chloride	13.5	12.8
Hydrochloric acid,			Succinonitrile	10.1	20.8
31.5%	13.0	16.6	Sulfur dioxide	15.2	7.1
Iodobenzene	12.8	15.9	Sulfuric acid, 110%	7.2	27.4
Isobutyl alcohol	7.1	18.0	Sulfuric acid, 100%	8.0	25.1
Isobutyric acid	12.2	14.4	Sulfuric acid, 98%	7.0	24.8
Isopropyl alcohol	8.2	16.0	Sulfuric acid, 60%	10.2	21.3
Isopropyl bromide	14.1	9.2	Sulfuryl chloride	15.2	12.4
Isopropyl chloride	13.9	7.1	Tetrachloroethane	11.9	15.7
Isopropyl iodide	13.7	11.2	Thiophene	13.2	11.0
Kerosene	10.2	16.9	Titanium tetrachloride	14.4	12.3
Linseed oil, raw	7.5	27.2	Toluene	13.7	10.4
Mercury	18.4	16.4	Trichloroethylene	14.8	10.5
Methanol, 100%	12.4	10.5	Triethylene glycol	4.7	24.8
Methanol, 90%	12.3	11.8	Turpentine	11.5	14.9
Methanol, 40%	7.8	15.5	Vinyl acetate	14.0	8.8
Methyl acetate	14.2	8.2	Vinyl toluene	13.4	12.0
Methyl acrylate	13.0	9.5	Water	10.2	13.0
Methyl *n*-butyrate	13.2	10.3	*meta*-Xylene	13.9	10.6
Methyl chloride	15.0	3.8	*ortho*-Xylene	13.5	12.1
Methyl ethyl ketone	13.9	8.6	*para*-Xylene	13.9	10.9

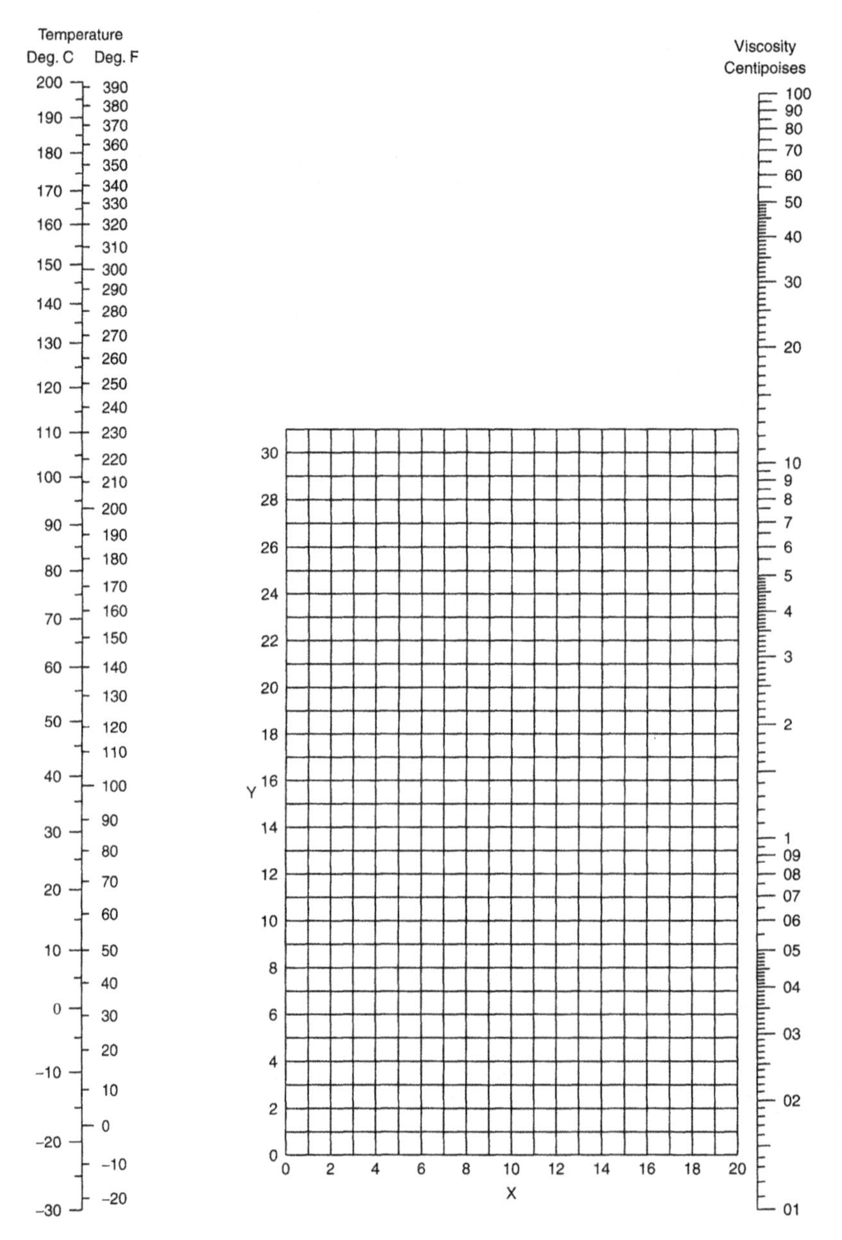

Figure B-1. Liquid viscosities. A centipoise is equal to 0.001 kg/m sec. (Reproduced with permission from reference 8. Copyright 1997, American Chemical Society.)

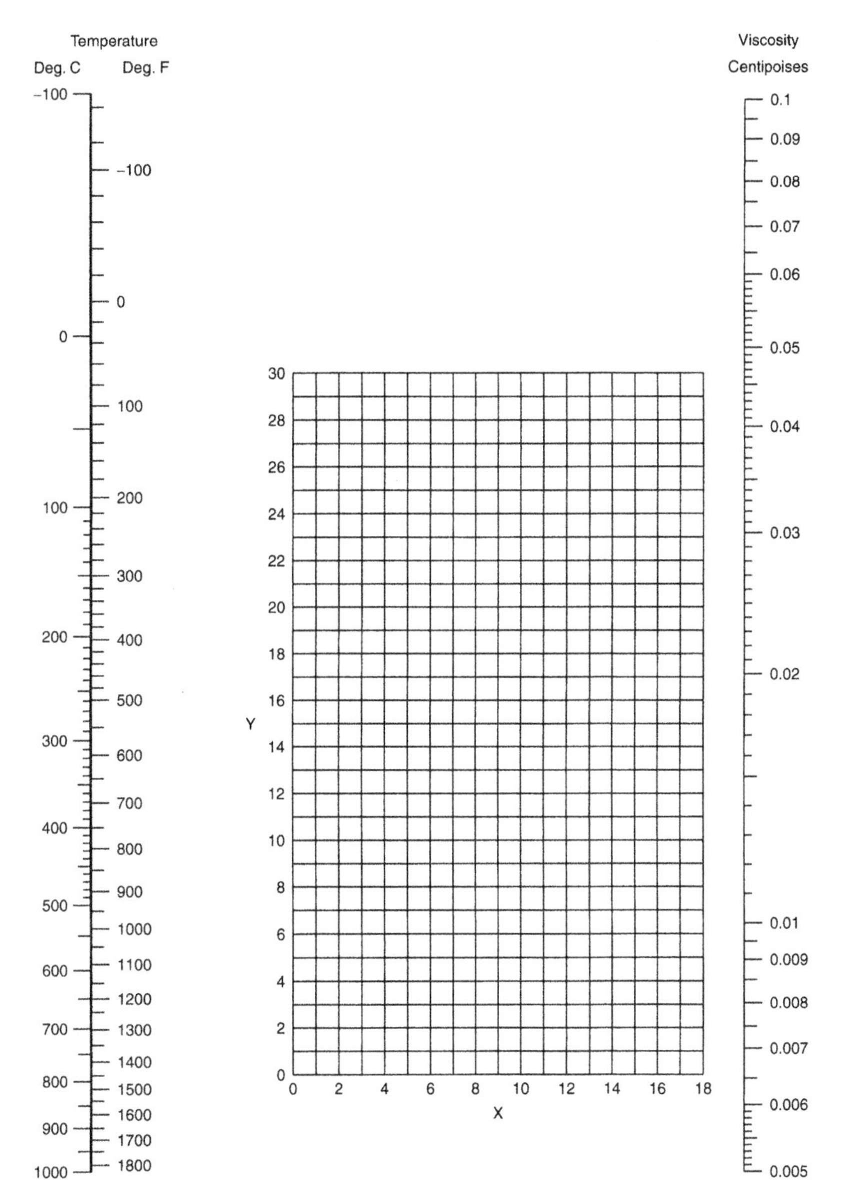

Figure B-2. Gas viscosities. A centipoise is equal to 0.0 kg/m sec. (Reproduced with permission from reference 8. Copyright 1997, American Chemical Society.)

Table B-2 Coordinates for Use in Nomograph (Figure B-2)

No.	Gas	*X*	*Y*	No.	Gas	*X*	*Y*
1	Acetic acid	7.7	14.3	29	Freon 113	11.3	14.0
2	Acetone	8.9	13.0	30	$3H_2(+)4N_2$	11.2	17.2
3	Acetylene	9.8	14.9	31	Helium	10.9	20.5
4	Air	11.0	20.0	32	Hexane	8.6	11.8
5	Ammonia	8.4	16.0	33	Hydrogen	11.2	12.4
6	Argon	10.5	22.4	34	Hydrogen bromide	8.8	20.9
7	Benzene	8.5	13.2	35	Hydrogen chloride	8.8	18.7
8	Bromine	8.9	19.2	36	Hydrogen cyanide	9.8	11.9
9	Butene	9.2	13.7	37	Hydrogen iodide	9.0	21.3
10	Butylene	8.9	13.0	38	Hydrogen sulfide	8.6	18.0
11	Carbon dioxide	9.5	18.7	39	Iodine	9.0	18.1
12	Carbon disulfide	8.0	16.0	40	Mercury	5.3	22.9
13	Carbon monoxide	11.0	20.0	41	Methane	9.9	15.5
14	Chlorine	9.0	18.4	42	Methyl alcohol	8.5	15.6
15	Chloroform	8.9	15.7	43	Nitric oxide	10.9	20.5
16	Cyanogen	9.2	15.2	44	Nitrogen	10.6	20.0
17	Cyclohexane	9.2	12.0	45	Nitronyl chloride	8.0	17.6
18	Ethane	9.1	14.5	46	Nitrous oxide	8.8	19.0
19	Ethyl acetate	8.5	13.2	47	Oxygen	11.0	21.3
20	Ethyl alcohol	9.2	14.2	48	Pentane	7.0	12.8
21	Ethyl chloride	8.5	15.6	49	Propane	9.7	12.9
22	Ethyl ether	8.9	13.0	50	Propyl alcohol	8.4	13.4
23	Ethylene	9.5	15.1	51	Propylene	9.0	13.8
24	Fluorine	7.3	23.8	52	Sulfur dioxide	9.6	17.0
25	Freon 11	10.6	15.1	53	Toluene	8.6	12.4
26	Freon 12	11.1	16.0	54	2,3,3-Trimethylbutane	9.5	10.5
27	Freon 21	10.8	15.3	55	Water	8.0	16.0
28	Freon 22	10.1	17.0	56	Xenon	9.3	23.0

Additional material on viscosity (effect of pressure, mixtures, dense systems) can be found in references 18–20.

Table B-3 Key for Figure B-3

		Temperature Range (°C)	
No.	Liquid	From	To
29	Acetic acid, 100%	0	80
32	Acetone	20	50
52	Ammonia	−70	50
26	Amyl acetate	0	100
37	Amyl alcohol	−50	25
30	Aniline	0	130
23	Benzene	10	80
27	Benzyl alcohol	−20	30

Table B-3 (*continued*)

No.	Liquid	Temperature Range (°C) From	To
10	Benzyl chloride	−30	30
49	Brine (25%$CaCl_2$)	−40	20
51	Brine (25% NaCl)	−40	20
44	Butyl alcohol	0	100
2	Carbon disulfide	−100	25
3	Carbon tetrachloride	10	60
8	Chlorobenzene	0	100
4	Chloroform	0	50
21	Decane	−80	25
6A	Dichloroethane	−30	60
5	Dichloromethane	40	50
15	Diphenyl	80	120
16	Diphenyl oxide	0	200
22	Diphenylmethane	30	100
16	Dowtherm A	0	200
24	Ethyl acetate	−50	25
42	Ethyl alcohol, 100%	30	80
46	Ethyl alcohol, 95%	20	80
50	Ethyl alcohol, 50%	20	80
1	Ethyl bromide	5	25
13	Ethyl chloride	−30	40
36	Ethyl ether	−100	25
7	Ethyl iodide	0	100
25	Ethylbenzene	0	100
39	Ethylene glycol	−40	200
2A	Freon 11 (CCl_3F)	−20	70
6	Freon 12 (CCl_2F_2)	−40	15
4A	Freon 21 ($CHCl_2F$)	−20	70
7A	Freon 22 ($CHClF_2$)	−20	60
3A	Freon 113 (CCl_2FCClF_2)	20	70
38	Glycerol	−40	20
28	Heptane	0	60
35	Hexane	−80	20
48	Hydrochloric acid 30%	20	100
41	Isoamyl alcohol	10	100
43	Isobutyl alcohol	0	100
47	Isopropyl alcohol	−20	50
31	Isopropyl ether	−80	20
40	Methyl alcohol	−40	20
13A	Methyl chloride	−80	20
14	Naphthalene	90	200
12	Nitrobenzene	0	100
34	Nonane	−50	25
33	Octane	−50	25
3	Perchlorethylene	−30	140
45	Propyl alcohol	−20	100
20	Pyridine	−50	25

(*continued overleaf*)

Table B-3 *(continued)*

No.	Liquid	Temperature Range (°C) From	To
11	Sulfur dioxide	−20	100
9	Sulfuric acid 98%	10	45
23	Toluene	0	60
53	Water	10	200
18	*meta*-Xylene	0	100
19	*ortho*-Xylene	0	100
17	*para*-Xylene	0	100

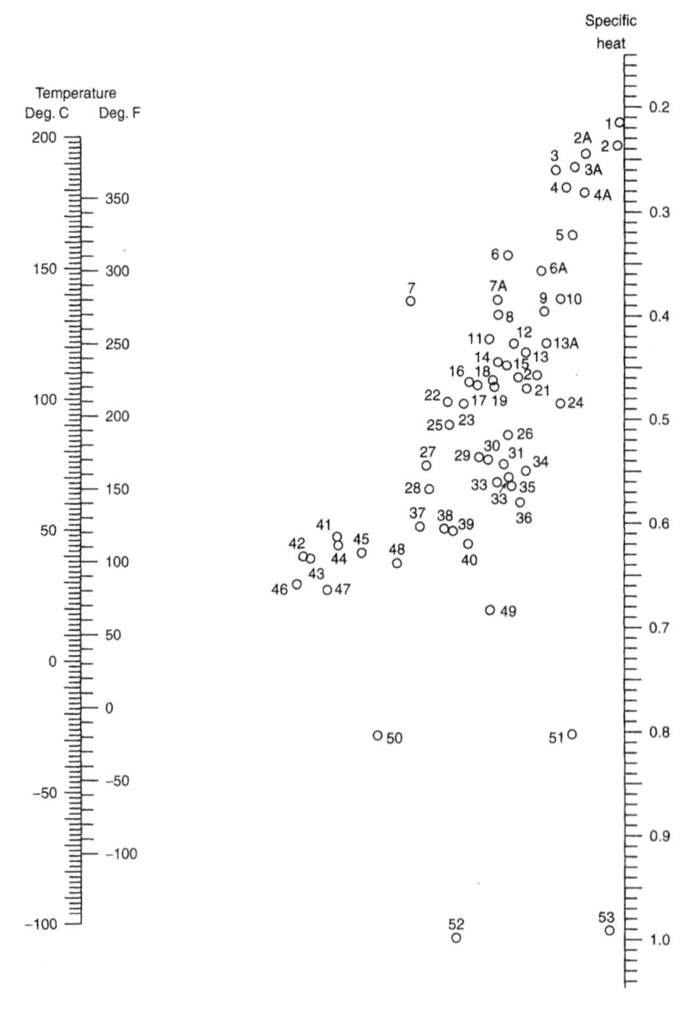

Figure B-3. Liquid specific heat data. Specific heats can be expressed as P.c.u./lb (°C), Btu/lb (°F), or calories/g. (Reproduced with permission from reference 8. Copyright 1997, American Chemical Society.)

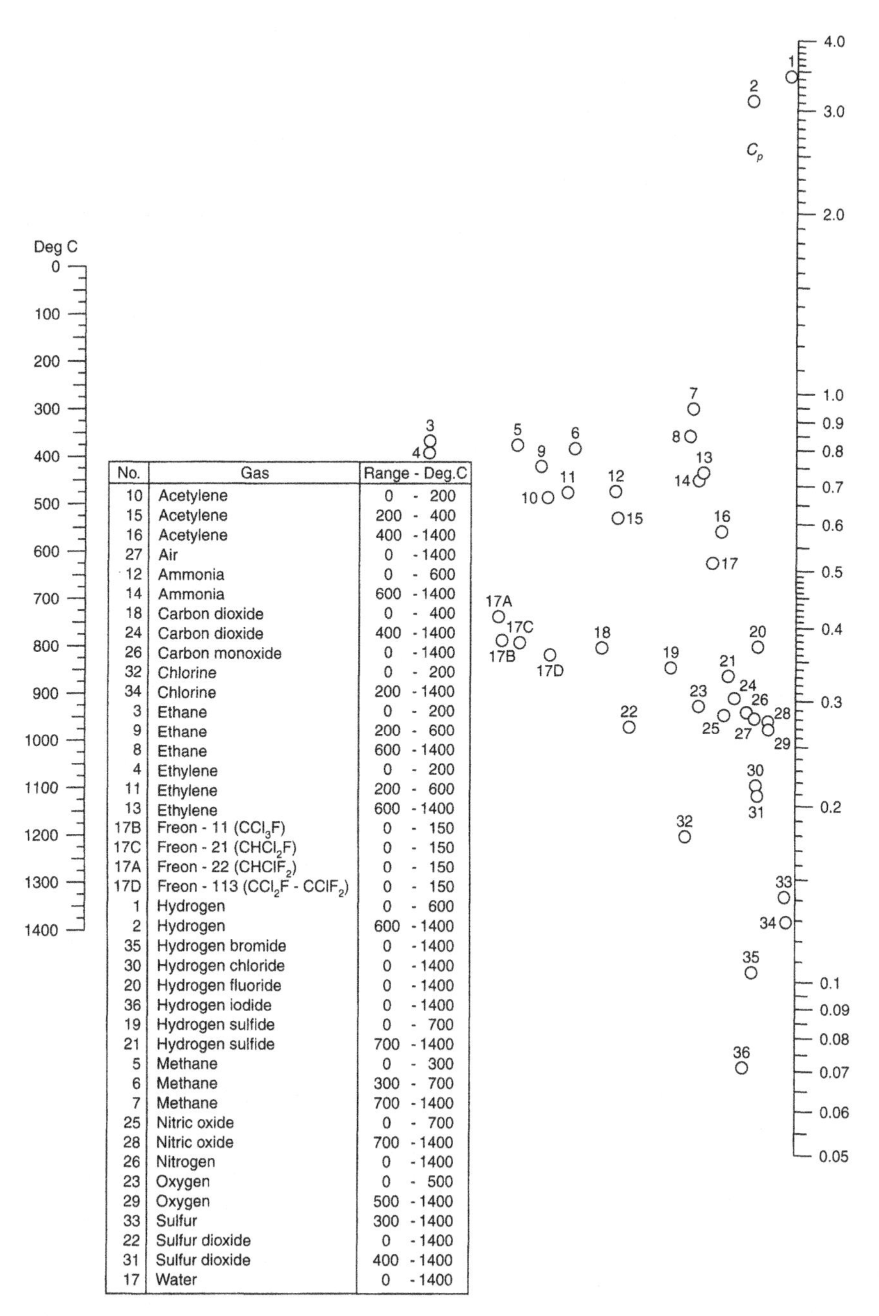

No.	Gas	Range - Deg.C
10	Acetylene	0 - 200
15	Acetylene	200 - 400
16	Acetylene	400 - 1400
27	Air	0 - 1400
12	Ammonia	0 - 600
14	Ammonia	600 - 1400
18	Carbon dioxide	0 - 400
24	Carbon dioxide	400 - 1400
26	Carbon monoxide	0 - 1400
32	Chlorine	0 - 200
34	Chlorine	200 - 1400
3	Ethane	0 - 200
9	Ethane	200 - 600
8	Ethane	600 - 1400
4	Ethylene	0 - 200
11	Ethylene	200 - 600
13	Ethylene	600 - 1400
17B	Freon - 11 (CCl_3F)	0 - 150
17C	Freon - 21 ($CHCl_2F$)	0 - 150
17A	Freon - 22 ($CHClF_2$)	0 - 150
17D	Freon - 113 (CCl_2F - $CClF_2$)	0 - 150
1	Hydrogen	0 - 600
2	Hydrogen	600 - 1400
35	Hydrogen bromide	0 - 1400
30	Hydrogen chloride	0 - 1400
20	Hydrogen fluoride	0 - 1400
36	Hydrogen iodide	0 - 1400
19	Hydrogen sulfide	0 - 700
21	Hydrogen sulfide	700 - 1400
5	Methane	0 - 300
6	Methane	300 - 700
7	Methane	700 - 1400
25	Nitric oxide	0 - 700
28	Nitric oxide	700 - 1400
26	Nitrogen	0 - 1400
23	Oxygen	0 - 500
29	Oxygen	500 - 1400
33	Sulfur	300 - 1400
22	Sulfur dioxide	0 - 1400
31	Sulfur dioxide	400 - 1400
17	Water	0 - 1400

Figure B-4. Gas specific heat data. Specific heats can be expressed as P.c.u./lb (°C), Btu/lb (°F) or calories/g (°C) (6).

Table B-4 Prandtl Numbers for Gases (at 1 atm, 100°C) (7.8)

Gas	$C_P\mu/k$
Air, hydrogen	0.69
Ammonia	0.86
Argon	0.66
Carbon dioxide, methane	0.75
Carbon monoxide	0.72
Helium	0.71
Nitric oxide, nitrous oxide	0.72
Nitrogen, oxygen	0.70
Steam (low pressure)	1.06

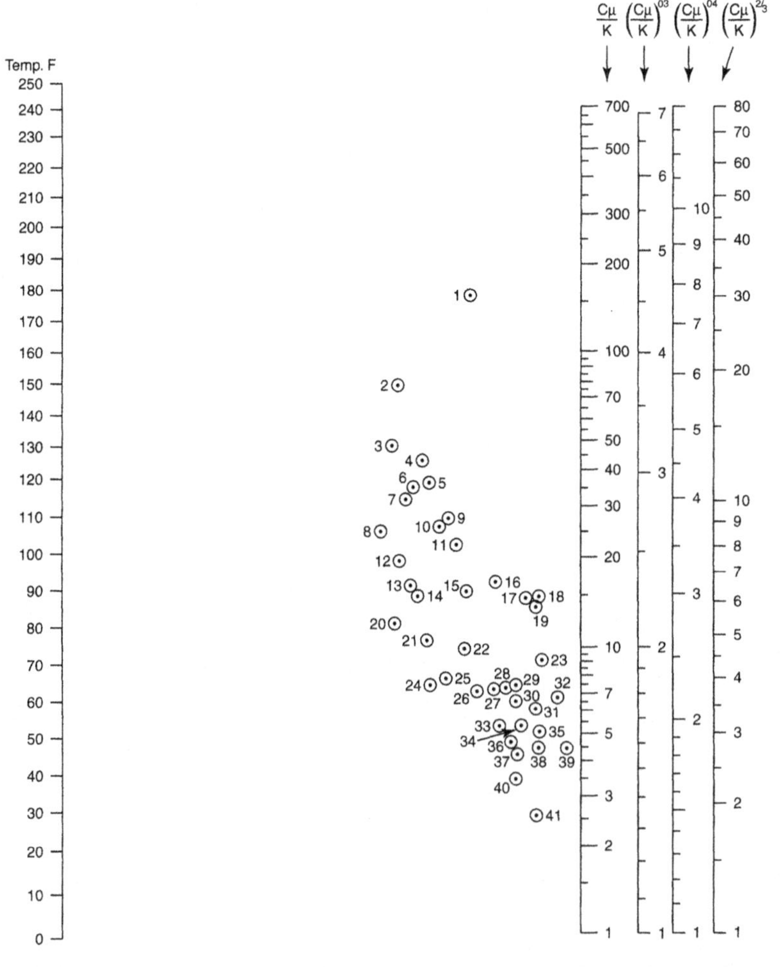

Figure B-5. Liquid Prandtl numbers. (Reproduced with permission from reference 8. Copyright 1997, American Chemical Society.)

Table B-5 Key for Figure B-5

No.	Liquid	Range (°F)
19	Acetic acid, 100%	50–140
14	Acetic acid, 50%	50–194
36	Acetone	14–176
21	Ammonia, 26%	14–230
18	Amyl acetate	32–104
6	Amyl alcohol	86–212
7	Aniline	14–248
28	Benzene	32–194
25	Brine, 25% $CaCl_2$	−4–176
20	Brine, 25% NaCl	−4–122
9	Butyl alcohol	14–230
41	Carbon disulfide	14–212
29	Carbon tetrachloride	32–176
23	Chlorobenzene	32–194
33	Chloroform	32–176
32	Ethyl acetate	32–140
17	Ethyl alcohol, 100%	14–212
16	Ethyl alcohol, 95%	50–158
12	Ethyl alcohol, 50%	50–176
40	Ethyl bromide	14–104
37	Ethyl ether	14–158
39	Ethyl iodide	14–176
1	Ethylene glycol	32–122
6	Glycerol, 50%	32–158
31	Heptane	14–140
34	Hexane	68–140
22	Hydrochloric acid, 30%	50–176
5	Isoamyl alcohol	50–230
10	Isoamyl alcohol	32–212
27	Methyl alcohol, 100%	14–176
13	Methyl alcohol, 40%	14–176
15	Nitrobenzene	68–212
35	Octane	14–122
36	Pentane	14–122
11	Propyl alcohol	86–176
2	Sulfuric acid, 111%	63–176
3	Sulfuric acid, 98%	50–194
4	Sulfuric acid, 60%	50–212
30	Toluene	14–230
24	Water	50–212
26	Xylene	14–122

Table B-6 Liquid Specific Gravities (6)

$$\text{Specific gravity} = \frac{\text{Density of material at indicated temperature}}{\text{Density of liquid water at 4°C}}$$

Density of liquid water at 4°C = 1.000 g/cm^3 = 62.3 lb/ft^3 = 1000.0 kg/m^3

Pure Liquid	Formula	Temperature (°C)	Specific Gravity
Acetaldehyde	CH_3CHO	18	0.783
Acetic acid	CH_3CO_2H	0	1.067
		30	1.038
Acetone	CH_3COCH_3	20	0.792
Benzene	C_6H_6	20	0.879
n-Butyl alcohol	$C_2H_5CH_2Ch_2OH$	20	0.810
Carbon tetrachloride	CCl_4	20	1.595
Ethyl alcohol	CH_3CH_2OH	10	0.798
		30	0.791
Ethyl ether	$(CH_3CH_2)_2O$	25	0.708
Ethylene glycol	CH_2OH CH_2OH	19	1.113
Glycerol	CH_2OH CHOH CH_2OH	15	1.264
		30	1.255
Isobutyl alcohol	$(CH_3)_2CHCH_2OH$	18	0.805
Isopropyl alcohol	$(CH_3)_2CHOH$	0	0.802
		30	0.777
Methyl alcohol	CH_3OH	0	0.810
		20	0.792
Nitric acid	HNO_3	10	1.531
		30	1.495
Phenol	C_6H_5OH	25	1.071
n-Propyl alcohol	$CH_3CH_2CH_2OH$	20	0.804
Sulfuric acid	H_2SO_4	10	1.841
		30	1.821
Water	H_2O	4	1.000
		100	0.958

Table B-7 Thermal Conductivity of Gases (6)

Gas	Temperature (°F)	Btu/(hr)(ft^2)(°F/ft)[a]
Air	32	0.0140
	212	0.0183
	392	0.0226
Ammonia	32	0.0128
	122	0.0157
Carbon dioxide	32	0.0085
	212	0.0133
Chlorine	32	0.0043

Table B-7 *(continued)*

Gas	Temperature (°F)	Btu/(hr)(ft^2)(°F/ft)[a]
Hydrogen	32	0.100
	212	0.129
Methane	32	0.0175
	122	0.0125
Nitrogen	32	0.0140
	212	0.0180
Oxygen	32	0.0142
	212	0.0185
Sulfur dioxide	32	0.0050
	212	0.0069
Water vapor	200	0.0159
	600	0.0256

[a] Btu/(hr)(ft^2)(°F/ft) × 1.7307 = J/(sec)(m^2)(°K/m).

Table B-8 Thermal Conductivity of Metals (6)

	k AS Btu/(h) (ft^2)(°F/ft)[a]		
Metal	At 32°F	At 212°F	At 572°F
Aluminum	117	119	133
Brass (70-30)	56	60	66
Cast iron	32	30	26
Copper	224	218	212
Lead	20	19	18
Nickel	36	34	32
Silver	242	238	
Steel (mild)		26	25
Tin	36	34	
Wrought iron		32	28
Zinc	65	64	59

[a] Btu/(hr)(ft^2)(°F/ft) × 1.7307 = J/(sec)(m^2)(°K/m).

Table B-9 Thermal Conductivity of Liquids (6)

Liquid	Temperature (°F)	Btu/(hr)(ft^2) (°F/ft)[a]
Acetic acid, 100%	68	0.099
50%	68	0.20
Acetone	86	0.102
	167	0.095

(continued overleaf)

Table B-9 ***(continued)***

Liquid	Temperature (°F)	Btu/(hr)(ft^2)(°F/ft)[a]
Benzene	86	0.092
	140	0.087
Ethyl alcohol, 100%	68	0.105
	122	0.087
40%	68	0.224
Ethylene glycol	32	0.153
Glycerol, 100%	68	0.164
	212	0.164
40%	68	0.259
n-Heptane	86	0.081
Kerosene	68	0.086
Methyl alcohol, 100%	68	0.124
	122	0.114
40%	68	0.234
n-Octane	86	0.083
Sodium chloride brine, 25%	86	0.330
Sulfuric acid, 90%	86	0.210
30%	86	0.300
Toluene	86	0.086
Water	32	0.320
	200	0.392

[a]Btu/(hr)(ft^2)(°F/ft) × 1.7307 = J/(sec)(m^2)(°K/m).

Table B-10 Thermal Conductivity of Nonmetallic Solids (6)

Material	Temperature (°F)	Btu/(hr)(ft^2)(°F/ft)[a]
Asbestos–cement boards	68	0.43
Bricks:		
Building	68	0.40
Fire clay	392	0.58
	1832	0.95
Sil-O-Cel	400	0.042
Calcium carbonate (natural)	86	1.3
Calcium sulfate (building plaster)	77	0.25
Celluloid	86	0.12
Concrete (stone)		0.54
Cork board	86	0.025
Felt (wool)	86	0.03

Table B-10 (*continued*)

Material	Temperature (°F)	Btu/(hr)(ft^2)(°F/ft)[a]
Glass (window)		0.3-0.61
Rubber (hard)	32	0.087
Wood (across grain):		
Oak	59	0.12
Maple	122	0.11
Pine	59	0.087

[a] Btu/(hr)(ft^2)(°F/ft) × 1.7307 = J/(sec)(m^2)(°K/m).

Additional material on thermal conductivity (pressure effect, mixtures) is presented in references 21–26.

Table B-11 Gaseous System Diffusion Coefficients

System	T(°K)	$D_{AB}P$ (obs.) $\frac{cm^2}{sec}$ atm.	Reference
Air–carbon dioxide	276.2	0.142	27
Air–carbon dioxide	317.3	0.177	27
Air–ethanol	313	0.145	28
Air–helium	276.2	0.624	27
Air–helium	317.2	0.765	27
Air–helium	346.2	0.902	27
Air–hexane	294	0.0800	29
Air–hexane	328	0.0930	29
Air–pentane	294	0.071	29
Air–pentane	391	0.0985	29
Air–water	313	0.288	28
Ammonia–diethyl ether	288.3	0.0999	30
Ammonia–diethyl ether	337.5	0.137	30
Argon–ammonia	254.7	0.150	31
Argon–ammonia	333	0.253	31
Argon–carbon dioxide	276.2	0.133	27
Argon–helium	276.2	0.646	27
Argon–helium	288	0.696	32
Argon–helium	298	0.729	33
Argon–helium	418	1.398	32
Argon–hydrogen	242.2	0.562	34
Argon–hydrogen	295.4	0.83	35
Argon–hydrogen	448	1.76	35

(*continued overleaf*)

Table B-11 (*continued*)

System	T(°K)	$D_{AB}P$ (obs.) $\frac{cm^2}{sec}$ atm.	Reference
Argon–hydrogen	628	3.21	35
Argon–hydrogen	806	4.86	35
Argon–hydrogen	958	6.81	35
Argon–hydrogen	1069	8.10	35
Argon–krypton	273	0.119	36
Argon–methane	298	0.202	32
Argon–neon	273	0.276	36
Argon–sulfur dioxide	263	0.077	28
Argon–xenon	195	0.158	32
Argon–xenon	194.7	0.508	36
Argon–xenon	329.9	0.137	36
Argon–xenon	378	0.178	32
Carbon dioxide–carbon dioxide	194.8	0.0516	37
Carbon dioxide–carbon dioxide	312.8	0.125	37
Carbon dioxide–helium	276.2	0.531	27
Carbon dioxide–helium	298	0.612	38
Carbon dioxide–helium	299	0.611	39
Carbon dioxide–helium	346.2	0.765	27
Carbon dioxide–helium	498	1.414	38
Carbon dioxide–nitrogen	298	0.167	40
Carbon dioxide–nitrogen	299	0.171	39
Carbon dioxide–nitrous oxide	194.8	0.531	37
Carbon dioxide–nitrous oxide	312.8	0.128	37
Carbon dioxide–oxygen	293.2	0.153	41
Carbon dioxide–sulfur dioxide	263	0.064	28
Carbon dioxide–sulfur dioxide	473	0.195	28
Carbon dioxide–water	307.2	0.198	42
Carbon dioxide–water	328.6	0.257	42
Carbon dioxide–water	352.3	0.245	43
Carbon monoxide–carbon monoxide	194.7	0.109	44
Carbon monoxide–carbon monoxide	373	0.323	44
Carbon monoxide–nitrogen	194.7	0.105	44
Carbon monoxide–nitrogen	373	0.318	44
Ethane–*n*-hexane	294	0.0375	29
Ethylene–water	328.4	0.233	43
Helium–benzene	423	0.61	38
Helium–*n*-butanol	423	0.587	38
Helium–ethanol	423	0.821	38
Helium–*n*-hexanol	423	0.469	38
Helium–methane	298	0.675	32
Helium–methanol	423	1.032	38
Helium–neon	242.2	0.792	34
Helium–neon	341.2	1.405	34

Table B-11 *(continued)*

System	T(°K)	$D_{AB}P$ (obs.) $\frac{cm^2}{sec}$ atm.	Reference
Helium–nitrogen	298	0.687	38
Helium–nitrogen (He trace)	298	0.73	45
Helium–nitrogen (N_2 trace)	298	0.688	45
Helium–oxygen	298	0.729	38
Helium–n-pentanol	423	0.507	38
Helium–i-propanol	423	0.677	38
Helium–n-propanol	423	0.676	38
Helium–water	307.1	0.902	43
Helium–water	352.4	1.121	43
Hydrogen–acetone	296	0.424	28
Hydrogen–ammonia	263	0.57	28
Hydrogen–ammonia	273	0.745	46
Hydrogen–ammonia	298	0.783	28
Hydrogen–ammonia	358	1.093	28
Hydrogen–ammonia	473	1.86	28
Hydrogen–ammonia	533	2.149	46
Hydrogen–benzene	311.3	0.404	47
Hydrogen–cyclohexane	288.6	0.319	47
Hydrogen–methane	288	0.694	32
Hydrogen–nitrogen	294	0.763	46
Hydrogen–nitrogen	298	0.784	28
Hydrogen–nitrogen	358	1.052	28
Hydrogen–nitrogen	573	2.417	46
Nitrogen–piperidine	315	0.403	47
Hydrogen–pyridine	318	0.437	47
Hydrogen–sulfur dioxide	285.5	0.525	28
Hydrogen–sulfur dioxide	473	1.23	28
Hydrogen–thiophene	302	0.400	47
Hydrogen–water	307.1	0.915	42
Hydrogen–water	328.5	1.121	43
Hydrogen–xenon	341.2	0.751	34
Methane–water	352.3	0.356	43
Neon–krypton	273	0.223	48
Nitrogen–ammonia	298	0.230	28
Nitrogen–ammonia	358	0.328	28
Nitrogen–benzene	311.3	0.102	47
Nitrogen–cyclohexane	288.6	0.731	47
Nitrogen–piperidine	315	0.0953	47
Nitrogen–sulfur dioxide	263	0.104	28
Nitrogen–water	307.5	0.256	43
Nitrogen–water	328.9	0.313	42
Nitrogen–water	349.1	0.354	42

(continued overleaf)

Table B-11 (*continued*)

System	T(°K)	$D_{AB}P$ (obs.) $\frac{cm^2}{sec}$ atm.	Reference
Nitrogen–water	352.1	0.359	43
Oxygen–benzene	311.3	0.101	47
Oxygen–carbon tetrachloride	296	0.749	28
Oxygen–cyclohexane	288.6	0.746	47
Oxygen–piperidine	315	0.0953	47
Oxygen–water	352.3	0.352	43

Multicomponent systems are treated in references 9 and 50-53.

Table B-12 Diffusion Coefficients (Aqueous Solutions, Infinite Dilution)

Solute	T(°C)	D^0_{AB}(exp) $\frac{cm^2}{sec} \times 10^5$	Reference
Helium	25	6.3	54
Hydrogen	25	4.8	54
Oxygen	25	2.41	54
Carbon dioxide	25	2.00	54
Ammonia	12	1.64	55
Nitrous oxide	25	2.67	56
Chlorine	25	1.25	57
Propylene	25	1.44	54
Benzene	25	1.09	58
Methyl alcohol	15	1.26	59
Ethyl alcohol	10	0.84	59
	15	1.00	55
	25	1.24	59
n-Propyl alcohol	15	0.87	59
i-Propyl alcohol	15	0.87	59
n-Butyl alcohol	15	0.77	59,55
i-Butyl alcohol	15	0.77	59
	20	0.792	60
	20	0.84	61
i-Amyl alcohol	15	0.69	55
Allyl alcohol	15	0.90	55
Ethylene glycol	20	1.04	60
1,2-Propylene glycol	20	0.88	60
Glycerol	15	0.72	55
	20	0.825	60
Mannitol	20	0.56	55
	20	0.673	62

Table B-12 (*continued*)

Solute	T(°C)	D^0_{AB}(exp) $\frac{cm^2}{sec} \times 10^5$	Reference
Benzyl alcohol	20	0.82	60
Acetic acid	20	1.19	61
Oxalic acid	20	1.53	55
Tartaric acid	15	0.61	55
Benzoic acid	25	1.21	63
Glycine	25	1.055	62
Ethyl acetate	20	1.00	61
Acetone	15	1.22	59
	20	1.16	61
	25	1.28	64
Furfural	20	1.04	61
Acetamide	15	0.96	55
Urea	20	1.20	55
	25	1.378	62
Urethane	15	0.80	55
Diethylamine	20	0.97	61
Aniline	20	0.92	61
Acetonitrile	15	1.26	55
Pyridine	15	0.58	55
Water	25	2.44	65

Table B-13 Diffusion Coefficients (Infinite Dilution)

Solute	Solvent	Temperature (°C)	$D^0_{AB} \times 10^5$ (obs.) $\frac{cm^2}{sec}$	Reference
Acetic acid	Acetone	25	3.31	66
Benzoic acid	Acetone	25	2.62	66
Formic acid	Acetone	25	3.77	66
Water	Aniline	20	0.70	61
Acetic acid	Benzene	25	2.09	66
Benzoic acid	Benzene	25	1.38	66
Bromobenzene	Benzene	7.5	1.45	67
Carbon tetrachloride	Benzene	25	1.92	68
Carbon tetrachloride	Benzene	20	1.76	69
Cinnamic acid	Benzene	25	1.12	66
Ethylene chloride	Benzene	7.5	1.77	67
Ethanol	Benzene	15	2.25	59
Formic acid	Benzene	25	2.28	66
Methanol	Benzene	25	3.82	70
Methyl iodide	Benzene	7.5	2.06	67

(*continued overleaf*)

Table B-13 (*continued*)

Solute	Solvent	Temperature (°C)	$D^0_{AB} \times 10^5$ (obs.) $\frac{cm^2}{sec}$	Reference
Naphthalene	Benzene	7.5	1.19	67
1,2,4-Trichlorobenzene	Benzene	7.5	1.34	67
Acetone	*iso*-Butanol	20	0.74	61
Acetone	Carbon tetrachloride	20	1.86	61
Benzene	Chlorobenzene	20	1.25	69
Acetone	Chloroform	15	2.36	59
Benzene	Chloroform	15	2.51	59
Ethanol	Chloroform	15	2.20	59
Ethyl ether	Chloroform	15	2.07	59
Carbon tetrachloride	Cyclohexane	25	1.49	69
Allyl alcohol	Ethanol	20	0.98	55
iso-Amyl alcohol	Ethanol	20	0.81	55
Azobenzene	Ethanol	20	0.74	55
Bromoform	Ethanol	20	0.97	71
Camphor	Ethanol	20	0.70	55
Carbon dioxide	Ethanol	17	3.2	55
Glycerol	Ethanol	20	0.51	55
Iodine	Ethanol	25	1.32	66
Iodobenzene	Ethanol	20	1.00	55
Pyridine	Ethanol	20	1.10	55
Urea	Ethanol	12	0.54	55
Water	Ethanol	25	1.132	72
Acetic acid	Ethyl acetate	20	2.18	61
Water	Ethylene glycol	20	0.18	60
Water	Glycerol	20	0.0083	60
Carbon tetrachloride	*n*-Hexane	25	3.70	73
n-Hexane	*n*-Hexane	25	4.21	74
Toluene	*n*-Hexane	25	4.21	66
Mercury	Mercury	234.3	1.63	75
Tin	Mercury	30	1.60	61
Water	*u*-Propanol	15	0.87	59
Water	1,2-Propylene glycol	20	0.075	60
Tin	Tin	485	5.9	76
Acetic acid	Toluene	20	2.00	61
Acetic acid	Toluene	25	2.26	66
Acetone	Toluene	20	2.93	61
Benzoic acid	Toluene	20	1.74	61
Benzoic acid	Toluene	25	1.49	66
Chlorobenzene	Toluene	20	2.06	69
Diethylamine	Toluene	20	2.36	61
Ethanol	Toluene	15	3.00	59
Formic acid	Toluene	25	2.65	66

Table B-14 Mutual Diffusion Coefficients of Inorganic Salts in Aqueous Solutions (55)

Solute	T (°C)	Concentration (g mole/liter)	$D_{AB} \times 10^5$ [(cm^2/sec) × 10^5]
HCl	12	0.1	2.29
H_2SO_4	20	0.25	1.63
HNO_3	20	0.05	2.62
		0.25	2.59
NH_4Cl	20	1.0	1.64
H_3PO_4	20	0.25	0.89
$HgCl_2$	18	0.25	0.92
$CuSO_4$	14	0.4	0.39
$AgNO_3$	15	0.17	1.28
$CoCl_2$	20	0.1	1.0
$MgSO_4$	10	0.4	0.39
$Ca(OH)_2$	20	0.2	1.6
$Ca(NO_3)_2$	14	0.14	0.85
LiCl	18	0.05	1.12
NaOH	15	0.05	1.49
NaCl	18	0.4	1.17
		0.8	1.19
		2.0	1.23
KOH	18	0.01	2.20
		0.1	2.15
		1.8	2.19
KCl	18	0.4	1.46
		0.8	1.49
		2.0	1.58
KBr	15	0.046	1.49
KNO_3	18	0.2	1.43

Table B-15 Diffusion Coefficients in Solid Polymers (79)

D_{AB} Values in $\frac{\text{cm}^2}{\text{sec}} \times 10^{-6}$

Polymer	H_E	H_2	O_2	CO_2	CH_4
1 Polyethylene terephthalate (glassy crystalline)	1.7		0.0036	0.00054	0.00017
2 Polycarbonate (Lexan)		0.64	0.021	0.0048	
3 Polyethylene, density 0.964	3.07		0.170	0.124	0.057
4 Polyethylene density 0.914	6.8		0.46	0.372	0.193
5 Polystyrene	10.4	4.36	0.11	0.058	
6 Butyl rubber	5.93	1.52	0.81	0.058	
7 Polychloroprene (neoprene)		4.31	0.43	0.27	

(continued overleaf)

Table B-15 (*continued*)

D_{AB} Values in $\frac{cm^2}{sec} \times 10^{-6}$

Polymer	H_e	H_2	O_2	CO_2	CH_4
8 Natural rubber	21.6	10.2	1.58	1.10	0.89
9 Silicone rubber, 10 percent filler (extrapolated)	53.4	67.1	17.0		
10 Polypropylene, isotactic	19.5	2.12			
11 Polypropylene, atactic	41.6	5.7			
12 Polyethyl methacrylate	44.1		0.11	0.030	
13 Butadiene-acrylonitrile (Perbunan)	11.7	4.5	0.43	0.19	
14 Polybutadiene		9.6	1.5	1.05	
15 Polyvinyl acetate (glassy)	9.52	2.10	0.051		0.0019

Table B-16 Diffusivities in Molten and Thermally Softened Polymers (80–87)

Gas	Diffusivities			
	Polyethylene	Polypropylene	Polyisobutylene	Polystyrene
Helium	17.09	10.51	12.96	
Argon	9.19	7.40	5.18	
Krypton			7.30	
Monochlorodifluormethane	4.16	4.02		
Methane	5.50		2.00	0.42
Nitrogen	6.04	3.51	2.04	0.348
Carbon dioxide	5.69	4.25	3.37	0.39

Table B-17 Values of Heats of Diffusion (E_D) and Diffusion Temperature Coefficients (D_0) for Molten, Thermally Softened Polymers (88–91)

$$D = D_0 \exp(-E_d/RT)$$

Polymer	Gas	E_D (kcal/mole)	$D_0 \times 10^5$ (cm^2/sec)	Temperature Range (°C)	References
Polyethylene	N_2	2.0	53.414	125–188	83–86
	CO_2	4.4	688.135	188–224	80–82
Polypropylene	CO_2	3.0	111.768	188–224	80–82
Polystyrene	H_2	10.1	218.064	120–188	87
	N_2	9.6	21.119	119–188	87
	CH_4	3.6	21.24	125–188	83–86

APPENDIX C

Table C-1 Equations of Continuity

Rectangular coordinates (x, y, z):

$$\frac{\partial \rho}{\partial t} + \frac{\partial}{\partial x}(\rho v_x) + \frac{\partial}{\partial y}(\rho v_y) + \frac{\partial}{\partial z}(\rho v_z) = 0$$

Cylindrical coordinates (r, θ, z):

$$\frac{\partial \rho}{\partial t} + \frac{1}{r}\frac{\partial}{\partial r}(\rho r v_r) + \frac{1}{r}\frac{\partial}{\partial \theta}(\rho v_\theta) + \frac{\partial}{\partial z}(\rho v_z) = 0$$

Spherical coordinates (r, θ, ϕ):

$$\frac{\partial \rho}{\partial t} + \frac{1}{r^2}\frac{\partial}{\partial r}(\rho r^2 v_r) + \frac{1}{r\sin\theta}\frac{\partial}{\partial \theta}(\rho v_\theta \sin\theta) + \frac{1}{r\sin\theta}\frac{\partial}{\partial \phi}(\rho v_\phi) = 0$$

Source: Adapted from reference 1.

Table C-2 Equation of Motion in Cylindrical Coordinates

In terms of τ:

r-component[a]

$$\rho\left(\frac{\partial v_r}{\partial t} + v_r\frac{\partial v_r}{\partial r} + \frac{v_\theta}{r}\frac{\partial v_r}{\partial \theta} - \frac{v_\theta^2}{r} + v_z\frac{\partial v_r}{\partial z}\right) = -\frac{\partial p}{\partial r} - \left(\frac{1}{r}\frac{\partial}{\partial r}(r\tau_{rr}) + \frac{1}{r}\frac{\partial \tau_{r\theta}}{\partial \theta} - \frac{\tau_{\theta\theta}}{r} + \frac{\partial \tau_{rz}}{\partial z}\right) + \rho g_r$$

θ-component[b]

$$\rho\left(\frac{\partial v_\theta}{\partial t} + v_r\frac{\partial v_\theta}{\partial r} + \frac{v_\theta}{r}\frac{\partial v_\theta}{\partial \theta} + \frac{v_r v_\theta}{r} + v_z\frac{\partial v_\theta}{\partial z}\right) = -\frac{1}{r}\frac{\partial p}{\partial \theta} - \left(\frac{1}{r^2}\frac{\partial}{\partial r}(r^2\tau_{r\theta}) + \frac{1}{r}\frac{\partial \tau_{\theta\theta}}{\partial \theta} + \frac{\partial \tau_{\theta z}}{\partial z}\right) + \rho g_\theta$$

(*continued overleaf*)

Table C-2 (*continued*)

$$z\text{-component}\quad \rho\left(\frac{\partial v_z}{\partial t}+v_r\frac{\partial v_z}{\partial r}+\frac{v_\theta}{r}\frac{\partial v_z}{\partial \theta}+v_z\frac{\partial v_z}{\partial z}\right)=-\frac{\partial p}{\partial z}$$
$$-\left(\frac{1}{r}\frac{\partial}{\partial r}(r\tau_{rz})+\frac{1}{r}\frac{\partial \tau_{\theta z}}{\partial \theta}+\frac{\partial \tau_{zz}}{\partial z}\right)+\rho g_z$$

In terms of velocity gradients for a Newtonian fluid with constant ρ and μ:

$$r\text{-component}\quad \rho\left(\frac{\partial v_r}{\partial t}+v_r\frac{\partial v_r}{\partial r}+\frac{v_\theta}{r}\frac{\partial v_r}{\partial \theta}-\frac{v_\theta^2}{r}+v_z\frac{\partial v_r}{\partial z}\right)=-\frac{\partial p}{\partial r}$$
$$+\mu\left[\frac{\partial}{\partial r}\left(\frac{1}{r}\frac{\partial}{\partial r}(rv_r)\right)+\frac{1}{r^2}\frac{\partial^2 v_r}{\partial \theta^2}-\frac{2}{r^2}\frac{\partial v_\theta}{\partial \theta}+\frac{\partial^2 v_r}{\partial z^2}\right]+\rho g_r$$

$$\theta\text{-component}\quad \rho\left(\frac{\partial v_\theta}{\partial t}+v_r\frac{\partial v_\theta}{\partial r}+\frac{v_\theta}{r}\frac{\partial v_\theta}{\partial \theta}+\frac{v_r v_\theta}{r}+v_z\frac{\partial v_\theta}{\partial z}\right)=-\frac{1}{r}\frac{\partial p}{\partial \theta}$$
$$+\mu\left[\frac{\partial}{\partial r}\left(\frac{1}{r}\frac{\partial}{\partial r}(rv_\theta)\right)+\frac{1}{r^2}\frac{\partial^2 v_\theta}{\partial \theta^2}+\frac{2}{r^2}\frac{\partial v_r}{\partial \theta}+\frac{\partial^2 v_\theta}{\partial z^2}\right]+\rho g_\theta$$

$$z\text{-component}\quad \rho\left(\frac{\partial v_z}{\partial t}+v_r\frac{\partial v_z}{\partial r}+\frac{v_\theta}{r}\frac{\partial v_z}{\partial \theta}+v_z\frac{\partial v_z}{\partial z}\right)=-\frac{\partial p}{\partial z}$$
$$+\mu\left[\frac{1}{r}\frac{\partial}{\partial r}\left(r\frac{\partial v_z}{\partial r}\right)+\frac{1}{r^2}\frac{\partial^2 v_z}{\partial \theta^2}+\frac{\partial^2 v_z}{\partial z^2}\right]+\rho g_z$$

Source: Adapted from reference 1.

Table C-3 Equation of Motion in Spherical Coordinates

In terms of τ:

$$r\text{-component}\quad \rho\left(\frac{\partial v_r}{\partial t}+v_r\frac{\partial v_r}{\partial r}+\frac{v_\theta}{r}\frac{\partial v_r}{\partial \theta}+\frac{v_\phi}{r\sin\theta}\frac{\partial v_r}{\partial \phi}-\frac{v_\theta^2+v_\phi^2}{r}\right)$$
$$=-\frac{\partial p}{\partial r}-\left(\frac{1}{r^2}\frac{\partial}{\partial r}(r^2\tau_{rr})+\frac{1}{r\sin\theta}\frac{\partial}{\partial \theta}(\tau_{r\theta}\sin\theta)\right.$$
$$\left.+\frac{1}{r\sin\theta}\frac{\partial \tau_{r\phi}}{\partial \phi}-\frac{\tau_{\theta\theta}+\tau_{\phi\phi}}{r}\right)+\rho g_r$$

$$\theta\text{-component}\quad \rho\left(\frac{\partial v_\theta}{\partial t}+v_r\frac{\partial v_\theta}{\partial r}+\frac{v_\theta}{r}\frac{\partial v_\theta}{\partial \theta}+\frac{v_\phi}{r\sin\theta}\frac{\partial v_\theta}{\partial \phi}+\frac{v_r v_\theta}{r}-\frac{v_\phi^2\cot\theta}{r}\right)$$
$$=-\frac{1}{r}\frac{\partial p}{\partial \theta}-\left(\frac{1}{r^2}\frac{\partial}{\partial r}(r^2\tau_{r\theta})+\frac{1}{r\sin\theta}\frac{\partial}{\partial \theta}(\tau_{\theta\theta}\sin\theta)+\frac{1}{r\sin\theta}\frac{\partial \tau_{\theta\phi}}{\partial \phi}\right.$$
$$\left.+\frac{\tau_{r\theta}}{r}-\frac{\cot\theta}{r}\tau_{\phi\phi}\right)+\rho g_\theta$$

$$\phi\text{-component}\quad \rho\left(\frac{\partial v_\phi}{\partial t}+v_r\frac{\partial v_\phi}{\partial r}+\frac{v_\theta}{r}\frac{\partial v_\phi}{\partial \theta}+\frac{v_\phi}{r\sin\theta}\frac{\partial v_\phi}{\partial \phi}+\frac{v_\phi v_r}{r}+\frac{v_\theta v_\phi}{r}\cot\theta\right)$$
$$=-\frac{1}{r\sin\theta}\frac{\partial p}{\partial \phi}-\left(\frac{1}{r^2}\frac{\partial}{\partial r}(r^2\tau_{r\phi})+\frac{1}{r}\frac{\partial \tau_{\theta\phi}}{\partial \theta}+\frac{1}{r\sin\theta}\frac{\partial \tau_{\phi\phi}}{\partial \phi}\right.$$
$$\left.+\frac{\tau_{r\phi}}{r}+\frac{2\cot\theta}{r}\tau_{\theta\phi}\right)+\rho g_\phi$$

Table C-3 ***(continued)***

In terms of velocity gradients for a Newtonian fluid with constant ρ and μ:

r-component $$\rho\left(\frac{\partial v_r}{\partial t}+v_r\frac{\partial v_r}{\partial r}+\frac{v_\theta}{r}\frac{\partial v_r}{\partial \theta}+\frac{v_\phi}{r\sin\theta}\frac{\partial v_r}{\partial \phi}-\frac{v_\theta^2+v_\phi^2}{r}\right)$$

$$=-\frac{\partial p}{\partial r}+\mu\left(\frac{1}{r^2}\frac{\partial^2}{\partial r^2}(r^2 v_r)+\frac{1}{r^2\sin\theta}\frac{\partial}{\partial\theta}\left(\sin\theta\frac{\partial v_r}{\partial\theta}\right)+\frac{1}{r^2\sin^2\theta}\frac{\partial^2 v_r}{\partial\phi^2}\right)+\rho g_r$$

θ-component $$\rho\left(\frac{\partial v_\theta}{\partial t}+v_r\frac{\partial v_\theta}{\partial r}+\frac{v_\theta}{r}\frac{\partial v_\theta}{\partial \theta}+\frac{v_\phi}{r\sin\theta}\frac{\partial v_\theta}{\partial \phi}+\frac{v_r v_\theta}{r}-\frac{v_\phi^2\cot\theta}{r}\right)$$

$$=-\frac{1}{r}\frac{\partial p}{\partial\theta}+\mu\left(\frac{1}{r^2}\frac{\partial}{\partial r}\left(r^2\frac{\partial v_\theta}{\partial r}\right)+\frac{1}{r^2}\frac{\partial}{\partial\theta}\left(\frac{1}{\sin\theta}\frac{\partial}{\partial\theta}(v_\theta\sin\theta)\right)+\frac{1}{r^2\sin^2\theta}\frac{\partial^2 v_\theta}{\partial\phi^2}\right.$$

$$\left.+\frac{2}{r^2}\frac{\partial v_r}{\partial\theta}-\frac{2\cos\theta}{r^2\sin^2\theta}\frac{\partial v_\phi}{\partial\phi}\right)+\rho g_\theta$$

φ-component $$\rho\left(\frac{\partial v_\phi}{\partial t}+v_r\frac{\partial v_\phi}{\partial r}+\frac{v_\theta}{r}\frac{\partial v_\phi}{\partial \theta}+\frac{v_\phi}{r\sin\theta}\frac{\partial v_\phi}{\partial \phi}+\frac{v_\phi v_r}{r}+\frac{v_\theta v_\phi}{r}\cot\theta\right)$$

$$=-\frac{1}{r\sin\theta}\frac{\partial p}{\partial\phi}+\mu\left(\frac{1}{r^2}\frac{\partial}{\partial r}\left(r^2\frac{\partial v_\phi}{\partial r}\right)+\frac{1}{r^2}\frac{\partial}{\partial\theta}\left(\frac{1}{\sin\theta}\frac{\partial}{\partial\theta}(v_\phi\sin\theta)\right)+\frac{1}{r^2\sin^2\theta}\frac{\partial^2 v_\phi}{\partial\phi^2}\right.$$

$$\left.+\frac{2}{r^2\sin\theta}\frac{\partial v_r}{\partial\phi}+\frac{2\cos\theta}{r^2\sin^2\theta}\frac{\partial v_\theta}{\partial\phi}\right)+\rho g_\phi$$

Source: Adapted from reference 1.

APPENDIX REFERENCES

1. R. B. Bird, W. E. Stewart, and E. N. Lightfoot, *Transport Phenomena*, John Wiley and Sons, New York (1960).
2. S. J. Kline, *Similitude and Approximation Theory*, McGraw-Hill, New York (1965).
3. P. W. Bridgman, *Dimensional Analysis*, Harvard University Press, Cambridge, MA (1921).
4. E. Buckingham, *Phys. Rev.* **4**(4), 345 (1914).
5. E. Buckingham, *Nature* **95**, 66 (1915).
6. M. S. Peters, *Elementary Chemical Engineering*, McGraw-Hill, New York (1984).
7. W. H. McAdams, *Heat Transmission*; McGraw-Hill, New York (1954).
8. R. G. Griskey, *Chemical Engineering for Chemists*, American Chemical Society, Washington, D.C. (1997).
9. J. O. Hirschfelder, C. F. Curtis, and R. B. Bird, *Molecular Theory of Gases and Liquids*, John Wiley and Sons, New York (1954).
10. E. A. Mason, *J. Chem. Phys.* **32**, 1832 (1960).
11. K. A. Kobe and R. E. Lynn, Jr., *Chem. Rev.* **52**, 117 (1952).
12. F. D. Rossini, editor, *A.P.I. Res. Proj.* **44**; Pittsburgh, PA (1952).
13. O. A. Hougen and K. M. Watson, *Chemical Process Principles*, Vol. III, John Wiley and Sons, New York (1947).
14. E. J. Owens and G. Thodos, *AIChE J.* **3**, 454 (1957).
15. J. O. Hirschfelder, R. B. Bird, and E. L. Spotz, *Chem. Rev.* **44**, 205 (1949).
16. S. Glasstone, K. J. Laidler, and H. Eyring, *Theory of Rate Processes*, McGraw-Hill, New York (1941).
17. J. F. Kincaid, H. Eyring, and A. W. Stern, *Chem. Revs.* **28**, 301 (1941).
18. O. A. Uyehara and K. M. Watson, *Nat. Petrol. News Tech. Section* **36**, 764 (1944).
19. N. L. Carr, R. Kobayashi, and D. B. Burroughs, *Am. Inst. Min. Met. Engs.* **6**, 47 (1954).
20. C. R. Wilke, *J. Chem. Phys.* **17**, 550 (1949).
21. A. Eucken, *Physik Z.* **14**, 324 (1913).
22. P. W. Bridgman, *Proc. Am. Acad. Arts Sci.* **59**, 141 (1923).
23. L. Lorenz, *Pogg. Ann.* **147**, 429 (1872).
24. G. Wiedemann and R. Franz, *Annu. Phys. Chem.* **89**, 530 (1853).
25. E. J. Owens and G. Thodos, *AIChE J.* **3**, 454 (1957).
26. J. M. Lenoir, W. A. Junk, and E. W. Comings, *Chem. Eng. Prog.* **49**, 539 (1953).
27. J. N. Holsen and M. R. Strunk, *Ind. Eng. Chem. Fund.* **3**, 163 (1964).
28. E. A. Mason and L. Monchick, *J. Chem. Phys.* **36**, 2746 (1962).
29. L. T. Carmichael, B. H. Sage, and W. N. Lacey, *AIChE J.* **1**, 385 (1955).
30. B. N. Srivastava and I. B. Srivastava, *J. Chem. Phys.* **38**, 1183 (1963).
31. B. N. Srivastava and I. B. Srivastava, *J. Chem. Phys.* **36**, 2616 (1962).
32. A. J. Carswell and J. C. Stryland, *Can. J. Phys.* **41**, 708 (1963).
33. S. L. Seager, L. R. Geerston, and J. C. Giddings, *J. Chem. Eng. Data* **8**, 168 (1963).
34. R. Paul and I. B. Srivastava, *J. Chem. Phys.* **36**, 1621 (1961).

35. A. A. Westenberg and G. Frazier, *J. Chem. Phys.* **36**, 3499 (1962).
36. I. Amdur and T. F. Schatzki, *J. Chem. Phys.* **27**, 1049 (1957).
37. I. Amdur, J. Ross, and E. A. Mason, *J. Chem. Phys.* **20**, 1620 (1952).
38. S. L. Seager, L. R. Geerston, and J. C. Giddings, *J. Chem. Eng. Data* **8**, 168 (1963).
39. R. E. Walker, N. DeHaas, and A. A. Westenberg, *J. Chem. Phys.* **32**, 1314 (1960).
40. R. E. Walker and A. A. Westenberg, *J. Chem. Phys.* **29**, 1139 (1958).
41. R. E. Walker and A. A. Westenberg, *J. Chem. Phys.* **32**, 436 (1960).
42. W. L. Crider, *J. Am. Chem. Soc.* **78**, 924 (1956).
43. F. A. Schwartz and J. E. Brow, *J. Chem. Phys.* **19**, 640 (1951).
44. I. Amdur and L. M. Shuler, *J. Chem. Phys.* **38**, 188 (1963).
45. A. A. Westenberg and R. E. Walker, *Thermodynamic and Transport Properties of Gases, Liquids and Solids*, ASME Symposium, McGraw-Hill, New York (1959).
46. D. S. Scott and K. E. Cox, *Can. J. Chem. Eng.* **38**, 201 (1960).
47. G. H. Hudson and J. C. McCoubrey, *Ubbelohde Trans. Faraday Soc.* **56**, 1144 (1960).
48. B. N. Srivastava and K. P. Srivastava, *J. Chem. Phys.* **30**, 984 (1959).
49. T. R. Marrero and E. A. Mason, *J. Phys. Chem. Ref. Data* **1**, 3 (1972).
50. C. F. Curtiss and J. O. Hirschfelder, *J. Chem. Phys.* **17**, 550 (1949).
51. C. R. Wilke, *Chem. Eng. Prog.* **46**, 95 (1950).
52. E. R. Gilliland, *Absorption and Extraction*, T. K. Sherwood, editor 1st ed.; McGraw-Hill, New York (1937), p. 11.
53. H. L. Toor, *AIChE J.* **3**, 198 (1957).
54. J. E. Vivian and C. J. King, *AIChE J.* **10**, 220 (1964).
55. *International Critical Tables*, McGraw-Hill, New York (1926-1930).
56. J. F. Davidson and E. J. Cullen, *Trans. Inst. Chem. Eng. (London)* **35**, 51 (1957).
57. J. Kramus, R. A. Douglas, and R. M. Ulmann, *Chem. Eng. Sci.* **10**, 190 (1959).
58. G. A. Ratcliff and N. J. Reid, *Trans. Inst. Chem. Eng. (London)* **39**, 423 (1961).
59. P. A. Johnson and A. L. Babb, *Chem Rev.* **56**, 387 (1956).
60. F. H. Garner and P. J. M. Marchant, *Trans. Inst. Chem. Eng. (London)* **39**, 397 (1961).
61. J. B. Lewis, *J. Appl. Chem.* **5**, 228 (1955).
62. L. G. Longworth, *J. Phys. Chem.* **58**, 770 (1954).
63. S. Y. Chang, M. S. Thesis, M.I.T. Cambridge, MA (1959).
64. D. K. Anderson, J. R. Hall, and A. L. Babb, *J. Phys. Chem.* **62**, 404 (1958).
65. R. A. Robinson and R. H. Stokes, *Electrolyte Solutions*, Second edition, Academic Press, New York (1959).
66. P. Chang and C. R. Wilke, *J. Phys. Chem.* **59**, 592 (1955).
67. A. E. Stearn, E. M. Irish, and H. Eyring, *J. Phys. Chem.* **44**, 981 (1940).
68. J. K. Horrocks and E. McLaughlin, *Trans. Faraday Soc.* **58**, 1357 (1962).
69. C. S. Caldwell and A. L. Babb, *J. Phys. Chem.* **60**, 14, 56 (1956).
70. R. R. Cram and A. W. Adamson, *J. Phys. Chem.*
71. H. S. Taylor, *J. Chem. Phys.* **6**, 331 (1938).
72. B. R. Hammond and R. H. Stokes, *Trans. Faraday Soc.* **49**, 890 (1953).
73. B. R. Hammond and R. H. Stokes, *Trans. Faraday Soc.* **51**, 1641 (1955).

74. D. W. McCall and D. C. Douglas, *Phys. Fluids* **2**, 87 (1959).
75. N. H. Nachtrieb and J. Petit, *J. Chem. Phys.* **24**, 746 (1956).
76. C. A. Ma and R. A. Swalin, *J. Chem. Phys.* **36**, 3014 (1962).
77. D. F. Othmer and M. S. Thakar, *Ind. Eng. Chem.* **45**, 589 (1953).
78. C. R. Wilke and P. Chang, *AIChE J.* **1**, 264 (1955).
79. J. Crank and G. S. Park, editors, *Diffusion in Polymers*, Academic, New York (1968), Chapter 2.
80. R. G. Griskey and P. L. Durill, *AIChE J.* **12**, 1147 (1966).
81. R. G. Griskey and P. L. Durill, *AIChE J.* **15**, 106 (1967).
82. R. G. Griskey, *Modern Plastics* **54**(6), 158 (1977).
83. J. L. Lundberg, M. B. Wilk, and M. J. Huyett, *J. Appl. Phys.* **31**(6), 1131 (1960).
84. J. L. Lundberg, M. B. Wilk, and M. J. Huyett, *J. Polymer Sci.* **57**, 275 (1962).
85. J. L. Lundberg, M. B. Wilk, and M. J. Huyett, *Ind. Eng. Chem. Fund.* **2**, 37 (1963).
86. J. L. Lundberg, E. J. Mooney, and C. E. Rodgers, *J. Polymer Sci.* A-2, **7**, 947 (1969).
87. D. M. Newitt and K. E. J. Weale, *Chem. Soc. (London)*, 1541 (1948).
88. R. M. Barrier, *Diffusion in and through Solids*, Macmillan, New York (1941).
89. W. Jost, *Diffusion in Solids, Liquids and Gases*, Academic Press, New York (1960).
90. P. G. Shewman, *Diffusion in Solids*, McGraw-Hill, New York (1963).
91. J. P. Stark, *Solid State Diffusion*, John Wiley and Sons, New York (1976).
92. R. G. Griskey, *Chemical Engineer's Portable Handbook*, McGraw-Hill, New York (2000).

INDEX